Classification of Genetic Variation in Animals

Classification of Genetic Variation in Animals

Edited by **Dominic Fasso**

R CALLISTO
REFERENCE

New York

Published by Callisto Reference,
106 Park Avenue, Suite 200,
New York, NY 10016, USA
www.callistoreference.com

Classification of Genetic Variation in Animals
Edited by Dominic Fasso

International Standard Book Number: 978-1-63239-113-1 (Hardback)

Printed in the United States of America.

Contents

Preface

This book is a compilation of various researches conducted by experts which focus on the genetic variation in animals. It discusses the scale of genetic variation present in animals. The genetic diversity exhibited by molecular markers gets extensive interest because of the utility of this information in breeding and conservation programs. In this theory, molecular markers give important data. The increasing availability of highly sophisticated molecular markers produces a detailed analysis and evaluation of genetic diversity in animals, and recognition of genes influencing reasonably significant qualities. The objective of the book is to offer a glance into the dynamic procedure of genetic variation in animals by presenting the opinion of experts who are engaged in the generation of new initiatives and techniques employed for the assessment of genetic diversity. The book should prove helpful for students and experts in the field of genetic protection.

The researches compiled throughout the book are authentic and of high quality, combining several disciplines and from very diverse regions from around the world. Drawing on the contributions of many researchers from diverse countries, the book's objective is to provide the readers with the latest achievements in the area of research. This book will surely be a source of knowledge to all interested and researching the field.

In the end, I would like to express my deep sense of gratitude to all the authors for meeting the set deadlines in completing and submitting their research chapters. I would also like to thank the publisher for the support offered to us throughout the course of the book. Finally, I extend my sincere thanks to my family for being a constant source of inspiration and encouragement.

Editor

Part 1

Molecular Phylogenetics

Genetic Characterization of Romanian Local Breeds Using Microsatellite Markers

Georgescu Sergiu Emil and Costache Marieta
University of Bucharest, Faculty of Biology
Department of Biochemistry and Molecular Biology, Splaiul Independentei, Bucharest
Romania

1. Introduction

The rapid evolution of civilization led to a loss in the genetic diversity of domestic animals. Due to the need to exploit the highly productive breeds, with production characteristics features constantly improved in time, the less productive local breeds have been neglected, changed by crossbreeding or even replaced. This practice has become extremely dangerous because it endangers the possibility of future improvement and development of domestic animals. In the future, such practice may affect the durability of genetic resources conservation and improvement because, together with the populations' disappearance, the species genetic pool is critically reduced.

During the last years, the issue of conserving the genetic diversity as a component of the conservation of the environment has been raised at an international level. In this respect, one of the main aspects of scientific research activities is conserving the biodiversity of local breeds, especially those of economic interest. FAO (Food and Agriculture Organization) included the issue of conservation, evaluation and use of animal genetic resources in its fields of interest four decades ago. In this context, a list of domestic breeds threatened with extinction has been published (World Watch List for Domestic Animal Diversity).

Some of the most important molecular markers used to study genomes are VNTR (Variable Number Tandem Repeat) and SNP (Single Nucleotide Polymorphism). They can be highlighted through various techniques such as: RFLP (Restriction Fragment Length Polymorphism) analysis, SSCP (Single-Strand Conformation Polymorphism), genotyping fluorescent labeled fragments or sequencing. Microsatellites are useful markers for population genetic studies because they offer advantages which are particularly appropriate for conservation projects: they are widely available and exhibit a high degree of polymorphism. In addition, it is assumed they are neutral to selection, since the observed genetic diversity is constituted as the consequence of two forces: genetic drift and mutation.

Worldwide, there is a special concern regarding the evaluation and genetic characterization of local populations, the so-called rare animal breeds. Most of the studies performed refer to the breeds' genetic characterization and the identification of the phylogenetic relations among them. Due their characteristics (high polymorphism, a higher power of discrimination comparative to other genetic markers, codominant Mendelian inheritance, easily amplified by PCR), microsatellites prove themselves very useful for genetic characterization of farm animals. During recent years, a whole series of studies was performed regarding horses (Canon et al., 2000, Tozaki et al., 2003, Aberle et al., 2004), bovines (Del Bo et al., 2001, Kim et al., 2002, Mateus et al., 2004), swine (Li et al., 2000, Fabuel et al., 2004) and sheep (Arranz et al., 2001, Baumung et al., 2006).

Microsatellites are tandem repeats present in all the genomes of vertebrates. They are short repetitive sequences, of 2-9bp, dispersed throughout the entire genome. Their frequencies are increased in non-coding regions. It was noted that certain categories of repetitions are encountered with a much higher frequency, in which case repetition of the type (CA)n is the most popular. This class of repetitive elements shows a high polymorphism and is species-specific. Microsatellite is the term usually used when the length of the repeating unit is bellow 10bp, and the term of minisatellite is used when the repeating unit is between 10 and 100bp.

Microsatellites have the highest variability among DNA sequences originating in the genome, since their polymorphisms are derived from fragment length and not from primary sequence. The cause of variation between individuals of the same species at the microsatellites or minisatellites level consists in the different number of the repetitive unit. The rate of mutations in the minisatellite/microsatellite sequences is very high (about 10^{-4} per kb) and the frequency of exchange for each locus is assumed to be proportional to the minisatellite/microsatellite length. The high variability makes them especially useful for genomic mapping, because there is a high probability of individual variation in their alleles for each locus.

2. Study of microsatellite markers

2.1 Microsatellite analysis

The study of these markers implies the analysis of sets comprising over 10 microsatellites, each set being species-specific. Their analysis is done by individual genetic profile, determining the interrelations between different individuals and evaluating their allele frequency in populations. Microsatellite analysis involves the following steps: i) sample collection; ii) DNA extraction, purification and evaluation; iii) multiplex PCR amplification; iv) evaluation of amplified fragments by electrophoresis in polyacrylamide gels or by automatic genotyping using capillary electrophoresis and amplicons fluorescence detection.

The biological materials used consist in blood samples collected on anticoagulant or hairs. DNA extraction from biological samples can be done using specific kits or conventional processes.

The study of microsatellite markers will involve analysis of sets comprising 10 to 17 microsatellites, each set being species-specific. Their analysis is done by individual genetic profiling (genetic fingerprinting), determining the interrelations between different individuals and evaluating their allele frequency in populations. Microsatellite sets will be selected based on of relevance in terms of population analyses from the international databases. A microsatellite analysis test consists in extracting DNA, multiplex PCR amplification, capillary electrophoresis and fluorescence detection in the fragments amplified. In the case of the genotyping technique, first a multiplex PCR reaction is done using genomic DNA, followed by capillary electrophoresis and fluorescence detection in the resulting product. For the multiplex PCR reaction we use a combination of several primers with specific sequences of fragments of interest, resulting in simultaneous amplification of these areas. The method for amplification should use primers with hybridization temperatures similar to the DNA sequences of interest. The size of amplified fragments is identified using an automated genetic analyzer system using high-resolution separation by capillary electrophoresis. For each DNA marker alleles are detected and then genotyped by comparison with international standardized sets of markers. One of the primers of each microsatellite is marked with an available fluorescent dye to enable the multiplex analysis of markers in a single reaction. The resulting products are compared with molecular weight standards that ensure the accuracy and precision of determinations.

2.2 Statistical analysis

For a neutral marker, the degree of polymorphism is proportional to the mutational rate. Rates of mutations and their effects are important factors in the calculation of genetic distance based on data obtained from microsatellite analysis. Researchers can determine the period of time passed since the separation of two populations or measure the degree of alleles transfer between them by applying theoretical models of empirical data obtained. Thus, they are able to establish mathematical models that enable them to assess genetic diversity and phylogenetic relationships between different populations.

Genetic diversity is given by multiple alleles and genotypes established for a study group (population, species or species group). According to the Hardy-Weinberg equilibrium principle, allele frequencies and genotypes in a population remain constant - meaning that they are in equilibrium - from generation to generation, except in the case of outside influences. These influences could be controlled by mating, mutations, small populations, genetic drift and gene exchange. Hardy-Weinberg equilibrium is extremely important in conservation studies and genetic evolution. It provides basic information to identify random mating, mutations occurrence or inbreeding effects. Deviations from expected values for the Hardy-Weinberg equilibrium can have several causes, such as low population size, inbreeding, or presence of null alleles among populations that can lead to an excess of false homozygosity. Analysis of a larger number of loci can provide an accurate picture of genetic diversity because each locus will contain an independent history of the population which depends on the proportion of mutations, the genetic drift or migration.

Heterozygosity is one of the most important parameters that can give us information about diversity and even the history of a population. Values vary from 0 (absence of heterozygosity) and 1 (where a large number of alleles have the same frequency). Higher values of average heterozygosity are equal to high levels of genetic variation. Conversely, if the average heterozygosity is reduced, genetic diversity is also reduced.

Over time, as a result of human intervention, many populations of animals around the world have been fragmented. The impact of fragmentation on genetic diversity, inbreeding and extinction risk of these populations depends largely on the gene exchange between different occurring sub-populations. Subsequent to the fragmentation of a population, there are gradual differences between these sub-populations. The degree of differentiation between different sub-populations is directly correlated with inbreeding coefficients in both populations and interpopulations.

The inbreeding coefficient of the whole population (F_{IT} = Factor of Inbreeding in the Total population) can be divided into: i) inbreeding coefficient of individuals in relation to a sub-population which includes individuals (F_{IS} = Factor of Inbreeding relative to sub-population); ii) inbreeding coefficient due to differences between sub-populations, to the whole population at baseline (F_{ST} = Factor of Inbreeding Relative to Total Population). F_{ST} is decreased when the exchange of genes between subpopulations is large. If the exchange rate drops, F_{ST} grows and subpopulations are separated and distinguished from one another (Weir & Cockerham, 1984). Regarding F_{IS}, a positive value shows a deficit of heterozygosity. The deficit is even greater as the value obtained is higher, which means a high level of inbreeding. In general, the values obtained for F_{IS} vary between -1 (no inbreeding) to 1 (complete identity).

2.3 Construction of phylogenetic trees based on microsatellite frequencies

The analysis of phylogenetic relationships is based on the definition of that sequence of steps (algorithms) which can build the best phylogenetic tree. A phylogenetic tree is the graphic representation of the phylogeny of a group of organisms. To obtain phylogenetic trees the following algorithms must be followed: obtaining microsatellites data, comparing the data, selecting optimal phylogenetic methods, constructing and evaluating the trees.

To study the phylogenetic relationships between closely related species or populations microsatellites are the markers to be used. Based on their frequency, genetic distances are calculated and are subsequently used to build phylogenetic trees. Some of the genetic distances used are the following: the Nei standard distance (Nei, 1972), the Cavalli-Sforza distance (Cavalli-Sforza and Edwards, 1967) and the Reynolds genetic distance (Reynolds et al., 1983). The calculation of these three distances is based on similar assumptions that admit that the differences between populations are largely due to genetic drift.

Thus, the evaluation of Nei standard distance (DS) is performed based on two assumptions: i) all loci have the same neutral mutation rate and genetic variability is due, in equal proportions, to both mutations and genetic drift, and ii) population size remains constant in time. In this context, distance is expected to increase proportionally with time. The other two genetic distances are evaluated without considering the mutational process and based on the assumption that allele frequency differences are due to genetic drift only in a population whose size is not kept constant over time. Thus, distance does not increase proportionally with time, but with the ratio $1/N$ (where N is the population size).

After calculating the genetic distance, a phylogenetic tree can be built using several methods, including Neighbor-Joining or UPGMA (Unweighted Pair Group Method with Arithmetic Mean).

3. Genetic characterization of some Romanian local breeds using microsatellites

The rapid evolution of civilization has led to a loss of genetic diversity in farm animals. Apart from the exploitation of very productive breeds, with constantly improved production features, there is an increase tendency to overlook or to simply replace the less productive indigenous species. This situation endangers the possibilities for any future improvement and development of farm animals. In the future, such practice may affect the durability of the conservation and improvement of genetic resources because, together with the populations' disappearance, the species genetic pool is seriously reduced.

The assessment of genetic variability in domestic animals is an important issue for the preservation of genetic resources and the maintenance of protection of future breeding options, in order to satisfy the demands of changing market needs. Conservation policies of native breeds will depend on an increase of our knowledge about historic and genetic relationships among breeds, as well as on cultural factors.

During recent years, the issue of conserving the biodiversity as a component of the conservation of the environment has been raised at an international level. In this respect, one of the main aspects of scientific research activities is conserving the biodiversity of local breeds, especially those of economic interest. FAO included the issue of conservation, evaluation and use of animal genetic resources in its fields of interest four decades ago. In this respect, a list of domestic breeds threatened with extinction has been published (World Watch List for Domestic Animal Diversity). In Romania it was determined several domestic populations and breeds are threatened with extinction, needing urgent rescue and conservation measures. Such populations are the following: Grey Steppe (cattle), Tsurcana and Karabash (sheep), Mangalitsa and Bazna (swine) and Hucul (horses). These are the only local breeds for which steps have been taken in order to be saved until now, but they are still considered to be endangered.

The new genetics and molecular biology methodologies applied via the identification and genetic characterization of indigenous breeds will provide new possibilities to reclaim natural resources by genetically improving animal populations and by preserving biodiversity.

3.1 Genetic structure of Romanian Hucul horse inferred from microsatellite data

The equine genome is distributed on 31 pairs of autosomes and the X/Y sex chromosomes. What they have in common with other mammals is the fact that the majority of horse DNA is made up of repeated sequences, which do not encode proteins. The repeated sequences comprise many different types such as the minisatellite, microsatellite, SINE, LINE and telomeric amongst others (Bowling & Ruvinsky, 2000).

The horse microsatellites were characterized first by Ellegren et al., 1992 a,b, who isolated microsatellite sets with repetitive motifs CA/GT and showed that they are polymorphic, so that they could be used for paternity tests, genetic linkage studies and genetic diversity analysis. Subsequently, many studies have described the isolation of equine microsatellites and their use in studies of paternity and genetic linkage maps of horse population composition. In recent years, microsatellites were used for the analyses of different horse breeds around the world, including primitive horse breeds like the German draught horse (Aberle et al., 2004), the Spanish Celtic breeds (Canon et al., 2000), or the Norwegian breeds (Bjørnstad et al., 2000).

The Hucul breed is an indigenous equine race, a true descendant of the wild horses in Romania. The rusticity and adaptability to the environment traits are unique in the equine breeds in our country, and constitute a valuable genetic inheritance. The Hucul represents the only local horse breed and its origin is controversial. The Hucul could be considered a direct descendant of the Tarpan horse in the Northern segment of the Carpathians and it was consolidated as a breed starting with 1872, together with the founding of the Rădăuţi Stud. Five bloodlines were established and named after the foundation stallions: Goral, Hroby, Ouşor, Pietrosu and Prislop. Currently the Hucul horse is bred in stud farms in Slovakia, Romania, Hungary and Poland.

The name of these horses originates from the ethnic group, Hutsul, which originally bred them. Acclimatized in an area with low temperatures, in mountain areas, the Hucul is considered to be a component of the mountain ecosystems, a cultural landscape formed in hundreds of years by human interaction with nature.

The morphological characteristics of the Hucul horses make them similar to the Tarpan. It is a small horse, perfectly adjusted to mountainous areas, possessing great endurance in the wilderness, characterized by the ease with which it finds its food and extremely resistant to disease. As an adaptation trait, to the mountainous areas, these horses have very resistant hooves and do not require horseshoes.

In order to analyze the genetic diversity of the Hucul breed, a comparative study was initiated between four breeds of horses from Romania (Hucul, Arabian, Romanian Sport Horse and Thoroughbred), using a set of 12 microsatellite markers (Georgescu et al., 2008). This study included the analysis of heterozygosity, inbreeding, the Hardy-Weinberg equilibrium (HWE) test and breeds relationship. An UPGMA tree based on the Reynolds's genetic distance relating the four horse populations was built using the PHYLIP 3.5 software (Felsenstein, 1989).

We collected fresh blood samples from 240 randomly chosen individuals (60 from each breed) and the isolation of genomic DNA was performed with the Wizard Genomic DNA Extraction Kit (Promega). The amplification of the microsatellite loci (VHL20, HTG4, AHT4, HMS7, HTG6, HMS6, HTG7, HMS3, AHT5, ASB2, HTG10, and HMS2) was performed using StockMarks® for Horses Equine Genotyping Kit (AppliedBiosystems). PCR products were detected using an ABI Prism 310 DNA Genetic Analyzer (AppliedBiosystems) and the allele sizes were determined using GeneScan-500 LIZ Size Standard (AppliedBiosystems).

A total of 119 different alleles were detected for all the four analyzed species and the entire group of 12 microsatellites was polymorphic (Table 1).

Microsatellite	Thoroughbred	Arabian	Romanian Sport Horse	Hucul
VHL20	6	8	8	10
HTG4	7	3	4	5
AHT4	6	8	7	7
HMS7	5	5	6	6
HTG6	9	6	6	7
AHT5	8	7	5	7
HMS6	6	7	5	5
ASB2	11	8	8	7
HTG10	12	11	7	9
HTG7	7	7	3	5
HMS3	10	8	7	7
HMS2	5	9	8	5

Table 1. The numbers of alleles per locus for each breed (Georgescu et al., 2008).

Observed and expected heterozygosities per breed (Table 2) ranged from 0.662 and 0.676 (Hucul) to 0.759 (Thoroughbred) and 0.741 (Romanian Sport Horse), respectively (Georgescu et al., 2008).

Breed	H_O	H_E
Thoroughbred	0.759±0.09	0.720±0.108
Arabian	0.691±0.12	0.738±0.101
Romanian Sport Horse	0.709±0.064	0.741±0.07
Hucul	0.662±0.135	0.676±0.122

Table 2. Observed (H_O) and expected (H_E) heterozygosities of 12 microsatellites in four horse breeds from Romania (Georgescu et al., 2008).

	Thoroughbred	Arabian	Romanian Sport Horse	Hucul
Thoroughbred		0.0997	0.0878	0.1501
Arabian	0.108410		0.1116	0.1877
Romanian Sport Horse	0.096567	0.119528		0.0979
Hucul	0.157910	0.194924	0.105775	

Table 3. Fst estimates compared in pairs - above diagonal - and Reynolds's genetic distance - below diagonal- (Georgescu et al., 2008).

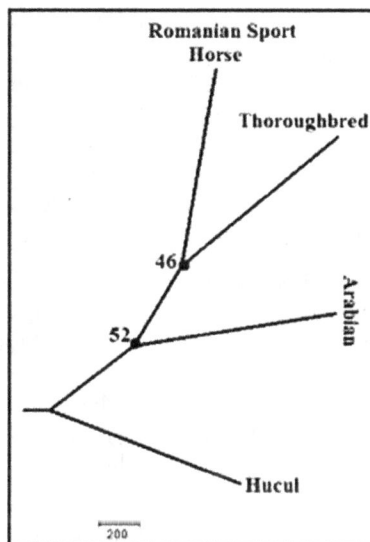

Fig. 1. Phylogenetic tree constructed based on Reynolds' genetic distance by UPGMA method. The numbers at the nodes are values for 1000 bootstrap replications (Georgescu et al., 2008).

The F_{ST} values ranged from 9.7% for the Romanian Sport Horse-Hucul pair to 1.8% for the Arabian-Hucul pair. The Reynold's genetic distance ranged from 0.096 to 0.194 (Table 3).

The results obtained by the Hardy-Weinberg test demonstrate that all the four horse populations are in equilibrium, without any digressions from it (Georgescu et al., 2008).

The phylogenetic tree obtained by the UPGMA method shows an early and clear divergence of the Hucul breed in comparison to the other three breeds analyzed (Figure 1). These data confirm the clear-cut divergence of the Hucul from the common branch (Georgescu et al., 2008). This was the first microsatellite-based genetic diversity study performed on the Hucul population in Romania.

3.2 Genetic diversity of Romanian cattle breeds based on microsatellites

The cattle genome consists of 29 pairs of autosomes and one pair of sex chromosomes. The total amount of DNA per cell is $6x10^{-12}g$, similar to other mammals. As in the case of all vertebrate genomes, the cattle genome is interspersed with numerous repetitive sequences with different structures and origins, such as microsatellites.

In the last decade, useful microsatellites studies have been published on the subject of European cattle breeds (MacHugh, 1998; Martin-Burriel *et al.*, 1999, 2007; Del Bo et al., 2001, Canon et al., 2001, Mateus *et al.*, 2004), African cattle breeds (Ibeagha-Awemu & Erhardt, 2005), and Asian cattle (Kim *et al.*, 2002; Mao *et al.*, 2007, 2008).

The last local cattle breed from Romania is the Grey Steppe which originates from the wild ancestor *Bos taurus primigenius* and is included in the Grey breed group encountered in various European countries (Ukrainian Grey, Hungarian Grey, Yugoslavian Grey Steppe, Greek Grey Steppe). Grey Steppe was spread on Romanian territory, except for the mountain areas, and had various ecologic species: the Moldavian, Transylvanian, Ialomiţa and Dobrudja varieties (Georgescu et al., 2009). The number of specimens registered a serious decline after the 1st World War. Grey Steppe has the following characteristics: robust-rough constitution, lively temper, strong tardiness and longevity, good fertility. It has very good qualities of rusticity, health, resistance to bad weather and diseases and has universal uses (traction, milk, meat). Currently, the Grey Steppe breed is on the brink of extinction, thus the genetic fund must be conserved.

The first study regarding genetic diversity and phylogenetic relationships of cattle breeds from Romania was carried out focusing on five populations: Grey Steppe, Romanian Spotted, Romanian Black Spotted, Romanian Brown and Montbeliarde, and was based on allelic frequencies of 11 microsatellite loci (Georgescu et al., 2009). This study included analyses of heterozygosity, inbreeding, the Hardy-Weinberg equilibrium test and breeds relationship. Also, a Neighbor-Joining (NJ) tree based on Reynolds's genetic distance relating the five cattle populations was built using the PHYLIP 3.5 software (Felsenstein, 1989).

We collected fresh blood samples from 190 randomly chosen individuals and the isolation of genomic DNA was performed with the Wizard Genomic DNA Extraction Kit (Promega). The amplification of the microsatellite loci (TGLA227, BM2113, TGLA53, ETH10, SPS115, TGLA126, TGLA122, INRA23, ETH3, ETH225, and BM1824) was carried out using StockMarks® for Cattle Bovine Genotyping Kit (AppliedBiosystems). PCR products were detected using an ABI Prism 310 DNA Genetic Analyzer (AppliedBiosystems) and the size of alleles was determined using GeneScan-500 ROX Size Standard (AppliedBiosystems).

A total number of 125 distinct alleles were detected across the 11 microsatellites analyzed in the five cattle breeds (Table 4) and the number of alleles varied between four and 12 (Georgescu et al., 2009).

Locus	Romanian Spotted	Romanian Black Spotted	Romanian Brown	Montbeliarde	Grey Steppe
TGLA227	12	13	10	7	5
BM2113	9	11	11	5	8
TGLA53	11	11	8	12	8
ETH10	4	8	4	4	6
SPS115	8	5	5	4	6
TGLA126	7	6	5	4	7
TGLA122	8	12	9	8	5
INRA23	7	7	6	8	8
ETH3	7	8	9	7	7
ETH225	6	5	5	6	5
BM1824	7	7	6	7	5

Table 4. The number of alleles per locus in each population (Georgescu et al., 2009).

Observed and expected heterozygosities per breed (Table 5) ranged from 0.580 (Montbeliarde) and 0.711 (Romanian Spotted) to 0.690 (Romanian Brown) and 0.778 (Romanian Black Spotted), respectively. The Hardy-Weinberg equilibrium was tested for all breed combinations and the results obtained demonstrated that all the five bovine populations are in equilibrium (Georgescu et al., 2009).

Breed	H_O	H_E	MNA
Romanian Spotted	0.593±0.163	0.711±0.127	7.818±2.227
Romanian Black Spotted	0.641±0.151	0.778±0.083	8.454±2.841
Romanian Brown	0.690±0.140	0.746±0.088	7.181±2.358
Montbeliarde	0.580±0.239	0.725±0.159	6.636±2.419
Grey Steppe	0.687±0.216	0.763±0.054	6.363±1.286

Table 5. Observed (H_O) and expected (H_E) heterozygosities of 11 microsatellites in five Romanian cattle populations (Georgescu et al., 2009).

The F_{ST} values indicate that 7% of the total genetic variation could be explained by the breeds' differences and the remaining 93% correspond to differences among individuals. In Table 6 the Reynold's genetic distance was calculated ranging from 0.056757 to 0.131480, with the smallest distances for the pair Montbeliarde - Romanian Spotted and the largest distance between Romanian Brown and Romanian Spotted (Georgescu et al., 2009). The phylogenetic tree obtained using the Neighbor-Joining (NJ) method based on Reynolds' genetic distances (Figure 2) shows that the Grey Steppe breed is clearly distinct from the other four cattle populations, which form two distinct clusters (Georgescu et al., 2009).

The two clusters presented in Figure 2 are in agreement with the expected relationships between breeds. The data obtained from this study represent a starting point in the genetic characterization of the Grey Steppe breed and can be useful for the preservation of these natural resources on the brink of extinction.

	Romanian Spotted	Romanian Black Spotted	Romanian Brown	Montbeliarde	Grey Steppe
Romanian Spotted		0.0662	**0.1132**	*0.0356*	0.0660
Romanian Black Spotted	0.083546		0.0522	0.0675	0.0400
Romanian Brown	**0.131480**	0.071088		0.1054	0.0696
Montbeliarde	*0.056757*	0.087499	0.126578		0.0753
Grey Steppe	0.090666	0.065676	0.095277	0.102635	

Table 6. Fst estimates compared in pairs -above diagonal- and Reynolds's genetic distance - below diagonal- (Georgescu et al., 2009).

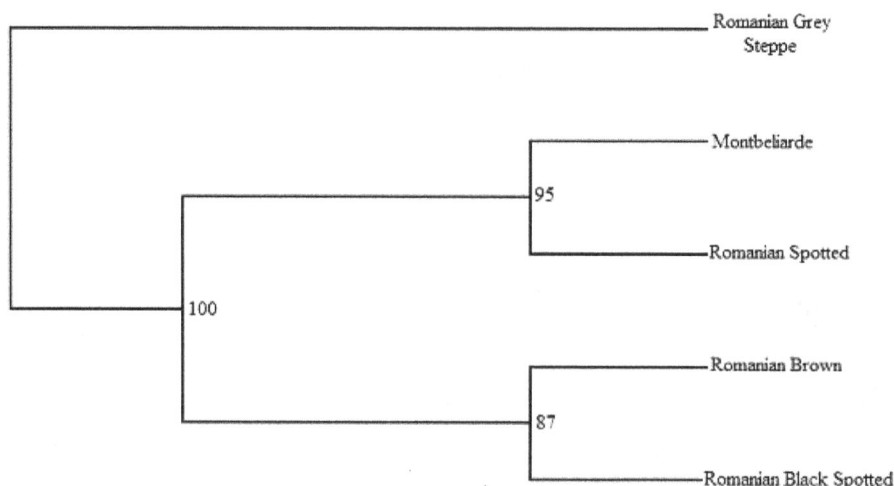

Fig. 2. NJ dendrogram of genetic relationships between the five Romanian cattle breeds. The numbers on the nodes are percentage bootstrap values in 1000 replications (Georgescu et al., 2009).

3.3 Genetic diversity of Mangalitsa pig population from Romania based on 10 microsatellites

In the last century, swine were included in improvement programs. Thus, over the years these programs have obtained breeds that satisfy the increasingly commercial requirements. But the selection methods used have involved a reduction of the numbers of animals in the population selected for breeding. Therefore, the consequence of selection for price efficiency is a reduction of the genetic variability of populations.

Several studies based on microsatellites focused on studying genetic diversity of pig breeds from Europe (Laval et al. 2000; Fabuel et al., 2004) and China (Li et al. 2000; Zhang et al., 2003), but only one such study was carried out in the case of pig breeds from Romania (Manea et al., 2009).

The Mangalitsa breed, although not originally from Romania, is still considered a local breed, since it has been bred in Romania for more than a hundred years. This breed is known for the remarkable quality of the meat, often used in traditional dishes.

Today, Mangalitsa is one of the swine breeds from Romania included on the FAO list of endangered species. The Mangalitsa breed is the only swine breed in our country dedicated to lard production. The prolificacy of the breed is good with the average of 5-6 piglets per farrowing. Subcutaneous fat (lard) and intramuscular fat obtained from the Mangalitsa breed have a lower content in "noxious" cholesterol than plant margarines.

One of the first study regarding phylogenetic relationships and genetic diversity of Romanian swine breeds based on microsatellite markers was carried out on seven populations: Synthetic Line-345 Peris (LS-345), Synthetic Line LSP-2000 (LSP-2000), Pietrain, Large White, Landrace, Mangalitsa and Wild Boar. This study included analysis of inbreeding and heterozygosity, breed relationships and Hardy-Weinberg equilibrium tests. Phylogenetic analysis between swine populations were performed using PHYLIP v3.5 software and a dendrogram was constructed using the Neighbor-Joining method (Manea et al., 2009).

Fresh blood samples from swine individuals, chosen at random, were collected and the isolation of genomic DNA was performed with Wizard Genomic DNA Extraction Kit (Promega). The amplification of the microsatellite loci (SW936, SO228, SO155, SW911, SO355, SW240, SW857, SO101, SO386, and SO005) was performed in two multiplex polymerase chain reactions using dye labeled primers (Manea et al., 2009). The combined PCR products were detected by capillary electrophoresis using an ABI Prism 310 DNA Genetic Analyzer (AppliedBiosystems) and the size of alleles was determined using GeneScan-500 LIZ Size Standard (AppliedBiosystems).

A total of 112 different alleles were identified for 10 microsatellites in all seven breeds analyzed (Table 7). The most polymorphic marker was SO005 with a total of 21 alleles, while microsatellite SO101 was the least polymorphic, showing only 7 alleles (Manea et al, 2009).

Locus	LS-345 Peris	LSP-2000	Pietrain	Large White	Landrace	Mangalitsa	Wild Boar	Total
SW936	8	9	7	8	6	7	4	12
SO155	7	7	6	6	5	4	4	8
SO228	6	5	6	6	5	6	4	9
SW911	4	4	6	5	3	5	3	9
SO355	6	6	6	7	6	6	1	9
SW240	8	7	6	8	7	9	5	14
SW857	8	9	7	7	7	7	4	10
SO101	4	4	3	5	4	4	5	7
SO386	9	7	3	8	5	7	8	13
SO005	11	13	6	14	8	8	8	21

Table 7. The numbers of alleles per locus in each breed (Manea et al., 2009).

As shown in Tabel 8, the range of heterozygosity of the analyzed markers in the seven evaluated pig breeds was between 0.5 and 0.699. Observed and expected heterozygosities ranged from 0.5 and 0.591 (Wild Boar) to 0.699 (Large White) and 0.765 (Pietrain). HWE was tested for all breed-combinations and the results demonstrated that all seven populations are in equilibrium (Manea et al., 2009).

Population	H_O	H_E	MNA
Synthetic Line-345 Peris	0.68±0.115	0.723±0.115	7.2±2.25
Synthetic Line LSP-2000	0.674±0.122	0.728±0.122	7.1±2.726
Pietrain	0.65±0.283	0.765±0.121	5.6±1.43
Large White	0.699±0.141	0.746±0.083	7.5±2.592
Landrace	0.626±0.159	0.706±0.083	5.7±1.337
Mangalitsa	0.651±0.138	0.616±0.138	6.3±1.636
Wild Boar	0.5±0.286	0.591±0.267	4.6±2.117

Table 8. Observed (H_O) and expected (H_E) heterozygosities and the mean number of alleles (MNA) of 10 microsatellites in seven Romanian swine populations (Manea et al., 2009).

Fst estimates compared in pairs and the Cavalli-Sforza's chord distance (Cavali-Sforza and Edwards, 1967) have been presented in Table 9. The Cavalli-Sforza genetic distance (D_C) ranged from 0.034, for LS-345-LSP-2000 pair, to 0.219, for Pietrain-Wild Boar pair. Also, a Neighbor-Joining tree of the seven pig populations (Figure 3) was constructed using Cavalli-Sforza's chord distances, based on the 10 microsatellite loci data (Manea et al., 2009). This dendrogram presents the phylogenetic relationships among the analyzed Romanian swine populations.

	LS-345	LSP-2000	Pietrain	Large White	Landrace	Mangalitsa	Wild Boar
LS-345		*0.018*	0.058	0.056	0.057	0.094	0.192
LSP-2000	*0.034*		0.025	0.057	0.055	0.113	0.169
Pietrain	0.067	0.046		0.067	0.067	0.133	0.203
Large White	0.056	0.056	0.084		0.036	0.105	0.176
Landrace	0.083	0.085	0.097	0.069		0.123	**0.203**
Mangalitsa	0.107	0.111	0.133	0.094	0.130		0.155
Wild Boar	0.198	0.173	**0.219**	0.170	0.199	0.146	

Table 9. Fst estimates compared in pairs -above diagonal- and Cavalli-Sforza's chord distances -below diagonal- (Manea et al., 2009).

As shown in Figure 3, with a bootstrap value of 100% Mangalitsa is the closest breed to the Wild Boar. At the same time, LS-345 and LSP-2000 together with the Pietrain breed, form a distinct cluster, which is normal because three different swine breeds participated in the formation of the LS-345 (Belgian Landrace, Duroc and Hampshire), while LSP-2000 was formed by crossing swine from the Synthetic Line-345 with the Pietrain breed (Manea et al., 2009).

This study was the first based on microsatellite markers for the genetic characterization of swine populations from Romania and the data obtained confirmed the ancient origin of the Mangalitsa. Also, this study contributed to an improved knowledge of the genetic diversity and phylogenetic relationships of pig breeds.

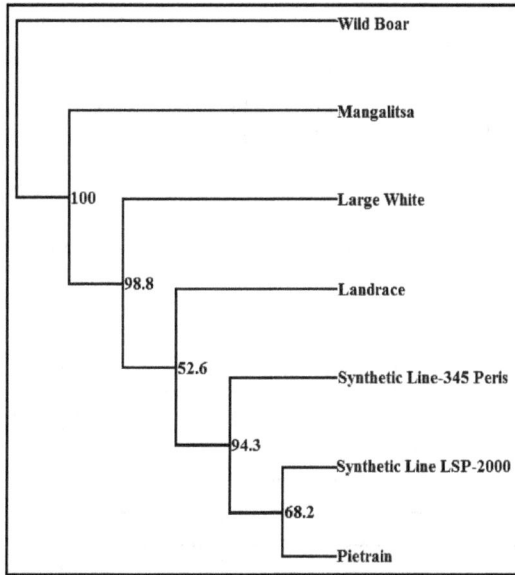

Fig. 3. Neighbor-Joining dendrogram of genetic relationships among seven Romanian swine breeds. The numbers on the nodes are percentage bootstrap values in 1000 replications (Manea et al., 2009).

3.4 Genetic diversity and phylogenetic relationships of four Romanian sheep breeds based on microsatellites

Today, the need to preserve local genetic resources is a global high priority and many measures to support primitive breeds have been established. Regarding sheep breeds, the maintenance of genetic diversity in livestock requires the adequate implementation of conservation measures, which should be based on complete information concerning the genetic characterization of the populations.

In the last decade, several studies based on microsatellites were performed on sheep breeds from Europe (Arranz et al. 2001; Stahlberger-Saitbekova et al., 2001, Rendo et al., 2004, Baumung et al., 2006) and Asia (Mukesh et al., 2006), but in Romania the first study was conducted in 2010 (Kevorkian et al., 2010).

The Karabash breed is an endangered race and it was saved by individual breeders, not through an organized effort. The Karabash together with Tsurcana represent the last two local sheep breeds from Romania. This particular sheep breed was observed due to some remarkable external characteristics, but mainly because of the high production and lamb earliness. The Karabash was mentioned for the first time in 1912 and it is considered that the two local sheep breeds, Tsigai and Tsurcana, contributed to its formation.

The aim of the first study, based on molecular markers, was to analyze the genetic diversity, variability and the phylogenetic relationships of four Romanian sheep breeds (Botoşani Karakul, Karabash, Palas Milk Line and Palas Meat Line), using 11 microsatellites. Blood samples from a total of 161 individuals from the four different sheep breeds were collected and the isolation of total DNA was performed with Wizard Genomic DNA Extraction Kit (Promega). Two PCR multiplex reactions were performed to amplify 11 microsatellites: 8-Plex reaction for OarCP20, OarCP34, MAF70, MAF214, MAF65, BM143, McM42, HSC and 3-Plex reaction for OarFC11, OarFCB20 and MAF33. The forward primers were labeled with a fluorescent compound. The combined PCR products were detected with ABI Prism 310 automated sequencer (Applied Biosystems), using the GeneScan-500 LIZ Size Standard (Kevorkian et al., 2010).

Table 10 presents the total number of allele identified in for the four breeds analyzed. All the 11 microsatellites were successfully amplified. MAF70 marker showed the highest number of alleles while OarCP20 was the least polymorphic (Kevorkian et al., 2010).

Locus	Milk Line Palas	Meat Line Palas	Karabash	Botoşani Karakul	Total
OarFCB 11	8	12	6	13	18
Oar FCB 20	10	9	7	13	20
Oar CP34	10	8	9	11	14
MAF 70	11	16	6	18	30
Oar CP20	6	7	3	9	11
MAF214	7	4	7	7	12
BM143	12	11	6	9	16
MAF33	7	3	7	15	17
McM42	5	4	10	10	17
MAF65	9	7	10	9	15
HSC	15	12	15	17	27

Table 10. Number of alleles per locus for each analyzed breed (Kevorkian et al., 2010).

Table 11 presents the results concerning the observed and expected heterozygosities for the four Romanian sheep breeds; these ranged from 0.580, respectively 0.670 (Karabash) to 0.720 (Meat Line Palas) and 0.790 (Botosani Karakul). HWE was tested for all breed-combinations and the results demonstrated that all four investigated populations are in equilibrium (Kevorkian et al., 2010).

Breed	H_o	H_e	MNA
Karabasch	0.580	0.670	7.8
Meat Line Palas	0.720	0.770	8.5
Milk Line Palas	0.590	0.740	9.2
Botoşani Karakul	0.670	0.790	11.6
Mean	0.640	0.740	9.275

Table 11. Mean number of alleles (MNA), mean observed heterozygosity (Ho) and mean expected heterozygosity (He) across 11 loci (Kevorkian et al., 2010).

The matrix of Nei's standard genetic distances and estimates of pairwise distances between breeds are presented in Table 12. The genetic distances between breeds ranged from 0.263

for Milk Line Palas and Meat Line Palas to 0.606 for Karabash and Meat Line Palas. The correspondent phylogenetic tree is presented in Figure 4 (Kevorkian et al., 2010).

	Botoşani Karakul	Karabash	Milk Line Palas	Meat Line Palas
Botoşani Karakul	-	0.0855	0.0803	0.0811
Karabash	0.381176	-	0.1063	0.1256
Milk Line Palas	0.412315	0.459566	-	0.0493
Meat Line Palas	0.448552	0.606219	0.263112	-

Table 12. Estimates of pairwise F_{ST} distances between the analyzed breeds (above diagonal) and Nei's standard genetic distances (below diagonal).

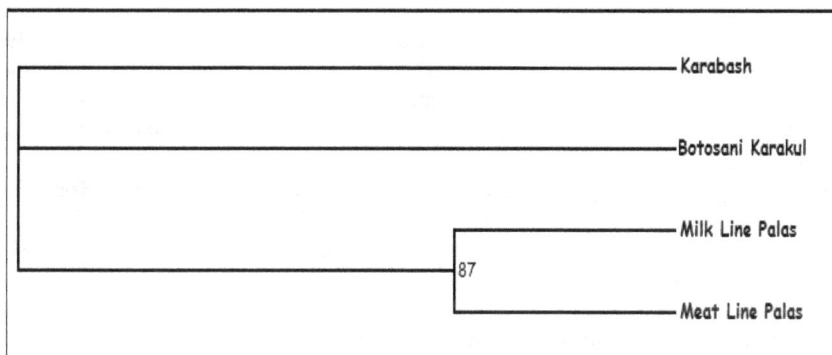

Fig. 4. Neighbor-Joining phylogenetic tree based on Nei's standard genetic distances for four Romanian sheep breeds (Kevorkian et al., 2010).

The resulting tree revealed that the most closely related breeds were the two synthetic lines. Although for the Karabash breed several clear difference are noticed (Kevorkian et al., 2010). This study was the first which has led to results regarding the characterization of genetic variability based on microsatellite markers of Romanian sheep breeds.

4. Conclusions

The need for characterizing the genetic variability of local breeds is one of the global priorities of scientific research and it is dictated both by the revaluation of practices in livestock breeding and by the conservation of genetic resources. The local breeds, perfectly adapted to the environment conditions in their foundation area, are preserved as gene pools for the next generations.

The originality of our studies is based on the fact that local breeds from Romania have never been studied and characterized using microsatellite markers. In addition these breeds were never analyzed in terms of phylogenetic relationships in comparison with other local breeds all over the world and their origins had not been clearly established. The research activities could open new horizons in clarifying issues concerning the origins of the Romanian local breeds and their kinship with other breeds.

5. Acknowledgments

This work was supported by the National Authority for Scientific Research, CNMP, grant PN II 52-124 "Technology for the improvement of the health status in sheep and goats by employing genetic markers" and grant PN II 52-163 "Economic and genetic optimization of sheep breeding programs in the context of European harmonization".

6. References

Aberle K.S., Hamann H., Drögemüller C., Distl O. (2004) Genetic Diversity in German Draught Horse Breeds Compared with a Group of Primitive, Riding and Wild Horses by Means of Microsatellite DNA Markers. *Animal Genetics* 35, pp. 270-277, ISSN 0268-9146.

Arranz J.J., Bayon Y., San Primitivo F. (2001) Differentiation among Spanish sheep breeds using microsatellites. *Genetics Selection Evolution*, 33, pp. 529–542, ISSN 0999-193X.

Baumung R., Cubric-Curik V., Schwend K., Achmann R. Solkner J. (2006) Genetic characterisation and breed assignment in Austrian sheep breeds using microsatellite marker information. *Journal of Animal Breeding and Genetics*, 123, pp. 265–271, ISSN 0931-2668.

Bjørnstad G, Gunby E., Røed K.H. (2000) Genetic Structure of Norwegian Horse Breeds. *Journal of Animal Breeding and Genetics*, 117, pp. 307-317, ISSN 0018-067X.

Bowling A.T., Ruvinsky A. (2000) The Genetics of the Horse. CABI Publishing, ISBN 0 85199 429 6, New York, USA.

Canõn J., Checa M.L., Carleos C., Vega-Plá J.L., Vallejo M., Dunner S. (2000) The Genetic Structure Of Spanish Celtic Horse Breeds Inferred From Microsatellite Data., *Animal Genetics*, 31, pp. 39-48, ISSN 0268-9146.

Cañón J., Alexandrino P., Bessa I., Carleos C., Carretero Y., Dunner S., Ferran N., Garcia D., Jordana J., Laloë D., Pereira A., Sanchez A., Moazami-Goudarzi K. (2001) Genetic diversity measures of local European beef cattle breeds for conservation purposes. *Genetics Selection Evolution*, 33, pp. 311–332, ISSN 0999-193X.

Cavalli-Sforza L.L. and Edwards A.W.F. (1967) Phylogenetic analysis: models and estimation procedures. *American Journal of Human Genetics*, 19, pp. 233–257, ISSN 0002-9297.

Del Bo L., Polli M., Longeri M., Ceriotti G., Looft C., Barre-Dirie A., Dolf G. and Zanotti M. (2001) Genetic diversity among some cattle breeds in the Alpine area. *Journal of Animal Breeding and Genetics*, 118, pp. 317-325, ISSN 0931-2668.

Fabuel E., Barragan C., Silio L., Rodriguez M.C., and Toro M.A. (2004) Analysis of genetic diversity and conservation priorities in Iberian pigs based on microsatellite markers. *Heredity*, 93, pp. 104-113, ISSN 0018-067X.

Kim K.S., Yeo J.S., Choi C.B. (2002) Genetic diversity of north-east Asian cattle based on microsatellite data. *Animal Genetics*, 33, pp. 201–204, ISSN 0268-9146.

Li K., Chen Y., Moran C., Fan B., Zhao S., Peng Z. (2000) Analysis of diversity and genetic relationships between four Chinese indigenous pig breeds and one Australian commercial pig breed. *Animal Genetics*, 31, pp. 322-325, ISSN 0268-9146.

Ellegren H., Andersson L., Johansson M., Sandberg K. (1992a) DNA fingerprinting in horses using simple (TG)n probe and its application to population comparisons. *Animal Genetics*, 23, pp. 1-9, ISSN 0268-9146.

Ellegren H., Johansson M., Sandberg K., Andersson L. (1992b) Cloning of highly polymorphic microsatellites in the horse. *Animal Genetics*, 23, pp. 133-142, ISSN 0268-9146.

Fabuel E., Barragan C., Silio L., Rodriguez M.C., Toro M.A. (2004) Analysis of genetic diversity and conservation priorities in Iberian pigs based on microsatellite markers. *Heredity*, 93, pp. 104-113, ISSN 0018-067X.

Felsenstein J. (1989) PHYLIP - Phylogeny Inference Package (Version 3.2). *Cladistics*, 5, pp. 164-166, ISSN 0748-3007.

Georgescu S.E., Manea M.A., Costache M. (2008) The genetic structure of indigenous Romanian Hucul horse breed inferred from microsatellite data. *Romanian Biotechnological Letters*, 13 (6), pp. 4030-4036, ISSN 1224-5984.

Georgescu S.E., Manea M.A., Zaulet, M., Costache M. (2009) Genetic diversity among Romanian cattle breeds with a special focus on the Romanian Grey Steppe Breed. *Romanian Biotechnological Letters*, 14 (1), pp. 4194-4200, ISSN 1224-5984.

Ibeagha-Awemu E.M. and Erhardt G. (2005) Genetic structure and differentiation of 12 African *Bos indicus* and *Bos taurus* cattle breeds inferred from protein and microsatellite polymorphisms. *Journal of Animal Breeding and Genetics*, 122 (1), pp. 12-20, ISSN 0931-2668.

Kevorkian S.E.M., Georgescu S.E., Manea M.A., Zaulet M., Hermenean A.O., Costache M. (2010) Genetic diversity using microsatellite markers in four Romanian autochthonous sheep breeds, *Romanian Biotechnological Letters*, 15 (1), pp. 5059-5065, ISSN 1224-5984.

Kim K.S., Yeo J.S., Choi C.B. (2002) Genetic diversity of north-east Asian cattle based on microsatellite data. *Animal Genetics*, 33, pp. 201–204, ISSN 0268-9146.

Laval C., Iannuccelli N., Legault C., Milan D., Groenen M.A., Giuffra E., Andersson L., Nissen P.H., Jorgensen C.B., Beeckmann P., Geldermann H., Foulley J. L., Chevalet C., Ollivier L. (2000) Genetic diversity of eleven European pig breeds. *Genetics Selection Evolution*, 32, pp. 187–203, ISSN 0999-193X.

Li K., Chen Y., Moran C., Fan B., Zhao S., Peng Z. (2000) Analysis of diversity and genetic relationships between four Chinese indigenous pig breeds and one Australian commercial pig breed. *Animal Genetics*, 31, pp. 322-325, ISSN 0268-9146.

MacHugh D.E., Loftus R.T., Cunningham P., Bradley D.G. (1998) Genetic Structure Of Seven European Cattle Breeds Assessed Using 20 Microsatellite Markers. *Animal Genetics*, 29, pp. 333-340, ISSN 0268-9146.

Manea M.A., Georgescu S.E., Kevorkian S., Costache M. (2009) Genetic diversity analyses of seven Romanian pig populations based on 10 microsatellites, *Romanian Biotechnological Letters*, 14 (6), pp. 4827-4834, ISSN 1224-5984.

Mao Y, Chang H, Yang Z, Zhang L, Xu M, Sun W, Chang G, Song G. (2007) Genetic structure and differentiation of three Chinese indigenous cattle populations. *Biochemical Genetics*, 45 (3-4), pp. 195-209, ISSN 0006-2928.

Mao Y, Chang H, Yang Z, Zhang L, Xu M, Chang G, Sun W, Song G, Ji D. (2008) The analysis of genetic diversity and differentiation of six Chinese cattle populations using microsatellite markers. *Journal of Genetics and Genomics*, 35, pp. 25-32, ISSN 1673-8527.

Mateus J.C., Penedo M.T., Alves V.C., Ramos M. and Rangel-Figueiredo T. (2004) Genetic diversity and differentiation in Portuguese cattle breeds using microsatellites. *Animal Genetics*, 35, pp. 106-113, ISSN 0268-9146.

Mukesh M., Sodhi M., Bhatia S. (2006) Microsatellite-based diversity analysis and genetic relationships of three Indian sheep breeds. *Journal of Animal Breeding and Genetics*, 123, pp. 258-264, ISSN 0931-2668.

Nei M. (1972) Genetic distance between populations. *The American Naturalist*, 106, pp. 283–292, ISSN 0003-0147.

Rendo F., Irondo B.M., Jugo L.I., Aguirre A., Vicario A., Estonba A. (2004) Tracking diversity and differentiation in six sheep breeds from the North Iberian Peninsula through DNA variation. *Small Ruminant Research*, 52, pp. 195–202, ISSN 0921-4488.

Reynolds J., Weir B.S. and Cockerham C. (1983) Estimation of the coancestry coefficient: basis for a short-term genetic distance. *Genetics*, 105, pp. 767-779, ISSN 0016-6731.

Tozaki T., Takezaki N., Hasegawa T., Ishida N., Kurosawa M., Tomita M., Saitou N., Mukoyama H. (2003) Microsatellite Variation in Japanese and Asian Horses and Their Phylogenetic Relationship Using a European Horse Outgroup. *Journal of Heredity*, 94, pp. 374-380, ISSN 0022-1503.

Stahlberger-Saitbekova N., Schlaèpfer J., Dolf G., Gaillard C. (2001) Genetic relationships in Swiss sheep breeds based on microsatellite analysis. *Journal of Animal Breeding and Genetics*, 118, pp. 379-387, ISSN 0931-2668.

Zhang G.X., Wang Z.G., Sun F.Z., Chen W.S., Yang G.Y., Guo S. J., Li Y.J., Zhao X.L., Zhang Y., Sun J., Fan B., Yang S.L., Li K. (2003) Genetic diversity of microsatellite loci in fifty-six Chinese native pig breeds. *Acta Genetica Sinica*, 30, pp. 225-233, ISSN 1673-8527.

Weir B.S., Cockerham C.C. (1984) Estimating F-Statistics for the Analysis of Population Structure. *Evolution*, 38, pp. 1358-1370, ISSN 0014-3820.

Genetic Characterization of Albanian Sheep Breeds by Microsatellite Markers

Anila Hoda[1] and Paolo Ajmone Marsan[2]
[1]Agricultural University of Tirana
[2]Università Cattolica del S. Cuore, Piacenza
[1]Albania
[2]Italy

1. Introduction

Albania is a Mediterranean country, located in West of Balkan Peninsula. Albanian farmers have a long tradition in sheep breeding. Sheep comprise one of the most important domestic livestock species in Albania and play an important role in the livelihood of local community since they are a good source of meat, milk and coarse wool (Dobi et al., 2006; Porcu and Markovic, 2005). There are three important local sheep breed in Albania: Bardhoka, Ruda and Shkodrane, which are also the object of this study. The genetic characterization of a breed is very important for the evaluation of genetic variability, which is an important element in conservation of genetic resources and for breeding strategies. Genetic characterization can be done by different classes of molecular markers, such as Restriction Fragment Length Polymorphisms (RFLP) (Abdel-rahman et al., 2010), Single Stranded Conformation Polymorphisms (SSCP) (Bastos et al., 2001), Random Amplified Polymorphic DNA (RAPD) markers (Jawasreh et al., 2011; Kantanen et al., 1995; Paiva et al., 2005; Qasim et al., 2011), Amplified Fragment Length Polymorphisms (AFLP) (Xiao et al., 2009), Single Nucleotide Polymorphisms (SNP) (Pariset et al., 2006a,b), and microsatellites.

Microsatellites are short tandem nucleotide repeats that are randomly distributed throughout eukaryotic genomes. The repeat units can range from two to six base pairs motifs (Tautz and Schlotterer, 1994). They are called also as, simple sequence repeats (SSR) (Tautz, 1989), short tandem repeats (STRs), (Edwards et al., 1991) or variable number tandem repeats (VNTR). Alleles at a specific locus can differ in the number of repeats. They are generally found in nuclear genome, usually in non-coding parts of genome. Microsatellites are "junk" DNA, and are selectively neutral (Li et al., 2002).

Microsatellite loci are often hypervariable with high mutation rates and therefore are highly polymorphic in most mammalian species (Weber, 1990, Jeffreys et al., 1994). Mechanisms of mutation are believed to be unequal crossover during recombination (Smith, 1976), polymerase slippage and especially slipped-strand mispairing during replication (Levinson and Gutman, 1987) resulting in the addition or loss of one or a small number of repeats. They are inherited co dominantly in Mendelian fashion and are relatively easy to score directly.

There is several mutation models considered for microsatellites. The infinite allele model (IAM), (Kimura and Crow, 1964) assumes that all new alleles are unique. Stepwise mutation model (SMM), (Kimura and Ohta, 1978), involve addition or deletion of one repeat. Mutations are also described as or as a combination of single and multiple steps by the two-phase mutation model (TPM), (Di Rienzo et al., 1994).

Microsatellite markers are currently the markers of choice for molecular genetic studies such as reconstruction of phylogenetic and relationships among populations (Bowcock et al., 1994; Forbes et al., 1995; MacHugh et al., 1997), determination of paternity and kinship analyses (Glowatzki-Mullis et al., 1995; Heyen et al., 1997; Luikart et al., 1999; Schlotterer, 2004) forensic studies (Edwards et al., 1992), linkage analysis (Francisco et al., 1996; Kappes et al., 1997; Mellersh et al., 1997), population structures (Arora and Bhatia, 2004; Bruford and Wayne, 1993).

Microsatellites are used, in livestock species for estimating genetic variation within and among breeds (Buchanan et al., 1994; Cronin et al., 2009; Diez-Tascon et al., 2000a; Dowling et al., 2008; Schmid et al., 1999; Saitbekova et al., 1999), for admixture studies (Alvarez et al., 2004; Freeman et al., 2004, 2006; MacHugh et al., 1997; Vicente et al., 2008) and for assigning individuals to breeds (Baumung et al., 2006; Cornuet et al., 1999; Maudet et al., 2002; Meadows et al., 2006; Troy et al., 2001).

Microsatellites as DNA markers are advantageous over many other markers as they are highly polymorphic, highly abundant, co-dominant inheritance, simply to analyze and easy to score, but nevertheless this type of marker has disadvantages such as null alleles, or size homoplasy (Schlotterer, 2004).

Microsatellites are the commonest markers used for genetic characterization of sheep breeds. Diez-Tascon et al., (2000b) studied genetic variability among six Merino populations using 20 microsatellites. Stahlberger-Saitbekova et al., (2001) analyzed genetic diversity between seven breeds from Swiss Alps and Mufflon, using ovine, caprine and bovine microsatellite markers. Arranz et al., (2001) have analyzed genetic variability of five Spanish sheep breed and Awassi by 18 microsatellites markers. Grigaliunaite et al., (2003) studied variability, paternity and possible bottleneck in 7 Baltic sheep breeds, using 15 microsatellite markers. Pariset et al., (2003) have used 11 microsatellites for genetic variation and inbreeding analysis in 17 flocks from Sarda breed in Viterbo province. Rendo et al. (2004) analyzed genetic variability of six autochthonous Nordeuropean sheep breeds based on 11 microsatellite markers. Alvarez et al., (2004) have used 14 microsatellite markers to analyze the relationship between Nordeuropean sheep breeds. Tapio et al., (2005) have used 25 microsatellites in 20 native and 12 imported North European sheep breeds in order to evaluate the importance of each breed for gene diversity. Baumung et al., (2006) have used 25 microsatellite markers for genetic characterization and breed assignment in 11 Austrian sheep breeds. Peter et al., (2007) have examined in a comprehensive study the genetic diversity of 57 European and Middle Eastern sheep breeds, using 31 microsatellites markers. Part of this study has been also the three Albanian sheep breeds considered here. Cinkulov et al., (2008) have used 15 microsatellite markers and mtDNA to estimate genetic variation in seven Pramenka types from West Balkan. Ligda et al., (2009) have used 28 microsatellite markers to analyze genetic diversity and differentiation in 8 Greek sheep breeds. Dalvit et al., (2008) have used 19 microsatellite markers for genetic characterization of 8 sheep breeds from Italy, Germany and Slovenia. Dalvit et al., (2009) have used 19 microsatellite markers for genetic variation and presence of breed substructure of four native sheep breeds from North Italy. Arora and Bhatia, (2004) have used 13 microsatellites to asses genetic effects of the population declines in Muzzafarnagri Sheep from India. Tapio et al., (2010) have used 20 microsatellite markers for genetic diversity and population structure of 52 sheep breed from three geographical regions Caucasus, Asia, and the eastern fringe of Europe.

The Food and Agriculture Organization (FAO) has proposed an integrated programme for the global management of genetic resources, Measurement of Domestic Animal Diversity (MoDAD) program, using panels of microsatellites for characterizing farm AnGR.

The genetic characterization of local sheep in Albania, for a long time has been very limited based mainly on blood or milk protein polymorphism and visible phenotypic profile (Zoraqi, 1991). Recently, in the frame of Econogene project (www.econogene.eu) these breeds are characterized at molecular level using several set of markers like microsatellite (Hoda et al., 2009; Peter et al., 2007), AFLP (Hoda et al., 2010), SNP (Hoda et al., 2011). The study was undertaken to characterize the genetic diversity, to evaluate the genetic relationship and structure of these local sheep breeds, using 31 microsatellite markers recommended by MoDAD/FAO.

2. Materials and methods

2.1 Sample collection and microsatellites

A total of 93 individuals representing 3 different Albanian sheep breeds were analyzed. The breeds were Bardhoka, Ruda and Shkodrane. For each breed, 31 unrelated individuals were selected. Sampling was carried out in mountainous area, where still have pure breed individuals, from ten to eleven flocks.

Bardhoka breed (Figure 1) is classified under the long tail group. Its origin is North/Northeast of Albania and Western part of Kosova as well. This is a sheep with triple productive profile, milk, lamb and wool. It has a good developed body and a strong skeleton. The head has strong mandibles, wide face and big ears. The legs are strong and with thick bones. Bardhoka has a totally white fleece/cover. A well developed udder is characteristic of the breed. It has good volume and well-developed teats, very appropriate for milking. Usually, ewes are polled while the rams are horned.

Fig. 1. Bardhoka sheep breed

Ruda (Figure 2) is triple purpose breed with half-fine wool belongs to the long tail group. This breed is part of Tsigaya family regarding to the wool quality and other zootechnic traits. It is originated and widespread in North-Eastern part of Albania. This breed is adapted to pastures in high altitude and for long distance transhumant. Animals have a well-developed body with long legs that is a characteristic for this breed. The animals are generally white but sometimes black ones can show up. Ewes are polled and rams have big horns. Animals are covered by a non-dense fleece; while their neck and abdomen is uncovered.

Fig. 2. Ruda sheep breed

Shkodrane sheep breed (Figure 3) belongs to the long tail group of a triple purpose use. Its origin is Northern Albania. The tendency is the reduction of population. Most of the crosses are made with Bardhoka breed aiming to increase the milk production and body weight features. Shkodrane is a small sheep, well adapted to poor and stony pastures of North Albania. It has low requirements for the feed and it is resistant towards cold and dry climate. The very long and coarse wool is typical for this breed. Characteristic of its exterior is the light brown pigmentation of the skin at legs and face. Ewes are polled while the rams are horned. This breed is estimated as "potentially endangered", and some efforts to establish conservation programs are in process.

Fig. 3. Shkodrane sheep breed

In Table 1 are shown some of phenotypic traits of Albanian sheep breeds (Dobi et al., 2006; Porcu and Markovic, 2006).

Breed	Color	Wool	Tail	Body weight (kg)		Withers Height (cm)		Milk Production (kg)	Use
				Male	Female	Male	Female		
Bardhoka	White,	Coarse	Long	60	40	70	60	150 - 200	Milk, wool, meat, fur
Ruda	White	Half-fine	Long	50	40	65	55	90	Milk, wool, meat, fur
Shkodrane	White, reddish face	Long, coarse	Long	42	36	55	52	130 - 150	Milk, meat, wool,

Table 1. Phenotypic traits of three Albanian sheep breeds.

2.2 DNA extraction and microsatellite analyses

Blood samples of 5 – 10 ml were collected in EDTA tubes and stored at –20°C. DNA was isolated according to standard phenol-chloroform extraction method. All samples were genotyped for 31 microsatellite markers according to the methodology explained in detail, by (Peter et al., 2007).

2.3 Data analysis

Allele frequencies, observed heterozygosity (Ho), expected heterozygosity (He) were estimated for 31 microsatellite markers using Genalex 6 program (Peakall and Smouse, 2006). Polymorphic information content (PIC) was estimated for all markers in all breeds using the Cervus software (Marshall, 2001).

Tests of genotype frequencies for deviation from Hardy-Weinberg equilibrium (HWE) as well as for linkage disequilibrium were carried out using Markov Chain Monte Carlo simulation (100 batches, 5000 iterations and a dememorization number of 10 000) implemented in the Genepop V.1.2 program(Raymond and Rousset, 1995).

The program FSTAT, (Nei, 1987), and estimation of Wright's fixation index (Weir and Cockerham, 1984). A significance test on the estimates for each microsatellite locus were obtained by constructing 95% and 99% confidence intervals based on the standard deviations estimated by jackknifing across populations using FSTAT (Goudet, 2001). Estimates of genetic variability for each breed (He, Ho), mean number of alleles were computed using GENETIX program (Belkhir et al., 2001). Gene flow (Wright, 1931) was calculated using the same program (Belkhir et al., 2001).

Cavalli-Sforzas Chord Distance D_C (Cavalli-Sforza and Edwards, 1967), Reynolds' D_R distance (Reynolds et al., 1983), Nei's D_A distance (Nei et al., 1983) and Neis Standard Distance (Nei, 1972) among breeds were computed using Populations program (Langella, 2002). In order to test the presence of correlations between different distance matrices, a Mantel test, modified by (Manly, 2007) was carried out with Programm FSTAT (Goudet, 2001).

The genetic distance of Reynolds (D_R) among breeds was used for the construction of UPGMA consensus tree (Felsenstein, 1993). Bootstrap (1000 replicates) resampling was performed to test the robustness of the dendrogram topology.

Genetic distances among individuals were estimated as the proportion of shared alleles (D_{PS}) using Populations program (Langella, 2002). Individual distances were represented by a neighbor-joining tree and depicted using software package TreeView version 1.6.6 (Page, 1996).

The analysis of population's structure by a clustering analysis based in Bayesian model was carried out by the program STRUCTURE (Pritchard et al., 2000). The program uses Markov Chain Monte Carlo method based on the "admixture model", where allelic frequencies were correlated, with "burn in period" and "period of data collection" of 300000 iterations. The samples were analyzed with K from 1 to 4, applying 5 independent running. Evanno's method (Evanno et al., 2005) was used to identify the appropriate number of clusters using the *ad hoc* statistic Δk, which is based on the second order rate of change of the likelihood function with respect to successive values of K.

To test for evidence of a recent genetic bottleneck, the program BOTTLENECK (Piry et al., 1999) was used. The program tests for departure from mutation drift equilibrium based on heterozygosity excess or deficiency. It compares heterozygosity expected (*He*) at Hardy–Weinberg equilibrium to the heterozygosity expected (*Heq*) at mutation drift equilibrium in the same sample, that has the same size and the same number of alleles. Wilcoxon signed rank test was used to test for heterozyosity excess under all three mutation models, infinite alleles (IAM), two-phase (TPM), and the step-wise mutation model (SMM).The method of graphical representation of mode-shift indicator, was also used for assessing distortion in allele frequency, indicative of possible bottleneck.

Nei genetic distance calculated from the allele data was plotted as PCA using GenAlEx program (Peakall and Smouse, 2006).

The Factorial Correspondence Analysis (FCA) is performed to visualize the relationships between individuals from different breeds and to test possible admixtures between the populations. FCA was computed using GENETIX program (Belkhir et al., 2001).

Geneclass2, (Paetkau et al., 1995), assuming a default allelic frequency of 0.001 and a threshold of 0.05. The assignment of individuals to the reference population was carried out using Bayesian approach (Rannala and Mountain, 1997). The "leave one out" procedure assignment was applied for the individuals. The confidence level was 99%.

The hierarchical analysis was carried out using analysis of molecular variance (AMOVA) implemented in the ARLEQUIN Ver. 3.0 package (Excoffier et al., 2005). AMOVA yields estimations of population structure at different levels of the specified hierarchy.

3. Results

3.1 Microsatellite markers

All markers were highly polymorphic. In table 2 are displayed the variability parameters of the investigated loci. A total number of 348 alleles were observed at 31 microsatellite loci. Except of SRCRSP4, all the markers displayed 5 or more alleles. The total number of alleles varied from 4 (SRCRSP5) to 20 (INRA63) with an overall mean of 11.33 alleles/locus. Mean number of alleles per locus ranged from 4 (SRCRSP5) to 13.33 (INRA63) having a pooled mean of 8.54. The effective number of alleles (N_E) ranged between 2.17 (SRCRSP9) to 7.7 (OARFCB20), with an overall mean of 4.57. PIC ranged from 0.495 (SRCRSP9) to 0.856 (OARFCB20). The within-breed deficit in heterozygosity, as evaluated by the F_{IS} parameter, ranged between −0.081 (SRCRSP9) to 0.458 (SRCRSP5) having a total mean of 0.041 for all loci. A very high contribute displayed the markers Oarae129 (0.368), Srcrsp5 (0.458) and Mcm527 (0.237) with (p<0.001) high significant F_{IS} values. F_{IT} values ranged from -0.044 (BM8125) to 0.5 (Srcrsp5). The global heterozygosity deficit (F_{IT}) was estimated 0.052 and global breed differentiation evaluated by F_{ST}, was estimated 0.011. The contribution of the

Locus	Allelic range (bp)	TNA	MNA	NE	PIC	AR	F_{IS}	F_{IT}	F_{ST}
BM1329	162-180	6	5.00	2.82	0.584	5.309	0.02	0.027	0.007
BM8125	112-126	7	6.33	3.37	0.657	6.148	-0.047	-0.044	0.003
DYMS1	163-199	14	10.67	7.32	0.827	11.515	-0.029	-0.007	0.022***
HUJ616	114-162	15	10.00	5.03	0.762	10.939	-0.025	-0.019	0.006***
OARAE129	136-162	7	5.33	3.12	0.606	5.023	0.368***	0.367***	-0.001
OARCP34	118-130	7	6.33	5.02	0.754	6.237	-0.017	-0.006	0.01*
OARCP38	120-136	8	5.67	2.26	0.518	5.667	-0.04	-0.043	-0.003
OARFCB128	98-128	9	7.33	4.63	0.740	7.192	0.083	0.085	0.003
OARHH47	121-155	13	9.67	5.43	0.761	9.93	0.163***	0.193***	0.036
OARVH72	125-141	9	8.33	7.12	0.829	8.481	0.047	0.05	0.003
MAF65	122-138	9	7.33	3.99	0.704	7.069	0.021	0.013	-0.008
INRA63	158-206	20	13.33	5.41	0.781	12.909	-0.025	-0.023	0.002
OARFCB20	92-118	13	11.00	7.70	0.834	10.854	0.056	0.071*	0.016*
OARJMP58	141-175	15	11.67	5.48	0.779	11.734	0.034	0.05	0.016**
OARJMP29	115-157	15	11.00	5.82	0.777	10.653	-0.082	-0.048	0.031***
OAFCB193	95-133	15	9.67	2.87	0.620	9.557	0.001	-0.001	-0.002
BM1824	169-175	5	5.00	3.82	0.684	4.978	0.013	0.016	0.002
MAF70	128-160	16	12.33	5.65	0.784	12.5	0.065	0.084*	0.021**
MAF209	110-140	12	9.67	5.11	0.762	10.422	-0.023	-0.005	0.018**
OAFCB304	149-189	16	10.67	4.05	0.701	11.06	0.089	0.107*	0.02***
SRCRSP9	108-128	10	6.33	2.17	0.490	6.023	-0.089	-0.096	-0.007
ILSTS5	186-211	10	7.00	2.86	0.590	7.573	-0.026	0.008	0.033**
ILSTS11	271-286	9	6.67	4.61	0.736	7.163	0.017	0.02	0.003
ILSTS28	127-169	13	10.00	6.33	0.815	9.777	-0.05	-0.052	-0.002
MAF214	180-265	15	10.67	4.19	0.712	10.613	-0.08	-0.054	0.024*
SRCRSP5	146-154	4	4.00	2.56	0.510	3.916	0.458***	0.5***	0.077***
SRCRSP1	126-140	8	6.67	2.86	0.583	6.328	0.116	0.121*	0.006
MAF33	116-143	12	9.33	4.27	0.723	9.263	0.048	0.043	-0.005
MCM140	161-192	13	10.33	5.97	0.804	10.439	0.111*	0.107*	-0.005
OAFCB226	118-156	14	10.67	5.28	0.778	10.73	-0.064	-0.058	0.006
MCM527	159-179	9	6.67	4.66	0.736	6.93	0.237***	0.243***	0.007
Mean/breed			11.23	8.54	4.57	0.72		0.041***	0.052***	0.011***

Table 2. Fragment sizes, total number of alleles per locus (TNA), Mean number of alleles (MNA), effective number of alleles (NE), polymorphic information content (PIC), Allelic richness (AR), Wright's F-statistics (FIT, FIS, FST) for each locus and all loci in three Albanian sheep breeds.

microsatellite markers for breed differentiation was estimated by the significance of the F_{ST} statistics. Only twelve markers had significant F_{ST} values and therefore contributed to breed differentiation. F_{ST} values ranged from -0.008 (MAF65) to 0.077 (SRCRSP5). The overall estimates for F-statistics were significantly (p < 0.05) different from zero.

In table 3, is shown Nei genetic diversity for 31 markers H_T. The values for observed heterozygosity ranged from 0.315 (SRCRSP5) to 0.891 (ILSTS28), with an overall mean value of H_O of 0.72, while the values of expected heterozygosity ranged from 0.543 (SRCRSP9) to 0.865 (OARFCB20). The mean value of Nei gene diversity, H_T was 0.75. The diversity within

Loci	H_O	H_S	H_T	D_{ST}	G_{ST}
BM1329	0.632	0.645	0.648	0.003	0.008
BM8125	0.739	0.705	0.707	0.001	0.003
DYMS1	0.88	0.856	0.868	0.012	0.021
HUJ616	0.822	0.802	0.805	0.003	0.006
OARAE1	0.432	0.686	0.685	-0.001	-0.001
OARCP3	0.812	0.8	0.805	0.006	0.01
OARCP3	0.583	0.563	0.562	-0.001	-0.003
OARFCB	0.705	0.774	0.775	0.001	0.002
OARHH4	0.67	0.801	0.821	0.02	0.037
OARVH7	0.821	0.862	0.864	0.002	0.003
MAF65	0.742	0.758	0.754	-0.004	-0.008
INRA63	0.839	0.818	0.819	0.001	0.002
OARFCB	0.75	0.775	0.794	0.019	0.036
OARJMP	0.785	0.813	0.822	0.009	0.016
OARJMP	0.882	0.815	0.832	0.018	0.031
OAFCB1	0.628	0.635	0.634	-0.001	-0.002
BM1824	0.731	0.741	0.742	0.001	0.002
MAF70	0.763	0.816	0.828	0.011	0.021
MAF209	0.817	0.799	0.808	0.01	0.018
OAFCB3	0.677	0.747	0.757	0.01	0.02
SRCRSP	0.591	0.543	0.541	-0.002	-0.007
ILSTS5	0.656	0.641	0.655	0.014	0.033
ILSTS1	0.774	0.786	0.787	0.002	0.003
ILSTS2	0.891	0.848	0.847	-0.001	-0.002
MAF214	0.814	0.753	0.765	0.013	0.025
SRCRSP	0.315	0.582	0.614	0.032	0.076
SRCRSP	0.576	0.652	0.654	0.002	0.005
MAF33	0.733	0.771	0.769	-0.003	-0.005
MCM140	0.747	0.84	0.838	-0.003	-0.005
OAFCB2	0.864	0.811	0.814	0.004	0.006
MCM527	0.599	0.786	0.79	0.004	0.007
Overall	0.718	0.749	0.755	0.006	0.012

Table 3. Nei's genetic diversity for each loci across all populations.

breeds (H_S) was 0.749 and diversity between breeds (D_{ST}) of 0,006. The diversity within breeds relative to the diversity of the whole population, G_{ST} value was 0.011. This value is similar with F_{ST} value.

Except of BM1824, all the markers displayed private alleles. A total of 89 private alleles were found, but only 10 private alleles had a frequency higher than 5% (Table 4).

Breed	Locus	Allele	Freq
Bardhoka	BM1329	180	0.050
Bardhoka	DYMS1	175	0.100
Bardhoka	HUJ616	156	0.083
Ruda	OARJMP58	165	0.065
Ruda	OAFCB304	179	0.081
Ruda	MAF33	124	0.065
Ruda	OAFCB226	152	0.081
Shkodrane	DYMS1	177	0.065
Shkodrane	HUJ616	154	0.067
Shkodrane	MAF209	136	0.065

Table 4. List of private alleles with frequency higher than 5%.

3.2 Genetic variation

The genetic variability for each breed was studied, regarding mean number of alleles and allelic richness, (Table 5). All breeds have similar mean number of alleles. The lowest value was displayed by Shkodrane breed. Overall estimate of allelic richness was 8.61. Within breed mean expected heterozygosity varied from 0.74 in Shkodrane to 0.77 in Ruda having an overall mean value of 0.75. Mean estimate of observed heterozygosity overall breed and loci was 0.72. Bardhoka and Shkodrane showed a significant deficit of heterozygotes ($p<0.05$), while Ruda showed a nonsignificant excess of heterozygotes. Mean value of inbreeding coefficient (F_{IS}) was 0.04. Deviations from Hardy-Weinberg equilibrium were significant for 6 out 93 loci breed combinations ($p<0.01$). The microsatellites SRCRSP5 and OARAE129 showed deviations in Bardhoka and Shkodrane. OARJMP29 was deviating only in Bardhoka and MCM27 was deviating only in Shkodrane. Ruda showed deviations in none of the markers.

3.3 Genetic differentiation

Polymorphism information content (PIC) in three Albanian sheep breeds ranged from 0.690 (Shkodrane) to 0.722 (Ruda). The breeds showed poor genetic differentiation, where F_{ST} index was equal to 0.011. Also, the average G_{ST} values over all loci was 0.011 indicating that a 1.1% of total genetic variation corresponded to differences among populations, whereas 99% was explained by difference among individuals.

Breed	MNA	H_O(SD)	H_E(SD)	AR	F_{IS}	PIC	NPA	HWE
Badhoka	8.58	0.71(0.015)	0.76(0.014)	8.35	0.065***	0.712	24	3
Ruda	8.84	0.75(0.014)	0.77(0.019)	8.54	0.022	0.722	38	1
Shkodrane	8.19	0.71(0.015)	0.74(0.018)	7.97	0.037*	0.690	27	3
Total	8.54	0.72(0.13)	0.75(0.09)	8.61	0.041	0.708		

Table 5. Mean number of alleles (MNA), mean observed (H_O) and expected (H_E) heterozygosity, allelic richness (AR), within-breed heterozygote deficiency (F_{IS}), number of private alleles (NPA), polymorphism information content (PIC) and number of loci not in the Hardy- Weinberg equilibrium at P < 0.05, for each breed across 31 loci.

The correlations between different distance matrices, D_R, D_A, D_C and D_S were tested by Mantel-Test modified by (Manly, 2010). High significant correlations were obtained between different matrices. The highest correlations were observed between Nei's D_A, Distance and Cavalli-Sforza, D_C, Distance (0,999912, p < 0,01) and the lowest between Reynolds, D_R, Distance and Cavalli-Sforza, D_C, Distance (0.673549, p < 0,01) (Table 6).

	D_A	D_S	D_R
D_C	0.999912**	0.796600**	0.673549**
D_A		0.804546**	0.683290**
D_S			0.983366**

Table 6. Corelations between distance matrix, D_C, D_A, D_S and D_R and their significance (** p < 0,01).

D_R genetic distance between each pair of populations are shown in Table 7. Distance matrix was used to build NJ phylogenetic tree. The smallest distance was between Bardhoka and Shkodrane. In figure 4 is presented the UPGMA tree based on Reynolds distance. Bootstrapping tested the robustness of the tree. Bootstrap values ranged from 41 to 100 indicating rather strong topology of the phylogenetic tree. Pairwise F_{ST} values between each pair of three sheep breeds were very low, reflecting a poor genetic differentiation breeds.

	Bardhoka	Ruda	Shkodrane
Bardhoka		0.012 (21)	0.010 (24)
Ruda	0.032 29		0.011 (23)
Shkodrane	0.031 28	0.029 28	

Table 7. Reynold's D_R genetic distance matrix (below diagonal), pairwise F_{ST} distance between breeds (above diagonal) and gene flow (Nm) (in bracket)

The low degree of genetic differentiation found between Albanian sheep breeds is supported by high level of gene flow (Nm, number of migrants per generation) between breeds (Table 7). Similar values of gene flow between populations are observed, but the highest value is observed between Bardhoka and Shkodrane.

The program BOTTLENECK (Piry et al., 1999) was used to investigate the hypothesis of a recent bottleneck. Wilcoxon sign-rank test under three mutations models IAM, TPM and SMM and shift mode test were used to find out recent bottleneck (heterozygosity excess) in the three Albanian sheep breeds. The heterozygosity excess obtained (Table 8) were nonsignificant ($P < 0.5$) under all the models in all sheep populations. These results were consistent with the normal L-shaped distribution of allele frequency in all populations (Figure 5). The results obtained here, demonstrate that the null hypothesis of mutation–drift equilibrium was fulfilled in these breeds.

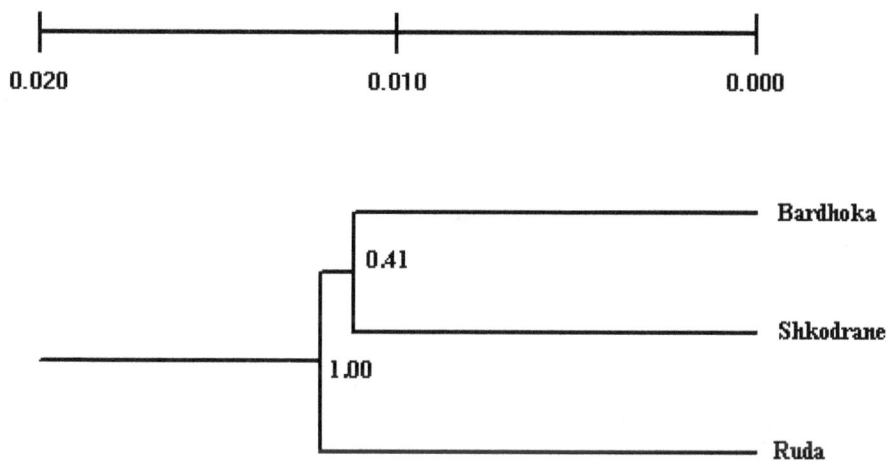

Fig. 4. Dendrogram of Reynolds genetic distance between Albanian sheep breeds by UPGMA algorithm

Breed	Mutation model		
	IAM	TPM	SMM
Bardhoka	0.995	0.795	0.999
Ruda	0.999	0.618	0.999
Shkodrane	0.951	0.909	0.999

Table 8. Bottleneck analysis for Albanian sheep breeds using Wilcoxon rank test under three mutation models

An AMOVA analysis was carried out to analyze the variation within and between breeds. The AMOVA revealed that percentage of variation among populations was 1.18% and within populations was 98.82% (Hoda et al., 2009). Variance components among population were highly significant ($p<0.001$). SRCRSP5 marker contributed the highest variability (8.42%) among populations. ILSTS11 and OARVH72 contributed the lowest variability (0.34% and 0.39% respectively).

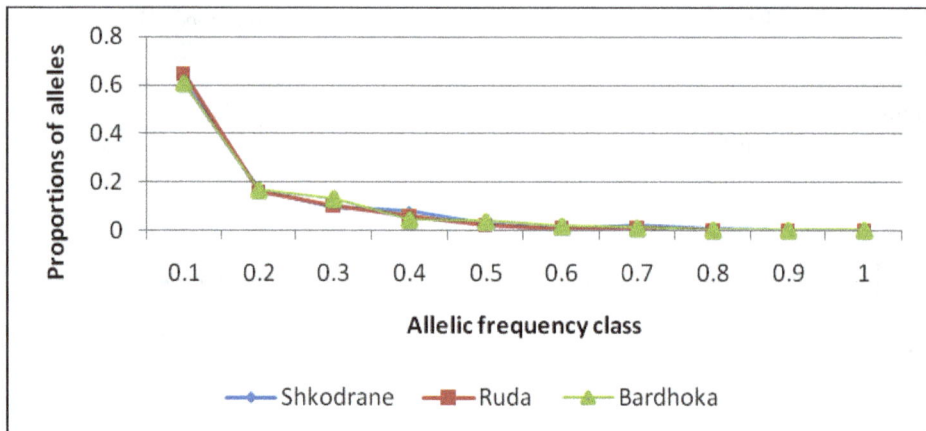

Fig. 5. Graphical representation of proportions of alleles and their distribution in three Albanian sheep breeds

3.4 Population structure

The program STRUCTURE (Falush et al., 2003; Pritchard et al., 2000) was run 5 times independently, with K ranging from 1 to 4, in order to choose the appropriate value of K. The results of the analyses with Structure are summarized in Table 9. We have used two methods to estimate the best K value: Pritchard and Evanno methods. According to Pritchard method, the average likelihood values Ln Pr(X | K) for each K, were plotted against K, in order to choose the optimal K value. The likelihood values reaches a maximum at K= 2 and afterwards decreased rapidly. Also the variance reaches the lowest value at K = 2. The results of this method are shown also previously (Hoda et al., 2009). Evanno method (Evanno et al., 2005) was applied and was calculated, an *ad hoc* statistic based on the second order rate of change of the likelihood function with respect to K. This statistic peaked at K = 2 (Table 9) (Figure 6) indicating strong support for 2 groups. Graphic representation of the estimated membership coefficients to the clusters for each individual, (K= 2), is shown previously (Hoda et al., 2009). The results of structure analysis show a high level of breed admixture.

K	Ln Pr(X | K)]	ΔK
1	-10080.2	
2	-10073.9	13.30326
3	-10198	0.702013
4	-10172.7	

Table 9. Estimated posterior probabilities [Ln Pr(X | K)] for different numbers of inferred clusters (K) and ΔK statistic.

The tree of individuals based on proportion of shared alleles D_{AS}, obtained through Neighbor-joining algoritm (figure 7) showed that animals from three different breeds are mixed together.

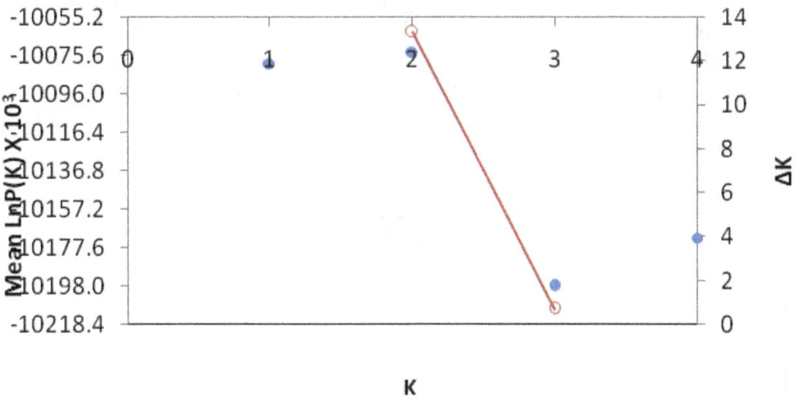

Fig. 6. Results of the STRUCTURE analysis showing mean Ln P(D) and ΔK values.

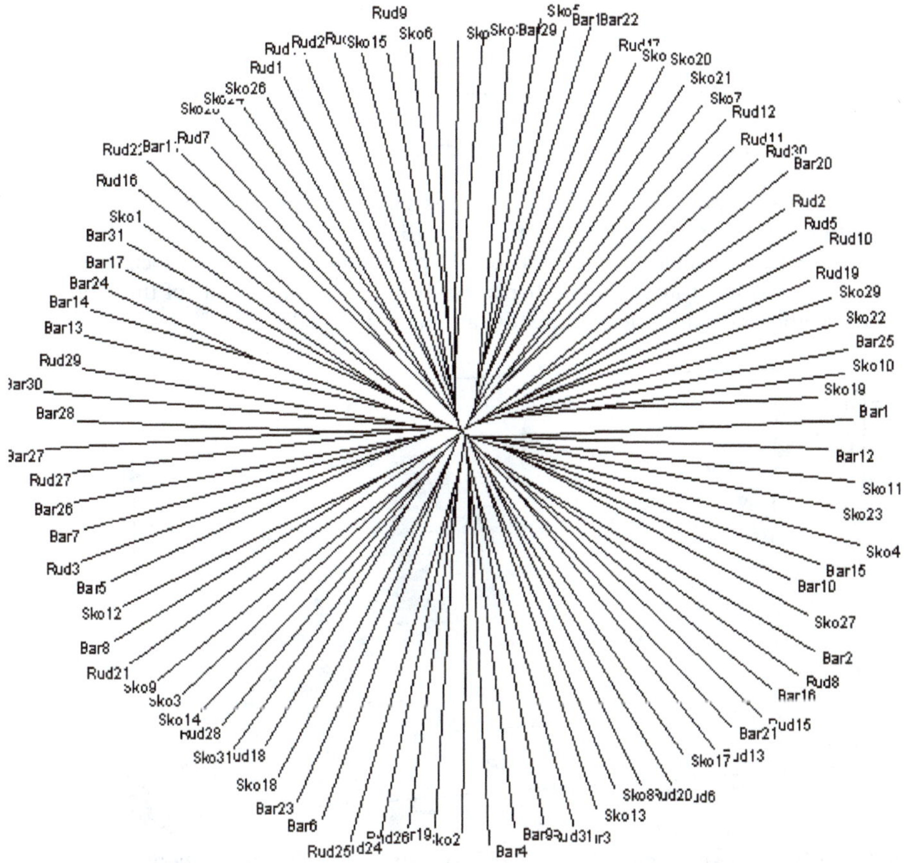

Fig. 7. Dendrogram of allele sharing distances between each individual of three sheep breeds

Relationship of populations based on Nei's genetic distance was performed by frequency Principle Component Analysis (PCA). The PCA was carried out with Genalex (Peakall and Smouse, 2006). In the Principal Component Analyses, the first and second axis accounted for 58.7, and 41.3% of the total inertia respectively. As is shown in figure 8, the first axis separate Ruda from two other breeds and the second axis separate Shkodrane from other breeds.

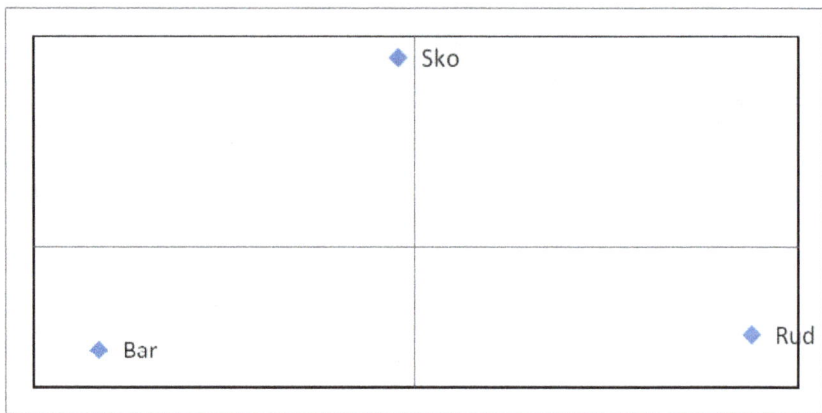

Fig. 8. Diagram of PCA based on Nei's genetic distance

The Factorial Correspondence Analysis (FCA) is performed to visualize the relationships between the individual using Genetix program (Belkhir et al., 2001), It is a multivariate method of analysis. Allele frequencies, of all populations and at all loci, are used as variables. The results of analysis are displayed in figure 9. Individuals of three breeds are grouped together indicating clear admixture among individuals.

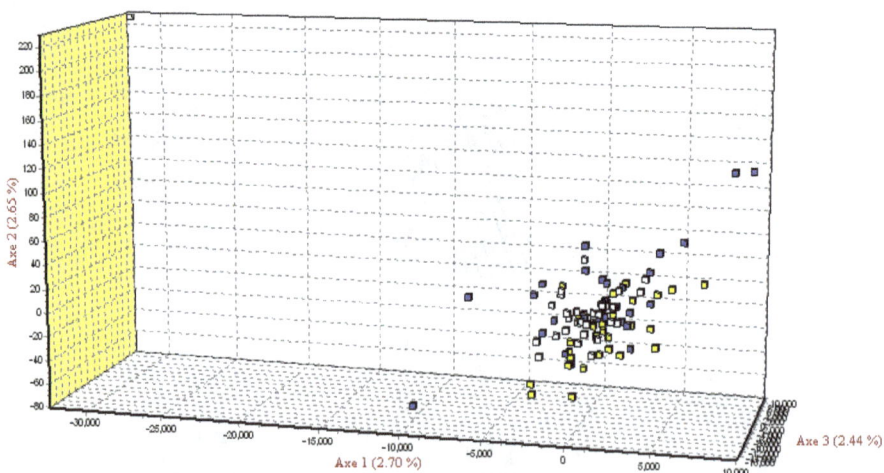

Fig. 9. Factorial Correspondence Analysis (FCA) results showing the relationship between all of the individuals analyzed in the study.

Breed	Direct					Simulation				
	D_C	D_S	D_A	Freq	Bayes	D_C	D_S	D_A	Freq	Bayes
Bardhoka	51.61	61.29	54.84	70.97	64.52	6.45	6.45	6.45	12.90	19.35
Ruda	58.06	64.52	54.84	64.52	67.74	19.35	12.90	25.81	22.58	29.03
Shkodrane	83.87	67.74	83.87	70.97	64.52	0.00	3.23	3.23	3.23	6.45
Total	64.52	64.52	64.52	68.82	65.59	8.60	7.53	11.83	12.90	18.28

Table 10. Percentage of individuals from each sheep breed correctly assigned to their reference population by likelihood and genetic distance methods.

Likelihood and genetic distance based methods were used for a direct assignment of and for an exclusion analysis of individuals to their reference population. The likelihood based methods are used also previously, (Hoda et al., 2009). Table 10 shows the results of the assignment test obtained through different methods. By direct assignment of individuals to their reference populations, the best scores were obtained with frequency method (68.82%). The highest rate of excluded animals out of the 10000 simulated individuals was obtained using Bayes theorem (18.28%). By direct assignment, using the genetic distance based methods, the highest percentage of correctly assigned individuals was obtained by Nei's genetic distances D_A and D_S. The highest number of correctly assigned animals, for all methods were from Shkodrane.

4. Discussions

The local breeds analyzed in this study are the most important Albanian sheep breeds that are reared on small familiar farms in extensive or semi-intensive systems. The aim of this study was to examine the genetic diversity within and between them using 31microsatellite analysis.

The number of alleles observed at a locus is an indication of genetic variability at that locus. FAO have recommended that microsatellite loci for genetic diversity studies should have more than four alleles. The total number of alleles per locus ranged from 4 (SRCRSP5) to 20 (INRA63), with a mean value of 11.23, indicating that all the microsatellite loci were sufficiently polymorphic and were appropriate to analyze diversity. This value was smaller than values found for four Romanian sheep breeds, by 11 microsatellite markers (17.9, Kevorkian et al., 2010)), Italian native sheep breeds, by 19 microsatellite markers (15.4, Dalvit et al., 2009)), or European sheep, by 23 microsatellite markers (19.9, Handley et al., 2007)), but were higher than values provided for Gentile di Puglia sheep breed, by 19 microsatellite markers (9.68, d'Angelo et al., 2009).

Takezaki and Nei, (1996) have determined that gene diversity should be in the range of 0.3 to 0.8 in the populations, in order that markers to be useful for measuring genetic variation. Gene diversity for each breed ranged from 0.74 to 0.77, with an average value of 0.75. This confirmed that these markers were appropriate for measuring genetic variation.

The polymorphic information content (PIC) is a parameter indicative of the degree of informativeness of a marker. All markers have PIC values higher than 0.5, indicating that are highly informative. Only SRCRSP9 had a PIC value close to 0.5 (0.490). This confirm again that the set of microsatellite markers were effective for genetic diversity estimation in Albanian sheep breeds.

The mean allele number (allele diversity) provides a reasonable indicator of the levels of variability present within a breed. The range of allele diversity measure was 8.19 – 8.84 indicated high level of genetic variability of Albanian sheep breed. The allele diversity estimates obtained in this study are comparable with those reported for Italian native sheep breeds (7.1 – 9.6, Dalvit et al., 2009), for Baltic sheep (3.93 – 8.33, Grigaliunaite et al., 2003), for Brazilian sheep breeds (4.22 – 8.39, Paiva et al., 2005), for Austrian sheep (6.19 – 10.7, Baumung et al., 2006), in Romanian sheep (7.8-11.6, Kevorkian et al., 2010), for five Bulgarian sheep breeds (6.3 – 8.6, Kusza et al., 2010), in Six Indian Sheep Breeds, (7.72 – 9.56, Arora et al., 2010).

The allelic richness had an average value of 8.61 alleles per breed. That is higher than what is observed in Northern European sheep breeds (6.98, Tapio et al., 2005), in European sheep breeds (7.5,), in Romanian sheep (4.82, KEVORKIAN et al., 2010), in West Balkan pramenka sheep types (7.9, Cinkulov et al., 2008), and in native Italian sheep breeds (7.3, Dalvit et al., 2009).

The genetic diversity values found in Albanian sheep breeds ranged from 0.74 to 0.77, with a mean value of 0.75. The values of gene diversity estimates were comparable with values detected for Baltik (0.57 – 0.76, Grigaliunaite et al., 2003), Austrian (0.67 – 0.78, Baumung et al., 2006), Bulgarian (0.73 – 0.80, Kusza et al., 2010), Romanian (0.67 – 0.79, Kevorkian et al., 2010), Italian native breeds (0.70 – 0.80, Dalvit et al., 2009). These gene diversity estimates are smaller than those detected by for Egyptian sheep breeds (0.81 – 0.86, El Nahas et al., 2010), Iranian sheep breeds (0.83, Nanekarani et al., 2010), but higher than values found for Indian sheep breeds (0.59-0.65, Mukesh et al., 2006), Slovak Tsigai populations (0.46 – 0.61, Kusza et al., 2009), West Balkan Pramenka sheep types (0.739 – 0.830, Cinkulov et al., 2008).

Estimates of observed heterozygosity confirm the high level of diversity evidenced in the Albanian sheep breeds. The average observed heterozygosity was 0.72. Overall heterozygosity estimates are comparable with what found in Swiss sheep breeds (Stahlberger-Saitbekova et al., 2001), Austrian sheep breeds (Baumung et al., 2006), Spanish breeds (Alvarez et al., 2004) and Alpine sheep breeds (Dalvit et al., 2008), Sarda sheep flocks (Pariset et al., 2003).

The comparison of average observed and expected heterozygosity values did not show great differences between breeds. All breeds showed smaller observed than expected heterozygosities (Table 3). Bardhoka and Shkodrane showed on 3 loci out of 31, significant deviation from Hardy Weinberg proportion, but Ruda showed no deviations from Hardy Weinberg proportion. The observed heterozygosity (Ho) of microsatellite loci was always larger than 0.50 (Table 3), except of Oarae129 (0.432) and SRCRSCP5 (0.315). Most of the loci showed the heterozygote deficit as also depicted by positive F_{IS} value (Table 2).

Several factors can contribute to less than expected heterozygosity in a population. One reason might be inbreeding, i.e. mating between relatives. In case of inbreeding, the deficit affects all or most of the loci in a similar way. The number of loci, with a significant deficit of heterozygotes, is very small. The farmers do efforts to avoid as much as possible, the breeding between relatives, trying not to use the rams from their own flock. Other factor that can also cause a deficit of heterozygotes in the population might be the presence of "null alleles" (non-amplifying alleles). This may cause a afalse observation of homozygotes excess. Peter et al., (2005) have indicated the presence of null alleles in locus OarAe 129. Finally, the presence of population substructure within the breed may lead to a Wahlund effect, since the animals were sampled from many small flocks. For all breeds sampling was carried out in 10 – 11 small flocks.

The high values of heterozygosities and allelic richness obtained in this study confirm that native breeds of sheep represent an important reservoir of genetic diversity, even though the level of differentiation among closely located breeds is small. This is in accordance with the prediction of (Handley et al., 2007) of a higher within-breed diversity and lower genetic differentiation in southern than in northern European breeds. Peter et al., (2007), observed a higher genetic diversity of Middle Eastern, Turkish, Greek, Albanian and Romanian sheep breeds compared with northwestern European breeds.

Grigaliunaite et al., (2003) showed that when unique allele has a frequency below 0.1 it might be an allele that is present in several populations at low frequency and could be found also in other breeds, if greater fraction of the total population would be screened. A high number of private alleles were observed in Albanian sheep breeds and none of them had frequency higher than 10%. Peter, (2005) in the study of 57 European and Middle-Eastern sheep breeds, including also the Albanian sheep breed considered here, showed the presence of 2 private alleles in Bardhoka, only one in Ruda and none to Shkodrane. Therefore all private allele data, obtained here are not informative

The studied breeds showed a poor, but significant genetic differentiation (0.011), which is very low compared to those from other genetic diversity studies, e.g. 18.3% for Indian sheep (Mukesh et al., 2006), 13.3% Slovak Tcigai populations (Kusza et al., 2009), 8.2% Romanian breeds (Kevorkian et al., 2010), 8.3% Bulgarian breeds (Kusza et al., 2009), 8% Austrian (Baumung et al., 2006), 6.1% Six Indian Sheep Breeds, (Arora et al., 2010), 5.7% for Alpine sheep (Dalvit et al., 2008) and European and Middle-Eastern breeds including also the Albanian sheep breeds (Peter et al., 2007), 5% for Pramenka types (Cinkulov et al., 2008), 4.6% for Ethiopian sheep (Gizaw et al., 2007), 3.7% in three Egyptian sheep (El Nahas et al., 2010) and Manchega sheep (Calvo et al., 2006). Our results are similar to those reported by (Nanekarani et al., 2010) for pelt sheep breeds of Iran (0.018). Low Fst value (0.29) have found also (Calvo et al., 2006) for Portugese native sheep, northern Spanish (F_{ST} = 0.029, (Rendo et al., 2004). The low genetic differentiation of Albanian sheep breed displayed here is in concordance with the results obtained with AFLP markers (Hoda et al., 2010) and SNP markers (Hoda et al., 2011).

An analysis of Nei genetic distance indicated that the three Albanian sheep breeds are closely related. The pairwise F_{ST} value of 0.05 implies moderate differentiation between breeds (Hartl, 1980). The pairwise F_{GT} values provided here between all pairs of the tested breeds are less than 0.05, indicating a low differentiation between Albanian sheep breeds. The degree of genetic differentiation was poor and had similar values between all pair of breeds. This is supported by the high level of gene flow (Nm) between breeds.

All the genetic distance measures employed to estimate inter-breed closeness showed low levels of distances between the sheep breeds. The smallest distance was observed between Shkodrane and Bardhoka, that is in concordance with results obtained previously using

AFLP markers (Hoda et al., 2010). The pattern of clustering observed with the allele-sharing distance measures (D_{AS}) among individual animals reflected the admixture of individuals coming from different breeds. This is in accordance also with the model of clustering of the same individuals, using Jaccard's similarity coefficients matrix based on AFLP data (Hoda et al., 2010). Results for the correct assignment of individuals to their reference origin, using different methods are shown in Table 10. Low percentage of correctly assigned individuals is found for all breeds. The results of assignment test can be used to identify pure breed individuals that might be used in the breeding programs in the near future. The low percentage of correctly assigned individuals to their reference population reflect also the high gene flow and intermixing of gene pool between the breeds and suggest that they are genetically very close. Low percentage of correctly assignment was obtained also by SNP markers (Hoda et al., 2011).The assignment tests, Factorial Correspondence Analysis (FCA) and Structure analysis showed a high degree of genetic similarity between individuals of three breeds and high level of breed admixture. The results of bottleneck analysis revealed that the sheep breeds have not undergone any recent bottleneck, i.e, any recent reduction in the effective population size and are at mutation drift equilibrium.

The results obtained here reflect sheep management in Albania. Sheep farming is an important activity for the farmers in Albania. Typically, the farms are small having 20 – 30 individuals with one ram. Management system is extensive or semi-extensive. The animals graze on natural grasses from morning till evening, without any supplement feed. They provide an important source of milk, meat and wool, mainly for family consumption. Product marketing and processing is limited and difficult due to the low rural socio economic level, poor infrastructure and investments. There is no breeding program for these sheep breeds. The mating is natural. In most of the cases there is only one ram per flock that breed all the ewes in the flock. The rams and ewes are housed and grazed together thereby no controlled mating is practiced at farmer's level. The rams are selected by the farmer, trying to avoid the use of males from their own flock. Usually the farmer buys the rams in the farm animal market, or from neighbor farms without any information or control of their origin, resulting in mating without parentage control. The lack of herd book, until nowadays, probably has facilitate the gene flow and the admixture of the breeds resulting to a low level of genetic differentiation.

Based on the results of this study, but also of previous studies by AFLP and SNPs markers we may conclude that Albanian sheep breed are important reservoir of genetic diversity, have a low level of differentiation and high level of admixture. All this results may be used and help in starting a breeding strategy and policy involving the decision on crossbreeding or pure breeding.

5. Conclusions

Traditionally, classifications of Albanian sheep breeds were based on visible phenotypic traits and productive traits. Characterization at molecular level using different set of markers was made possible in frame of Econogene project.

Molecular characterization using a huge set of microsatellite markers showed that Albanian sheep breed have more within breed variation than between breed variation.

All microsatellite markers have more than 4 alleles and a high level of gene diversity and high PIC values, indicating that were sufficiently polymorphic and were appropriate to analyze diversity.

Genetic distances between breed were small. The pairwise FST values were small and similar between all breeds. A high level of gene flow was detected between breeds. All these data show a poor level of genetic differentiation.

Factorial Correspondence Analysis (FCA) and Structure analysis showed a high degree of genetic similarity between individuals of three breeds and high level of breed admixture.

The results of bottleneck analysis revealed that the sheep breeds have not undergone any recent bottleneck, i.e, any recent reduction in the effective population size and are at mutation drift equilibrium.

Albanian sheep breed are important reservoir of genetic diversity, have a low level of differentiation and high level of admixture.

All this results may be used and help in starting a breeding strategy and policy involving the decision on crossbreeding or pure breeding.

6. Acknowledgements

This work has been supported by the EU Econogene contract QLK5-CT-2001-02461. The content of the publication does not represent necessarily the views of the Commission or its services.

7. References

Abdel-rahman, S. M.; A. F. El-nahas, S. A. Hemeda, S. A. El-fiky, and S. M. Nasr. 2010. Genetic Variability among Four Egyptian Sheep Breeds Using Random Amplified Polymorphic Dna (RAPD) and PCR-RFLP Techniques. Journal of Applied Sciences Research 6: 1-5.

Alvarez, I. et al. 2004. Genetic relationships and admixture among sheep breeds from Northern Spain assessed using microsatellites. Journal of animal science 82: 2246.

Arora, R.; and S. Bhatia. 2004. Genetic structure of Muzzafarnagri sheep based on microsatellite analysis. Small ruminant research 54: 227-230.

Arora, R.; S. Bhatia, and B. P. Mishra. 2010. Genetic Variation and Relationship of Six Indian Sheep Breeds Adapted to the Northwestern Arid Zone of Rajasthan. Biochemical Genetics: 1-9.

Arranz, J. J.; Y. Bayón, and F. S. Primitivo. 2001. Differentiation among Spanish sheep breeds using microsatellites. Genetics Selection Evolution 33: 1-14.

Bastos, E.; A. Cravador, J. Azevedo, and H. Guedes-Pinto. 2001. Single strand conformation polymorphism (SSCP) detection in six genes in Portuguese indigenous sheep breed" Churra da Terra Quente". BIOTECHNOLOGIE AGRONOMIE SOCIETE ET ENVIRONNEMENT 5: 7-16.

Baumung, R.; V. Cubric-Curik, K. Schwend, R. Achmann, and J. Slkner. 2006. Genetic characterisation and breed assignment in Austrian sheep breeds using microsatellite marker information. Journal of Animal Breeding and Genetics 123: 265-271.

Beckmann, J. S.; and J. L. Weber. 1992. Survey of human and rat microsatellites. Genomics 12: 627-631.

Belkhir, K.; P. Borsa, J. Goudet, L. Chikhi, and F. Bonhomme. 2001. GENETIX, logiciel sous WindowsTM pour la génétique des populations. Laboratoire genome et populations, CNRS UPR 9060.

Bowcock, A. M. et al. 1994. High resolution of human evolutionary trees with polymorphic microsatellites. Nature 368: 455-457.

Bruford, M. W.; and R. K. Wayne. 1993. Microsatellites and their application to population genetic studies. Current Opinion in Genetics Development 3: 939-943.

Buchanan, F. C.; S. M. Galloway, and A. M. Crawford. 1994. Ovine microsatellites at the OarFCB5, OarFCB19, OarFCB20, OarFCB48, OarFCB129 and OarFCB226 loci. Animal genetics 25: 60-60.

Calvo, J. H.; J. A. Bouzada, J. J. Jurado, and M. Serrano. 2006. Genetic substructure of the Spanish Manchega sheep breed. Small Ruminant Research 64: 116-125.

Cavalli-Sforza, L. L.; and A. W. Edwards. 1967. Phylogenetic analysis. Models and estimation procedures. American Journal of Human Genetics 19: 233.

Cinkulov, M. et al. 2008. Genetic diversity and structure of the West Balkan Pramenka sheep types as revealed by microsatellite and mitochondrial DNA analysis. Journal of Animal Breeding and Genetics 125: 417-426.

Cornuet, J. M.; S. Piry, G. Luikart, A. Estoup, and M. Solignac. 1999. New methods employing multilocus genotypes to select or exclude populations as origins of individuals. Genetics 153: 1989.

Cronin, M. A.; L. A. Renecker, and J. C. Patton. 2009. Genetic variation in domestic and wild elk (Cervus elaphus). Journal of animal science 87: 829.

d'Angelo, F. et al. 2009. Genetic variability of the Gentile di Puglia sheep breed based on microsatellite polymorphism. Journal of animal science 87: 1205.

Dalvit, C. et al. 2008. Genetic diversity and variability in Alpine sheep breeds. Small Ruminant Research 80: 45-51.

Dalvit, C.; M. De Marchi, E. Zanetti, and M. Cassandro. 2009. Genetic variation and population structure of Italian native sheep breeds undergoing in situ conservation. Journal of animal science 87: 3837.

Di Rienzo, A. et al. 1994. Mutational processes of simple-sequence repeat loci in human populations. Proceedings of the National Academy of Sciences 91: 3166.

Diez-Tascon, C.; R. P. Littlejohn, P. Almeida, and A. M. Crawford. 2000a. Genetic variation within the Merino sheep breed: analysis of closely related populations using microsatellites. Animal Genetics 31: 243-251.

Dobi, P.; A. Hoda, and E. a. K. Sallaku, V. 2006. Racat autoktone te" bage"tive te" imta (Native breeds of small ruminant species). Tirana, Albania, Dajti 2000.

Dowling, D. K.; U. Friberg, and J. Lindell. 2008. Evolutionary implications of non-neutral mitochondrial genetic variation. Trends in Ecology Evolution 23: 546-554.

Edwards, A. L.; H. A. Hammond, L. Jin, C. T. Caskey, and R. Chakraborty. 1992. Genetic variation at five trimeric and tetrameric tandem repeat loci in four human population groups. Genomics 12: 241-253.

Edwards, A.; A. Civitello, H. A. Hammond, and C. T. Caskey. 1991. DNA typing and genetic mapping with trimeric and tetrameric tandem repeats. American Journal of Human Genetics 49: 746.

El Nahas, S. M. et al. 2010. Analysis of genetic variation in different sheep breeds using microsatellites. African Journal of Biotechnology 7.

Evanno, G.; S. Regnaut, and J. Goudet. 2005. Detecting the number of clusters of individuals using the software STRUCTURE: a simulation study. Molecular Ecology 14: 2611-2620.

Excoffier, L.; G. Laval, and S. Schneider. 2005. Arlequin (version 3.0): an integrated software package for population genetics data analysis. Evolutionary bioinformatics online 1: 47.

Falush, D.; M. Stephens, and J. K. Pritchard. 2003. Inference of population structure using multilocus genotype data: linked loci and correlated allele frequencies. Genetics 164: 1567.

Felsenstein, J. 1993. PHYLIP (phylogeny inference package), version 3.5 c. .

Forbes, S. H.; J. T. Hogg, F. C. Buchanan, A. M. Crawford, and F. W. Allendorf. 1995. Microsatellite evolution in congeneric mammals: domestic and bighorn sheep. Molecular Biology and Evolution 12: 1106.

Francisco, L. V.; A. A. Langsten, C. S. Mellersh, C. L. Neal, and E. A. Ostrander. 1996. A class of highly polymorphic tetranucleotide repeats for canine genetic mapping. Mammalian Genome 7: 359-362.

Freeman, A. R. et al. 2004. Admixture and diversity in West African cattle populations. Molecular Ecology 13: 3477-3487.

Freeman, A. R.; D. G. Bradley, S. Nagda, J. P. Gibson, and O. Hanotte. 2006. Combination of multiple microsatellite data sets to investigate genetic diversity and admixture of domestic cattle. Animal Genetics 37: 1-9.

Gizaw, S.; J. A. Van Arendonk, H. Komen, J. J. Windig, and O. Hanotte. 2007. Population structure, genetic variation and morphological diversity in indigenous sheep of Ethiopia. Animal Genetics 38: 621-628.

Glowatzki-Mullis, M.; C. Gaillard, G. Wigger, and R. Fries. 1995. Microsatellite-based parentage control in cattle. Animal Genetics 26: 7-12.

Goudet, J. 2001. FSTAT, a program to estimate and test gene diversities and fixation indices (version 2.9. 3). .

Grigaliunaite, I. et al. 2003. Microsatellite variation in the baltic sheep breeds. Vet Zootech 1: 66-73.

Handley, L. J. et al. 2007. Genetic structure of European sheep breeds. Heredity 99: 620-631.

Heyen, D. W. et al. 1997. Exclusion probabilities of 22 bovine microsatellite markers in fluorescent multiplexes for semiautomated parentage testing. Animal Genetics 28: 21-27.

Hoda, A.; G. Hykaj, L. Sena, and E. Delia. 2011. Population structure in three Albanian sheep breeds using 36 single nucleotide polymorphisms. Acta Agriculturae Scand Section A 61: 12-20.

Hoda, A.; P. Ajmone-Marsan, G. Hykaj, and Econogene Consortium. 2010. Genetic diversity in albanian sheep breeds estimated by AFLP markers. Albanian j. agric. sci. 9: 23-29.

Hoda, A.; P. Dobi, and G. Hyka. 2009. Population structure in Albanian sheep breeds analyzed by microsatellite markers. Livestock Research for Rural Development 21.

Jawasreh, K. et al. 2011. Genetic relatedness among Jordanian local Awassi lines Baladi, Sagri, and Blackface and the black Najdi breed using RAPD analysis. Genomics and Quantitative Genetics: 31-36.

Jeffreys, A. J. et al. 1994. Complex gene conversion events in germline mutation at human minisatellites. Nature Genetics 6: 136-145.

Kantanen, J.; J. Vilkki, K. Elo, and A. Maki-Tanila. 1995. Random amplified polymorphic DNA in cattle and sheep: application for detecting genetic variation. Animal Genetics 26: 315-320.

Kappes, S. M. et al. 1997. A second-generation linkage map of the bovine genome. Genome Research 7: 235.

Kevorkian, S. E.; S. E. Georgescu, M. Adina, M. Z. Manea, and A. O. Hermenean. 2010. Genetic diversity using microsatellite markers in four Romanian autochthonous sheep breeds. Romanian Biotechnological Letters 15: 5060.

Kimura, M.; and J. F. Crow. 1964. The number of alleles that can be maintained in a finite population. Genetics 49: 725.

Kimura, M.; and T. Ohta. 1978. Stepwise mutation model and distribution of allelic frequencies in a finite population. Proceedings of the National Academy of Sciences 75: 2868.

Kusza, S. et al. 2009. Study of genetic differences among Slovak Tsigai populations using microsatellite markers. Czech Journal of Animal Science 54: 468-474.

Kusza, S. et al. 2010. Microsatellite analysis to estimate genetic relationships among five bulgarian sheep breeds. Genetics and Molecular Biology 33: 51-56.

Langella, O. 2002. POPULATIONS 1.2. 28. Population genetic software (individuals or populations distances, phylogenetic trees). CNRS, France.

Levinson, G.; and G. A. Gutman. 1987. Slipped-strand mispairing: a major mechanism for DNA sequence evolution. Molecular biology and evolution 4: 203.

Li, Y. C.; A. B. Korol, T. Fahima, A. Beiles, and E. Nevo. 2002. Microsatellites: genomic distribution, putative functions and mutational mechanisms: a review. Molecular Ecology 11: 2453-2465.

Ligda, C.; J. Altarayrah, and A. Georgoudis. 2009. Genetic analysis of Greek sheep breeds using microsatellite markers for setting conservation priorities. Small Ruminant Research 83: 42-48.

Luikart, G. et al. 1999. Power of 22 microsatellite markers in fluorescent multiplexes for parentage testing in goats (Capra hircus). Animal Genetics 30: 431-438.

MacHugh, D. E.; M. D. Shriver, R. T. Loftus, P. Cunningham, and D. G. Bradley. 1997. Microsatellite DNA variation and the evolution, domestication and phylogeography of taurine and zebu cattle (Bos taurus and Bos indicus). Genetics 146: 1071.

Manly, B. F. 2007. Randomization, bootstrap and Monte Carlo methods in biology. Chapman Hall/CRC.

Marshall, T. C. 2001. Cervus 2.0. Available from helios. bto. ed. ac. uk/evolgen.

Maudet, C.; G. Luikart, and P. Taberlet. 2002. Genetic diversity and assignment tests among seven French cattle breeds based on microsatellite DNA analysis. Journal of Animal Science 80: 942.

Meadows, J. et al. 2006. Globally dispersed Y chromosomal haplotypes in wild and domestic sheep. Animal Genetics 37: 444-453.

Mellersh, C. S. et al. 1997. A linkage map of the canine genome. Genomics 46: 326-336.

Mukesh, M.; M. Sodhi, and S. Bhatia. 2006. Microsatellite-based diversity analysis and genetic relationships of three Indian sheep breeds. Journal of Animal Breeding and Genetics 123: 258-264.

Nanekarani, S.; C. Amirinia, N. Amirmozafari, R. V. Torshizi, and A. A. Gharahdaghi. 2010. Genetic variation among pelt sheep population using microsatellite markers. African Journal of Biotechnology 9: 7437-7445.

Nei, M. 1972. Genetic distance between populations. American naturalist: 283-292.

Nei, M. 1987. Molecular evolutionary genetics. Columbia Univ Pr.

Nei, M.; F. Tajima, and Y. Tateno. 1983. Accuracy of estimated phylogenetic trees from molecular data. Journal of Molecular Evolution 19: 153-170.

Paetkau, D.; W. Calvert, I. Stirling, and C. Strobeck. 1995. Microsatellite analysis of population structure in Canadian polar bears. Molecular Ecology 4: 347-354.

Page, R. D. 1996. Tree View: an application to display phylogenetic trees on personal computers. Computer applications in the biosciences: CABIOS 12: 357.

Paiva, S. R. et al. 2005. Genetic variability among Brazilian sheep using microsatellites of. Proc. The Role of biotechnology, Turin, March: 5-7.

Pariset, L. et al. 2006a. Characterization of 37 breed-specific single-nucleotide polymorphisms in sheep. Journal of Heredity 97: 531.

Pariset, L. et al. 2006b. Characterization of single nucleotide polymorphisms in sheep and their variation as evidence of selection. Animal Genetics 37: 290-292.

Pariset, L.; M. C. Savarese, I. Cappuccio, and A. Valentini. 2003. Use of microsatellites for genetic variation and inbreeding analysis in Sarda sheep flocks of central Italy. Journal of Animal Breeding and Genetics 120: 425-432.

Peakall, R.; and P. E. Smouse. 2006. GENALEX 6: genetic analysis in Excel. Population genetic software for teaching and research. Molecular Ecology Notes 6: 288-295.

Peter, C. 2005. Molekulargenetische Charakterisierung von Schafrassen Europas und des Nahen Ostens
auf der Basis von Mikrosatelliten. PhD, Justus-Liebig-Universitaet Giessen.

Peter, C. et al. 2007. Genetic diversity and subdivision of 57 European and Middle-Eastern sheep breeds. Animal genetics 38: 37-44.

Peter, C.; E. M. Prinzenberg, and G. Erhardt. 2005. Null allele at the OarAE129 locus and corresponding allele frequencies in five German sheep breeds. Anim. Genet 36: 92.

Piry, S.; G. Luikart, and J. M. Cornuet. 1999. BOTTLENECK: a program for detecting recent effective population size reductions from allele data frequencies. Montpellier, France.

Porcu, K.; and B. and B. Markovic. 2005. Catalogue of West Balkan Pramenka Sheep Breed Types. Faculty of Agricultural Sciences and Food, Skopje, Republic of Macedonia.

Pritchard, J. K.; M. Stephens, and P. Donnelly. 2000. Inference of population structure using multilocus genotype data. Genetics 155: 945.

Qasim, M. et al. 2011. Estimation of Genetic Diversity in Sheep (Ovis aries) using Randomly Amplified Polymorphic DNA. International Journal of Animal and Veterinary Advances 3: 6-9.

Rannala, B.; and J. L. Mountain. 1997. Detecting immigration by using multilocus genotypes. Proceedings of the National Academy of Sciences 94: 9197.

Raymond, M.; and F. Rousset. 1995. GENEPOP (version 1.2): population genetics software for exact tests and ecumenicism. Journal of heredity 86: 248.

Rendo, F. et al. 2004. Tracking diversity and differentiation in six sheep breeds from the North Iberian Peninsula through DNA variation. Small Ruminant Research 52: 195-202.

Reynolds, J.; B. S. Weir, and C. C. Cockerham. 1983. Estimation of the coancestry coefficient: basis for a short-term genetic distance. Genetics 105: 767.

Saitbekova, N.; C. Gaillard, G. Obexer-Ruff, and G. Dolf. 1999. Genetic diversity in Swiss goat breeds based on microsatellite analysis. Animal Genetics 30: 36-41.

Schlotterer, C. 2004. The evolution of molecular markers—just a matter of fashion? Nature Reviews Genetics 5: 63-69.

Schmid, B. M.; N. Saitbekova, C. Gaillard, and G. Dolf. 1999. Genetic diversity in Swiss cattle breeds. Journal of Animal Breeding and Genetics 116: 1-8.

Smith, G. P. 1976. Evolution of repeated DNA sequences by unequal crossover. Science 191: 528.

Stahlberger-Saitbekova, N.; J. Schlpfer, G. Dolf, and C. Gaillard. 2001. Genetic relationships in Swiss sheep breeds based on microsatellite analysis. Journal of Animal Breeding and Genetics 118: 379-387.

Takezaki, N.; and M. Nei. 1996. Genetic distances and reconstruction of phylogenetic trees from microsatellite DNA. Genetics 144: 389.

Tapio, I. et al. 2005. Unfolding of population structure in Baltic sheep breeds using microsatellite analysis. Heredity 94: 448-456.

Tapio, M. et al. 2005. Native breeds demonstrate high contributions to the molecular variation in northern European sheep. Molecular ecology 14: 3951-3963.

Tapio, M. et al. 2010. Microsatellite-based genetic diversity and population structure of domestic sheep in northern Eurasia. BMC genetics 11: 76.

Tautz, D. 1989. Hypervariabflity of simple sequences as a general source for polymorphic DNA markers. Nucleic Acids Research 17: 6463.

Tautz, D.; and C. Schlotterer. 1994. Simple sequences. Current Opinion in Genetics Development 4: 832-837.

Troy, C. S. et al. 2001. Genetic evidence for Near-Eastern origins of European cattle. Nature 410: 1088-1091.

Vicente, A. A. et al. 2008. Genetic diversity in native and commercial breeds of pigs in Portugal assessed by microsatellites. Journal of animal science 86: 2496.

Weber, J. L. 1990. Human DNA polymorphisms based on length variations in simple-sequence tandem repeats. Genome analysis 1: 159-181.

Weir, B. S.; and C. C. Cockerham. 1984. Estimating F-statistics for the analysis of population structure. Evolution: 1358-1370.

Wright, S. 1931. Evolution in Mendelian populations. Genetics 16: 97.

Xiao, F.; Y. Fu, T. Shi, and J. Wang. 2009. Six Local Sheep Breeds AFLP Analysis of Genetic Diversity in Xinjiang. China Animal Husbandry Veterinary Medicine.

Zoraqi, G. 1991. Study of the genetic structure of "Bardhoka" and"Shkodrane" native sheep breeds, by polymorphic genetic markersand genetic relationship between breeds with productive traits. PhD, Agricultural University of Tirana, Albania.

3

Spatial Variation of Genetic Diversity in *Drosophila* Species from Two Different South American Environments

Luciana P. B. Machado, Daniele C. Silva,
Daiane P. Simão and Rogério P. Mateus
Universidade Estadual do Centro-Oeste – UNICENTRO
Brazil

1. Introduction

Currently, the main factor that generates natural habitats fragmentation is the use of land that results in reduced vegetation cover and in an asymmetric distribution of the remnants, which show different sizes and shapes (Didham et al., 1996). Usually, deforestation leads to fragmentation of continuous areas, generating the occurrence of vegetation islands that are isolated from each other by areas covered with grasslands or another type of culture (Lovejoy, 1980). In addition to deforestation, cyclic global climate change, such as glacial and interglacial periods with global warming, also has effects on the distribution of several morphoclimatic plant domains in South America, also resulting in habitat fragmentation (Ab'Saber, 2000).

In the southern and southeast region of Brazil there are two types of vegetation that are fragmented. They are the relicts of xerophytic vegetation and the Atlantic Forest domain. In the first case, there is in South America an extensive area in northeast-southwest axis, including the Caatinga, the Cerrado and Chaco, named "dry diagonal", which is located between the Amazon rainforest and the Atlantic Forest (Prado & Gibbs, 1993). The morphoclimatic areas of Caatinga and Chaco, along with the Caribbean coast of Colombia and Venezuela, have high density and diversity of cactus species (Hueck, 1972). However, adjacent areas, including the area of distribution of *Araucaria* forests, cacti are also detected in isolated populations, mainly in rocky leveling or associated with vegetation formations in sandy substrates (Manfrin & Sene, 2006). These populations of cacti are remnants of the xerophytic vegetation retraction in the interglacial periods (Ab'Saber, 2000). According to Ab'Saber (2000), the Caatinga and Chaco were connected at least four times during the Quaternary glaciations due to periods where the climate was dry and cold. In the interglacial periods, the climate became warm and wet, leading to the retraction of xerophytic vegetation and expansion of tropical forests. As cacti are good indicators of dry areas, their distribution is altered with climate change. Currently, the discontinuous distribution and size of the fragments of xerophytic vegetation are the result not only of paleoclimate cycles, but also of human action. In this context, it is expected that the *Drosophila* species associated with these cacti have followed the retractions and expansions of the dry areas of the paleoenvironment, and their population structures to reflect such changes.

In the second case, the Atlantic Forest domain includes a diverse mosaic of biomes (dense ombrophylous forest, mixed ombrophylous forest (*Araucaria* forest), deciduous and semi-deciduous seasonal forests, high altitude wet forests, northeastern enclaves, riparian forests) and associated ecosystems (high altitude grasslands, marshes, mangroves (Lino, 2002).

The mixed ombrophylous forest (MOF) includes the southern Brazil typical forests, with disjunctions in southeast and in neighboring countries, Paraguay and Argentina (Kozera et al., 2006). It is one of the most diverse biome in the world despite only few fragments of the original forest remains. This fact also makes it one of the most threatened biomes in the world, being considered a biodiversity hotspot and one of the priority areas for conservation (Myers et al., 2000). The remnants of this biome are located in the states of Rio Grande do Sul, Santa Catarina, Paraná, São Paulo and Minas Gerais of Brazil (Inoue et al., 1984). They are characterized by presenting high and dense phytophysionomy, being structurally composed of an upper stratum dominated by *Araucaria angustifolia* and a subforest composed of angiosperms and gymnosperms (Machado & Siqueira, 1980). In Brazil, until the beginning of the 20th century, the *Araucaria* forest was the predominant landscape of the south region, with an area of approximately 200,000 km². It is estimated that in 1870, the area covered by *Araucaria angustifolia* natural forests in the state of Paraná was approximately 73,780 km². By the year 1995, due to intense wood and other non-logging species exploitation, the area in Paraná was reduced to 2,594 km² (Sanquetta, 1999).

The semi-deciduous seazonal forest is also known as interior forest. Scattered remnants of this type of biome are found in the Brazilian plateaus, in the states of São Paulo, Paraná, Minas Gerais, Mato Grosso do Sul, Santa Catarina and Rio Grande do Sul. In the southern states of Brazil, it is often associated with the *Araucaria* forest. Some enclaves are also found in the Brazilian northeast. The semi-deciduous seazonal forest features are strongly determined by its continentality (Warren, 1996). Climate changes causes 20% to 50% of the trees to lose their leaves during the dry season. This is one of the most endangered biome within the Atlantic forest domain. What remains is confined to small and medium-sized fragments and very distant from each other, mostly of them located in protected areas (MMA-SBF, 2002). It is divided into lowland semi-deciduous seazonal forest, submountain semi-deciduous seasonal forest, and mountain semi-deciduous seazonal forest (Veloso & Góes-Filho, 1982). The ecological concept of this type of vegetation is conditioned to the dual climate seasonality: one tropical, with a time of intense summer rains followed by accentuated drought; and other subtropical, without dry season, but with physiological drought caused by the intense winter cold, with averages temperatures below 15°C (Rizzini, 1979).

The Atlantic Forest domain, as all other Brazilian biomes, is under strong fragmentation because of agricultural activities and the progressive growing the town surrounding. This drastic decrease of the vegetation cover make it as an area of extreme biological importance, priority in researches about species inventories and determination of genetic variability from species that are found in these areas. *Araucaria* forest (or Mixed Ombrophilous Forest) is one ecosystem of the Atlantic Forest biome. Since 2002, the Ministry of Environment of Brazil recommended the creation of an ecological corridor linking the remaining areas of Santa Catarina and Parana in order to avoid extinction of this ecosystem.

In this context, drosophilids can be used as bioindicators of environment quality and diversity. These animals have the capacity to reflect ecological changes because different species have different requirements regarding the quality of the environment (Ferreira & Tidon, 2005, Mateus et al., 2006), and most species have limited dispersal ability (Markow & Castrezana, 2000). More recently, the *Drosophila* genus has been the focus on biodiversity studies (van der Linde & Seventer, 2002; Tidon, 2006; Torres & Madi-Ravazzi, 2006, de Toni et al., 2007) because of its great diversity, morphologically and ecologically. In this sense, these insects are suitable for studies involving ecological, biogeographic and evolutionary approaches.

For the analysis of the xerophytic vegetation islands, the *Drosophila antonietae* species is a good model for study (Manfrin & Sene, 2006). On the other hand, for the analysis of the Atlantic forest biome, which includes *Araucaria* forest and semi-deciduous seazonal forest ecosystems, *Drosophila ornatifrons* was the species of choice because it is one of the prevailing species in the Atlantic forest (Sene et al., 1980; Tidon-Sklorz et al., 1994). Thus, this work aimed to characterize the genetic variability of two *Drosophila* species found in fragments of two different southern South America environments, one from the xerophytic vegetation islands of the "dry diagonal" (*D. antonietae*) and the other from Atlantic forest biome (*D. ornatifrons*), with the purpose to evaluate the effect of differetnt types of habitat fragmentation over this genetic feature.

2. Materials and methods

2.1 Species

Drosophila antonietae specimens were collected in three areas of *Cereus hildmaniannus* occurrence in the Iguassu river basin, located in the middle portion of the Parana-Uruguay rivers basin, named as Cantagalo-PR (25°25′00.0″ S, 52°04′14.9″ W), Rio do Poço/Guarapuava-PR (25°17′29.8″ S, 51°53′08.8″ W) and Segredo (25°46′27,2″ S, 52°06′55,6″ W), all in Parana State, Brazil. Following the protocol proposed by Mateus et al. (2005), thorax and abdomen were individually dry stored at -20°C and used for the allozymic analyses in another work (Lorenci et al., 2010). The respective heads were stored in 70% ethanol at -20°C and were used for DNA extraction, which were amplified for 7 microsatellite loci (AluRSAIanto-1, HaeIIIanto-2, HaeIIIanto-3, HaeIII400anto-4, HaeIII400anto-5, AluRSAIanto-6, AluRSAIanto-7), according to Machado et al. (2003).

Drosophila ornatifrons were collected in four Atlantic forest fragmented areas in the south and southeast regions of Brazil, two *Araucaria* forest fragments, and two semi-deciduous seazonal forest fragments, named as Parque das Araucárias/Guarapuava-PR - PA (25°23′36″ S, 51°27′19″ W), Salto São Francisco/Guarapuava-PR – SSF (25°03′49.1″ S, 51°17′29.8″ W), Cajuru/SP - CAJ (21°19′10.2″ S, 47°16′14.7″ W) and Sertãozinho/SP – SRT (21°09′07.8″ S, 48°04′58.0″ W), respectively. Following the protocol proposed by Silva et al. (2010), thorax and abdomen were individually dry stored at -20°C for the isoenzymatic analyses using six allozymic systems (EST, 1-GPDH, IDH, MDH, PGM and ME) according to Mateus and Sene (2003, 2007). The respective heads were stored in 70% ethanol at -20°C and will be used in further analyses (not performed yet) using DNA markers (microsatellite).

2.2 Population genetics analyses

The population genetic analyses were performed, using two different softwares. The allele frequency, mean heterozigosity (observed – Ho, and expected – He), polymorphic loci percentage, Wright's F statistics (Weir & Cockerham, 1984), genetic distance (Nei, 1972), Hardy-Weinberg equilibrium test (using the exact test with the conventional method of Monte Carlo and the Markov Chain test – 10 batches and 2,000 permutations per batch) and the UPGMA grouping analysis (using Nei's D) were obtained through the TFPGA (Miller, 1997) software. The genetic distance (Reynolds et al., 1983), the presence of exclusive allele and the Neighbor-Joining grouping analysis (Saitou e Nei, 1987) were performed using the GDA (Lewis e Zaykin, 2001) software. The correlation between *D. antonietae* populations genetic distances and both geographical (straight distances between populations) and ecological (distances between populations through the rivers) distances, according to Mateus et al. (2007), were tested using the Mantel test in the TFPGA software. In this test, the pairwise populations were compared and the genetic distances applyed were Reynolds et al. (1983), obtained in the GDA software.

The Fst values obtained were used for: (1) Classification of the genetic differentiation among populations (spatial analyses) using the qualitative guide proposed by Wright (1978) as 'low' (0 – 0,05), 'moderate' (0,05 – 0,15), 'high' (0,15 – 0,25) and 'very high' (> 0,25); (2) Gene flow estimate among populations using the formula described by Wright (1931), Fst = 1/(4 Nm + 1), where Nm represents the efective number of migrant gametes among populations per generation, and the verification of the balance between genetic drift and gene flow according to Kimura & Weiss (1964). In this case, it was assumed that the populations were in equilibrium according to the island model of gene flow, in which the equation is based.

3. Results and discussion

3.1 Environment 1: Xerophytic vegetation islands

The amplification of the seven microsatellite loci described by Machado et al. (2003) was conducted in 75 *Drosophila antonietae* specimens (25 from each natural population analyzed). The HaeIIIanto-2 locus was amplified only in four of these samples, two from Cantagalo and two from Rio do Poço. Thus, this locus was not used in the populational analyses. The allele numbers detected for each locus were: five in both AluIRSAIanto-1 and HaeIIIanto-3 loci; four in HaeIII400anto-4, AluRSAIanto-6 and AluRSAIanto-7; and three in HaeIII400anto-5.

Table 1 shows the allele frequencies for 7 loci in three populations analyzed. A locus was considered polymorphic when the most frequent allele did not have frequence above 95%.

The allele frequencies analysis (Table 1) demonstrated that all populations showed polymorphism in all loci. Exclusive alleles were found for the AluIRSAIanto-1 (allele 1 in Rio do Poço) and HaeIIIanto-3 (allele 5 in Rio do Poço) loci. All loci showed significant departure from the expected by the Hardy-Weinberg equilibrium in at least one population, with Rio do Poço (PR) presenting only one and Cantagalo and Segredo presenting four loci out of the Hardy-Weinberg expectations. These results indicated that the frequencies and the genetic diversity observer can not be maintained by recurrent mutation alone.

A very high within and among populations heterozygote deficiency were detected (Fis = 0.2561 and Fit = 0.2927), however they were not statistically different from zero (Table 2), confirming the observation that all mean expected heterozygosities (He) were higher than all mean observed heterozygosities (Ho) in all populations (Table 1). However, the overall mean observed heterozygosity, considering the three analysed populations, was 0.4614. This

Loci/Alleles	Rio do Poço	Cantagalo	Segredo
AluRSAIanto-1			
1	**0.24**	-	-
2	**0.37**	-	**0.42**
3	**0.27**	0.59	**0.14**
4	**0.04**	0.35	**0.33**
5	**0.07**	0.06	**0.11**
HaeIIIanto-3			
1	0.52	**0.48**	**0.43**
2	0.19	**0.29**	**0.28**
3	0.17	**0.17**	**0.22**
4	0.10	**0.06**	**0.07**
5	0.02	-	-
HaeIII400anto-4			
1	0.50	**0.37**	0.47
2	0.11	**0.07**	0.08
3	0.34	**0.41**	0.36
4	0.04	**0.07**	0.09
5	-	**0.02**	-
6	-	**0.06**	-
HaeIII400anto-5			
1	0.46	**0.48**	**0.40**
2	0.46	**0.48**	**0.34**
3	0.08	**0.04**	**0.26**
AluRSAIanto-6			
1	0.42	**0.07**	0.31
2	0.46	**0.60**	0.53
3	0.04	**0.19**	0.16
4	-	**0.14**	-
5	0.08	-	-
AluRSAIanto-7			
1	0.54	0.43	**0.10**
2	0.29	0.50	**0.32**
3	0.15	0.07	**0.46**
4	0.02	-	**0.12**
$P_{0.95}$	100	100	100
Ho	0.4926	0.5664	0.4132
He	0.6389	0.6112	0.6396

Table 1. Allelic frequencies for seven microsatellite loci in three populations of *Drosophila unlonietue*. Ho = mean observed heterozygosity; He = mean expected heterozygosity; $P_{0.95}$ = polymorphic loci percentage; numers in bold indicate loci that showed departure from the Hardy-Weinberg equilibrium.

value is higher than the mean observed heterozygositiy obtained for cactophilic (Ho = 0.087) and non cactophilic (Ho = 0.160) *Drosophila*, using allozyme data (revision from Zouros, 1973; Johnson, 1974; Barker & Mulley, 1976; Moraes & Sene, 2002). It was also higher than the previous calculated for *D. antonietae*, also with allozymes (Ho = 0.2242 – Mateus & Sene, 2003; Ho = 0.319 – Mateus & Sene, 2007), it was higher than the Ho obtained for 5 populations of this species by Machado et al. (2003) using microsatellite DNA (Ho = 0.2543), and even slightly higher than the Ho obtained for 10 populations of this species analyzed by Machado (2003) also using microsatellite DNA (Ho = 0.3835).

Locus	Fis	Fit	Fst
AluRSAIanto-1	0,4313	0,5303	0,1742
HaeIIIanto-3	0,4861	0,4749	-0,0217
HaeIII400anto-4	-0,3963	-0,4144	-0,0130
HaeIII400anto-5	0,3030	0,2991	-0,0056
AluRSAIanto-6	0,2379	0,2318	-0,0080
AluRSAIanto-7	0,4196	0,4889	0,1194
All loci	0,2561	0,2927	0,0491
I.C. 95% - minimum	-0,0301	-0,0295	-0,0142
- maximum	0,4296	0,4823	0,1135

Table 2. Wright F statistics for seven *loci* in three *Drosophila antonietae* populations.

According to Prout & Barker (1993), there are several possible reasons for a positive Fis: positive assortative mating, inbreeding resulting from sib mating, null alleles, and temporal Wahlund effect. Besides those, selection against heterozygotes is another possible cause. Kimura & Crow (1963) proposed that the Fis should be negative under a random mating system. Our results evidenced one locus with negative Fis (HaeIIanto-4). For all others, a variation between high and very high deficiency of heterozygotes within populations occurred for all populations (Table 2). These results diverge from what it is expected for endogamic populations as different Fis values were observed for each locus and inbreeding should affect all loci at the same rate. Furthermore, Mateus & Sene (2003) showed that *D. antonietae* do not display an inbreeding behavior analysing the allozymic pattern of flies emerged from different rotting cacti.

Wahlund effect can be a possible cause of heterozygote deficiency in populational genetics studies (Johnson e Black, 1984) and, in the present case, it can not be discarded. The most plausible scenario is that our samples are compoused by flies from different generations (temporal Wahlund effect). A spatial Wahlund effect is less possible as there is population structure and geographic isolation among *D. antonietae* populations.

Another plausible cause of heterozygote deficiency is the presence of null alleles, which could be quite commom in microsatellite samples (see Van Treuren, 1998; and McGoldrick et al., 2000, as examples). The fixation os a null allele is responsable for most of the failure in amplification experiments (Callan et al., 1993). Null alleles are represented by segments that do not amplify and segregates with other amplifying alleles, generating a false homozygote. The amplification of only two alleles for HaeIIIanto-2 locus in only four specimens (two from Rio do Poço and two from Cantagalo) is an evidence of null allele presence for this locus. According to Machado (2003), the presence of null alleles can explain the non amplification of *Drosophila buzzatii* microsatellite loci in *D. antonietae*. Machado et al. (2010), analyzing the same seven microssatellite loci of *D. antonietae*, detected the presence of null allele in the AluRSAlanto-6 locus and size homoplasy in four out of seven loci. However, they concluded that null allele and size homoplasy do not appear to represent significant problems for the population genetics analyses because the large amount of variability at microsatellite loci can compensate the low frequency of these problems in the populations investigated. In our case, the presence of null alleles was important as the HaeIIIanto-2 locus was excluded from the populational analyses.

Natural selection against heterozygotes is another event that could be generating the heterozygosity deficiency observed. However, the data present here do not allow the evaluation of such event as the fitness of the heterozygotes over time was not measured. In *D. antonietae*, Mateus & Sene (2003) verified the possible action of natural selection using temporal and spatial approaches of the allozymic variation. Later, Mateus & Sene (2007) pointed out that natural selection is a possible factor preventing genetic divergence among *D. antonietae* populations. Machado (2003) tested the occurrence of a hitchiking effect between 7 microsatellite and 10 allozymatic loci in *D. antonietae* and no correlation was found between these markers. However, the hitchiking hypothesis with another genetic system was not discarded.

Natural populations can present heterozygotes deficience because of assortative mating. For the *D. buzzatii* cluster, Machado et al. (2002), analysing courtship behavior, observed that males always court females no matter they are the same species or not. Kelly & Noor (1996) also observed the same pattern with other *Drosophila* species. Therefore, there are evidences that assortative mating is not occurring not only in the *D. buzzatii* cluster but in the *Drosophila* genus in general.

The Wright F statisctics (Table 2) also showed low and statistically not different from zero genetic differentiation among populations (Fst = 0.0491). The highest Reynolds et al. (1983) genetic distance (Table 3) was found between Cantagalo and Segredo (0.0725) and the lowest was between Cantagalo and Rio do Poço (0.0321). The Neighbor-Joining analysis using Reynolds et al. (1983) distances did not show any correlation between ecological and genetic distances among populations (Figure 1). Rio do Poço population, which is located in the head of the Cavernoso river and therefore is the first in the sequence of this river transection (frow headwater to the mouth), was grouped in the middle of Cantagalo (the second downstream) and Segredo (the third and the last before Cavernoso river flows into Iguassu river). The Mantel test did not result in a statistically sigficative correlation between genetic and ecologic distances (r = 0.63; p = 0.33) and genetic and geographic distances (r = 0.82; p = 0.32). As these populations are in the same river system and because no correlation was found among distances, it was not possible to assume that they were in regional equilibrium (Hutchison and Templeton, 1999) and thus the number of migrantes (*Nm*) was not calculated.

Populations	Rio do Poço	Cantagalo
Cantagalo	0.0321	ᴬᴬᴬᴬ
Segredo	0.0425	0.0725

Table 3. Reynolds *et al.* (1983) distances between all pairwise populations of *D. antonietae*.

The genetic differentiation (Fst) found for *D. antonietae* (Table 2) was lower than the observed in a microgeographic analysis realized with *D. mediopunctata* from *Araucaria* forest fragments in the Parana state, Brazil (Fst = 0.066 in the winter; Cavasini, 2009). However, it was higher than the Fst detected by Mateus & Sene (2003) for *D. antonietae* in the within population spatial and temporal allozyme variation approaches (Fst = 0.0355 and 0.0023, respectivelly). Considering a among population approach, the Fst obtained here was lower (almost half the value) than those obtained using allozymes (Fst = 0.0723; Mateus & Sene, 2007) and microsatellite (Fst = 0.0730; Machado, 2003). In the same way, Reynolds et al. (1983) genetic distances (Tabela 3) showed values consistents with the Fst, sugesting that despite the low genetic differentiation there is population structure among the populations studied.

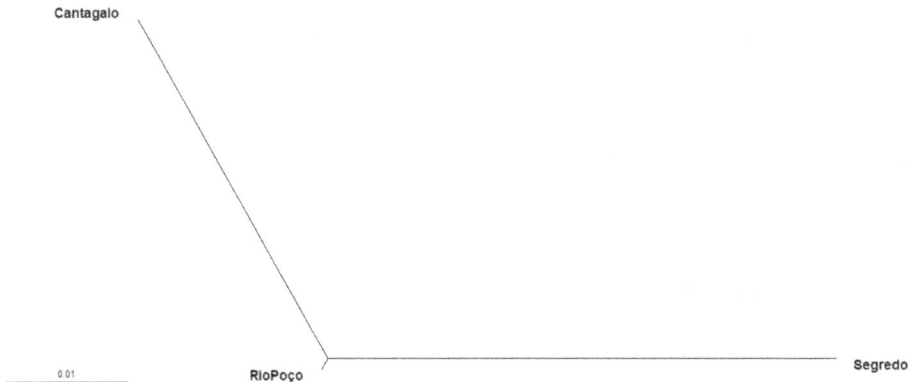

Fig. 1. Neighbor-Joining analysis for three *Drosophila antonietae* populations using Reynolds et al. (1983) genetic distances.

Gene flow is a factor that could decrease among population differentiation and the allele sharing could increase the genetic variability within populations. All that is possible considering that population size is not as small as it should be to make the genetic drift to become an important force. As detected, there is high polymorphism and diversity within and low genetic differentiation among the analyzed populations. Gene flow was suggested by Monteiro & Sene (1995) as the promoter of the low morphological differentation in the aedagi of flies from geographic isolated *D. antonietae* populations. However, the geographic distances between populations of this species raises the question of the possibility of a real and current gene flow among them. Mateus & Sene (2007) with allozymes and Manfrin et al. (2001) and de Brito et al. (2002) with mDNA indicated a certain degree of differentiation among the populations from the high portion and from the low portion of the Parana-Uruguay rivers basin. Because of the association between *D. antonietae* and the cactus *Cereus hildmaniannus*, which is distributed mainly along the rivers basin, it has been proposed that there is gene flow among *Drosophila* populations through these rivers "corridors" following the cacti distribution, preventing genetic differentiation among them (Monteiro & Sene, 1995; Machado, 2003; Mateus & Sene, 2007). Our results did not support this hypothesis.

In this ecological specificity context, Manfrin & Sene (2006) argue that this cactophilic association of the *D. buzzattii* cluster is an important aspect of this group evolution. Cactophilic *Drosophila* populations might have followed the xerophitic vegetation expansions and retractions caused by the paleoclimate cyclic changes, with the actual distribution being a result of all the glaciation cycles that occured in the Terciary and

Quaternary periods (Bigarella et al., 1975; Ab'Saber, 2000), associated with recent fragmentation caused by human activities. These events might have acted as a vicariant agent contributing to this group evolutionary history (Sene et al., 1988). A possible hypothesis to explain this pattern is the historical maintenance of shared polymorphisms, as proposed by Mateus & Sene (2007). After each retraction, the remaining populations could have retained the ancestral population polymorphism through several non-exclusive factors: 1) by having large effective population size (Mateus & Sene, 2003), decreasing the genetic drift performance; 2) by a non-sufficient divergent time among populations (Manfrin & Sene, 2006; Mateus & Sene, 2007), and; 3) the microsatellite markers could be hitchhiking a natural selection on other genetic systems, with the exception of allozymes (Machado, 2003). Current or recent gene flow among populations is unlikely to be the explanation as the Mantel test performed here did not showed correlation between geographic positioning and genetic differentiation.

In spite of the low genetic differentiation among populations obtained, exclusive alleles were detected for two out of the seven evaluated loci. Cantagalo and Segredo were the most different populations regarding allele composition and also Reynolds et al. (1983) genetic distance. Rio do Poço and Cantagalo showed the lowest distance (Table 3). In the Neighbor-Joining grouping analysis Rio do Poço population was placed bewteen Cantagalo and Segredo populations, despite it is located upstream in the Cavernoso river (Figure 1). This absence of pattern between geographical positioning and genetic distances was corroborated by the Mantel test. These data seems to indicate that an incipient diversification is occuring in the populations analysed, process that could be in response to a regional adaptation or simply by stocastic events.

Thus, the results obtained using microsatellite loci diversity analyses of *D. antonietae* populations from the Iguassu river basin, which flows into the Parana river, are in agreement with other markers for this species, that is, high within population diversity and low among population differentiation, however with the presence of an ancipient diversification.

3.2 Environment 2: Atlantic forest fragments

The electrophoretic analyses of 9 isoenzymatic systems of 145 *Drosophila ornatifrons* specimens (58 from PA, 55 from SSF, 17 from CAJ and 15 from SRT) resulted in 7 loci and 46 alleles. The allele frequencies for all *D. ornatifrons* populations are presented in Table 4.

The allele numbers detected for each locus were: ten in *Est*; and six in *Gpdh*, *Idh*, *Mdh-1*, *Mdh-2*, *Me* and *Pgm*. A locus was considered polymorphic when the most frequent allele did not have frequence above 95%. Three out of four populations (PA, SSF e SRT) showed all loci polymorphic. The CAJ population showed only one monomorphic locus (*Mdh-1*). These results demonstrated that there is high allele diversity for this species, higher than most of the microstellite loci previous tested for this species (Laborda et al., 2009) and other species of this genus. For example, Mateus et al. (2008), analysing the allele diversity of 4 isoenzymatic systems (IDH, MDH, ME e1-GPDH) in 11 species of the *tripunctata* group (that belongs to the same *quinaria-tripuncata* section where the *guarani* group is found) observed higher number of alleles only for *Idh* and *Mdh-1*. Saavedra et al (1995) studied the polymorphism for 4 allozymatic loci (*Sod*, *Odh*, *Est-2* and *Est-3*) in natural populations from south of Brazil of *D. maculifrons*, another species os the *guarani* group, and also found high allele diversity (32 alleles), however lower than the one found here.

Loci/Alleles	PA	SSF	CAJ	SRT	Loci/Alleles	PA	SSF	CAJ	SRT
Mdh-1					*Est*				
1	0.10	0.23	1.00	0.07	1	0.02	0.02	-	-
2	0.42	0.09	-	0.63	2	-	0.06	-	-
3	0.22	-	-	0.17	3	0.07	0.12	-	0.04
4	0.22	0.39	-	0.13	4	0.10	0.14	0.04	0.43
5	0.02	0.29	-	-	5	0.22	0.15	0.23	0.21
6	0.01	-	-	-	6	0.14	0.18	0.18	0.21
Mdh-2					7	0.14	0.08	0.14	0.07
1	0.09	0.37	-	-	8	0.03	0.20	0.14	0.04
2	0.17	-	-	-	9	0.17	0.04	0.27	-
3	0.04	0.63		-	10	0.11	0.01	-	-
4	0.33	-	-	-	*Gpdh*				
5	0.30	-	-	-	1	0.04	-	-	-
6	0.07	-	-	-	2	0.06	-	-	-
Me					3	0.20	-	-	0.07
1	0.19	0.33	0.29	0.21	4	0.46	0.77	0.83	0.68
2	0.09	0.22	0.29	0.57	5	0.20	0.21	0.17	0.25
3	0.43	0.37	0.36	0.71	6	0.04	0.02	-	-
4	0.12	0.06	0.03	-	*Idh*				
5	0.12	0.01	0.03	0.14	1	0.36	0.36	0.21	0.15
6	0.05	0.01	-	-	2	0.20	0.37	0.46	0.11
Pgm					3	0.25	0.25	0.08	0.54
1	-	-	0.81	0.20	4	0.12	0.02	-	0.04
2	0.02	-	0.06	0.55	5	0.07	-	0.21	0.08
3	0.20	0.09	0.13	0.25	6	-	-	0.04	0.08
4	0.09	0.16	-	-	$P_{0.95}$	100	100	83	100
5	0.64	0.60	-	-	Ho	0.3609	0.4060	0.1489	0.2640
6	0.05	0.16	-	-	He	0.7326	0.6308	0.4881	0.6229

Table 4. Allelic frequencies for nine allozyme loci in four populations of *Drosophila ornatifrons*. Ho = mean observed heterozygosity; He = mean expected heterozygosity; $P_{0.95}$ = polymorphic loci percentage; numers in bold indicate loci that showed departure from the Hardy-Weinberg equilibrium.

The mean observed heterozygosity (Ho) was 0.3609 for PA, 0.4060 for SSF, 0.1489 for CAJ and 0.2640 for SRT. The mean expected hetreozygosity (He) was 0.7326 for PA, 0.6308 for SSF, 0.4881 for CAJ and 0.6229 for SRT. These values were higher than the Ho obtained for other *Drosophila* species (Zouros, 1973; Johnson, 1974; Barker & Mulley, 1976; Moraes & Sene, 2002; and Cavasini, 2009) and the São Paulo populations showed lower Ho than the Parana populations. This result could be due to the fact that *D. ornatifrons* is a species restricted to non-disturbed environments (Ferreira & Tidon, 2005), fact that could also explain the Ho similarity between semi-deciduos forest populations and between *Araucaria* forest populations.

All loci showed significant departure from the expected by the Hardy-Weinberg equilibrium in at least two populations: PA presented all nine polymorphic loci out of the Hardy-Weinberg expectations; SSF and SRT showed only *1-Gpdh* in equilibrium; and CAJ showed only *Pgm* in equilibrium. These results indicated that the frequencies and the genetic diversity observed can not be maintained by recurrent mutation alone and is expected in natural populations. According to Falconer & Mackay (1996) the changes in the allele frequencies in natural populations could be due to systematic process as mutation, natural selection, gene flow, inbreeding or a dispersive process as genetic drift. All these will be discussed here.

Mutation have a very little impact in the populational genetic diversity over time and could be disregarded. Natural selection caused by environmental factors is possible because several authors already reported that this genus is sensitive to environmental changes (Barker et al., 1986; Moraes, 2000; Mateus, 2001; Moraes & Sene, 2002; Mateus & Sene, 2003; Ferreira & Tidon, 2005; Cavasini, 2009). Several works suggested some association of the allozyme genetic variation with environment; however whether the variation is maintained by natural selection still remains as an important question in evolutionary biology (Lewontin, 1974; Nei, 1975; Kimura, 1983; Koehn et al., 1983). Zapata et al. (2000) detected that natural selection is an evolutionary force operating in the allozyme-chromossomic inversion association in *D. subobscura*. In *D. buzzatii*, the analysis of allozyme variation in colonizing populations in Australia and Spain suggested a significant role of natural selection shaping the allele frequency distribution for several loci (Barker & East, 1980; Barker et al., 1986; Rodriguez et al., 2000), which are strictly linked to chromossomal inversions rearrangements (Schaffer et al., 1993; Betrán et al., 1995; Rodriguez et al., 2000). In the *guarani* group would be necessary to verify the association between allozymes and chomossomic inversions that were already detected in some of its species (Brncic, 1953; Salzano, 1954).

In the present work was not possible to directly associate the genetic diversity with environmental changes because abiotic data were not collected to perform this type of analysis. However, the possible action of natural selection caused by environmental factor can not be discarded because the areas of collection show different characteristics, such as different sizes, vegetation (*Araucaria* forest in Parana and Seazonal semideciduos forest in São Paulo), conservation level, climate, altitude, average temperature and pluvisosity, which could result in different microhabitats and selection pressures.

Table 5 shows the results of the Wright F statistics. The Fis indicated a very high and statistically significant heterozygote deficiency both within (Fis = 0.4896) and among populations (Fit = 0.5621). All loci presented a very high heterozygote deficiency.

These values were higher than the heterozygote deficiency found for other species of the genus, such as *D. mediopunctata* (Cavasini, 2009), *D. antonietae* (Mateus & Sene, 2003, 2007, and the present work), *D. gouveai* (Moraes & Sene, 2002). According to Kimura & Crow (1963), a negative Fis is expected under random mating. There are several reasons to detect a positive Fis: null alleles, temporal Wahlund effects, selection against heterozygotes, assortative mating and inbreeding (Prout & Barker, 1993). The frequency of null alleles in allozymes is low and not sufficient to explain a positive Fis, remaining inbreeding, Wahlund effect and selection against heterozygotes as the most likely hypothesis. Inbreeding should affect all loci in the same way and similar Fis values are expected for all loci. However, the data presented here showed different values of Fis for each locus, which is inconsistent with inbreeding. Assortative mating seems unlikely because this behavior was never detected in

Drosophila (Kelly & Noor, 1996) and some works already described the existence of hybrids in the *guarani* group (King, 1947; Kastritsis, 1969). A Wahlund effect is possible if it is considered that more than one generation was sampled in each collection (overlapping generations). Selection against heterozygotes is difficult to be tested as discussed above.

Locus		Fis	Fit	Fst
Est		0,4753	0,4902	0,0284
Gpdh		0,2625	0,3007	0,0518
Idh		0,5322	0,5540	0,0465
Mdh-1		0,5407	0,6086	0,1480
Mdh-2		0,2611	0,5010	0,3246
Me		0.6729	0.6904	0.0534
Pgm		0,6246	0,7301	0,2811
All loci		0,4896	0,5621	0,1420
I.C. 95%	- minimum	0,3740	0,4764	0,0583
	- maximum	0,5810	0,6437	0,2338

Table 5. Wright F statistics for nine loci in four *Drosophila ornatifrons* populations.

The overall Fst (0.1420) indicated a moderate genetic differentiation and the existence of a population structure among the four analysed populations as already detected for other *Drosophila* species: *D. pavani* (Kojima et al., 1972), *D. mediopunctata* (Cavasini, 2009), *D. antonietae* (Mateus & Sene, 2007). This result was corroborated by the Nei (1972) genetic distance analyses (Table 6) that showed variation between 0.3145 (between PA and SSF) and 0.7166 (between PA and CAJ). The Nei (1972) genetic identities ranged between 0.4884 (between PA and CAJ) and 0.7302 (between PA and SSF). These results showed that a possible correlation between geographic and genetic distance exist for *D. ornatifrons* as the Parana populations (SSF and PA) presented the higher identity and the lower distance. However, this possible relation was not confirmed in the Neighbor-Joining analysis (Figure 2).

Populations	PA	SSF	CAJ	SRT
PA	-	0.3145	0.7166	0.4643
SSF	0.7302	-	0.4902	0.5343
CAJ	0.4884	0.6125	-	0.6642
SRT	0.6286	0.5861	0.5147	-

Table 6. Nei (1972) distances (above diagonal) and identities (below diagonal) between all pairwise populations of *D. ornatifrons*.

Through the Fst obtained for *D. ornatifrons* (0.1420), the effective number of migrants (Nm) was calculated as 1.51, indicating moderated levels of gene flow and genetic drift according to Kimura e Weiss (1964). However, the Nei (1972) distances and identities obtained were also an indication of the presence of more than one evolutionary lineage in the samples of this species. According to Avise & Smith (1997) and Thorpe (1983), populations of the same species tend to show Nei's identity above 0.9 and distance below 0.1, and congeneric species show identity between 0.25 and 0.85 and distances between 0.16 and 1.39. Similar results were found by Mateus et al. (2010) in two criptic species of the *buzzatii* cluster: *D. antonietae* and *D. gouveai*. In fact, a recent aedeagi analysis of flies identifed as *D. ornatifrons* resulted in at least three different aedagi present in the sample (N. P. Heinz, personal communication).

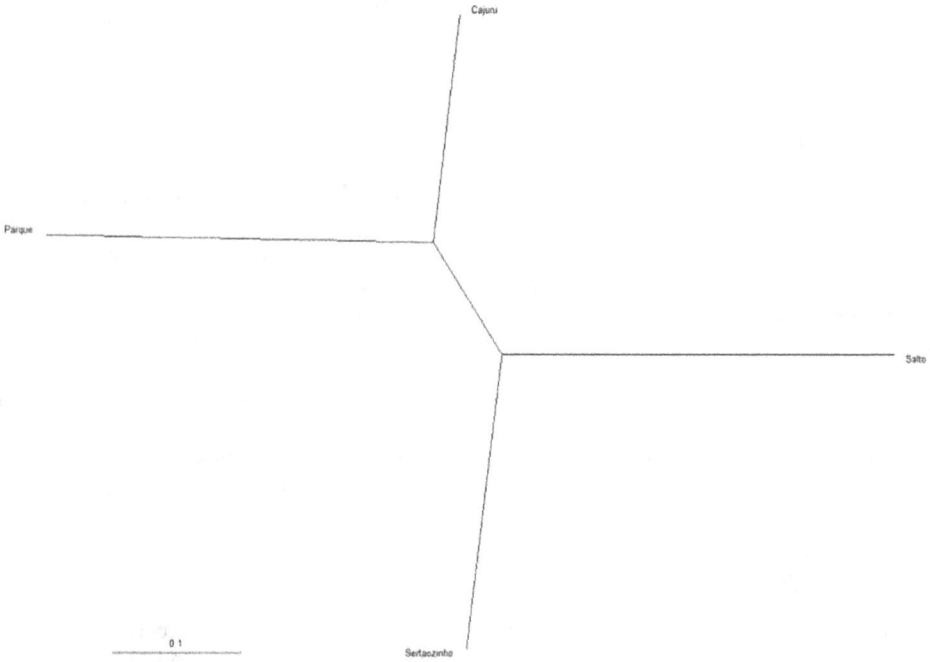

Fig. 2. Neighbor-Joining analysis for four *Drosophila ornatifrons* populations using Nei (1972) genetic distances.

There are several reasons that could explain the results obtained so far for *D. ornatifrons*. The characteristics of each collection area, such as type of vegetation, climate, altitude and mainly the conservation status. All four areas of collection worked here are conservation units that present different degrees of perturbation. It seems that the PA and SSF areas are bigger and better preserved than CAJ and SRT, which are smaller and relatively more isolated fragments. This could explain the moderate genetic differentiation among these populations. The spatial analysis of the allozymatic variability distribution for *D. ornatifrons* revealed that the PA (43 alleles and Ho = 0.3609) and SSF (33 alleles and Ho = 0.4060) populations showed higher genetic variability than the CAJ (22 alleles and Ho = 0.1489) and SRT (26 alleles and Ho = 0.2640). These results probably also reflect the characteristics of each fragment as described above, and also because *D. ornatifrons* is considered ecologicaly more restricted and should suffer more drastically with its habitat fragmentation. Nevertheless, the process of fragmentation of the Atlantic forest biome is relatively recent when compared to other biomes, which could lead to a greater and stronger inbalance over the genetic diversity distribution. This fact reinforces the urgency for new conservation program in these areas.

4. Conclusions

These results show that the fragmentation in each of the environments sampled in this work has differentiated effects over the *Drosophila* species analyzed. It seems that the xerophytic vegetation, which has a more ancient fragmentation process when compared to the Atlantic Forest, implies over the *Drosophila* species a more constant genetic diversity distribution. On the other hand, the *Drosophila* species that occurs in the Atlantic Forest biome seems to be under a more conspicuous pressure because of the fragmentation, maybe because this fragmentation is recent and lead to an imbalance greater and stronger over the genetic diversity distribution.

Our results also showed that the microsatellite loci diversity analysis of *D. antonietae* populations from the Iguassu river basin, which flows into the Parana river, are in agreement with other markers for this species, that is, high within population diversity and low among population differentiation, however with the presence of an ancipient diversification. Regarding *D. ornatifrons* populations, a high genetic differentiation was detected, which could due to the presence of the new species in the populations analyzed. However, new data should be added to corroborate this hypothesis as this is the first of a series of works on the genetic diversity distribution for this species. Currently, our laboratory is producing new data regarding the genetic diversity distribution for this species using other molecular and morphological data, besides a philogeographical approach. In the future, we hope to shed some light and better understand the effect of forest fragmentation on the genetic diversity distribution in *D. ornatifrons*.

5. Acknowledgments

We would like to thank Emanuele C. Gustani, Katiane dos Santos and Norbert P. Heinz for technical assistance. Funds were provided by: Fundação Araucária (Rogerio P. Mateus, grant numbers 103/2005, 231/2007 and 415/2009), FINEP (grant numbers 01.05.0420.00/2003 and 1663/2005), CAPES (Daiane P. Simão and Daniele C. Silva Master Fellowships) and UNICENTRO.

6. References

Ab'Saber, A. N., Spaces occupied by the expansion of dry climtes in South America during the Quaternary ice ages. Revista do Instituto Geológico, São Paulo. p.71-78, 2000.

Avise, J.C., Smith, M.H., 1977. Gene frequency comparisons between sunfish (Centrarchidae) populations at various stages of evolutionary divergence. Syst. Zool. 26, 319-335.

Barker, J. S. F., Mulley, J. C. Isozyme variation in natural populations of Drosophila buzzatii. Evolution, v. 30, p. 213-233, 1976.

Barker, J. S. F., East, P. D. Evidence for selection following perturbation of allozyme frequencies in natural population of *Drosophila*. Nature, v. 284, p. 166-168, 1980.

Barker, J. S. F., East, P. D., Weir, B. S. Temporal and microgeographic variation in allozyme frequencies in a natural population of *Drosophila buzzatii*. Genetics, v. 112, p. 577-611, 1986.

Betrán, E., Quezada-Díaz, J. E., Ruiz, A., Santos, M., Fontdevila, A. The evolutionary history of *Drosophila buzzatii*. XXXII. Linkage disequilibrium between allozymes and chromosome inversions in two colonizing populations. Heredity, v. 74, p. 188-199, 1995.

Bigarella, J. J., Andrade-Lima, D., Riehs, P. J. Considerações a respeito das mudanças paleoclimáticas na distribuição de algumas espécies vegetais e animais no Brasil. Anais da Academia Brasileira de Ciências, v. 41, p. 411-464, 1975.

Brncic, D. J. Chromosomal variation in natural populations of *Drosophila guaramunu*. Journal Molecular and General Genetics, v. 85, p. 1-11, 1953.

Callan, D. F., Thompson, A. D., Shen, Y., Phillips, H. A., Richards, R. I., Mulley, J. C., Sutherland, G. R. Incidence and origin of null alleles of the (AC)n microsatellite markers. Ann. J. Hum. Genet., v. 52, p. 922-927, 1993.

Cavasini, R. Aspectos ecológicos e genéticos no gênero *Drosophila* relacionados à fragmentação da Floresta de Araucária. Dissertação de Mestrado, Universidade Estadual do Centro-Oeste, Guarapuava/Paraná. 114p., 2009.

de Brito, R. O. A., Manfrin, M. H., Sene, F. M. Nested cladystic analysis of Brazilian populations of *Drosophila serido*. Mol. Phyl. Evol., v. 22, p. 131-143, 2002.

de Toni, D. C., Gottschalk, M. S., Cordeiro, J., Hofmann, P. P. R., Valente, V. L. S. Study of the Drosophilidae (Diptera) Communities on Atlantic Forest Islands of Santa Catarina State, Brazil. Neotropical Entomology, v. 36, p. 356-375, 2007.

Didham, R. K., Ghazoul, J., Stork, N. E., Davis, A. J.. Insects in fragmented forests: a functional approach. Trends in Ecology and Evolution, v. 11, p. 255-260, 1996.

Falconer, D. S.; Mackay, T. F. Introdução à genética quantitativa. Londres: Longman, 1996, 464 p.

Ferreira, L. B., Tidon, R. Colonizing potential of Drosophilidae (Insecta, Diptera) in environments with different grades of urbanization. Biodiversity and Conservation, v. 14, p. 1809–1821, 2005.

Hueck, K. As florestas da América do Sul: ecologia, composição e importância econômica. Polígono, São Paulo, 466 p., 1972.

Hutchison, D., Templeton, A. R. Correlation of pair-wise genetic and geographic distance measures inferring the relative influences of gene flow and drift on the distribution of genetic variability. Evolution, v. 53, p. 1898-1914, 1999.

Inoue, M. T., Roderjan, C. V., Kuniyoshi, S. Y. Projeto madeira do Paraná, FUPEF, Curitiba, 260p., 1984.

Jonhson, M. S., Black, R. The Wahlund effect and the geographical scale of variation in the intertidal limpet *Siphonaria* sp. Marine Biology, v. 79, p. 295-302, 1984.

Johnson, G. B. Enzyme polymorphism and metabolism. Science, v. 184, p. 28-37, 1974.

Kastritsis, C. D. The chromosomes of some species of the *guarani* group of *Drosophila*. J. Hered., v. 60, p. 50–57, 1969.

Kelly, J. K., Noor, M. A. F. Speciation by reinforcement: a model derived from studies of *Drosophila*. Genetics, v. 143, p. 1485-1497, 1996.

Kimura, M. The neutral theory of molecular evolution. Cambridge Univ. Press, Cambridge, 1983.

Kimura, M., Crow, J. F. The measurement of effective population number. Evolution, v. 17, p. 279-288, 1963.

Kimura, M., Weiss, G. H. The stepping-stone model of population structure and the decrease of genetic correlation with distance. Genetics, v. 49, p. 561-576, 1964.

King, J. C. Interespecific relationships within the *guarani* group of *Drosophila*. Evolution, v. 1, p. 143-153, 1947.

Koehn, R. K., Zera, A. J., Hall, J. G. Enzyme polymorphisms and natural selection. In: Nei, M., Koehn, R. K. (Eds.), Evolution of genes and proteins. Sinauer, Sunderland, p.115-136, 1983.

Kojima, K., Smouse, P.; Yang, S.; Nair, P. S.; Brncic, D. Isozyme frequency patterns in *Drosophila pavani* associated with geographical seasonal variables. Genetics, v. 72, p. 721-731, 1972.

Kozera, C., Dittrich, V. A. de O.; Silva, S. M. Composição florísica da Floresta Ombrófila Mista Montana do Parque Municipal do Barigüi, Curitiba, PR. Floresta, v.36, p.45-58, 2006.

Laborda, P. R., Klaczko, L. B., Souza, A. P. *Drosophila mediopunctata* microsatellites. II: Cross-species amplification in the *tripunctata* group and other *Drosophila* species. Conservation Genet. Resour., v. 1, p. 281-296, 2009.

Lewis, P. O., Zaykin, D. Genetic data analysis: computer program for the analysis of allelic data, version 1.0 (d16c). Free program distributed by the authors over the Internet. from http://lewis.eeb.uconn.edu/lewishome/software.html, 2001.

Lewontin, R. C. The genetic basis of evolutionary change. Columbia Univ. Press, New York, 1974.

Lino, C. F. Os Domínios da Mata Atlântica. In: www.rbma.org.br. 2003. Acessado em 30/06/2011.

Lorenci, M., Simão, D. P., Machado, L. P. B. and Mateus, R. P. Genetic variability analysis of two natural populations of Drosophila antonietae (Diptera; Drosophilidae). Drosophila Information Service, v. 93, p. 178-182, 2010.

Lovejoy, T. E. Foreword. In: Soulé, M. E., Wilcox, B. A. (eds.). Conservation biology: an evolutionary-ecological perspective, Sinauer Associates, Sunderland, p.5-9, 1980.

Machado, L. P. B. Descrição e análise de loci de microssatélites em populações naturais da espécie cactofílica Drosophila antonietae (DIPTERA; DROSOPHILIDAE). Tese de Doutorado. FMRP. Universidade de São Paulo, 104p., 2003.

Machado, L. P. B., Castro, J. P., Ravazzi, L. M. Avaliation of the Courtship and of the Hybrid Male Sterility Among Drosophila buzzatii Cluster Species (Diptera, Drosophilidae). Revista Brasileira de Biologia, v. 62, p. 601-608, 2002.

Machado, L. P. B., Manfrin, M. H., Silva-Junior, W. A., Sene, F. M. Microsatellite loci in the cactophilic species Drosophila antonietae (Diptera; Drosophilidae). Mol. Ecol. Notes, v. 3, p. 159-161, 2003.

Machado, L. P. B., Mateus, R. P., Sene, F. M., Manfrin, M. H. Microsatellite allele sequencing in population analyses of the South American cactophilic species Drosophila antonietae (Diptera: Drosophilidae). Biological Journal of the Linnean Society,v. 100, p. 573–584, 2010.

Machado, S. A., Siqueira, J. D. P. Distribuição natural da Araucaria angustifolia (Bert.). In: Ktze, O. (Ed.), IUFRO Meeting on Forestry Problems of the Genus Araucaria: Forestry problems of the genus Araucaria. FUPEF, Curitiba, p.4-9, 1980.

Manfrin, M. H., Sene, F. M. Cactophilic Drosophila in south America: a model for evolutionary studies. Genetica, v. 126, p. 57-75, 2006.

Manfrin, M. H., de Brito, R. O. A., Sene, F. M., Systematics and evolution of the Drosophila buzzatii cluster (Diptera: Drosophilidae) using mtDNA. Ann. Entomol. Soc. Am., v. 94, p. 333-346, 2001.

Markow, T. A., Castrezana, S. Dispersal in cactophilic Drosophila. Oikos, v.89, p. 378–386, 2000.

Mateus, R. P. Variação isoenzimática em populações naturais de Drosophila antonietae (Diptera; Drosophilidae). Tese de Doutorado- USP- Ribeirão Preto, 2001.

Mateus, R. P., Sene, F. M. Temporal and spatial allozyme variation in the South American cactophilic Drosophila antonietae (Diptera, Drosophilidae). Biochemical Genetics, v. 41, p. 219-233, 2003.

Mateus, R. P., Sene, F. M. Population genetic study of allozyme variation in natural populations of Drosophila antonietae (Insecta, Diptera). Journal of Zoological Systematics and Evolutionary Research, v. 45, p. 136-143, 2007.

Mateus, R. P., Machado, L. P. B., Sene, F. M. Differential body expression of isoenzymatic loci in adults of the cactophilic species Drosophila antonietae (Diptera: Drosophilidae). DIS, v. 88, p. 46-48, 2005.

Mateus, R. P, Buschini, M. L. T., Sene, F. M. The Drosophila community in xerophytic vegetations of the upper Parana–Paraguay River Basin. Braz J Biol, v. 66, p. 719–729, 2006.

Mateus, R. P., Machado, L. P. B., Cavasini, R., Gustani, E. C. Isoenzymatic analysis of South American species of the Drosophila tripunctata group (Diptera, Drosophilidae). Dros. Inf. Serv., v. 91, p. 53-56, 2008.

Mateus, R. P., Machado, L. P. B., Moraes, E. M., Sene, F. M. Allozymatic divergence between border populations of two cryptic species of the Drosophila buzzatii cluster species (Diptera: Drosophilidae). Biochem. Syst. Ecol., v. 38, p. 410-415, 2010.

McGoldrick, D. J., Hedgecock, D., English, L. J., Baoprasertkul, P., Wars, R. D. The transmission of microsatellite alleles in Australian and North American stocks of the Pacific oyster (Crassostrea gigas), selection and null alleles. Journal of Shellfish Research, v. 19, p. 779-788, 2000.

Miller, M. P. Tools for population genetic analyses - TFPGA - 1.3: A Windows program for the analysis of allozyme and molecular population genetic data. Computer software distributed by author, 1997.

MMA/SBF – Ministério do Meio Ambiente/Secretaria de Biodiversidade e Floresta. Biodiversidade brasileira: avaliação e identificação de áreas e ações prioritárias para conservação, utilização sustentável e repartição dos benefícios da biodiversidade nos biomas brasileiros. Brasília, 52p., 2002.

Monteiro, S. G., Sene, F. M. Estudo morfométrico de populações de Drosophila serido das regiões central e sul do Brasil. Revista Brasileira de Genética, v. 18 (suplemento), p.283, 1995.

Moraes, E. M. Estrutura de reprodução em uma população isolada de Drosophila sp. B, Dissertação de Mestrado, Depto. de Genética, Faculdade de Medicina, USP, Ribeirão Preto, 2000.

Moraes, E. M., Sene, F. M. Breeding structure in an isolated cactophilic Drosophila population from sandstone table hill. J. Zool. Syst. Evol. Res., v. 40, p. 123-128, 2002.

Myers, N., Mittermeier, R. A., Mittermeier, C. G., Fonseca, G. A. B., Kent, J. Biodiversity hotspots for conservation priorities. Nature, v.403, p.853–858, 2000.

Nei, M. (1972). Genetic distance between populations. Am. Nat. v.1065, p.283.

Nei, M. Molecular population genetics and evolution. American Elsevier, New York, 1975.

Prado, D. E., Gibbs, P. E. Patterns of species distributions in the dry seasonal forest South America. Annals of the Missouri Botanic Garden, v. 80, p. 902-927, 1993.

Prout, T., Barker, J. S. F. F statistics in Drosophila buzzatii: selection, population size and inbreeding. Genetics, v.134, p. 369-375, 1993.

Reynolds, J., Weir, B. S., Cockerham, C. C. Estimation of the coancestry coefficient: basis for a short-term genetic distance. Genetics, v.105, p.767-779, 1983.

Rizzini, C. T. Tratado de Fitogeografia do Brasil: Aspectos Sociológicos e Florísticos. HUCITEC, São Paulo, 2 vol., 374p., 1979.

Rodriguez, C., Piccinali, R., Levy, E., Hasson, E. Contrasting population genetic structures using allozymes and the inversion polymorphism in Drosophila buzzatii. J. Evol. Biol., v. 13, p. 976-984, 2000.

Saavedra, C. C. R., Valente, V. L. S., Napp, M. An ecological/genetic approach to the study of enzymatic polymorphisms in Drosophila maculifrons. Revista Brasileira de Genética, v. 18, p. 147-164, 1995.

Saitou, N., Nei, M. The neighbor-joining method: A new method for reconstructing phylogenetic tree. Mol. Biol. Evol., v. 4, p. 406-425, 1987.

Salzano, F. M. Chromosomal Relations in Two Species of *Drosophila*. The American Naturalist, v. 88, p. 399-405, 1954.

Sanquetta, C. R. Pinheiro do Paraná: lendas e realidades. FUPEF, Curitiba, 1999.

Schafer, D. J., Fredline, D. K., Knibb, W. R., Green, M. M., Barker, J. S. F. Genetics and linkage mapping of *Drosophila buzzatii*. J. Hered., v. 84, p. 188-194, 1993.

Sene, F. M., Val, F. C., Vilela, C. R., Pereira, M. A. Q. R. Preliminary data on the geographical distribution of Drosophila species within morphoclimatic domains of Brazil. Papeis Avulos de Zoologia; v. 33, p. 315-326, 1980.

Silva, D. C., dos Santos, K., Machado, L. P.B. and Mateus, R. P. Allozymatic activity in samples prepared for morphometric and molecular analyses in two species of the Drosophila guarani group (Diptera: Drosophilidae). Drosophila Information Service, v. 93, p.182-185, 2010.

Thorpe, J.P., 1983. Enzyme variation, genetic distance and evolutionary divergence in relation to levels of taxonomic separation. In: Oxford, G.S., Rollinson, D. (Eds.), Protein Polymorphism: Adaptive and Taxonomic Significance. Academic Press, New York, p.131-152.

Tidon, R. Relationships between drosophilids (Diptera, Drosophilidae) and the environment in two contrasting tropical vegetations. Biological Journal of the Linnean Society, v. 87, p. 233-247, 2006.

Tidon-Sklorz, R., Vilela, C. R., Sene, F. M.; Pereira, M. A. Q. R. The genus *Drosophila* in the Serra do Cipó. Revista Brasileira de Entomologia, v. 38, p. 627-637, 1994.

Torres, F. R., Madi-Ravazzi, L. Seasonal variation in natural populations of Drosophila spp. (Diptera) in two woodlands in the State of São Paulo, Brazil. Iheringia Série Zoologia, v. 96, p. 437-444, 2006.

van der Linde, K., Sevenster, J. G. Drosophila diversity over a disturbance gradient. Proc. Exp. App. Entomol., v. 13, p. 51–56. 2002.

van Treuren, R. Estimating null allele frequencies at a microsatellite locus in the oystercatcher (*Haematopus ostralegus*). Molecular Ecology, v. 7, p. 1413–1417, 1998.

Veloso, H.P., Góes-Filho, L. Fitogeografia Brasileira: Classificação Fisionômico-Ecológica da Vegetação. Bol. Téc. Projeto RADAMBRASIL. Sér. Vegetação N° 1, 1982.

Warren, D. A Ferro e Fogo: A História e a Devastação da Mata Atlântica Brasileira. Companhia das Letras, São Paulo, 484p., 1996.

Weir, B. S., Cockerham, C. C. Estimating F-statistics for the analysis of population structure. Evolution, v. 38, p. 1358-1370, 1984.

Wright, S. Evolution in Mendelian populations. Genetics, v. 16, p. 97-159, 1931.

Wright, S. Evolution and the genetics of population, vol. 4. Variability within and among natural populations. Univ. Chicago Press, Chicago, 1978.

Zapata, C., Alvarez, G., Rodriguez-Trelles, F., Maside, X. A long-term study on seasonal changes of gametic disequilibrium between allozymes and inversions in *Drosophila subobscura*. Evolution, v. 54, p. 1673-1679, 2000.

Zouros, E. Genic differentiation associated with the early stages of speciation in the mulleri subgroup of Drosophila. Evolution, v. 27, p. 601-621, 1973.

4

Molecular Approaches for the Study of Genetic Diversity in Microflora of Poultry Gastrointestinal Tract

Alireza Seidavi

Department of Animal Science, Rasht Branch, Islamic Azad University, Rasht
Iran

1. Introduction

Livestock production currently comprises approximately 40 per cent of the gross value of the world's agricultural produce. At the end of the 20th century, approximately 74 per cent of poultry meat and 68 per cent of eggs were produced in the industrial sector. It is predicted that the consumption, per capita, of poultry products will increase to 17.2 kg by 2020. These needs will require extensive scientific and technological development in many areas.

Because chicken protein plays such a huge role in the supply of human nutritional needs, much scientific research has been carried out to improve the productivity of these animal foodstuffs. The productivity of poultry, like all living birds, is influenced by genetics, the environment and interactions between these two factors. Accordingly, besides coherent and targeted programs that seek to increase the genetic potential of broilers, extensive research on environmental conditions, such as nutrition, physiology, hygiene, control of diseases and management improvement, must be carried out to improve productivity.

One of the most important factors affecting broiler productivity is the gut microbial flora that can play a very significant positive and negative role in the final yield of broilers. Our limited knowledge of the role of bacteria in the digestive tract of birds is largely derived from the little information obtained on the composition of chicken intestine microflora. Significant studies on the microbial flora of the digestive tract of poultry began in 1970, and currently, the microflora is being increasingly studied by researchers worldwide.

Microorganisms, especially bacteria, have considerable effects on the immune, nutritional and physiological processes of the host. Birds suffering from infections of harmful bacteria will cause heavy losses in the poultry herds and also in human communities. Investigation of the microbial flora of the gastrointestinal tract has a significant role to play in ensuring the health and safety of poultry products submitted for human consumption. Further, one of the most reliable methods for the quality control of poultry products is a survey of microorganisms in poultry herds (Skanseng et al., 2006).

Recently, the detection, differentiation and identification of microorganisms have been accomplished using methods such as phenotypic measurement, biochemical assays, immunological assays and molecular methods. However, methods that are based on phenotypic and biochemical assays to identify microorganisms are typically vague and inaccurate (Settanni and Corsetti, 2007). Although phenotypic identification remains standard and is a commonly used method for identifying most bacteria, it is very cumbersome, requires excessive time, exceptional technical ability and proper technical standards to obtain accurate results (Rossello-Mora and Amann, 2001).

Moreover, it is difficult to process a large number of samples. Typically, at least 10 tests are carried out to identify and differentiate all of the species within a sample. Rapid and accurate diagnosis of pathogens allows the correct course of action to be taken and also enables better understanding of the pathogen epidemiology. Microbiological culturing takes time (at least 24-48 hours for microbial growth), and the methods are not specific and do not have the required precision; hence, the current effectiveness of these methods is very limited (Miyashita et al., 2004). Indeed, because these methods cannot be used for non-culturable bacteria and can only be used for culturable bacteria, our knowledge of microbial flora in the digestive tract of poultry remains inaccurate and incomplete. Most bacteria, due to their unknown requirements for growth, cannot be grown *in vitro*.

Consequently, many researchers have sought to develop methods that would allow the rapid and accurate identification of useful and harmful microorganisms. Certainly, commercial identification kits, such as API, and automated identification systems, such as VITEK, allow the identification of many bacteria to a species level, thus enabling the user to identify the bacteria relatively quickly. However, many bacteria cannot be detected with these conventional methods. This may be due to the specific needs of the bacteria, for example, the Lactic Acid Bacteria (LAB) (Ampe et al., 1999). Further, because existing commercial systems on the market have been developed to primarily identify human pathogenic bacteria, they may not accurately detect and identify bacteria from other sources, such as animal bacteria and food product bacteria (Settanni and Corsetti, 2007). It is impossible to identify new bacterial species and strains using these commercial systems.

With the development of methods based on molecular techniques, the speed and accuracy of diagnosis of many poultry diseases has been increased, leading to a reduced time to diagnosis and a consequent reduction in treatment costs. Ultimately, these methods will reduce and prevent the incidence of disease in the herd.

Among the molecular methods, the restriction fragment length polymorphism (RFLP) of total genomic DNA (generated by non-PCR techniques) method belongs to the first generation of molecular methods that were widely used for differentiating microorganisms (Rosselló-Mora and Amann, 2001). Southern blot gel electrophoresis (SBGE) was also among the first-generation methods that were widely used to identify the microflora. Today, second-generation molecular methods are used and include PCR-based methods, such as PCR-RFLP and randomly amplified polymorphic DNA (RAPD), for the detection and differentiation of bacterial isolates. Because the resulting profiles on ethidium bromide-stained gels do not contain specific bacterial sequence information, they must be compared with profiles of reference bacteria. If this method is used in isolation, it will not relate samples with reference species because it relies on the existence of specific PCR products and produce from a specific beginner primer pair (Rosselló-Mora and Amann, 2001). The

RAPD-PCR method has been successfully implemented to identify different microorganisms (Raclasky et al., 2006).

Targeted gene amplification has been increasingly studied by researchers as a valid method to identify bacterial species (Settanni et al., 2005). At present, many specific PCRs for identifying bacterial species have been confirmed and are used for identifying bacterial preparations. One of the most common PCR-based methods for identifying microorganisms is denaturing gradient gel electrophoresis (DGGE). This informative method can differentiate between polymorphic gene sequences. The data are obtained using an acrylamide gel with a denaturing gradient (Muyzer et al., 1993). Molecular or genetic methods for the identification of bacteria, in addition to or indeed instead of phenotypic assays, provide more sensitive and more specific detection. These methods decrease the errors that result from subjective interpretation of morphological and biological bacterial characteristics. Essentially, bacterial DNA remains almost unchanged throughout the life cycle and even while exposed to environmental stresses. Therefore, molecular methods for the identification of bacteria (and other microorganisms) that target genomic DNA are being rapidly developed and are of increasing significance. Recent advances in PCR methods that allow rapid and accurate diagnosis of a broad range of bacteria, such as molecular genetic methods, have become a key method for detecting microorganisms. It is many years since 16S ribosomal RNA (rRNA) sequencing was used as the main tool for determining phylogenetic relationships between bacteria. The molecular specificity of this method allowed the determination of phylogenetic relationships for both the detection and identification of bacteria used in clinical laboratories. Studies have now shown that sequencing determination is a valid method for identifying bacteria with slow, unusual or difficult growth requirements; moreover, such bacteria were inadequately identified using the bacterial culture methods (Zhu and Joerger, 2003). Recent developments in ribosomal DNA- and RNA-based molecular methods allow the identification of different bacterial populations from environmental samples without culturing. The PCR method for determining bacterial diversity in the population of samples is used. Available reports indicate that there is a good correlation between PCR-based methods and culture methods for those bacteria that can be grown in culture.

Accordingly, molecular methods have been used to detect bacteria in the digestive tract of broiler chickens reared in commercial conditions. Although molecular methods also have limitations, they can be referred to error possibility in separating, amplification and DNA simulations in some bacteria and special sequences; but nevertheless, this method provides the ability to investigate the diversity of microbial flora in a particular sample. Data obtained using molecular methods will allow us to easily compare different isolates of poultry digestive systems without requiring subjective comparisons of biochemical features. Gastrointestinal microbial ecology is the study of the prevalence and diversity of existing microorganisms, their activities and their relationships with each other and the host animal (competitive and coherence effects).

The increase in information on human gut microbial ecology is due to three main factors: 1) development of anaerobic culture techniques; 2) use of laboratory rodents to understand the relationship between intestinal bacteria and the host; and 3) use of animal models lacking in specific microbes or a microbial flora (Savage, 2001).

Although the ability to culture digestive system bacteria is relatively high (10-50% of total species, Vaughan et al., 2000), only a part of the total species are cultured.

The main reasons underlying this partial culturing include an absence of required bacterial growth substances, the choice of medium required, the stress of the culturing process, the need to remove oxygen to maintain anaerobic conditions and the difficulty of simulating interactions with other bacteria and the host tissues.

Solutions to address these problems in culture methods are required. During the last decade, there have been notable increases in the use of 16S ribosomal RNA methods to determine bacterial population diversity (Vaughan et al., 2000). Nucleic acid sequence comparisons of isolates from bacterial ecosystems can be used to determine the molecular characteristics and classification of the bacteria and also to predict their evolutionary relationships.

Molecular technology provides information on nucleic acid sequences. Using these sequences, microorganisms can be identified within a specific environment and further, their function, their performance and their importance or role in a specific environment can be evaluated.

The molecular ecology of microbial flora using molecular technology - mainly based on information related to nucleic acid sequences - microorganisms identified in a specific environment, tasks and their performance is evaluated and the importance or role in an environment where residents are being evaluated.

The author of this chapter has performed many experiments on the molecular detection of poultry gastrointestinal tract microflora, has published many papers in notable journals and has also presented many papers in international conferences on this subject (Seidavi., 2008; 2009a; 2009b; 2009c; Seidavi & Chamani., 2010; Seidavi & Qotbi., 2009a; 2009b; Seidavi et al., 2007; 2008a; 2008b; 2008c; 2008d; 2008e; 2008f; 2008g; 2008h; 2008i; 2008j; 2008k; 2009a; 2009b; 2009c; 2009d; 2010a; 2010b; 2010c; 2011; Mirhosseini et al., 2008a; 2008b; 2009a; 2009b; 2009c; 2010). This chapter describes the molecular approaches that may be used for the study of genetic diversity in poultry gastrointestinal tract microflora.

2. Different methods used for the detection and identification of microbial flora

Several molecular genetics-based methods for the detection and identification of microbial flora in different samples have been developed, which have greater accuracy and are faster than classical methods.

2.1 Polymerase Chain Reaction (PCR)

Many of the molecular based polymerase chain reaction methods are used for the detection of microbial flora. Using specific primers, the PCR amplifies a specific sequence of DNA, thus confirming the presence of microorganisms. The principles and details of the polymerase chain reaction have been well described by various sources.

2.1.1 Multiplex PCR (mPCR)

One of the branches of applied molecular microbiology includes the monitoring and control of existing microorganisms in natural ecosystems (Settanni and Corsetti, 2007). Classical technologies and mono-PCR are insufficient for studying several complex species simultaneously, including microbial flora, and the use of multiple PCR reactions is highly desirable. Today, the use of mPCR for the rapid identification of multiple isolates is a useful tool for studying the structure of the microbial population and the dynamics of microbial communities, such as changes in the microbial population during fermentation or in response to environmental changes. Additionally, mPCR is semi-quantitative because it can be used to estimate bacterial concentrations as the threshold for identification is very low (Settanni and Corsetti, 2007). Today, the mPCR method is used in different applied science fields to accelerate and direct the identification and differentiation of microorganisms with or without prior isolation and culture.

2.1.2 Technical aspects and molecular targets of mPCR

In order to ensure specificity of the system (a unique target sequence for each primer pair), it is necessary to use protocol designs with longer melting (T_m) steps than in typical PCR. Additionally, starter DNA sequences that bind the primers or each other should be avoided because denaturing these sequences will reduce their availability for amplification.

The magnesium concentration in the mPCR is an important factor that affects the efficiency of the reaction (McPherson and Moller, 2000). In general, the $MgCl_2$ concentration in mPCRs is higher than the concentration used in typical PCR reactions. Depending on the number of bacteria that must be recognised, mPCR generally requires a single amplification reaction (usually 4-5 bacteria) or a two-stage amplification reaction (usually 5-6 or more bacteria).

For the molecular diagnosis, identification and classification of bacteria (Grahn et al., 2003), mPCR generally targets the 16S rRNA genes. This gene is widely used to understand the phylogenetic relationships between bacteria (Rosselló-Mora and Amann, 2001). However, sometimes the 16S gene sequence is insufficient to identify related species (Torriani et al., 2001), and therefore, other genes should be considered in the mPCR design.

2.1.3 The role of mPCR in the study of microorganisms

The mPCR method is widely used for the identification and separation of microorganisms, and these are some of the reasons underlying its ability to provide a simple fingerprint of the bacterial population. For this method, specific DNA sequences for each bacterium are necessary to obtain an individual and a single band for each bacterium. The sizes of these specific bands should be different to allow straightforward band location. Using the sample DNA as the template, and provided no cross-amplification is observed, the system can be used for total DNA extracted from the sample. Thus, DNA fragments of amplified products of desired bacteria can be observed in the agarose gel. Ultimately, this method provides a fingerprint that is specific for the sample analysed (a series of bands related to the bacteria in the mPCR).

The validity of the mPCR measurement is based on the availability of desired bacteria DNA. The primer mix should include a cross-specific primer that produces a PCR product for all bacteria to be identified in the experiment (Zarlenga and Higgins, 2001).

2.2 Sequencing of SSU rDNA clone libraries with small subunits

rRNA libraries of the small Subunits are necessary to identify all bacteria in a specific environment. The sequenced SSU rRNA genes have become a standard method for the identification of isolates, such that the complete description of microbial populations is impossible without information on SSU rRNA sequences (Zoetendal et al., 2004). At present, over 79,000 16S rRNA sequences are available in the DNA database, more than any other gene (http://rdp.cme.msu.edu/html). rRNA sequences can be obtained from the SSU rRNA sequences directly or from their coding genes (SSU rDNA) by reverse transcription (RT) or conventional PCR. In practice, SSU rRNA sequences are determined from rDNA clone libraries rather than cDNA libraries. Following this determination, amplified sequences are determined and compared with the SSU rDNA sequences stored in the database (http://www.ncbi.nlm.nih.gov/BLAST/ and http://rdp.cme.msu. edu / html) and then phylogenetic analysis is conducted (Cole et al., 2003). SSU rRNA sequence clone libraries from human faeces (Suau et al., 1999), colon and ileum (Wang et al., 2003) and oral cavity (Paster et al., 2001) have shown that a considerable number of living bacteria had not been discovered in previous studies. Similarly, studies of the gastrointestinal tract of various animal species have been reported (Daly et al., 2001; Gong et al., 2002), and microflora identified (Zoetendal et al., 2004). It is also important to estimate the amount of actual variation in the SSU rDNA clone libraries. This estimate depends on how many operational taxonomy units (OTUs) are known. Unfortunately, the OTU has not yet been standardised and sequences within an OTU differ by up to 5%, which makes it difficult to compare between clone libraries (Martin, 2002).

In traditional methods, bacteria were categorised using phenotypic characteristics. Following the development of technologies based on nucleic acids, SSU rDNA sequences were developed as a standard tool for the phylogenetic classification of bacteria.

It should also be stated that SSU rDNA sequences that are stored in the DNA databases relate to a small part of the total bacterial isolates. Further, new microbiological methods, such as the analysis of the isolate cello-bio or butyrate production, have allowed the isolation of increasing numbers of previously unknown bacteria (Zeotendal et al., 2003). Thus, an accurate estimation of the ability to kill bacteria in the digestive system and the elimination of these ambiguities will be possible.

Although sequencing of amplified SSU rDNA clone fragments provides important information for the identification of non-cultivable bacteria, these data are not quantitative, and cloning procedures and PCR are not infallible. Several PCR methods have been proposed to minimise these errors. However, SSU rDNA clone libraries have made major contributions to the understanding of gastrointestinal microbial flora.

2.3 Fingerprinting methods SSU rDNA: DGGE, TGGE, TTGE, SSCP and T-RFLP

Because of the effort required and the cost of cloning and sequencing of SSU rDNA sequences of the microbial flora population, several fingerprinting methods have been developed that appear ideal for investigating changes in microbial populations, for example comparing different parts of the gastrointestinal tract of different animals.

Denaturing gradient gel electrophoresis (DGGE) was used to study the ecology of the diverse bacteria in marine ecosystems (Muyzer et al., 1993). In this study, a microbial ecosystem was investigated using DGGE, temperature gradient gel electrophoresis (TGGE) and temporal temperature gradient gel electrophoresis (TTGE). Other methods include the analysis of microbial populations for single strand conformation polymorphisms (SSCPs) and terminal-restriction fragment length polymorphisms (T-RFLPs). DGGE, TGGE and TTGE are based on the specific melting behaviours of amplified sequences. The secondary structure of single-stranded DNA and SSCP-based T-FRLP is based on the specific targets of restriction enzymes. Interestingly, prior to their use in microbial ecology, these methods (except for T-FRLP) were used in clinical research, which demonstrates their efficacy. Today, with the development of statistical software packages, we can calculate the similarity indices and conduct cluster analyses of SSU rDNA profiles. Thus, these fingerprinting methods are very useful and allow the analysis and study of microbial populations over time (at all ages) and also allow study of the response of animals to diet. Several articles that review the details of fingerprinting methods have been published (Vaughan et al., 2000; Muyzer et al., 1993; Konstantinov et al., 2002).

DGGE, TGGE and TTGE analyses of SSU rDNA have been successful in determining the characteristics of human (Tannok et al., 2000; Seksik et al., 2003), cow (Kocherginskaya et al., 2001), dog (Simpson et al., 2002), rodent (Deplancke et al., 2000; McCracken et al., 2001) and chicken (Van der Wielen et al., 2002b; Zhu et al., 2002) intestinal bacteria populations. DGGE or TGGE methods are sufficiently sensitive to detect bacteria as a per centage of the total bacterial population (Zeotendal et al., 1998). The T-RFLP method has also proven useful for obtaining fingerprints of gastrointestinal microbial flora (Nagashima et al., 2003).

In these studies, the role of environmental factors on the microbial population, such as disorders, physiological conditions, the part of the digestive system under investigation and the host animal species, has been investigated.

The stability of the bacterial ecosystem is directly related to its diversity. Reducing the diversity of the bacterial population decreases its stability. Fluorescent *in situ* hybridisation (FISH) and the bacterial population structure of the dominant TGGE in human faeces of healthy adults showed that the population composition remained relatively stable over time (Franks et al., 1998). Previous studies, based on culture techniques, have shown that in the case of human faecal microbial flora, the population changes usually occur in newborns and in the elderly (Hopkins et al., 2001), with similar changes expected for animals. More recent studies using molecular methods have confirmed these findings (Schwiertz et al., 2003). More molecular studies are required to confirm the relationship between microbes and gastrointestinal diseases in humans and animals.

For example, individual faecal microbial populations are affected by unstable Crohn's disease (Seksik et al., 2003). The important point here is to understand why bacterial population changes cause disease (or vice versa). Another study on the bacterial population in the ileum of piglets showed that there is an inverse relationship between bacterial diversity and sensitivity to pathogenic Clostridium bacteria (Deplank et al., 2002). The microbial flora of the ileum in neonates that were fed by their mothers showed a lower diversity and Clostridium density than neonates that were fed through intestine by the parents.

Comparison of faecal samples by TGGE fingerprinting in the adult human has demonstrated that the constituent dominant bacterial populations depend on the animal host (Zeotendal et al., 1998). This finding was also noted in the stool samples of other individuals (Tannock et al., 2000) and in other animals, such as dogs, chickens and mice (Zhu et al., 2002; Toivanen et al., 2001; Vaahtovuo et al., 2003). These data show the dependence of the bacterial population on the host animal's digestive system. This dependence is a general phenomenon and is not restricted to a particular animal species. This effect could be due to the considerable influence of the animal host genotype on the bacterial population that for example, we can refer to Meta- neo-gene existing in digestive system in some vertebrate and invertebrate animals (Heksetin and Van Alen, 1996). This hypothesis has been investigated in humans by comparing the DGGE profiles of adult humans with relatives (people without kinship compared with identica twins, Zeotendal et al., 2001). The similarity between the DGGE profiles of mangos twins was significantly greater than between non-relatives, demonstrating the effect of genetic structure on the composition of microbial flora. Thus, the genetic makeup of the host animal, such as poultry, affects the microbes that interact with the host (the microbial flora) (Hooper et al., 2002).

These important findings indicate that specific effects on the host animal's intestinal microbial flora cannot be ignored. The ecology of the microbial flora of the digestive system, which varies in each part of the tract, is complex. Early studies that were based on fingerprinting demonstrated these differences in pigs (Simpson et al., 1999). The bacterial population in the colon mucosa is uniformly distributed; however, there are no significant differences with the faecal bacterial population (Zeotendal et al., 2001). There are a limited number of comparative and adaptive studies of different parts of the digestive tract of poultry and other animals. The observations discussed here show that faecal samples from other parts of the digestive system do not necessarily reflect the microbial flora population.

2.4 Diversity microarrays

DNA microarrays are a new, well-characterised method for the molecular identification of samples of environmental bacteria, and they include Biochips, gene chips or DNA chips. Typically, DNA microarrays consist of fragments of DNA that are covalently bound to several glass surfaces and that are available for hybridisation. DNA microarray technology has also been optimised for use in the studies of bacterial diversity in many ecosystems (Loy et al., 2002; El Fantrossi et al., 2003). The two main difficulties of DNA microarray analysis are hybridisation and quantitative signal determination. El Fantrossi et al. (2003) demonstrated that specific and nonspecific hybridisation can distinguish the differential thermal curve for each probe-goal pair. Initial efforts to develop DNA microarray studies of the gastrointestinal microbial ecosystem have been completed, and the technology appears promising (Wilson et al., 2002). Undoubtedly, further development of DNA microarray technology will occur, and the technology will be used in the study of the ecology of the gastrointestinal tract.

2.5 Non-SSU rRNA-based profiling

SSU rRNA-based profiles of several other methods, such as cellular fatty acid composition (Toivanen et al., 2001; Vaahtovuo et al., 2003) or C + G DNA content (Apajalahti et al., 2002), have been used successfully to investigate changes in bacterial populations of the gastrointestinal tract. However, compared with SSU rRNA-based methods, these methods provide no phylogenetic information.

3. Different methods of investigation and quantification of microbial flora

3.1 Various methods of quantitative determination of SSU rDNA and SSU rRNA

Although PCR is the most sensitive method for the identification of sequences that have very low densities in the environment, many factors influence the amplification reaction, and fingerprinting methods alone cannot provide quantitative information to researchers (Von Wintzingerode et al., 1997). However, it is possible to determine the density of the SSU rDNA in the PCR reaction.

3.2 The RT-PCR method

Competitive PCR or the RT-PCR method for quantitative determination of the desired products can determine the amount of mRNA in human cells, which was the initial purpose of this method (Wang et al., 1989). In this method, a specific standard DNA fragment, at different concentrations, is added to a target (desired) product and amplified by PCR. The difference in the amount of the target and the standard is then quantified on an agarose gel. Using competitive PCR, the SSU rDNA of several bacterial species was determined in samples of cow rumen (Koike and Kobayashi, 2001, Reily et al., 2002).

3.3 Quantitative determination of amplified fragments in TGGE profiles and integration of quantitative PCR and constant-denaturant capillary electrophoresis (CDCE)

A similar method (to 3.2 above) for the quantitative measurement of individual components of the amplified fragments of TGGE profiles is used (Felske et al., 1998). The advantage of this method is that the amplified sample fragments and amplified standard fragments are similar, and differences between them can be identified by melting behaviour. Notably, when quantitative PCR and constant-denaturant capillary electrophoresis (CDCE) protocols were combined, similar results were obtained (Lim et al., 2001).

3.4 The most probable number PCR method

The Most Probable Number (MPN) PCR approach for determining the amount of SSU rDNA in environmental samples has been used successfully for human stool samples (Wang et al., 1996). The principles of this method are the same as the bacterial MPN count. Thus, DNA is diluted to a very low concentration and, using primers specific for a particular group or bacterial species, the DNA is then used as a PCR template.

This method is relatively quick and is suitable for determining the main groups of bacteria present, but it is not useful for the analysis of complex populations of species.

3.5 The real time PCR method

This method has received much recent attention and has been used to determine the characteristics of different samples of human and neonate pig gut and to successfully determine rumen samples (Huijsdens et al., 2002; Malinen et al., 2003). Although the efficiency of real time PCR for complex bacterial populations requires further study, this method can be used to study very low numbers of bacteria (which is very difficult with other methods) and accordingly, its future is bright.

This method is based on the accurate and sensitive identification and quantitation of fluorescence, which shows an increased signal in proportion to the amount of PCR product. To identify more than a series of explicit microorganisms, recommended specific primers were used.

3.6 The dot blot electrophoresis method

The Electrophoresis Dot Blot (Blot spot or point) method is used to calculate the concentration of a population-specific 16S rRNA in a mixture. For this method, the total RNA is isolated and then filtered onto a Dot or Slot Blot (Blot fractured) and labelled with oligonucleotide probes. The relative rRNA concentration of hybridised material can be calculated by dividing the concentration of the general probe by the concentration of the specific probe (after normalisation of the rRNA signal with a control strain).

This method has been used to quantify rRNA from samples of human, horse and rumen (Sghir et al., 2000; Marteau et al., 2001; Daly and Shirazi - Beechey, 2003). Because PCR has no relation to other amplification methods, quantification determination is more accurate, and therefore, this method has a good reputation and is widely used. Recently, all of the data obtained from more than 700 probes was published, and the data are available in an online website (www.probebase.net), which facilitates the search for rRNA probes of microorganisms at the levels of family, genus and species.

Population studies have been conducted to compare the performance of quantitative PCR, Dot Blot or hybridisation; however, this type of population determination is relative (Rigottier-Gois et al., 2003a). The SSU rRNA and ribosomal density per cell are variable and depend on the species of bacteria, the growth stage and the activity level. Accordingly, Dot Blot electrophoresis can provide information on the number of bacteria, especially with regard to bacterial growth during culture. Similarly, genome size and the 16s rRNA gene copy number varies between different bacterial genomes, thus preventing accurate conclusions from these data.

3.7 Fluorescent hybridisation *in vivo*

To determine the amount of bacterial cells in environmental samples without the conventional culturing method, the FISH method may be used. In this method, oligonucleotide probes are used to target SSU rRNA. The combination of SSU rRNA probe hybridisation and epifluorescence, using light microscopy, confocal laser microscopy, or flow cytometry, allows the direct examination of a single bacterial population. This method can determine the relative prevalence of particular groups or genera of bacteria. FISH is being used more and more frequently to study the bacterial composition of the digestive system, and currently, probes for bacteria belonging to different genera, such as bifidobacterium, Streptococcus, Lactobacillus, Collineslla, Eubacteriuom, Fozobacteriuom, Clostridium, Veillonella, Fibrobacter, and Rominokokus, have been described (Harmsen and Welling, 2002). To facilitate enumeration, automatic methods for FISH quantification and computer programs that analyse the images have been generated (Jensen et al., 1999). Lastly, this is the best method for counting bacteria in the digestive tract. However, thus far, the FISH method has primarily been used for determining (Amann et al., 1990) the major bacterial groups in human faeces.

3.8 Cytometry

Recently, the efficacy of the flow cytometry method for the determination of faecal bacteria has been demonstrated (Rigottier-Gois et al., 2003a; 2003b; Zeotendal et al., 2002). Statistical analysis has shown that count results from two microscopic and cytometry methods are similar (Zeotendal et al., 2002). In the future, it is possible that both FISH and cytometry will be used to categorise bacteria without culturing. Although these bacteria are not alive, they can be used in molecular genetic studies. Limitations of the FISH method include the requirement for 16S rDNA sequence information from the database and the limited number of probes that can be used in each analysis. The FISH method depends on the permeability of the bacterial cells, the availability of target products and the construct number in each cell.

3.9 Densitometry

Densitometry measurements have many applications, especially for the measurement of SDS-PAGE protein gels (Zhang et al., 2007, Bromage and Kattari, 2007), but also for quantification of PCR products including mtDNA (Enzmann et al., 1999) and bacterial DNA (Amit-Romach et al., 2004) and many researchers have reported using this method.

4. Gene expression methods in microorganisms

Identifying the characteristics of the digestive system microbial flora is the first step in the study of this ecosystem because such data provide limited information on microorganism-microorganism and microorganism-host interactions. Given the complexity of microbial flora in the digestive system and the limited ability of many forms of bacterial culture, it is clear that determining the function of all of these microorganisms will be very difficult. However, some technical developments in bacterial analysis provide a good outlook for future research.

4.1 Gene expression analysis using RT-PCR in microorganisms

Functional gene expression is a suitable method for determining the activity of bacteria in an ecosystem. Using the RT-PCR method, Deplancke et al. (2000) identified mRNA expression of adenosine-5–phosphor soleplate in different parts of the digestive tract of rats. The RT-PCR method was also used to examine the effect of Helicobacter pylori infection on the expression of four genes in mouse and human gastric mucosa (Rokbi et al., 2001). The technology "gene expression in vivo" (IVET) is also a method for analysing gene expression in living organisms.

4.2 Gene expression evaluation in microorganisms using IVET

IVET allows the identification of promoters that are activated when bacteria are exposed to certain environmental conditions (Rainey and Preston, 2000). This method is typically used to study gene expression of pathogenic microorganisms. However, it was used recently to identify gene promoters that were activated when exposed to lactobasirus in the mouse digestive system (Walter et al., 2003). Interestingly, the expression of three genes was associated with the establishment of Lactobacillus in the gastrointestinal tract.

4.3 Gene expression evaluation in microorganisms using microorganism DNA

Despite the use of complete genomic sequences, genomic comparisons and DNA microarrays for studying transcription in microorganisms, these methods remain in the early stages of development and are expensive. Thus, researchers are faced with obstacles and could instead use the genomic sequences of well-known cultured bacteria.

4.4 Gene expression evaluation using bacterial artificial chromosome and subtractive hybridisation

New methods for research into the function of genes, especially gene function in non-cultivated microorganisms, include the use of bacterial artificial chromosomes (BACs) and subtractive hybridisation. BAC-related cloning of large DNA fragments (more than one hundred pounds) is possible. Because environmental DNA libraries from soil (Liles et al., 2003) and sea (Beja et al., 2000) have been established, it is possible to use these libraries for the study of the diversity and metabolic potential of these complex ecosystems. Thus, it is possible to associate genes with SSU rRNA gene function, without culturing.

4.5 Gene expression evaluation in microorganisms using marked substitute materials

The use of substitute materials, marked with an isotope, provides another method for evaluating the performance of a specific microorganism in a complex population. Environmental samples are marked and grown with the substitute materials that contain environmental isotopes (stable or radioactive), and the microorganisms are then identified using extracted DNA or rRNA. (Boschker et al., 2001; Polz et al., 2003).

4.6 Gene expression evaluation in microorganisms using micro autoradiography and FISH composition

Autoradiography and FISH composition are an additional successful method that generates phylogenetic information (Ouverney and Fuhrman, 1999). This method appears to be very promising. You should be able to remember some microorganism between different isotopes (Londry and Des Marais, 2003). However, it is clear that using and developing new methods such as those mentioned above will allow the examination of the normal microorganism interactions (microorganism-microorganism and microorganism-host).

5. Conclusion

Due to several difficulties, traditional and non-molecular based techniques do not provide the required quality for acceptance as a standard reference method for the identification and study of bacteria. Thus, researchers have disagreed over these methods, and the data have not been presented in any standardised or reliable format. These issues indicate the necessity for new and updated techniques. Microbial culture methods can theoretically detect a small number of live bacteria in the gastrointestinal tract of chickens. However, in many cases, it is necessary to use selective media and conduct enrichment. In practice, the minimum required number of bacteria for identification by culture methods is much greater than with molecular methods (Bjerrum et al., 2006). Due to their specific requirements, in current media, many bacteria show slow growth or sometimes no growth. Therefore, their identification by culture methods is difficult and sometimes impossible. Many other studies of bacterial detection have reported on the superiority of the polymerase chain reaction over culture methods. One reason underlying this superiority is that PCR can detect target sequences and is not related to the growth of target cells, while culture-based methods are correlated with the growth of target bacteria.

In this regard, many researchers, including Kothary and Babu (2001), Bjerrum et al. (2006) and Apajalahti et al. (2004), have stated that in many cases, the number of bacteria in the digestive tract of poultry is very low, and it is therefore necessary to use highly sensitive methods for the detection of these bacteria. However, due to the use of specific media for the accurate identification and elimination of false negative bacteria during culturing, several enrichments of these samples can be carried out. Nevertheless, false positive results, due to the growth of non-target bacteria that have similar growth requirements to the desired bacteria, must always be considered. A typical solution for this issue is the use of a specific medium for the bacterium being studied. When a very low population for a specific bacterial species in a sample exists, the resulting inhibitory effects on the polymerase chain reaction must be overcome through the use of selective media (Jofre et al., 2005; Rossen et al., 1992; Al-Soud and Radstrom, 2000; Al-Soud and Radstrom, 2001; Lantz et al., 2000). Li et al. (2005) believed that negative results could occur in a polymerase chain reaction even with medium-associated bacterial enrichment. The growth requirements and the appropriate media for growth of many bacteria remain unknown.

With the development of specific polymerase chain reactions for the detection of bacteria in the digestive tract of chickens, separately or simultaneously, new horizons develop in the field of poultry science research. In recent years, many studies have been conducted related to nutrition and health of poultry instead of using traditional methods in such a process approach to research in animal science is very attractive. Accordingly, investigations of the effects of dietary components, processed diets, antibiotics, and proboscis and diet enzymes on the gastrointestinal microbial flora of poultry and related research using these methods will be initiated. Indeed, poultry veterinary medicine had recently been leaning toward using these new methods and fields of research in this area. In addition, portions of the digestive system of birds used in this study included the microbial flora of the state organs, such as the crop, and by-products, such as carcasses and eggs and even the surrounding environment, such as the bedding, water and rations. Bacterial presence can be studied using these methods. Furthermore, using such methods, the microbial flora of poultry strains can be compared, and the differences between indigenous and industrial (modified) strains may be investigated. Individual variation in the microbial flora of the digestive system will benefit from the introduction of flora. In the near future, researchers may even investigate breeding herds guided by the microbial flora of poultry, which is one of the traits that influences the health and economic performance of chickens.

6. Acknowledgment

This study was supported by the Islamic Azad University, Rasht Branch, Rasht, Iran. The author wishes to sincerely thank Dr. S.Z. Mirhosseini and Dr. M. Shivazad for their comments. The author also acknowledges the kind advice of Dr. M. Chamani, Dr A.A. Sadeghi, and Mr. R. Pourseify for valuable comments and helpful assistance.

7. References

Al-Soud, W.A. & Radstrom, P. (2000). Effects of amplification facilitators on diagnostic PCR in the presence of blood, feces, & meat. *J. Clin. Microbiol.* Vol.38, pp. 4463-4470.

Al-Soud, W.A. & Radstrom, P. (2001). Purification and characterization of PCR-inhibitory components in blood cells. *J. Clin. Microbiol.* Vol.39, pp. 485-493.

Amann, R.I.; Krumholz, L. & Stahl, D.A. (1990). Fluorescent-oligonucleotide probing of whole cells for determinative, phylogenetic, & environmental studies in microbiology. *J. Bacteriol.* Vol.172, pp. 762-770.

Amit-Romach, E.; Sklan, D. & Uni, Z. (2004). Microflora ecology of the chicken intestine using 16S ribosomal DNA primers. *Poult. Sci.* Vol.83, pp. 1093-1098.

Ampe, F.; ben Omar, N.; Moizan, C.; Wacher, C. & Guyot, J.P. (1999). Polyphasic study of the spatial distribution of microorganisms in Mexican pozol, a fermented maize dough, demonstrates the need for cultivation-independent methods to investigate traditional fermentations. *Appl. Environ. Microbiol.* Vol.65, pp. 5464-5473.

Apajalahti, J.; Kettunen, A. & Graham, H. (2004). Characteristics of the gastrointestinal microbial communities, with special reference to the chicken. *World's Poult. Sci. J.* Vol.60, pp. 223-232.

Apajalahti, J.H.; Kettunen, H.; Kettunen, A,; Holben, W.E.; Nurminen, P.H.; Rautonen, N. & Mutanen, M. (2002). Culture-independent microbial community analysis reveals that inulin in the diet primarily affects previously unknown bacteria in the mouse ceacum. *Appl. Environ. Microbiol.* Vol.68, pp. 4986-4995.

Beja, O.; Suzuki, M.T.; Koonin, E.V.; Aravind, L.; Hadd, A.; Nguyen, L.P.; Villacorta, R.; Amjadi, M.; Garrigues, C.; Jovanovich, S.B.; Feldman, R.A. & DeLong, E.F. (2000). Construction and analysis of bacterial artificial chromosome libraries from marine microbial assemblage. *Environ. Microbiol.* Vol.2, pp. 516-529.

Bjerrum, L.; Engberg, R.M.; Leser, T.D.; Jensen, B.B.; Finster, K. & Pedersen, K. (2006). Microbial community composition of the ileum and cecum of broiler chickens as revealed by molecular and culture-based techniques. *Poult. Sci.* Vol.85, pp. 1151-1164.

Bonnet, R.; Suau, A.; Dore, J.; Gibson, G.R. & Collins, M.R. (2002). Differences in rDNA libraries of faecal bacteria derived from 10- and 25-cycle PCRs. *Int. J. Syst. Evol. Microbiol.* Vol.52, pp. 757-763.

Boschker, H.T.; de Graaf, W.; Koster, M.; Meyer-Reil, L. & Cappenberg, T.E. (2001). Bacterial populations and processes involved in acetate and propionate consumption in anoxic brackish sediment. *FEMS Microbiol. Ecol.* Vol.35, pp. 97-103.

Bromage, E.S. & Kaattari, S.L. (2007). Simultaneous quantitative analysis of multiple protein species within a single sample using standard scanning densitometry. *J. Immunol. Met.* Vol.323, pp. 109-113.

Cole, J.R.; Chai, B.; Marsh, T.L.; Farris, R.J.; Wang, Q.; Kulam, S.A.; Chandra, S.; McGarrell, D.M.; Schmidt, T.M.; Garrity, G.M. & Tiedje, J.M. (2003). The ribosomal database project (RDP-II): previewing a new autoaligner that allows regular updates and the new prokaryotic taxonomy. *Nucleic Acids Res.* Vol.31, pp. 442-443.

Daly, K. & Shirazi-Beechey, S.P. (2003). Design and evaluation of group-specific oligonucleotide probes for quantitative analysis of intestinal ecosystems: their application to assessment of equine colonic microflora. *FEMS Microbiol. Ecol.* Vol.44, pp. 243-252.

Daly, K.; Stewart, C.S.; Flint, H.J. & Shirazi-Beechey, S.P. (2001). Bacterial diversity within the equine large intestine as revealed by molecular analysis of cloned 16S rRNA genes. *FEMS Microbiol. Ecol.* Vol.38, pp, 141-151.

Deplancke, B.; Hristova, K.R.; Oakley, H.A.; McCracken, V.J.; Aminov, R.I.; Mackie, R.I. & Gaskins, H.R. (2000). Molecular ecological analysis of the succession and diversity of sulfate-reducing bacteria in the mouse gastrointestinal tract. *Appl. Environ. Microbiol.* Vol.66, pp. 2166-2174.

Deplancke, B.; Vidal, O.; Ganessunker, D.; Donovan, S.M.; Mackie, R.I. & Gaskins, H.R. (2002). Selective growth of mucolytic bacteria including *Clostridium perfringens* in a neonatal piglet model of total parenteral nutrition. *Am. J. Clin. Nutr.* Vol.76, pp. 1117-1125.

El Fantroussi, S.; Urakawa, H.; Bernhard, A.E.; Kelly, J.J.; Noble, P.A.; Smidt, H.; Yershov, G.M. & Stahl, D.A. (2003). Direct profiling of environmental microbial populations by thermal dissociation analysis of native rRNAs hybridized to oligonucleotide microarrays. *Appl. Environ. Microbiol.* Vol.69, pp. 2377-2382.

Enzmann, H.; Wiemann, C.; Ahr, H.J. & Schluter, G. (1999). Damage to mitochondrial DNA induced by the quinolone Bay y 3118 in embryonic turkey liver. *Mutation Res.* Vol.425, pp. 213-224.

Felske, A.; Akkermans, A.D.L. & de Vos, W.M. (1998). Quantification of 16S rRNAs in complex bacterial communities by multiple competitive reverse transcription-PCR in temperature gradient gel electrophoresis fingerprints. *Appl. Environ. Microbiol.* Vol.64, pp. 4581-4587.

Franks, A.H.; Harmsen, H.J.M.; Raangs, G.C.; Jansen, G.J.; Schut, F. & Welling, G.W. (1998). Variations of bacterial populations in human feces measured by fluorescent *in situ* hybridization with group-specific 16S rRNAtargeted oligonucleotide probes. *Appl. Environ. Microbiol.* Vol.64, pp. 3336-3345.

Grahn, N.; Olofsson. M.; Ellnebo-Svedlund, K.; Monstein, H.J. & Jonasson, J. (2003). Identification of mixed bacterial DNA contamination in broadrange PCR amplification of 16S rDNA V1 and V3 variable regions by pyrosequencing of cloned amplicons. *FEMS Microbiol. Lett.* Vol.219, pp. 87-91.

Hackstein, J.H.P. & Van Alen, T.A. (1996). Fecal methanogens and vertebrate evolution. *Evolution.* Vol.50, pp. 559-572.

Harmsen, H.J.M.; Raangs, G.C.; He, T.; Degener, J.E. & Welling, G.W. (2002). Extensive set of 16S rRNA-based probes for detection of bacteria in human feces. *Appl. Environ. Microbiol.* Vol.68, pp. 2982-2990.

Hooper, L.V.; Midtvedt, T. & Gordon, J.I. (2002). How host-microbial interactions shape the nutrient environment of the mammalian intestine. Annu. Rev. Nutr. Vol.22, pp. 283-307.

Hopkins, M.J.; Sharp, R. & Macfarlane, G.T. (2001). Age and disease related changes in intestinal bacterial populations assessed by cell culture, 16S rRNA abundance, & community cellular fatty acid profiles. *Gut.* Vol.48, pp. 198-205.

Huijsdens, X.W.; Linskens, R.K.; Mak, M.; Meuwissen, S.G.M.; Vandenbroucke-Grauls, C.M.J.E. & Savelkoul, P.H.M. (2002). Quantification of bacteria adherent to gastrointestinal mucosa by real-time PCR. *J. Clin. Microbiol.* Vol.40, pp. 4423-4427.

Jansen, G.J.; Wildeboer-Veloo, A.C.; Tonk, R.H.; Franks, A.H. & Welling, G.W. (1999). Development and validation of an automated, microscopy- based method for enumeration of groups of intestinal bacteria. *J. Microbiol. Methods.* Vol.37, pp. 215-221.

Jofre, A.; Martin, B.; Garrigaa, M.; Hugas, M.; Pla, M.; Rodriguez-Lazaro, D. & Aymerich, T. (2005). Simultaneous detection of *Listeria monocytogenes* and *Salmonella* by multiplex PCR in cooked ham. *Food Microbiol.* Vol.22, pp. 109-115.

Kocherginskaya, S.; Aminov, R.I. & White, B.A. (2001). Analysis of the rumen bacterial diversity under two different diet conditions using denaturing gradient gel electrophoresis, random sequencing, & statistical ecology approaches. *Anaerobe* Vol.7, pp. 19-134.

Koike, S. & Kobayashi, Y. (2001). Development and use of competitive PCR assays for the rumen cellulolytic bacteria: *Fibrobacter succinogenes, Ruminococcus albus* and *Ruminococcus flavefaciens. FEMS Microbiol. Lett.* Vol.204, pp. 361-366.

Kong, R.Y.C.; Lee, S.K.Y.; Law, T.W.F.; Law, S.H.W. & Wu, R.S.S. (2002). Rapid detection of six types of bacterial pathogens in marine waters by multiplex PCR. *Water Res.* Vol.36, pp. 2802-2812.

Konstantinov, S.R.; Fitzsimons, N.; Vaughan, E.E. & Akkermans, A.D.L. (2002). From composition to functionality of the intestinal microbial communities. In: Probiotics and Prebiotics: Where Are We Going? (Tannock, G.W.; Ed.), Caister Academic Press, London, UK, pp.59-84.

Kothary, M.H. & Babu, U.S. (2001). Infectious dose of foodborne pathogens in volunteers: a review. *J. Food Safety.* Vol.21, pp. 49-73.

Lantz, P.G.; Al-Soud, W.A.; Knutsson, R.; Hahn-Hagerdal, B. & Radstrom, P. (2000). Biotechnical use of polymerase chain reaction for microbiological analysis of biological samples. *Biotec. Annu. Rev.* Vol.5, pp. 87-130.

Li, Y.; Zhuang, S. & Mustapha, A. (2005). Application of a multiplex PCR for the simultaneous detection of *Escherichia coli* O157:H7, *Salmonella* and *Shigella* in raw and ready-to-eat meat products. *Meat Sci.* Vol.71, pp. 402-406.

Liles, M.R.; Manske, B.F.; Bintrim, S.B.; Handelsman, J. & Goodman, R.M. (2003). A census of rRNA genes and linked genomic sequences within a soil metagenomic library. *Appl. Environ. Microbiol.* Vol.69, pp. 2684-2691.

Lim, E.L.; Tomita, A.V.; Thilly, W.G. & Polz, M.F. (2001). Combination of competitive quantitative PCR and denaturant capillary electrophoresis for high-resolution detection and enumeration of microbial cells. *Appl. Environ. Microbiol.* Vol.67, pp. 3897-3903.

Londry, K.L. & Des Marais, D.J. (2003). Stable carbon isotope ractioning by sulfate-reducing bacteria. *Appl. Environ. Microbiol.* Vol.69, pp. 2942-2949.

Loy, A.; Lehner, A.; Lee, N,; Adamczyk, J.; Meier, H.; Ernst, J.; Schleifer, K.H. & Wagner, M. (2002). Oligonucleotide microarray for 16S rRNA genebased detection of all recognized lineages of sulfate-reducing prokaryotes in the environment. *Appl. Environ. Microbiol.* Vol.68, pp. 5064-5081.

Malinen, E.; Kassinen, A.; Rinttila, T. & Pavla, A. (2003). Comparison of real-time PCR with SYBR Green I or 5-nuclease assays and dot-blot hybridization with rDNA-targeted oligonucleotide probes in quantification of selected faecal bacteria. *Microbiol.* Vol.149, pp. 269-277.

Marteau, P.; Pochart, P.; Dore, J.; Bera-Maillet, C.; Bernallier, A. & Corthier, G. (2001). Comparative study of bacterial groups within the human cecal and fecal microbiota. *Appl. Environ. Microbiol.* Vol.67, pp. 4939-4942.

McCracken, V.J.; Simpson, J.M.; Mackie, R.I. & Gaskins, H.R. (2001). Molecular ecological analysis of dietary and antibiotic-induced alterations of the mouse intestinal microbiota. *J. Nut.* Vol.131, pp. 1862-1870.

McPherson, M.J. & Møller, S.G. (2000). Reagents and instrumentation. PCR. BIOS Scientific Publishers Ltd, Oxford, pp.23-60.

Mirhosseini, S.Z.; Seidavi, A.R.; Shivazad, M.; Chamani, M.; Sadeghi, A.A. & Pourseify, R. (2008a). Application of a duplex PCR approach for the specific and simultaneous detection of *Clostridium* spp. and *Lactobacillus* spp. in broiler gastrointestinal tract. *Indian Journal of Animal Nutrition.* Vol. 25, pp. 83-92.

Mirhosseini, S.Z.; Seidavi, A.R.; Shivazad, M.; Chamani, M.; Sadeghi, A.A. & Pourseify, R. (2009a). Detection of *Salmonella* spp. in Gastrointestinal Tract of Broiler Chickens by Polymerase Chain Reaction. *Kafkas Universitesi Veteriner Fakultesi Dergisi.* Vol.15, pp. 965-970.

Mirhosseini, S.Z.; Seidavi, A.R.; Shivazad, M.; Chamani, M.; Sadeghi, A.A. & Pourseify, R. (2010). Detection of *Clostridium* spp. and its relation to different ages and gastrointestinal segments as measured by molecular analysis of 16S rRNA genes. *Brazilian Archives of Biology and Technology.* Vol.53, pp. 69-76.

Mirhosseini, S.Z.; Seidavi, A.R.; Shivazad, M.; Chamani, M.; Sadeghi, A.A. & Pourseify, R. (2008b). Optimization of PCR for *Clostridium* spp. bacteria detection in gastrointestinal tract of broilers. *Proceedings of the 3rd Congress on Animal Science.* 15-16 October 2008. Faculty of Agriculture, Ferdowsi University of Mashhad, Mashhad, Iran, pp. 32.

Mirhosseini, S.Z.; Seidavi, A.R.; Shivazad, M.; Chamani, M.; Sadeghi, A.A. & Pourseify, R. (2009b). Application of a duplex PCR approach for the specific and simultaneous detection of *Bifidobacterium* spp. and *lactobacillus* spp. in duodenum, jejunum, ileum and cecum of broilers. *Proceedings of the British Society of Animal Science (BSAS), Annual Conference.* 30 March-1 April 2009, Southport, UK. pp. 207.

Mirhosseini, S.Z.; Seidavi, A.R.; Shivazad, M.; Chamani, M.; Sadeghi, A.A. & Pourseify, R. (2009c). Relative frequency alteration of *Escherichia coli* in broiler intestine. *Proceedings of the British Society of Animal Science (BSAS), Annual Conference.* 30 March-1 April 2009, Southport, UK. pp. 209.

Miyashita, N.; Saito, A.; Kohno, S.; Yamaguchi, K.; Watanabe, A.; Oda, H.; Kazuyama, Y. & Matsushima, T. (2004). Multiplex PCR for the simultaneous detection of *Chlamidia pneumoniae, Mycoplasma pneumoniae* and *Legionella pneumophila* in community-acquired pneumonia. *Respir. Med.* Vol.98, pp. 542-550.

Muyzer, G.; de Waal, E.C. & Uitterlinden, A.G. (1993). Profiling of complex microbial populations by denaturing gradient gel electrophoresis analysis of polymerase chain reaction-amplified genes coding for 16S rRNA. *Appl. Environ. Microbiol.* Vol.59, pp. 695-700.

Nagashima, K.; Hisada, T.; Sato, M. & Mochizuki, J. (2003). Application of new primer-enzyme combinations to terminal restriction fragment length polymorphism profiling of bacterial populations in human feces. *Appl. Environ. Microbiol.* Vol.69, pp. 1251-1262.

Ouverney, C.C. & Fuhrman, J.A. (1999). Combined microautoradiography- 6S rRNA probe technique for determination of radioisotope uptake by pecific microbial cell types *in situ. Appl. Environ. Microbiol.* Vol.65, pp. 1746-1752.

Polz, M.F.; Bertillson, S.; Acinas, S.G. & Hunt, D. (2003). A(r)ray of ope in analysis of the function and diversity of microbial communities. *Biol. Bull.* Vol.04, pp. 196-199.

Raclasky, V.; Trtkova, J.; Ruskova, L.; Buchta, V.; Bolehovska, R.; Vackova, M. & Hamal, P. (2006). Primer R108 performs best in the RAPD strain typing of three *Aspergillus* species frequently isolated from patients. *Folia Microbiol.* Vol.51, pp. 136-140.

Rainey, P.B. & Preston, G.M. (2000). *In vivo* expression technology strategies: valuable tools for biotechnology. *Curr. Opin. Biotechnol.* Vol.11, pp. 440-444.

Reilly, K.; Carruthers, V.R. & Attwood, G.T. (2002). Design and use of 16S ribosomal DNA-directed primers in competitive PCRs to enumerate proteolytic bacteria in the rumen. *Microb. Ecol.* Vol.43, pp. 259-270.

Rigottier-Gois, L.; Le Bourhis, A.G.; Gramet, G.; Rochet, V. & Dore, J. (2003a). Fluorescent hybridization combined with flow cytometry and hybridization of total RNA to analyse the composition of microbial communities in human faecal samples using 16S rRNA probes. *FEMS Microbiol. Ecol.* Vol.43, pp. 237-245.

Rigottier-Gois, L.; Rochet, V.; Garrec, N.; Suau, A. & Dore, J. (2003b). Enumeration of *Bacteroides* species in human faeces by fluorescent *in situ* hybridization combined with flow cytometry using 16S rRNA probes. *Syst. Appl. Microbiol.* Vol.26, pp. 110-118.

Rokbi, B.; Seguin, D.; Guy, B.; Mazarin, V.; Vidor, E.; Mion, F.; Cadoz, M. & Quentin-Millet, M.J. (2001). Assessment of *Helicobacter pylori* gene expression within mouse and human gastric mucosae by real-time reverse transcriptase PCR. *Infect. Immun.* Vol.69, pp. 4759-4766.

Rosselló-Mora, R. & Amann, R. (2001). The species concept for prokaryotes. *FEMS Microbiol. Rev.* Vol.25, pp. 39-67.

Rossen, L.; Noskov, P.; Holmstrom, K. & Rasmussen, O.F. (1992). Inhibition of PCR by components of food samples, microbial diagnostic assays and DNA-extraction solution. *Int. J. Food Microbiol.* Vol.17, pp. 37-45.

Savage, D.C. (2001). Microbial biota of the human intestine: a tribute to some pioneering scientists. *Curr. Issues Intest. Microbiol.* Vol.2, pp. 1-15.

Schwiertz, A.; Gruhl, B.; Lobnitz, M.; Michel, P.; Radke, M. & Blaut, M. (2003). Development of intestinal bacterial composition in hospitalized preterm infants in comparison with breast-fed, full-term infants. *Pediatr. Res.* Vol.54, pp. 393-399

Seidavi, A.R. & Chamani, M. (2010). Determining of Relative Population of *Clostridium* spp. Bacteria in Duodenum, Jejunum, Ileum and Cecum of Broilers at Various Ages using Densitometry Technique based on 16S rDNA Approach. Veterinary Research Bulletin, *Islamic Azad University, Garmsar Branch*, Vol.6, pp. 1-10.

Seidavi, A.R. & Qotbi, A. (2009a). Measurment of *Clostridium* spp. relative population in different segments of poultry's intestine by means of densitometry technique based on PCR. *Abstract Book of the 1st National Congress of Veterinary Laboratory Sciences.* 1-2 December 2009, Tehran, Iran.

Seidavi, A.R. & Qotbi, A. (2009b). Estimation of relative population of *Escherichia coli* in duodenum, jejunum, ileum and cecum of poultry and their alterations at various ages. *10th Iranian Congress of Microbiology, Ilam University of Medical Sciences.* April 21-23, 2009. Ilam, Iran. pp. 114.

Seidavi, A.R. (2008). Investigation on relative population of *Lactobacillus* spp. in broiler gut at various ages using densitometry technique. Abstract Book of 14th National and 2nd *International Conference of Biology.* 19-21 August 2008. University of Tehran, Iran, pp. 357.

Seidavi, A.R. (2009a). Application of genomic densitometry for calculating the relative population of *Escherichia coli* in the intestine of broiler chicks. *Veterinary Journal of Islamic Azad University, Tabriz Branch.* Vol.3, pp. 411-420.

Seidavi, A.R. (2009b). Investigation on Possibility of Quantification of Flora in Poultry Guts using Densitometry Technique. *Proceedings of 5th International Poultry Conference.* 10-13 March 2009. Taba, Egypt. pp. 1423.

Seidavi, A.R. (2009c). Investigation on relative population of *Salmonella* spp. bacteria in duodenum, jejunum, ileum and cecum of poultry using densitometry technique. *Proceedings of the British Society of Animal Science (BSAS), Annual Conference.* 30 March-1 April 2009, Southport, UK. pp. 208.

Seidavi, A.R.; Mirhosseini, S.Z.; Shivazad, M.; Chamani, M. & Sadeghi, A.A. (2008a). Optimizing multiplex polymerase chain reaction method for specific, sensitive and rapid detection of *Salmonella* sp.; *Escherichia coli* and *Bifidobacterium* sp. in chick gastrointestinal tract. *Asian Journal of Animal and Veterinary Advances.* Vol.3, pp. 230-235.

Seidavi, A.R.; Mirhosseini, S.Z.; Shivazad, M.; Chamani, M. & Sadeghi, A.A. (2008b). The development and evaluation of a duplex PCR detection of *Bifidobacterium* spp. and *Lactobacillus* spp. in duodenum, jejunum, ileum and cecum of broilers. *Journal of Rapid Methods and Automation in Microbiology.* Vol.16, pp. 100-112.

Seidavi, A.R.; Mirhosseini, S.Z.; Shivazad, M.; Chamani, M.; Sadeghi, A.A. & Pourseify, R. (2010a). Detection and investigation of *Escherichia coli* in contents of duodenum, jejunum, ileum and cecum of broilers at different ages by PCR. *Asia Pacific Journal of Molecular Biology and Biotechnology.* Vol.18, pp. 321-326.

Seidavi, A.R.; Mirhosseini, S.Z.; Shivazad, M.; Chamani, M.; Sadeghi, A.A. & Pourseify, R. (2007). Effect of five different components on optimization of PCR for detection of *Clostridium* spp. bacteria in broiler gastrointestinal tract. *Veterinary Journal of Islamic Azad University, Garmsar Branch.* Vol.3, pp. 153-162.

Seidavi, A.R.; Mirhosseini, S.Z.; Shivazad, M.; Chamani, M.; Sadeghi, A.A. & Pourseify, R. (2008c). PCR for detection of *Bifidobacterium* spp in broiler intestine. *Online Journal of Veterinary Research.* Vol.12, pp. 38-46.

Seidavi, A.R.; Mirhosseini, S.Z.; Shivazad, M.; Chamani, M.; Sadeghi, A.A. & Pourseify, R. (2011). A polymerase chain reaction assay for simple and rapid diagnosis of *Campylobacter* spp. in broiler chicken gastrointestinal tract. *Annals of Biological Research.* Vol.2, pp. 145-152.

Seidavi, A.R.; Mirhosseini, S.Z.; Shivazad, M.; Chamani, M.; Sadeghi, A.A. & Pourseify, R. (2008d). Development and assessment of a polymerase chain reaction assay as measured by molecular analysis of 16S rDNA genes for simple and rapid detection of *Lactobacillus* spp. in broiler chicken gastrointestinal tract. *10th Iranian Genetics Congress.* 21-23 May 2008. Razi Halls Center, Tehran, Iran, pp. 412.

Seidavi, A.R.; Mirhosseini, S.Z.; Shivazad, M.; Chamani, M.; Sadeghi, A.A. & Pourseify, R. (2008e). Alteration of relative frequency *Bifidobacterium* spp. bacteria in duodenum, jejunum, ileum and cecum of broilers. *Proceedings of the 3rd Congress on Animal Science.* 15-16 October 2008. Faculty of Agriculture, Ferdowsi University of Mashhad, Mashhad, Iran, pp. 165.

Seidavi, A.R.; Mirhosseini, S.Z.; Shivazad, M.; Chamani, M.; Sadeghi, A.A. & Pourseify, R. (2008f). Simultaneous detection of *Lactobacillus* spp. and *Clostridium* spp. bacteria in broiler intestine using a specific duplex PCR. *Proceedings of the 3rd Congress on Animal Science.* 15-16 October 2008. Faculty of Agriculture, Ferdowsi University of Mashhad, Mashhad, Iran, pp. 313.

Seidavi, A.R.; Mirhosseini, S.Z.; Shivazad, M.; Chamani, M.; Sadeghi, A.A. & Pourseify, R. (2008g). Introduction of Polymerase Chain Reaction (PCR) approach for simultaneous detection of *Bifidobacterium* spp.; *Salmonella* spp.; and *Escherichia coli*, in broilers gastrointestinal tract. *Proceedings of the 6th National Biotechnology Congress of IR Iran*. 13-15 August 2009, Tehran, Iran.

Seidavi, A.R.; Mirhosseini, S.Z.; Shivazad, M.; Chamani, M.; Sadeghi, A.A. & Pourseify, R. (2010b). Relative frequency of *Lactobacillus* spp. in various segments of broiler intestine at different ages. *2nd International Veterinary Poultry Congress*. February 20-21, 2010. Tehran, Iran. pp. 1.

Seidavi, A.R.; Mirhosseini, S.Z.; Shivazad, M.; Chamani, M.; Sadeghi, A.A. & Pourseify, R. (2008h). A PCR-based strategy for simple and rapid detection of *Bifidobacterium* spp. in duodenum, jejunum, ileum and cecum of broiler chickens at three different ages. *Book of Proceedings of 1st Mediterranean Summit of WPSA: Advances and Challenges in Poultry Science*. Porto Carras, Chalkidiki, Greece, 7-10 May 2008. Page 260-265; World Poultry Science Journal, Vol 64 (Supplement 1), pp. 31-32.

Seidavi, A.R.; Mirhosseini, S.Z.; Shivazad, M.; Chamani, M.; Sadeghi, A.A. & Pourseify, R. (2008i). Detection of *Escherichia coli* at different ages and gastrointestinal segments of broilers as measured by molecular analysis of 16S rRNA genes. *Book of Proceedings of 1st Mediterranean Summit of WPSA: Advances and Challenges in Poultry Science*. Porto Carras, Chalkidiki, Greece, 7-10 May 2008. Page 266-271; World Poultry Science Journal, Vol 64 (Supplement 1), pp. 32.

Seidavi, A.R.; Mirhosseini, S.Z.; Shivazad, M.; Chamani, M.; Sadeghi, A.A. & Pourseifi, R. (2008j). Application of a conventional PCR analysis for detection of *Salmonella* spp. in gastrointestinal of Iranian chickens at different ages. *Book of Abstract of XXIII (23rd) Worlds Poultry Congress, Brisbane, Australia*, 30 June- 4 July, 2008; World Poultry Science Journal, Vol 64 (Supplement 2), pp. 472.

Seidavi, A.R.; Mirhosseini, S.Z.; Shivazad, M.; Chamani, M.; Sadeghi, A.A. & Pourseify, R. 2009a. Development and Assessment of A Polymerase Chain Reaction Approach Based on 16S rDNA Analysis for Simple and Rapid Detection of *Clostridium* spp. in Broiler Chicken Gastrointestinal Tract. *Proceedings of 5th International Poultry Conference*. 10-13 March 2009. Taba, Egypt. pp. 1138-1147.

Seidavi, A.R.; Mirhosseini, S.Z.; Shivazad, M.; Chamani, M.; Sadeghi, A.A. & Pourseify, R. (2009b). Detection of *Campylobacter* spp. and its relation to different ages and Broiler Gastrointestinal Segments as Measured by Molecular Analysis of 16S rRNA Genes. *Proceedings of 5th International Poultry Conference*. 10-13 March 2009. Taba, Egypt. pp. 1424-1432.

Seidavi, A.R.; Mirhosseini, S.Z.; Shivazad, M.; Chamani, M.; Sadeghi, A.A. & Pourseify, R. (2009c). Investigation on relative frequency and alteration of *Clostridium* spp. in poultry intestine. *Proceedings of the 60th Annual Meeting of the European Association for Animal Production (EAAP)*. 24-27 August 2009, Barcelona, Spain. pp. 340.

Seidavi, A.R.; Mirhosseini, S.Z.; Shivazad, M.; Chamani, M.; Sadeghi, A.A. & Pourseify, R. (2010c). Study of Relative Frequency Alteration of Bacteria in Chickens Intestine. *Book of Abstracts of XIIIth (13rd) European Poultry Conference, Tours, France*, 23-27 August, 2010; Page 428; World's Poultry Science Journal, Vol.66 (Supplement).

Seidavi, A.R.; Qotbi, A. & Chamani, M. (2008k). Determination of Relative Population of *Bifidobacterium* spp. in Broiler Gut Using Densitometry Technique Based Molecular Genetics. Iranian Journal of Dynamic Agriculture, Islamic Azad University, *Varamin- Pishva Branch*. Vol.5, pp. 229-243.

Seidavi, A.R.; Qotbi, A. & Chamani, M. (2009d). Investigation on Frequency of *Lactobacillus* spp. Bacteria in Duodenum, Jejunum, Ileum and Cecum of Poultry using Densitometry Technique. *Iranian Journal of Veterinary Sciences.* Vol.6, pp. 747-755.

Seksik, P.; Rigottier-Gois, L.; Gramet, G.; Sutren, M.; Pochart, P.; Marteau, P.; Jian, R. & Dore, J. (2003). Alterations of the dominant faecal bacterial groups in patients with Crohn's disease of the colon. *Gut.* Vol.52, pp. 237-242.

Settanni, L. & Corsetti, A. (2007). The use of multiplex PCR to detect and differentiate food- and beverage-associated microorganisms: A review. J. Microbiol. Methods. Vol.69, pp. 1-22.

Settanni, L.; Van Sinderen, D.; Rossi, J. & Corsetti, A. (2005). Rapid differentiation and *in situ* detection of 16 sourdough *Lactobacillus* species by multiplex PCR. *Appl. Environ. Microbiol.* Vol.71, pp. 3049-3059.

Sghir, A.; Gramet, G.; Suau, A.; Rochet, V.; Pochart, P. & Dore, J. (2000). Quantification of bacterial groups within human fecal flora by oligonucleotide probe hybridization. *Appl. Environ. Microbiol.* Vol.66, pp. 2263-2266.

Simpson, J.M.; Martineau, B.; Jones, W.E.; Ballam, J.M. & Mackie, R.I. (2002). Characterization of fecal bacterial populations in canines: effects of age, breed and dietary fiber. *Microb. Ecol.* Vol.44, pp. 186-197.

Simpson, J.M.; McCracken, V.J.; White, B.A.; Gaskins, H.R. & Mackie, R.I. (1999). Application of denaturing gradient gel electrophoresis for the analysis of the porcine gastrointestinal microbiota. *J. Microbiol. Methods.* Vol.36, pp. 167-179.

Skanseng, B.S.; Kaldhusdal, M. & Rudi, K. (2006). Comparison of chicken gut colonisation by the pathogens *Campylobacter jujeni* and *Clostridium perfringens* by real-time quantitative PCR. *Mol. Cell. Probes.* Vol.20, pp. 269-279.

Suau, A.; Bonnet, R.; Sutren, M.; Godon, J.J.; Gibson, G.R.; Collins, M.D. & Dore, J. (1999). Direct analysis of genes encoding 16S rRNA from complex communities reveals many novel molecular species within the human gut. *Appl. Environ. Microbiol.* Vol.65, pp. 4799-4807.

Tannock, G.W.; Munro, K.; Harmsen, H.J.M.; Welling, G.W.; Smart, J. & Gopal, P.K. (2000). Analysis of the fecal microflora of human subjects consuming a probiotic product containing *Lactobacillus rhamnosus* DR20. *Appl. Environ. Microbiol.* Vol.66, pp. 2578-2588.

Toivanen, P.; Vaahtovuo, J. & Eerola, E. (2001). Influence of major histocompatibility complex on bacterial composition of fecal flora. *Infect. Immun.* Vol.69, pp. 2372-2377.

Torriani, S.; Felis, G.E. & Dellaglio, F. (2001). Differentiation of *Lactobacillus plantarum*, L. *pentosus*, & *L. paraplantarum* by recA gene sequence analysis and multiplex PCR assay with recA gene-derived primers. *Appl. Environ. Microbiol.* Vol.67, pp. 3450-3454.

Vaahtovuo, J.; Toivanen, P. & Eerola, E. (2003). Bacterial composition of murine fecal microflora is indigenous and genetically guided. *FEMS Microbiol. Ecol.* Vol.44, pp. 131-136.

Van der Wielen, P.W.J.J.; Lipman, L.J.A.; van Knapen, F. & Biesterveld, S. (2002). Competitive exclusion of *Salmonella enterica* serovar Enteritidis by *Lactobacillus crispatus* and *Clostridium lactatifermentans* in a sequencing fed-batch culture. *Appl. Environ. Microbiol.* Vol.68, pp. 555-559.

Vaughan, E.E.; Schut, F.; Heilig, C.H.J.; Zoetendal, E.G.; de Vos, W.M. & Akkermans, A.D.L. (2000). A molecular view of the intestinal ecosystem. *Curr. Issues Intest. Microbiol.* Vol.1, pp. 1-12.

Von Wintzingerode, F.; Goebel, U.B. & Stackebrandt, E. (1997). Determination of microbial diversity in environmental samples: pitfalls of PCR-based rRNA analysis. *FEMS Microbiol. Rev.* Vol.21, pp. 213-229.

Walter, J.; Heng, N.C.K.; Hammes, W.P.; Loach, D.M.; Tannock, G.W. & Hertel, C. (2003). Identification of *Lactobacillus reuteri* genes specifically induced in the mouse gastrointestinal tract. *Appl. Environ. Microbiol.* Vol.69, pp. 2044-2051.

Wang, A.M.; Doyle, M.V. & Mark, D.F. (1989). Quantification of mRNA by the polymerase chain reaction. *Proc. Nat. Acad. Sci. U.S.A.* Vol.86, pp. 9717-9721.

Wang, X.; Heazlewood, S.P.; Krause, D.O. & Florin, T.H.J. (2003). Molecular characterization of the microbial species that colonize human ileal and colonic mucosa by using 16S rDNA sequence analysis. *J. Appl. Microbiol.* Vol.95, pp. 508-520.

Wilson, K.H.; Wilson, W.J.; Radosevitch, J.L.; DeSantis, T.Z.; Viswanathan, V.S.; Kuczmarski, T.A. & Andersen, G.L. (2002). High-density microarray of small-subunit ribosomal DNA probes. *Appl. Environ. Microbiol.* Vol.68, pp. 2535-2541.

Zarlenga, D.S. & Higgins, J. (2001). PCR as a diagnostic and quantitative technique in veterinary parasitology. Vet. Parasitol. Vol.101, pp. 215-230.

Zhang, H.; Xu, J.; Wang, J.; Mnghebilige, M.; Sun, T.; Li, H.; & Guo, M. (2007). A survey on chemical and microbiological composition of kurut, naturally fermented yak milk from Qinghai in China. *Food Control.* doi:10.1016/j.foodcont.2007.06.010.

Zhu, X.Y. & Joerger, R.D. (2003). Composition of microbiota in content and mucus from cacae of broiler chickens as measured by fluorescent *in situ* hybridization with group-specific 16S rRNA-targeted oligonucleotide probes. *Poult. Sci.* Vol.82, pp. 1242-1249.

Zhu, X.Y.; Zhong, T.; Pandya, Y. & Joerger, R.D. (2002). 16S rRNA-based analysis of microbiota from the caecum of broiler chickens. *Appl. Environ. Microbiol.* Vol.68, pp. 124-137.

Zoetendal, E.G.; Akkermans, A.D.L. & de Vos, W.M. (1998). Temperature gradient gel electrophoresis analysis from human fecal samples reveals stable and host-specific communities of active bacteria. *Appl. Environ. Microbiol.* Vol.64, pp. 3854-3859.

Zoetendal, E.G.; Akkermans, A.D.L.; Akkermans van-Vliet, W.M.; de Visser, J.A.G.M. & de Vos, W.M. (2001). The host genotype affects the bacterial community in the human gastrointestinal tract. *Microb. Ecol. Health Disease.* Vol.13, pp. 129-134.

Zoetendal, E.G.; Collier, C.T.; Koike, S.; Mackie, R.I. & Gaskins, H.R. (2004). Molecular Ecological Analysis of the Gastrointestinal Microbiota: A Review. *J. Nutr.* Vol.134, pp. 465-472.

Zoetendal, E.G.; Plugge, G.M.; Akkermans, A.D.L. & de Vos, W.M. (2003). *Victivallis vadensis* gen. nov.; sp. nov.; a sugar-fermenting anaerobe from human faeces. *Int. J. Syst. Evol. Microbiol.* Vol.53, pp. 211-215.

Zoetendal, E.G.; von Wright, A.; Vilponnen-Salmela, T.; Ben-Amor, K.; Akkermans, A.D.L. & de Vos, W.M. (2002). Mucosa-associated bacteria in the human gastrointestinal tract are uniformly distributed along the colon and differ from the community recovered from feces. *Appl. Environ. Microbiol.* Vol.68, pp. 3401-3407.

5

Interspecific and Intraspecific Genetic Diversity of *Thunnus* Species

Mei-Chen Tseng[1], Chuen-Tan Jean[2], Peter J. Smith[3] and Yin-Huei Hung[1]
[1]Department of Aquaculture
National Pingtung University of Science & Technology, Pingtung
[2]Department of Physical Therapy
Shu Zen College of Medicine and Management, Kaohsiung
[3]Museum Victoria, Melbourne Victoria
[1,2]Taiwan, Republic of China
[3]Australia

1. Introduction

The genus *Thunnus* contains eight valid species with a fossil record extending back to the Middle Eocene, about 40 million years ago (Carrol, 1988, Benton, 1993). Collette (1978) distinguished tunas into the temperate subgenus *Thunnus* (bluefin tuna group) and the tropical subgenus *Neothunnus* (yellowfin group) by the presence or absence of a central heat exchanger. The yellowfin group contains blackfin (*T. atlanticus*), longtail (*T. tonggol*), and yellowfin (*T. albacares*) tunas. Albacore (*T. alalunga*), bigeye (*T. obesus*), northern bluefin (*T. thynnus & T. orientalis*), and southern bluefin (*T. maccoyii*) tunas are members of the bluefin group. *T. obesus* shares approximately the same number of characters with each group; however, it is classified as a member of the subgenus *Thunnus* since the character states consist of adaptations to life in colder environments (Collette, 1978).

Finnerty & Block (1995) explored *Thunnus* systematics using a portion of the *cytochrome (Cyt)* *b* gene. However, only five of eight tuna species were analyzed, and results were insufficient to draw conclusions about relationships within the genus *Thunnus*. Alvarado-Bremer et al. (1997) constructed phylogenetic relationships among tunas from a portion of the mitochondrial (mt) DNA control region. A Neighbor-joining tree supported monophyletic origins for the temperate subgenus *Thunnus* and of the tropical subgenus *Neothunnus*, except for bigeye tuna because it was difficult to place in either subgenus. This result is consistent with allozyme data suggesting that the bigeye tuna has a greater similarity to yellowfin and blackfin tunas than to temperate tunas (Sharp & Pirages, 1978; Elliott & Ward, 1995). Alvarado-Bermer et al. (1997), Ward (1995), and Chow & Kishino (1995) suggested that mtDNAs of albacore and Pacific bluefin tunas share a common ancestry. Nevertheless, Sharp & Pirages (1978) and Chow & Kishino (1995) indicated that albacore tunas are highly divergent from all other tunas, suggesting that the *Thunnus* subgenus is not a monophyletic group. Elliott and Ward (1995) reported that bluefin tunas were much closer to the yellow tuna than to albacore and bigeye tunas, and albacore was the most divergent species in the genus. Alvarado-Bremer et al. (1997) inferred that introgression of albacore mtDNA into the Pacific bluefin tuna occurred a long time ago.

Chow et al. (2006) examined intra- and interspecific nucleotide sequence variations of the first internal transcribed spacer (*ITS1*) of ribosomal (r)DNA among all *Thunnus* species. Their report supported mitochondrial introgression between species and contradicted the morphological subdivision of the genus into the two subgenera, *Neothunnus* and *Thunnus*. The cladogram constructed from the *ITS1* indistinctly resolved phylogenetic relationships among three tropical tunas (*T. albacare, T. tonggol,* and *T. atlanticus*). The *ITS1* and mtDNA ATCO sequence data both supported the monophyletic status of the yellowfin tuna group and indicated that these tropical tunas are recently derived taxa; nevertheless, *T. thynnus* and *T. orientalis* shared almost identical *ITS1* sequences, while having distinct mtDNA. These molecular data suggested that intermittent speciation events occurred in species of the genus *Thunnus*. Consequently, relationships among closely related *Thunnus* taxa remain unresolved.

The Pacific bluefin tuna *T. orientalis* is a migratory pelagic fish of high commercial value in world fisheries (Ward et al., 1995). This species is primarily distributed in the North Pacific Ocean, from the Gulf of Alaska to southern California and Baja California and from Sakhalin Island in the southern Sea of Okhotsk to the northern Philippines; the tuna is also found in the South Pacific Ocean around Australia, the Galapagos Islands, the Gulf of Papua, and New Zealand (Bayliff, 1994, Collette & Nauen, 1983, Collette & Smith, 1981, Smith et al., 2001; Ward et al., 1995). Pacific bluefin tuna spawn between Japan and the Philippines, in the Sea of Japan south of Honshu (Chen et al., 2006; Itoh, 2009, Tanaka et al., 2006, Tanaka et al., 2007); larvae are transported northwards towards Japan by the Kuroshio Current, and juveniles are found in waters near Japan. Some young fish migrate east as far as the western coast of North America, and are presumed to return to the western Pacific to breed (Bayliff, 1991, 2001). The Pacific bluefin tuna is believed to become sexually mature at about 5 years of age and to have a maximum lifespan of 25 years (Bayliff, 1994; Ueyanagi, 1975). Surprisingly little is known about the stock structure and population biology of *T. orientalis*. Currently there are no data on the population genetics of *T. orientalis* stocks exploited in the northern and southern Pacific Ocean.

In the present study, we collected complete *Cyt b* sequences from all eight *Thunnus* species, described interspecies genetic variations of the *Cyt b* gene, and explored their phylogeny. Results indicated that *Cyt b* is a good tool for tuna identification. The *Thunnus* phylogeny did not conform to the two-subgenera classification pattern. Another aim of this research was to investigate whether genetic differentiation appears between *T. orientalis* samples from Taiwan and New Zealand waters. Results suggested that these two samples shared high genetic homogeneity.

2. Materials and methods

2.1 Tuna sampling

2.1.1 Eight *Thunnus* tuna collection

All five *Thunnus* species *T. obesus, T. tonggol, T. alalunga, T. albacares,* and *T. orientalis* and an outgroup specimen *Katsuwonus pelamis* were collected from Taiwan waters. *Thunnus maccoyii* was obtained from the Indian Ocean by Taiwanese fishery observers in 2006 as described by Shiao et al. (2008). *Thunnus thynnus* was caught from Martinique waters of the eastern Caribbean Sea in 2008. Tissue specimens of *T. atlanticus* were supplied by the Florida Museum of Natural History, Gainesville, FL, USA.

2.1.2 *Thunnus orientalis* sampling

Two sets of muscle tissue samples of *T. orientalis* were collected from the North and South Pacific Oceans. Forty specimens (T1~T40) were sampled from southeastern Taiwan waters (T) during 2006~2007 and 40 specimens (Z1~Z40) from New Zealand waters (Z) in 2001 by scientific observers on commercial fishing vessels (Fig. 1). Muscle tissue samples were preserved in 95% ethanol for DNA extraction.

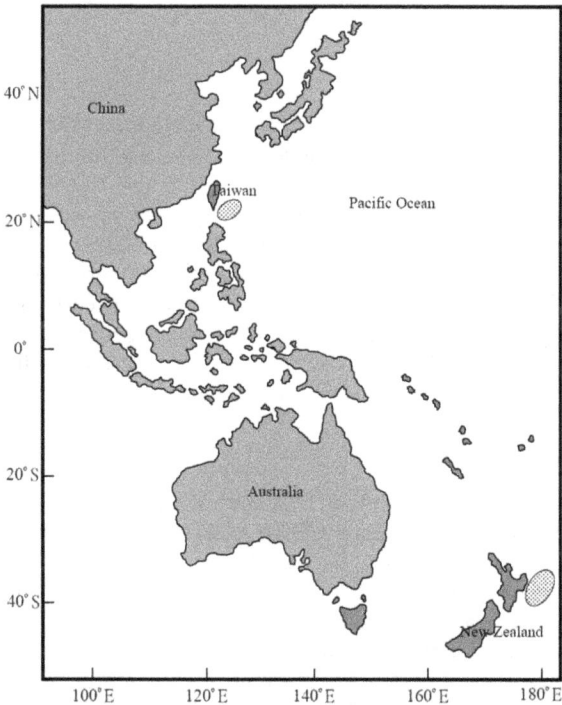

Fig. 1. Sampling locations of *Thunnus orientalis* in Taiwan and New Zealand waters.

2.2 DNA Cloning

2.2.1 DNA isolation and polymerase chain reaction (PCR) amplification

DNA was extracted as previously described (Kocher et al., 1989). Two sets of primers and amplification conditions were used to amplify the complete *Cyt b* gene and control region of mtDNA (Tseng et al., 2009, 2011). Amplification was performed in a BIO-RAD MJ Mini Gradient Thermal Cycler (Conmall Biotechnology, Singapore). Subsamples (at 10 µL) of amplified products were checked on 0.8% agarose gels to confirm the product sizes.

2.2.2 DNA purification, transformation, and sequencing

The remaining successful PCR products were purified from agarose gels using a DNA Clean/Extraction kit (GeneMark, Taichung, Taiwan). Isolated DNA fragments were subcloned into the pGEM-T easy vector (Promega, Madison, WI, USA) and transformed into the *Escherichia coli* JM109 strain. Plasmid DNA was isolated using a mini plasmid kit (Geneaid, Taichung, Taiwan). Twenty individuals from each of the Taiwanese and New Zealand samples of *T. orientalis* were randomly chosen for cloning followed by sequencing in an Applied Biosystems (ABI, Foster City, CA, USA) automated DNA sequencer 377 (vers. 3.3) with a Bigdye sequencing kit (Perkin-Elmer, Wellesley, MA, USA). The T7 and SP6 primers were used in the sequencing reaction, and PCR cycle parameters for sequencing were 35 cycles of 95°C for 30s, 50°C for 30s, and 72°C for 1 min.

2.3 Sequence analyses

2.3.1 *Cyt b* gene analysis

Interspecific genetic diversities of nucleotides were calculated using MEGA software (Tamura et al., 2007). GTR+I+G was the best-fitting model for DNA substitutions as determined by the Modeltest 3.7 program (Posada & Crandall, 1998) using the Akaike information criterion (AIC = 5091.39, gamma = 0.5623). The phylogenetic tree of nucleotide sequences was constructed using the Neighbor-joining (NJ) method (Nei & Kumar, 2000). The confidence of the clusters was assessed using a bootstrap analysis with 1000 replications. Genetic distances between amino acid sequences were estimated using Poisson correction methods (Nei & Kumar, 2000) in MEGA (Tamura et al., 2007). The phylogenetic tree of amino acid sequences was constructed by the NJ method with an anterior-branch test, and the confidence of the clusters was assessed by 1000 bootstrap replications.

2.3.2 Control region analysis

In total, 40 nucleotide sequences of the mtDNA control region from the two sets of samples (T and Z) were aligned with the program, CLUSTAL W (Thompson et al., 1994), and verified by eye. All 40 sequences were deposited in NCBI GenBank (with accession nos. JN631171~1190 and JN631211~1230). The number of polymorphic sites including substitutions and indels was estimated using Arlequin vers. 3.1 (Excoffier et al., 2010). Levels of inter- and intra-sample genetic diversity were quantified by indices of h_d and pairwise estimates of nucleotide divergence (d_{ij}), both among and within samples. The average number of nucleotide substitutions per site (π) and pairwise differences between samples were determined using the DnaSP program (Librado & Rozas, 2009). Intraspecific genetic distances were analyzed by Kimura's 2-parameter (K2P) model (Kimura, 1980). The best-fitting models of substitution were determined using MODELTEST 3.7 (Posada & Crandall, 1998) with the AIC. The TVM+I+G model with gamma = 0.724 and AIC = 5370.36 was chosen as the best-fitting model. Phylogenetic trees were constructed using the NJ method (Nei & Kumar, 2000), and the confidence of each node of the tree was tested by bootstrapping (Felsenstein, 1985) with 1000 replicates. Nodes with bootstrap values of >70 were significantly supported by a ≥95% probability (Hillis & Bull, 1993). A minimum spanning tree (MST) was computed from the matrix of pairwise distances between all pairs of 40 haplotypes using a modification of the algorithm described by Rohlf (1973). An exact test of genetic differentiation between samples was also estimated with F_{ST} (Raymond & Rousset, 1995) with 10^5 steps of a Markov chain.

2.4 Genotyping of microsatellites

Genetic variation in *T. orientalis* samples was estimated with six microsatellite loci: Tth14, -31, -185, -217, -226, and Ttho-4. The primers for these loci were developed by Clark et al. (2004) and Takagi et al. (1999). PCR cocktails consisted of ~1 ng genomic DNA, 10 pmol reverse primer, 10 pmol labeled forward primer, 25 mM dNTP, 0.05~0.1 mM $MgCl_2$, 10× buffer, and 0.5 U *Taq* polymerase (Takara Shuzo, Tokyo, Japan) made up to a 25-μL volume with Milli-Q water. Forward primers were labeled with the FAM, TAMRA, or HEX fluorescence markers. Polymerase chain reaction (PCR) amplifications were carried out in a BIO-RAD MJ Mini Gradient Thermal Cycler (Conmall Biotechnology, Singapore) programmed with the following schedule: 1 cycle of 95°C for 4 min, followed by 35 cycles of 94°C for 30s, annealing at 52~60°C for 30s, and 72°C for 30s. Each 5 μL of PCR production from three loci labeled with different fluorescence tags, was mixed and precipitated with 95% alcohol. Semiautomated genotyping was performed using a capillary MegaBACE-500 DNA analysis system (Amersham Biosciences, Piscataway, NJ, USA). Genotypes were scored with Genetic Profiler 1.5 (Amersham Biosciences). The overall genotype success rate was 100%. The size of each allele was checked by eye, and the Micro-checker software package (Van Oosterhout et al., 2004) was used to correct genotyping errors, such as non-amplified alleles, short allele dominance, and scoring of stutter peaks.

2.5 Microsatellite genotypic analyses

The total number of alleles (n_a), allelic frequencies, and Shannon index were estimated for each locus using the program POPGENE (Yang & Yeh, 1993). Observed (H_O) and expected (H_E) heterozygosities were calculated for each locus (Raymond & Rousset, 1995). Deviations from Hardy-Weinberg equilibrium (HWE) were examined by an exact test using GENEPOP (Raymond & Rousset, 1995). Genetic differentiation between samples was characterized using F_{ST} and R_{ST} values as implemented in ARLEQUIN 3.1 (Schneider et al., 2000). A factorial correspondence analysis was performed with GENETIX (Belkhir et al., 1996-2004) to plot multilocus genotypes of the two samples in two dimensions.

3. Results

3.1 Genetic variations and genealogy of the *Cyt b* gene

The mitochondrial *Cyt b* gene was completely sequenced for all eight tuna individuals examined. Of the 1141 aligned base pairs, 20 sites were singleton variable and 45 sites were parsimoniously informative. The *Cyt b* gene cloned from *T. tonggol* and homological sequence EF141181.1 shared an identical sequence. Some individual nucleotide compositions only appeared in specific species except *T. thynnus* (Fig. 2). For example *T. tonggol* could be individually distinguished from each other *Thunnus* tuna by the 45th, 81st, 321st, 360th, and 762nd sites of the *Cyt b* sequences. *T. albacares* could be particularly distinguished from each other *Thunnus* tuna by the 330th and 783rd sites.

```
[                                                                                            111111]
[                 1112222 3333333333 4444444555 5555666677 7777777777 7888888999 999000000]
[                 4680360237 0223556678 0144679012 2788012701 1345567788 9134459123 679022235]
[                 5312251249 3140470357 5747878495 8328281851 7810624739 2916987216 653803950]
T. thynnus       CTCCACCCCC ACAAACACGT ATCTCTATCC CTTCACTCGA CCTATCTCAT TTAGCCTTGG ACCCTGAGC
NC_004901.2      ...TG...TT ......G... ...C...... .....T.... .......... .......... .........
T. tonggol       T.T....... .T....T... G.......... .....T.... ..C.CTC.C. CC........ ......G..
EF141181.1       T.T....... .T....T... G.......... .....T.... ..C.CTC.C. CC........ ......G..
T. maccoyii      ........TT .........A. ...C.C..... .....T.... .......... ..GA...... G...C....
NC_014101.1      ........TT .........A. ...C...... .......... .......... ..G....... G...C....
T. alalunga      .C...T.... C.G....T.. .C.CTCG.TT .C...T..AG ...GC....C CCTAT.CCA. ...A..GAT
NC_005317.1      .C...T.... C.G....T.. .C.CTCG..T .C.T.T...G ...GC....C CCTAT.CCA. ...A..GAT
T. orientalis    .C...T.... C.G.G..T.. .CTCTCG.TT .CC..T...G T..GC..T.C CCTAT.CCA. ...A.AGAT
GU256524         .C...T.... C.G.G..T.. .CTCTCG.TT .C...T...G T..GC....C CCTAT.CCA. ...A.AGAT
T. albacares     .......... ...G...... ....T..C.. .......... ..C.C...T. C.........  ......G..
NC_014061.1      .......... ...G...... ......T.... .......... ..C.C...T. C.........  ......G..
T. obesus        .......T.T .....A...C ......C.... ...TGTCT.. .TC.C...C. C....T....  ......G..
NC_014059.1      ........T  .....A.T.. ......G... ...T.T.... .TC.C...C. C....T...A .TA...G..
T. atlanticus    ......T... .......... G...T..... T......... ..C.C.C.C. C.........  ......G..
```

Fig. 2. Total of 69 variable sites within 15 *Cyt b* sequences from eight tunas.

Numbers of different nucleotide between *Thunnus* species ranged from five (*T. alalunga* vs. GU256524) to 43 (*T. obesus* vs. *T. orientalis*), and interspecific K2P distances ranged 0.004~0.039 (Table 1). *Katsuwonus pelamis* had significant different nucleotide components from *Thunnus* tunas. The different nucleotide numbers between *K. pelamis* and *Thunnus* species ranged 122~128, and K2P distances ranged 0.117~0.123. Complete *Cyt b* sequences from all tuna species except *T. atlanticus* (wanting) were taken from NCBI (NC_014101.1, NC_004901.2, NC_005317.1, NC_014061.1, NC_014059.1, EF141181.1, and GU256524) and selected as reference sequences in the phylogenetic analysis. These reference sequences and eight congeneric sequences from the study were consistently clustered in the NJ tree. The tree contained two explicit clades, the first group consisted of *T. maccoyii*, *T. thynnus*, *T. atlanticus*, *T. albacares*, *T. obesus*, *T. tonggol*, and the second group included *T. alalunga* and *T. orientalis*, with extremely high bootstrapping (1000 replications) support (Fig. 3). All eight tuna individuals and seven reference sequences coded five different amino acid sequences with high conservation. The *Cyt b* gene sequence of our *T. maccoyii* specimen and its reference sequence as well as *T. orientalis* and its reference sequence coded identical amino acid sequences. One identical amino acid sequence was shared by nine specimens from five *Thunnus* species, *T. alalunga*, *T. thynnus*, *T. albacares*, *T. atlanticus*, and *T. tonggol*. Poisson correction distances between different amino acid sequences ranged 0.003~0.005. Different amino acid numbers between sequences ranged 1~2. All branches presented an insignificant cluster from eight *Thunnus* species with slightly high bootstrapping support suggesting that parallel evolution had occurred in the protein structure of the *Cyt b* gene.

3.2 D-Loop genealogy

In total, 52 mutations were shared among the samples, and no fixed differences occurred in particular samples. Forty-one mutations were polymorphic in sample T, but monomorphic in sample Z. In contrast, 49 mutations were polymorphic in sample Z, but monomorphic in sample T. The average number of nucleotide differences between these two samples was 19.79. The average number of nucleotide substitutions per site between these two samples (D_{xy}) was 0.023. Nucleotide ratios of the control region displayed a very high AT-rich composition (62%). A short double-repeat unit (TGCATGTGCAT) located at positions 47~57

	1	2	3	4	5	6	7	8	9	10	11	12	13	14	15
1. T. thunnus	*	7	17	16	14	10	7	31	30	36	34	8	7	9	122
2. NC_004901.2	0.006	*	20	19	18	9	8	34	33	39	37	15	14	16	128
3. T. tonggol	0.015	0.018	*	11	19	21	22	38	35	43	41	16	15	16	126
4. T. maccoyii	0.014	0.017	0.01	*	18	22	21	35	32	40	38	15	14	15	127
5. NC_014101.1	0.012	0.016	0.017	0.016	*	22	21	35	34	40	38	13	12	11	122
6. T. alalunga	0.009	0.008	0.019	0.02	0.02	*	3	32	31	37	35	18	17	19	127
7. NC_005317.1	0.006	0.007	0.02	0.019	0.019	0.003	*	35	34	40	38	15	14	16	127
8. T. orientalis	0.028	0.031	0.034	0.032	0.032	0.029	0.032	*	3	7	5	31	30	32	122
9. GU256524	0.027	0.030	0.032	0.029	0.031	0.028	0.031	0.003	*	8	6	30	29	31	123
10. T. albacares	0.033	0.035	0.039	0.036	0.036	0.033	0.036	0.006	0.007	*	2	36	35	37	128
11. NC_014061.1	0.031	0.033	0.037	0.034	0.034	0.032	0.034	0.004	0.005	0.002	*	34	33	35	126
12. T. obesus	0.007	0.013	0.014	0.013	0.012	0.016	0.013	0.028	0.027	0.032	0.031	*	1	6	123
13. NC_014059.1	0.06	0.012	0.013	0.012	0.011	0.015	0.012	0.027	0.026	0.032	0.030	0.001	*	5	122
14. T. atlanticus	0.008	0.014	0.014	0.013	0.010	0.017	0.014	0.029	0.028	0.033	0.032	0.005	0.004	*	123
15. K. pelamis	0.117	0.123	0.121	0.122	0.117	0.122	0.122	0.117	0.118	0.123	0.121	0.118	0.117	0.118	*

Table 1. Numbers of different nucleotides between tuna species (upper diagonal) and K2P genetic distances of nucleotide sequences (below the diagonal) from *cytochrome b* gene sequences of eight *Thunnus* species and the outgroup *Katsuwonus pelamis*.

nt was observed in six individuals of *T. orientalis*. In total, 133 mutant sites containing 128 transitions, 14 transversions, and 14 indels were observed among the 40 haplotypes.

Long fragments, 865~876 base pairs (bp), containing the entire mtDNA control region were successfully amplified. Following alignment, the consensus nucleotide sequence for all specimens of *T. orientalis* was 894 bp long. Forty different sequences were found in the 40 individuals from T and Z samples with a very high haplotype diversity (100%). The number of different nucleotides ranged 2 (T5 vs. T6) to 31 (T1 vs. Z28) (Fig. 4). The nucleotide diversity (d_{ij}) among sequences ranged 0.002 (T9 vs. Z19) to 0.040 (T1 vs. Z29) with an average of 0.023 ± 0.008 (Table 2).

K2P genetic distances among different haplotypes ranged 0.002~0.042 with an overall average of 0.023 ± 0.003. The average number of nucleotide differences among all specimens was 16.879 ± 2.068. The d_{ij} between these two samples was 0.023 ± 0.002. Genetic distances ranged 0.002 (T5 vs. T6) to 0.035 (T1 vs. T19), with an average of 0.021 in sample T, and ranged 0.003 (Z19 vs. Z33) to 0.035 (Z13 vs. Z29) with an average of 0.02 in sample Z. The topology of the NJ tree was estimated using the entire dataset with a non-significant geographical group (Fig. 5). Specimens from both sets of samples were scattered throughout the NJ tree. The pairwise F_{ST} value between these two samples was 0.009 (p = 0.244). These results indicated that no specimens of *T. orientalis* exhibited genetic differentiation. The MST was computed from a matrix of pairwise distances between all pairs of haplotypes of *T. orientalis* (Fig. 6). Two central haplotypes were found in the MST (T15 and T11), and most of haplotypes were located at the tip of the MST.

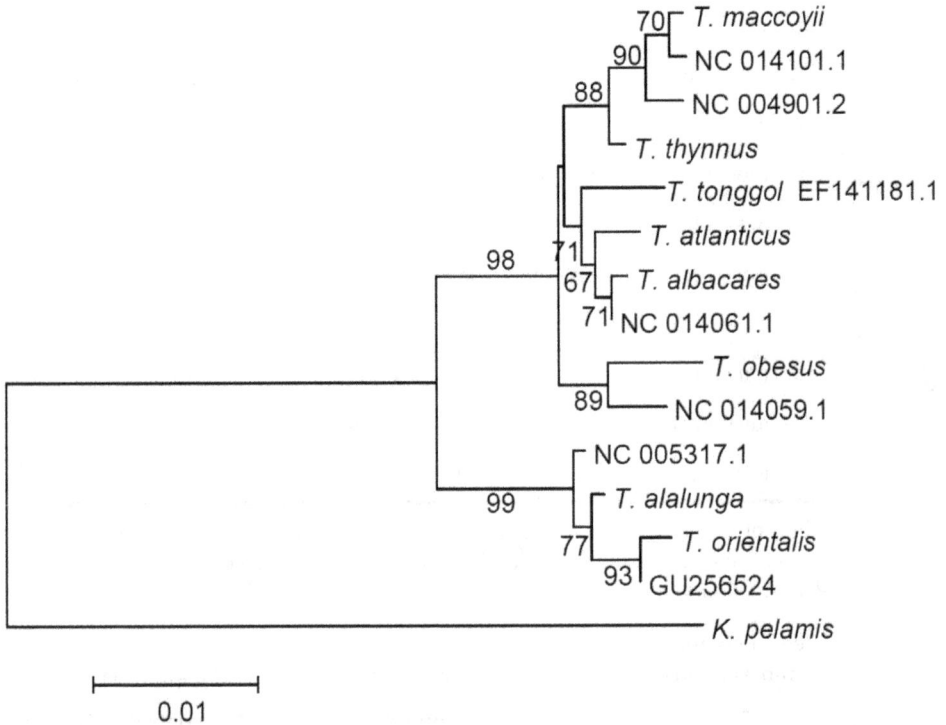

Fig. 3. Neighbor-joining tree constructed by 16 *Cyt b* gene sequences from eight *Thunnus* species and the outgroup *Katsuwonus pelamis*.

3.3 Microsatellite diversity

All six microsatellite loci were scored in these two sets of samples of *T. orientalis*, which had high levels of polymorphism (Fig. 7). The number of alleles per locus ranged 4 (Tth 185) to 19 (Tth 217) with an average of 10.50 ± 5.32 (n = 80). Shannon's information index (I) estimated for different loci ranged 0.906~2.165 (mean, 1.501 ± 0.410) in sample T, and 0.693~2.544 (mean, 1.632 ± 0.624) in sample Z. Mean numbers of alleles (na) per locus were 7.833 ± 3.833 for sample T and 9.833 ± 5.565 for sample Z (Table 3).

```
[                     11111 1111111111 1111122222 2222222222 2222222222]
[                     468901234 4456666789 9999900000 1111112234 4445556667]
[                     9424496541 2682357671 3678901239 0124582613 5692360574]
T1                    TGTTAATCAC CCCACTTAGA CTTATTCTTT TCTGTCAAAA TCCTACATCT
T5                    .......... T.T....... .......... C......... ..........
T9                    .A.....TG. T.TC..C..G .....CT... ...TC....G ..A...G...
T11                   C......T.. T.T...C..G .......... C........G ........T.
T17                   .AC....T.. T.T..CCG.G .....CT... C..T.T...G ........T.
T22                   .....G.T.. T.T......G T.C..C.... C...C....G .......CT.
T25                   .A.....T.. T.T...C..G T.C..CT.C. C.CCC....G .......T.
Z8                    .A.....T.. T.T...C..G T....CTC.. C..CC..G.G C.A.G..CT.
Z13                   .......T.. T.T...G.. .......... C...C..... ..........
Z24                   .......T.. T.T...C..G .....CT... C..A.....G .......T.
Z28                   .A.....T.. T.T.T.C... .....CT... C..CC....G C.......T.
Z34                   .A.....T.. T.T...C..G .....CT... C..C.T...G ........T.
Z36                   .A.....TG. T.TC..C..G .....CT... CT.TC....G .TAC....T.
Z39                   .......T.. TTT......G .......... C...C....G C.......TC
T. alalunga  .A..CG.A.T T.T....... TCC....CCC .T.T.TG..G ..A..T..T.
T. thynnus   .A.ATGCA.T T.T.....A. .CCTC..CCC .T.T.T..GG C.A.....T.


[                     2223333333 3334444444 4555555555 5666666666 677777888]
[                     8990001133 7780112345 8000223679 9014445667 906788356]
[                     6470141267 7813160957 3129195971 5431245123 060716648]
T1                    TGACAAAGCT CGTCGTTGTT ACCGATATTA TGATATATAC TTCGTCTAT
T5                    .......... .......... ...A....C. .......... .........
T9                    ...TG..... .A........ ...AG.GC.. .......G... ....T...
T11                   ...TG..... .A........ ...A..GC.G .......... ....TC..
T17                   C..TGG.... T......A.C G..A...C.. .......... ....T...
T22                   C...G..... .A...C.... ...A..GC.. .......... ....T...
T25                   ....G..... T......A.C ...A...C.. ....G..... ....T...
Z8                    ..GT...... TA...C.A.. G..A..GC.. .......... ....T...
Z13                   ...T...... ...T...C. ...A...C.. .......... ..........
Z24                   ...TG..... ......C... G..A...C.G .......... .....T..C
Z28                   .AGT...A.. TA.T...A.. G..A.CGC.G ...C.C.... C...CT...
Z34                   C..TG..... T......A.C G..A...C.. .......... ..AA.T...
Z36                   ...TG..... .A..A..... ...AG.GC.G .......... C....T...
Z39                   ...T...... .......... G..A...... .AG.....T .....T.G.
T. alalunga  .A.T..T.TC TA.T..CA.. .TTA..G.CG .......CG. .A...T...
T. thynnus   .A.T..T.TC TACT...A.. .TTA..G.C. C......... .A...T...
```

Fig. 4. Variable sites within 16 selected D-loop sequences from 14 Pacific bluefin tuna specimens and two outgroups.

	T1	T5	T9	T11	T17	T22	T25	Z8	Z13	Z24	Z28	Z34	Z36	Z39	T. alalunga	T. thynnus
T1	*	5	24	18	27	21	25	31	11	20	33	25	30	30	41	43
T5	0.006	*	23	15	24	18	22	28	8	17	30	22	27	16	38	40
T9	0.028	0.027	*	16	21	19	20	20	21	17	26	20	10	25	42	44
T11	0.021	0.018	0.019	*	19	13	19	21	15	10	23	17	16	17	38	42
T17	0.032	0.029	0.025	0.022	*	22	15	21	22	14	25	7	23	24	43	45
T22	0.025	0.021	0.022	0.015	0.026	*	16	18	18	15	28	20	21	20	39	43
T25	0.030	0.026	0.024	0.022	0.018	0.019	*	18	22	16	24	12	22	24	40	44
Z8	0.037	0.033	0.024	0.025	0.025	0.021	0.021	*	26	18	18	18	22	24	40	42
Z13	0.013	0.009	0.025	0.018	0.026	0.021	0.026	0.031	*	17	26	22	25	16	41	43
Z24	0.024	0.020	0.020	0.012	0.016	0.018	0.019	0.021	0.020	*	22	12	17	17	43	47
Z28	0.040	0.036	0.031	0.027	0.030	0.033	0.029	0.021	0.031	0.026	*	22	24	28	42	44
Z34	0.030	0.026	0.024	0.020	0.008	0.024	0.014	0.021	0.026	0.014	0.026	*	22	22	42	44
Z36	0.036	0.032	0.012	0.019	0.027	0.025	0.026	0.026	0.030	0.020	0.028	0.026	*	27	42	46
Z39	0.022	0.019	0.030	0.020	0.029	0.024	0.029	0.028	0.019	0.020	0.033	0.026	0.032	*	45	45
T. alalunga	0.050	0.046	0.051	0.046	0.053	0.047	0.049	0.049	0.050	0.052	0.051	0.051	0.051	0.055	*	17
T. thynnus	0.052	0.048	0.053	0.051	0.055	0.052	0.053	0.051	0.052	0.057	0.053	0.053	0.056	0.055	0.020	*

Table 2. Numbers of different nucleotides of the mtDNA D-loop region (above the diagonal) and K2P genetic distances of nucleotide sequences (below the diagonal) among tuna specimens.

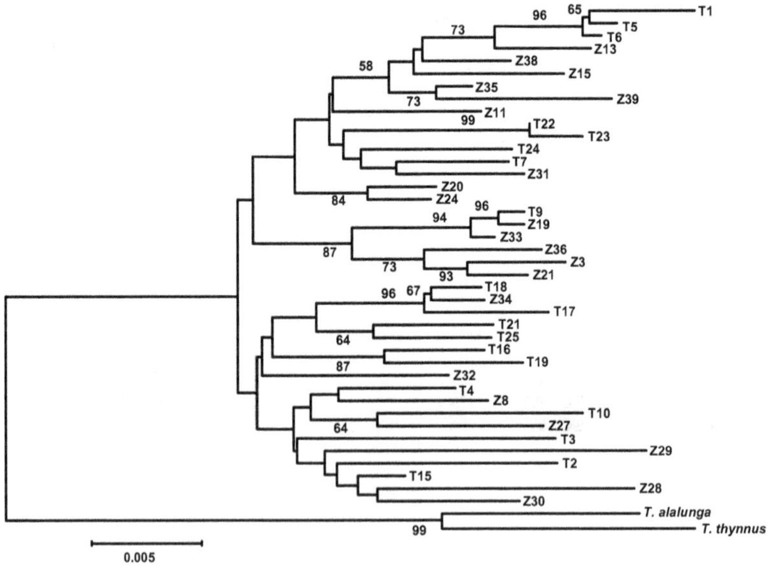

Fig. 5. Neighbor-joining tree constructed with 40 D-loop sequences from 20 individuals each from Taiwan and New Zealand waters. *Thunnus alalunga* and *T. thynnus* were selected as outgroups in the tree.

Four private alleles were found in sample T and 16 private alleles in sample Z. At the Tth-217 locus, there was a greater number of private alleles in sample T than in the other loci. The value of H_O over six loci ranged 0.425 (Tth-31) to 0.725 (Tth-217), with an average of 0.633 ± 0.127 in sample T and ranged 0.2 (Tth-185) to 0.75 (Ttho-4, Tth-27, and Tth-226), with an average of 0.613 ± 0.217 in sample Z. Values of H_E over six loci ranged 0.531 (Tth-185) to 0.872 (Tth-217), with an average of 0.712 ± 0.109 in sample T and ranged 0.461 (Tth-185) to 0.911 (Tth-217), with an average of 0.730 ± 0.154 in sample Z. Overall mean H_O and H_E values for the six loci were 0.623 ± 0.167 and 0.722 ± 0.130 (Table 3).

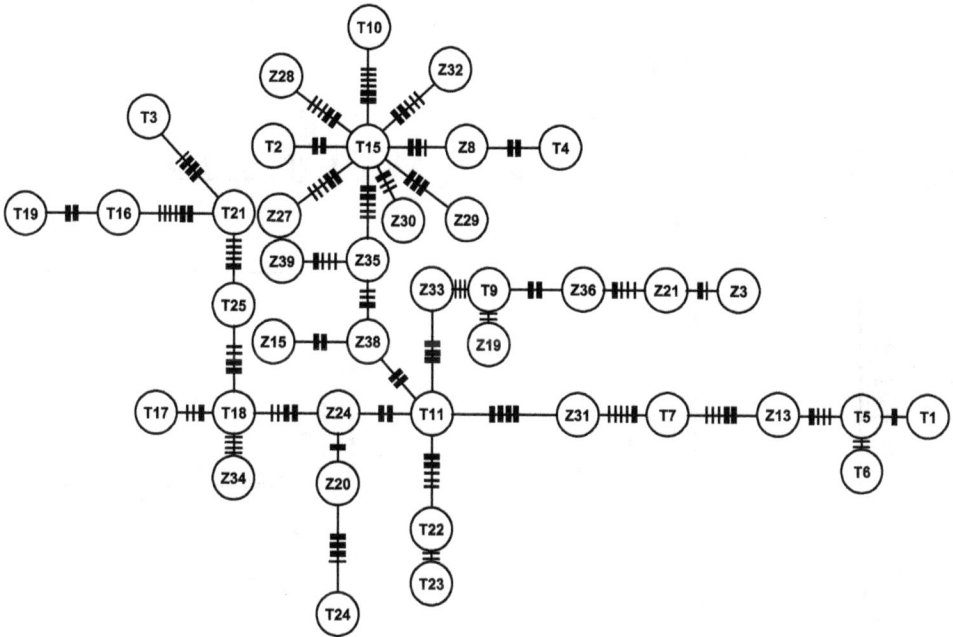

Fig. 6. Minimum spanning tree constructed from D-loop data.

Nei's genetic identity and genetic distance between the two samples were 0.978 and 0.222. Shannon's information index ranged 0.818 (Tth-185) to 2.435 (Tth-217) with an average of 1.609 ± 0.528. Permutation tests for linkage disequilibrium among the six loci for the two samples revealed only very slight disequilibrium for the entire dataset ($D_{IT}^2 = 0.009$). The genetic differentiation index, F_{ST}, ranged 0.001 (locus Tth-14) to 0.016 (locus Tth-31) within loci and averaged 0.076. F_{ST} and R_{ST} values between these two samples were 0.003 ($p = 0.243$) and 0.019 ($p = 0.099$), respectively. A factorial correspondence analysis showed that these two samples had distributions that almost completely overlapped to a great extent (Fig. 8).

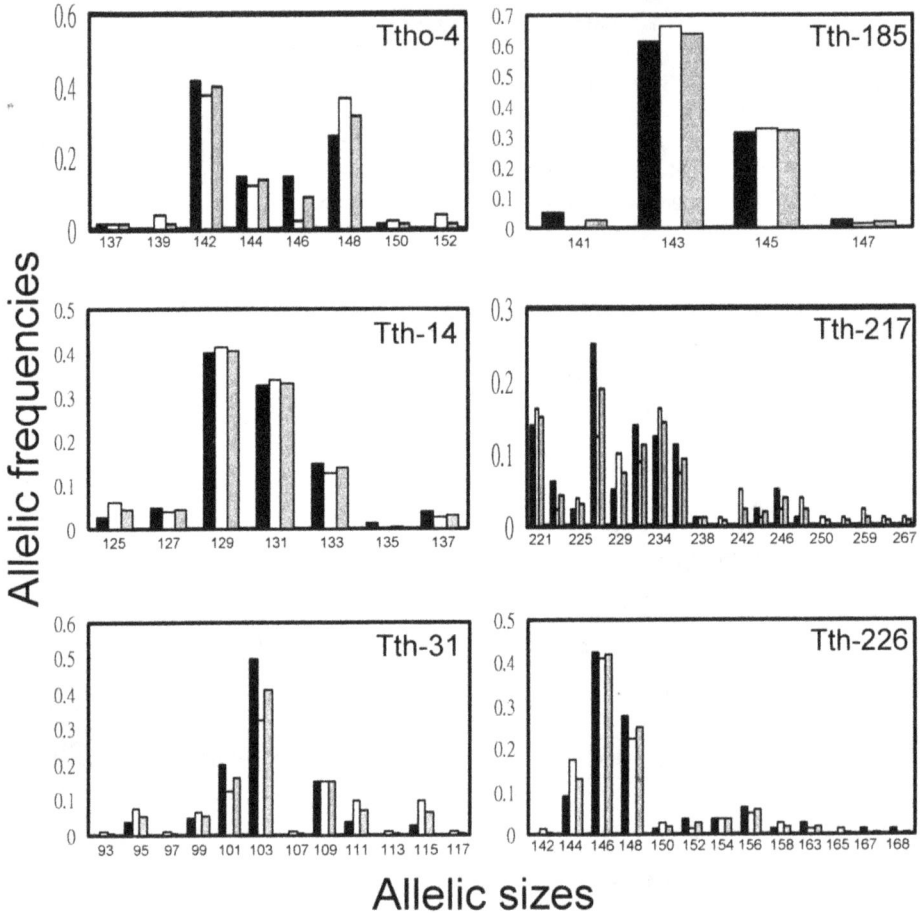

Fig. 7. Allelic sizes and frequencies from six microsatellite loci. White, Taiwanese samples; black, New Zealand samples; gray, overall sample.

4. Discussion

4.1 Interspecific divergence

In the past, tuna species identification was attempted using several different nuclear and mitochondrial markers. However, those results produced various conclusions, until Viñas & Tudela (2009) used a validated methodology for genetic identification of *Thunnus* species. They indicated that the combination of two genetic markers, one mitochondrial *CR* and another nuclear *ITS 1*, allowed full discrimination among all eight tuna species. In this study, we found that the *Cyt b* gene is also a well-established marker to distinguish *Thunnus* species, and it has been widely used in taxonomic studies of marine fishes in general (Johns & Avise, 1998). In a previous report, Block et al. (1993) constructed a phylogeny of tunas, bonitos, mackerels, and billfish based on a partial *Cyt b* gene sequence (about 515 bp in

Samples Loci	Taiwan sample $N = 40$	New Zealand sample $N = 40$	Overall $N = 80$
Ttho-4			
HWE test	n.s.	n.s.	n.s.
na/ne	6/3.517	8/3.426	8/3.566
I	1.395	1.481	1.495
H_O	0.6	0.750	0.675
H_E	0.725	0.717	0.724
Tth14			
HWE test	n.s.	n.s.	n.s.
na/ne	7/3.415	6/3.272	7/3.35
I	1.436	1.38	1.419
H_O	0.7	0.675	0.688
H_E	0.716	0.703	0.706
Tth31			
HWE test	n.s.	*	*
na/ne	7/3.140	12/5.745	12/4.27
I	1.441	2.012	1.785
H_O	0.575	0.55	0.563
H_E	0.690	0.836	0.771
Tth185			
HWE test	n.s.	*	*
na/ne	4/2.101	3/1.836	4/1.965
I	0.906	0.693	0.818
H_O	0.425	0.200	0.313
H_E	0.531	0.461	0.494
Tth217			
HWE test	n.s.	n.s.	n.s.
na/ne	12/7.191	19/10	19/8.945
I	2.165	2.544	2.435
H_O	0.725	0.75	0.738
H_E	0.872	0.911	0.894
Tth226			
HWE test	n.s.	n.s.	n.s.
na/ne	11/3.675	11/3.888	13/3.824
I	1.663	1.682	1.705
H_O	0.775	0.75	0.763
H_E	0.737	0.752	0.743
Total no. of alleles	47	59	63
Mean I per locus	1.501 ± 0.410	1.632 ± 0.624	1.609 ± 0.529
Mean na per locus	7.833 ± 3.833	9.833 ± 5.565	10.50 ± 5.32
Mean ne per locus	3.840 ± 1.735	4.695 ± 2.888	4.32 ± 2.396
Mean H_O per locus	0.633 ± 0.127	0.613 ± 0.217	0.623 ± 0.167
Mean H_E per locus	0.712 ± 0.109	0.730 ± 0.154	0.722 ± 0.130

* Significant at the 5% level; n.s., not significant.

Table 3. All information of *I* (Shannon index), *na* (observed allelic numbers), *ne* (effective allelic numbers), H_O, H_E, and test of Hardy-Weinberg equilibrium (HWE). * Indicates a significant deviation from HWE.

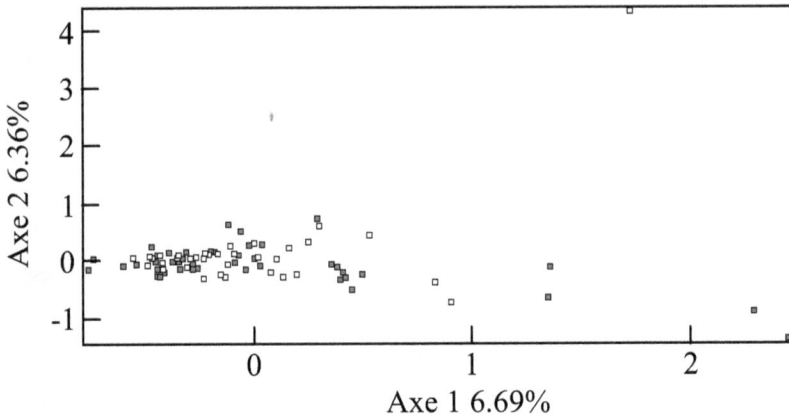

Fig. 8. Distribution of the factorial correspondence analysis between twoPacific bluefin tuna specimems from Taiwan and New Zealand waters. Gray squares, specimems from Taiwan; white squares, specimems from New Zealand waters.

length). The marker supplied important evolutionary information among these Scombridae fishes. Tseng et al. (2011) successfully distinguished the three bluefin tunas *T. orientalis*, *T. maccoyii*, and *T. thynnus* using the full-length *Cyt b* gene (1141 bp). Consequently, for this study, the entire *Cyt b* gene sequence could identify all eight *Thunnus* species.

Some specific nucleotide sites are valid for species identification of *Thunnus* tunas. Moreover, all eight subcloned and seven reference sequences had significant clusters among species, suggesting that the *Cyt b* gene is a useful marker for distinguishing *Thunnus* species. However, the identification of hybrid individuals is still difficult when a single genetic marker, such as *Cyt b*, is used. This implies that when one needs to verify equivocal tuna specimens, both nuclear and mitochondrial DNA should simultaneously be applied for identification.

Thunnus phylogeny was clarified on the NJ tree of the *Cyt b* gene. Two significant groups were present on the NJ tree with high bootstrapping values. Results indicated that *T. alalunga* and *T. orientalis* are sister species and share a recent common maternal ancestor, which is consistent with previous reports from Alvarado-Bermer et al. (1997), Ward (1995), and Chow & Kishino (1995). *Thunnus thynnus* and *T. maccoyii* have a closer relationship as do *T. atlanticus* and *T. albacares* with high boostrapping value support in the NJ tree. Moreover, the NJ tree supports monophyly for the tropical yellowfin group, but not for the temperate bluefin group. This is similar to results from Elliott & Ward's (1995) report which indicated that some bluefin tunas were much closer to the yellow tuna than to albacore and bigeye tunas. In summary, these data suggest that the phylogeny of *Thunnus* species does not fit into the current two-subgenera *Thunnus* and *Neothunnus* classification pattern.

Block et al. (1993) used a partial sequence of the *Cyt b* gene to draw an NJ tree and suggested that *T. alalunga* is the presumed most primitive species. In our study, two groups in the *Cyt b* genealogy seemed to have evolved in parallel. It is difficult to infer which species is the most ancestral species in *Thunnus* according to our NJ tree. The difference between a previous report by Block et al. (1993) and our results may have resulted from various sampling and sequence completeness. However, in this study we sequenced the entire *Cyt b* gene from all *Thunnus* species. Chow et al. (2003) used the mtDNA ATCO sequence to analyze *Thunnus* phylogeny and suggested that *T. alalunga* and *T. orientalis* should be sister species which was consistent with our results. According to these results, we suggest that bluefin tunas had a more-divergent origin.

4.2 Intraspecific divergence

The mtDNA CR of *T. orientalis* was AT-rich and similar to those of many other fishes. Short tandem repeat sequences were reported in the CR of some fishes (Ishikawa et al. 2001), but not in *Thunnus* species, except in *T. orientalis* (Alvarado-Bremer et al. 1997). In *T. orientalis*, double repeated sequences were present in 6/40 specimens and indicated that tandem repeat sequences may be a derived character in *Thunnus* species. Most of the central haplotypes in the MST were from sample T, while most individuals from sample Z were located at the terminus. Most of the haplotypes were located at the apex of the tree which is considered to be the result of adaptive radiation.

There is plenty of evidence to show a lack of differentiation in *T. orientalis* between Taiwanese and New Zealand waters. For one thing, the genetic distance of CR between the T and Z samples was equal to the mean genetic distances within samples. What is more, the CR had a lower Fst between the two samples with an insignificant probability. One final point, the NJ topology for the CR did not show a clear geographical grouping, and all haplotypes from samples T and Z were scattered within the NJ tree and MST. It should be concluded, from what was said above, that the null hypothesis of specimens from Taiwanese and New Zealand waters being taken from a single population cannot be rejected.

The mean H_O and H_E values were very similar between the T and Z samples. Low and non-significant genetic differentiation indices of Fst and Rst between these two samples indicated that individuals of *T. orientalis* belong to the same population, and no individuals could be assigned to a specific group according to the factorial correspondence analysis. Both mtDNA and microsatellite results led to the conclusion that the lack of genetic differentiation between the T and Z samples of *T. orientalis* is consistent with a "one-stock" hypothesis. In other tuna species DNA studies have revealed little intra-specific divergence within, but significant divergence between ocean basins. In bigeye tuna (*T. obesus*) DNA markers showed no divergence among population samples from the western Pacific Ocean (Chiang et al., 2006), but high divergence among Atlantic and Indo-Pacific populations (Chow et al., 2000; Chiang et al., 2006; 2008; Martinez et al., 2006). Similarly, in albacore tuna (*T. alalunga*) no genetic differences were detected among samples from the Northwest Pacific Ocean (Wu et al., 2009), but significant differences were reported among Atlantic and Indo-Pacific populations (Vinas et al., 2004). In the other bluefin tuna no spatial genetic structure was detected among samples of southern bluefin tuna (*T. maccoyii*) from the Indian and Pacific Oceans (Grewe et al., 1997), as expected for a species with a single spawning ground in the Java Sea. In contrast, in the Atlantic bluefin tuna (*T. thynnus*) no spatial genetic heterogeneity was detected among samples in the eastern Atlantic Ocean (Pujolar et al., 2003), but populations from the Gulf of Mexico and the Mediterranean were genetically distinct (Boustany et al., 2008). In recent decades, many wild organisms have experienced many environmental and harvesting effects that have led to population declines. For example, the abundance of juveniles of the freshwater eel *Anguilla japonica* Temminck & Schlegel, 1846 has decreased since 1970 due to climate change and overfishing (Dekker, 2003). The spawning stock of Atlantic bluefin tuna in the western Atlantic declined by ~50% between 1970 and 2000 (Porch, 2005). Southern bluefin tuna *T. maccoyii* is now in a depleted state. Historically, the stock was exploited for more than 50 years, with total catches peaking at 81,750 tons in 1961. However, the spawning stock was estimated to be around 10,000~20,000 tons in the previous 10~20-year period with a historically and critically low level (CCSBT, 2010). Currently Pacific bluefin tuna exhibits a high level of genetic polymorphism and population states under mutation-drift equilibrium, but significant catch decreases have occurred in the last decade due to overexploitation. Therefore, it is necessary to monitor the genetic diversity of the Pacific bluefin tuna population over a long period of time in the future.

5. Conclusions

The *Cyt b* gene is a well-established marker to distinguish *Thunnus* species. The genealogy of *Cyt b* suggested that *T. orientalis* and *T. alalunga* are sister species with a high bootstrapping value and supported monophyly of the tropical yellowfin group, but not of the temperate bluefin group. The phylogeny of *Thunnus* species does not fit into the current two-subgenera *Thunnus* and *Neothunnus* classification pattern. The mtDNA *CR* of *T. orientalis* shows that a lack differentiation in *T. orientalis* from Taiwanese and New Zealand waters is consistent with a "one-stock" hypothesis.

6. Acknowledgements

We are extremely grateful to Dr. J. C. Shiao for supplying specimens from the North Pacific Ocean. We also deeply appreciate S. L. Lin, J. Q. Ling, and D. S. Hsiung for their help with laboratory work. We also thank the Florida Museum of National History-Genetic Resources Repository and Dr. Rob Robins for supplying muscle tissues of *Thunnus atlanticus* specimens.

7. References

Alvarado-Bremer, J. R., Naseri, I., & Ely, B. (1997). Orthodox and unorthodox phylogenetic relationships among tunas revealed by the nucleotide sequence analysis of the mitochondrial control region. *Journal of Fish Biology* 50, 540-554.

Bayliff, W. H. (1991). Status of northern bluefin tuna in the Pacific Ocean. In IATTC, world meeting on stock Assessment of Bluefin tunas: strengths and weaknesses. *Inter-American tropical Tuna Commission Special Report* 7, 29-88.

Bayliff, W. H. (1994). A review of the biology and fisheries for northern bluefin tuna, *Thunnus thynnus*, in the Pacific Ocean, *Interactions of Pacific Tuna Fisheries*, (Eds R. S. Shomura, J. Majkowski, and S. Langi.) Vol. 2. FAO Fisheries Technical Paper No. 336/2, pp. 244-295.

Bayliff, W. H. (2001). Status of bluefin tuna in the Pacific Ocean. *Inter-American Tropical Tuna Commission Stock Assessment Report* 1, 211-254.

Belkhir, K., Borsa, P., Chikhi, L., Raufaste, N. & Bonhomme, F. (1996-2004). GENETIX 4.05: logiciel sous Windows TM pour la génétique des populations. Laboratoire Génome, Populations, Interactions, CNRS UMR 5000, Université de Montpellier II, Montpellier (France).

Benton, M. J., ed. (1993) The Fossil Record 2. Chapman and Hall, London

Block, B. A., Finnerty, J. R., Stewart, A. F. & Kidd, J. (1993). Evolution of endothermy in fish: mapping physiological traits on a molecular phylogeny. *Science* 260, 210-214.

Boustany, A. M., Reeb, C. A. & Block, B. A. (2008) Mitochondrial DNA and electronic tracking reveal population structure of Atlantic bluefin tuna (*Thunnus thynnus*). *Marine Biology* 156, 13-24.

Carrol, R. L. (1988). Vertebrate Paleontology and Evolution. New York: W. H. Freeman and Co.

CCSBT (2010). Report on biology, stock status and Management of Southern bluefin tuna. *Attachment 12 of the Report of the Fifteenth Meeting of the Scientific Committee.*

Chen, K. S., Crone, P. & Hsu, C. C. (2006). Reproductive biology of female Pacific bluefin tuna *Thunnus orientalis* from south-western North Pacific Ocean. *Fisheries Science* 72, 985-994.

Chiang, H. C., Hsu, C. C., Lin, H. D., Ma, G. C., Chiang, T. Y. & Yang, H. Y. (2006) Population structure of bigeye tuna (*Thunnus obesus*) in the South China Sea, Philippine Sea and western Pacific Ocean inferred from mitochondrial DNA. *Fisheries Research* 79, 219-225.

Chiang, H. C., Hsu, C. C., Wu, G. C. C., Chank, S. K. & Yang, H. Y. (2008) Population structure of bigeye tuna (*Thunnus obesus*) in the Indian Ocean inferred from mitochondrial DNA. *Fisheries Research* 90, 305-312.

Chow, S., Nohara, K., Tanabe, T., Itoh, T., Tsuji, S., Nishikawa, Y., Ueyanagi, S. & Uchikawa, K. (2003) Genetic and morphological identification of larval and small juvenile tunas (Pisces, Scombridae) caught by a mid-water trawl in the western Pacific. *Bulletin of the Fishery Research Agency* 8, 1-14.

Chow, S., Nakagawa, T., Suzuki, N., Takeyama, H. & Matsunaga, T. (2006) Phylogenetic relationships among *Thunnus* species inferred from rDNA *ITS1* sequence. *Journal of Fish Biology* 68, 24-35.

Chow, S. & Kishino, H. (1995) Phylogenetic relationships between tuna species of the genus *Thunnus* (Scombridae: Teleostei): Inconsistent implications from morphology, nuclear and mitochondrial genomes. *Journal of Molecular Evolution* 41, 741-748.

Chow, S., Okamoto, H., Miyabe, N., Hiramatsu, K. & Barut, N. (2000) Genetic divergence between Atlantic and Indo-Pacific stocks of bigeye tuna (*Thunnus obesus*) and admixture around South Africa. *Molecular Ecology* 9, 221-227.

Clark, T. B., Ma, L., Saillant, E. & Gold, J. R. (2004). Microsatellite DNA markers for population-genetic studies of Atlantic bluefin tuna (*Thunnus thynnus*) and other species of genus *Thunnus*. *Molecular Ecology Notes* 4, 70.

Collette, B. B. (1978) Adaptations and systematics of the mackerels and tunas. In G. D. Sharp and A. E. Dizon (eds.), The physiological ecology of tunas, pp. 7-39, Academic Press, New York.

Collette, B. B. & Nauen, C. E. (1983). FAO species catalogue. Vol 2 Scombrids of the world. An annotates and illustrated catalogue of tunas, mackerels, bonitos and related species known to date. *FAO Fisheries Synopsis* 2, pp. 137.

Collette, B. B. & Smith, B. R. (1981). Bluefin tuna, *Thunnus thynnus orientalis*, from the Gulf of Papua. *Japanese Journal of Ichthyology* 28, 166-168.

Dekker, W. (2003). Worldwide decline of eel resources necessitates immediate action: Quebec Declaration of Concern. *Fisheries* 28, 28-30.

Elliott, N. G. & Ward, R. D. (1995) Genetic relationships of eight species of Pacific tuna (Teleostei, Scombridae) inferred from allozyme analysis. *Marine and Freshwater Research* 46, 1021-1032.

Excoffier, L. & Lischer, H. E. L. (2010). Arlequin suite ver 3.5: A new series of programs to perform population genetics analyses under Linux and Windows. *Molecular Ecology Resource* 10, 564-567.

Felsenstein, J. (1985). Confidence limits on phylogenies: an approach using the bootstrap. *Evolution* 39, 783-791.

Finnerty, J. R. & Block, B. A. (1995) Evolution of *cytochrome b* in the Scombroidei (Teleostei): molecular insights into billfish (Istiophoridae and Xiphiidae) relationships. *Fishery Bulletin* 93, 78-96.

Grewe, P. M, Elliott, N. G., Innes, B. H. & Ward, R. D. (1997) Genetic population structure of southern bluefin tuna (*Thunnus maccoyii*). *Marine Biology* 127, 555-561.

Hillis, D. M. & Bull, J. J. (1993). An empirical test of bootstrapping as a method for assessing confidence in phylogenetic analysis. *Systematic Biology* 42, 182-192.

Ishikawa, S., Aoyama, J., Tsukamoto, K., and Nishida, M. (2001). Population structure of the Japanese eel *Anguilla japonica* as examined by mitochondrial DNA sequencing. *Fisheries Science* 67, 246-253.

Itoh, T. (2009). Contributions of different spawning seasons to the stock of Pacific bluefin tuna *Thunnus orientalis* estimated from otolith daily increments and catch-at-length data of age-0 fish. *Nippon Suisan Gakkaishi* 75, 412-418.

Johns, G. C. & Avise, J. C. (1998). A comparative summary of genetic distances in the vertebrates from the mitochondrial *cytochrome b* gene. *Molecular Biology and Evolution* 15, 1481-1490.

Kimura, M. (1980). A simple method for estimating evolutionary rate of base substitutions through comparative studies of nucleotide sequences. *Journal of Molecular Evolution* 16, 111-120.

Kocher, T. D., Thomas, W. K., Meyer, A., Edwards, S. V., Pabo, S., Villablanca, F.X. & Wilson, A. C. (1989). Dynamics of mitochondrial DNA evolution in animals:

amplification and sequencing with conserved primers. *Proceedings of the National Academy of Sciences USA* 86, 6196-6200.

Librado, P. & Rozas, J. (2009). DnaSP v5: A software for comprehensive analysis of DNA polymorphism data. *Bioinformatics* 25, 1451-1452 | doi: 10.1093/bioinformatics/btp187.

Martinez, P., Gonzalez, E. G., Castilho, R. & Zardoya, R. (2006) Genetic diversity and historical demography of Atlantic bigeye tuna (*Thunnus obesus*). *Molecular Phylogenetics and Evolution* 39, 404-416.

Nei, M. & Kumar, S. (2000). 'Molecular Evolution and Phylogenetics'. Oxford University Press, New York.

Porch, C. E. (2005). "The sustainability of western Atlantic bluefin tuna: a warm-blooded fish in a hot-blooded fishery". *Bulletin of Marine Science* 76.2, 363-84.

Posada, D. & Crandall, K. A. (1998). Modeltest: testing the model of DNA substitution. *Bioinformatics* 14, 817-818.

Pujolar, J. M., Roldán, M. I. & Pla, C. (2003) Genetic analysis of tuna populations *Thunnus thynnus thynnus* and *T. alalunga*. *Marine Biology* 143, 613-621.

Raymond, M. & Rousset, F. (1995). An exact test for population differentiation. *Evolution* 49, 1280-1283.

Rohlf, F. J. (1973). Algorithm 76. Hierarchical clustering using the minimum spanning tree. *Computer Journal* 16, 93-95.

Sharp, G. D. & Pirages, S. (1978) The distribution of red and white swimming muscles, their biochemistry, and the biochemical phylogeny of selected scombrid fishes. In: Sharp GD, Dizon AE (eds) *The physiological ecology of tunas,* pp. 41-78, Academic Press, New York,.

Shiao J. C., Chang, S. K., Lin, Y. T. & Tzeng, W. N. (2008). Size and age composition of southern bluefin tuna (*Thunnus maccoyii*) in the central Indian Ocean inferred from fisheries and otolith data. *Zoological Studies* 47, 158-171.

Smith, P. J., Griggs, L. & Chow, S. (2001). DNA identification of Pacific bluefin tuna (*Thunnus orientalis*) in the New Zealand fishery. *New Zealand Journal of Marine and Freshwater Research* 35, 843-850.

Takagi, M., Okamura, T., Chow, S. & Taniguchi, N. (1999). PCR primers for microsatellite loci in tuna species of the genus *Thunnus* and its application for population genetic study. *Fisheries Science* 65, 571-576.

Tamura, K., Dudley, J., Nei, M. & Kumar, S. (2007). *MEGA4*: Molecular Evolutionary Genetics Analysis (MEGA) software version 4.0. *Molecular Biology and Evolution* 24, 1596-1599.

Tanaka, Y., Mohri, M. & Yamada, H. (2007). Distribution, growth and hatch date of juvenile Pacific bluefin tuna *Thunnus orientalis* in the coastal area of the Sea of Japan. *Fisheries Science* 73, 534-542.

Tanaka, Y., Satoh, K., Iwahashi, M. & Yamada, H. (2006). Growth-dependent recruitment of Pacific bluefin tuna *Thunnus orientalis* in the northwestern Pacific Ocean. *Marine Ecology Progress Series* 319, 225-235.

Thompson, J. D., Higgins, D. G. & Gibson, T. J. (1994). Clustal W: improving the sensitivity of progressive multiple sequence alignment through sequence weighting, position-specific gap penalties and weight matrix choice. *Nucleic Acids Research* 22, 4673-4680.

Tseng, M. C., Jean, C. T., Tsai, W. L. & Chen, N. C. (2009). Distinguishing between two sympatric *Acanthopagrus* species from Dapeng Bay, Taiwan, using morphometric and genetic characters. *Journal of Fish Biology* 74, 357-376.

Tseng, M. C., Shiao, J. C. & Hung, Y. H. (2011). Genetic identification of *Thunnus orientalis, T. thynnus*, and *T. maccoyii* by a *cytochrome b* gene analysis. *Environmental Biology of Fishes* 91, 103-115.

Ueyanagi, S. (1975). *Thunnus* commentary. In Fisheries in Japan: Tuna. Uichi Noda Publishers; Japan Marine Products Photo Materials Association. Tokyo.

Van Oosterhout, C., Hutchinson, W. F., Wills, D. P. M. & Shipley, P. (2004). MICRO-CHECKER software for identifying and correcting genotyping errors in microsatellite data. *Molecular Ecology Notes* 4, 535-538.

Viñas, J., Alvarado-Bremer, J. R. & Pla, C. (2004) Inter-oceanic genetic differentiation among albacore (*Thunnus alalunga*) populations. *Marine Biology* 145, 225-232.

Viñas, J. & Tudela, S. (2009). A validated methodology for genetic identification of tuna species (Genus *Thunnus*). *PLoS ONE* 4: e7606. doi:10.1371/journal.pone.0007606

Ward, R. D., Elliott, N. G. & Grewe, P. M. (1995). Allozyme and mitochondrial DNA separation of Pacific Northern bluefin tuna, *Thunnus thynnus orientalis* (Temminck and Schlegel), from Southern bluefin tuna, *Thunnus maccoyii* (Castelnau). *Marine and Freshwater Research* 46, 921-930.

Ward, R. D. (1995). Population genetics of tunas. *Journal of Fish Biology* 47 (supplement A): 259-280.

Wu, G. C. C., Chiang, H. C., Chen, K. S., Hsu, C. C. & Yang, H. Y. (2009). Population structure of albacore (*Thunnus alalunga*) in the Northwestern Pacific Ocean inferred from mitochondrial DNA. *Fisheries Research* 95, 125-131.

Yang, R. C. & Yeh, F. C. (1993). Multilocus structure in *Pinus contoria* Dougl. *Theoretical and applied genetics* 87, 568-576.

6

Molecular Markers and Genetic Diversity in Neotropical Felids

Alexeia Barufatti Grisolia[1] and Vanessa Roma Moreno-Cotulio[2]
[1]Universidade Federal da Grande Dourados, UFGD, FCBA, MS
[2]Universidade Federal de Alfenas, UNIFAL-MG, ICN, MG
Brazil

1. Introduction

The Class Mammalia has a taxonomic diversity including 5416 existing species or recently extinct and this number increases each year with the description of new species. This extraordinary group shows not only a wide diversity of species but also of forms, ecologies, physiologies, life histories and behaviour (Wilson & Reeder, 1993).

Felines are important representatives of the Class Mammalia and the 38 existing species of felids occur naturally in almost all areas of the world except in some insular regions as Australia, New Guinea and New Zealand, Japan, Madagascar, Oceania, Andes and some Caribbean islands (Nowak, 1999; Johnson et al., 2006). The Neotropical region (covers the southern part of North America, Central America and South America) is occupied by 10 recognized cat species that are divided into three clades evolutionarily distinct, which have been distinguished using a variety of molecular genetic techniques. The first clade, ocelote lineage [*Leopardus pardalis* (Linnaeus, 1758)] that comprises all the species of the genus *Leopardus*. The second clade, puma lineage [*Puma concolor* (Linnaeus, 1771)] that comprises the two species of the genus *Puma*. The third clade, *Panthera* lineage is represented in this region by jaguar [*Panthera onca* (Linnaeus, 1758)] (Johnson et al., 2006). There are eight of these species in Brazil: *Leopardus pardalis* (ocelot), *Leopardus wiedii* (margay), *Leopardus tigrinus* (oncilla), *Leopardus geoffroyi* (Geoffrey's cat), *Leopardus colocolo* (colocolo), *Puma yagouaroundi* (jaguarundi), *Puma concolor* (cougar) and *Panthera onca* (jaguar) (Oliveira, 1994). The *Oreailurus jacobita* (andean mountain cat) and *Lynchailurus colocolo* (pampas cat) are the other species of ocelot lineage that occur in other South American regions.

Mattern & McLennan (2000), studying felines phylogeny and speciation, came to a phylogenetic tree that confirms the formation of the main groups of the family Felidae (Figure 1).

The IUCN (International Union for Conservation of Nature) published in 2001 the scheme transcribed below (Figure 2) concerning the categories used to evaluate the conservation status of the animals or plants species and that is used to elaborate the "Red list of the Brazilian fauna threatened with extinction" (MMA, 2008).

Among the endangered species, there is a classification in three levels: *critically endangered (CR):* taxon runs extremely high risk of extinction in the wild in a immediate future;

endangered (EN): taxon is not critically in danger but runs high risk of extinction in the wild in a near future; and, *vulnerable (VU):* taxon does not fit in the categories critically endangered or endangered, but runs high risk of extinction in the wild in the medium term.

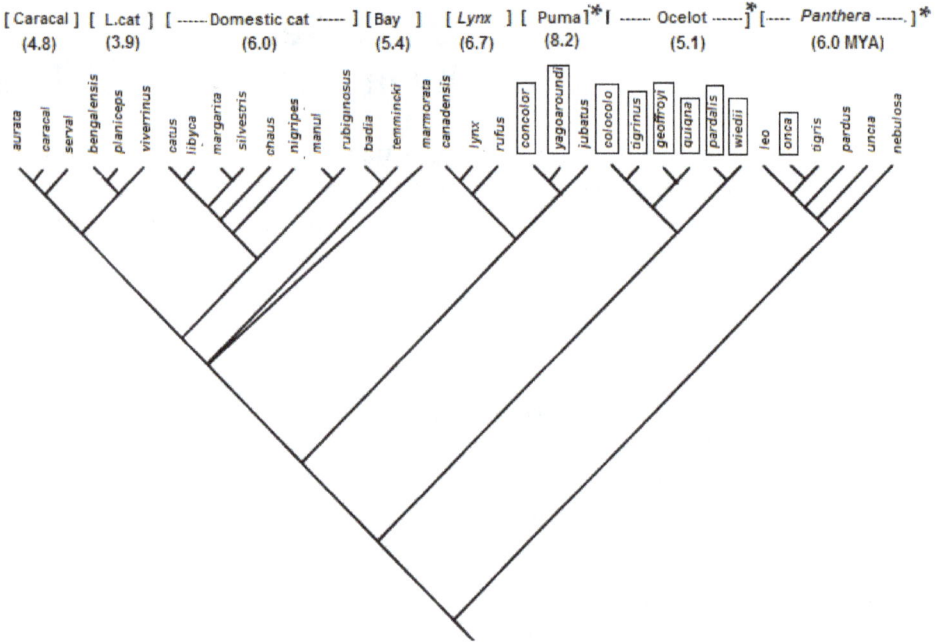

Fig. 1. Plylogenetic tree for Felidae based upon 1504 characters, ages of each lineage (MYA, million years ago) based upon amount of sequence divergence are written across the top of the tree. Only species names are indicated on the tree and Neotropical cats were framed. Adapted from Mattern & McLennan (2000). * Indicate clades of felids.

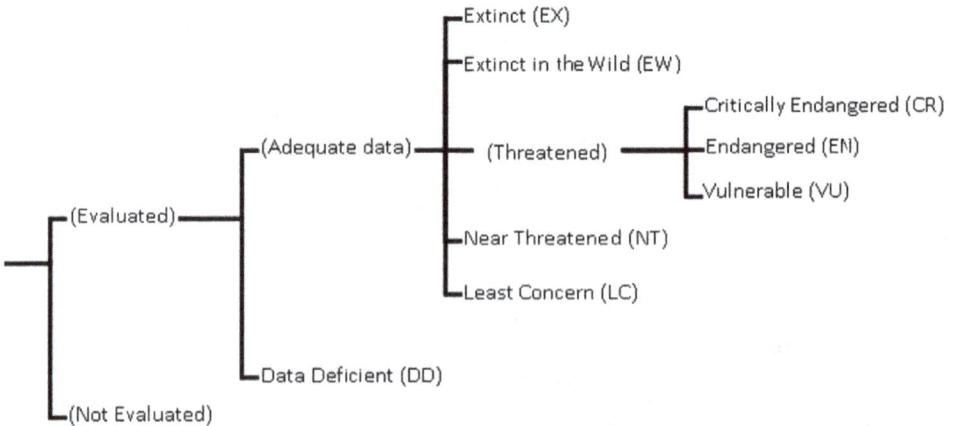

Fig. 2. Scheme for the classification of the species according to IUCN (2001).

All the neotropical felids are threatened with extinction on some level (IUCN, 2001), and this risk has increased mainly by anthropogenic changes that cause the reduction and fragmentation of the natural habitat of these animals, hunt and pollution, which significantly contribute to the decline and isolation of wild populations (Crawshaw Jr., 1997; Frankham et al., 2002; Schipper et al., 2008).

According to Frankham et al. (2008): "biodiversity is the variety of ecosystems, species, populations within species, as well as the genetic biodiversity existing within and between populations". One of the very promising research areas on the conservation of species threatened with extinction is the identification of "Hotspots" or areas with exceptional concentrations of endemic species, but with greatly reduced habitat. Among the 25 "Hotsposts" already indentified worldwide, the cerrado and the Atlantic rainforest of Brazil are included (Myers et al., 2000). The feline species that inhabit these areas where the deforestation is intense are particularly vulnerable. In the Brazilian rainforests, the ocelot (*Leopardus pardalis*), oncilla (*Leopardus tigrinus*), and margay (*Leopardus wiedii*), are among the felids that struggle to survive against the habitat destruction (Moreira et al., 2001).

The loss of biodiversity often culminates in the local extinction of species, reduction in the distribution and density of species, increasing the extinction risks. Therefore, the Conservation Biology has become an important science due to its multidisciplinary character, since it encompasses several other disciplines such as ecology and biology of populations, which allows accumulating data and information for the conservation and management of natural resources (Primack & Rodrigues, 2001; De Salle & Amato, 2004). Later and more recently the Conservation Genetics arose as an important area within the perspective of elaborating adequate management and conservation plans (Johnson et al., 2001; Perez-Sweeney et al., 2003).

The Conservation Genetics involves the use of genetic technologies to minimize the risks of extinction of endangered species. Among the topics studied in this area we can cite: (1) minimization of inbreeding and loss of genetic diversity; (2) identification of species or populations in risk due to reduced genetic diversity; (3) resolution of the structure of fragmented populations; (4) resolution of taxonomic uncertainties; (5) definition of management unities within the limits of the species; (6) detection of hybridization; (7) non-invasive sampling for genetic analyses; (8) definition of locations and choice of the best populations for reintroduction; (9) forensic analysis; (10) understanding the biology of the species (Frankham et al., 2002).

One of the main focuses of conservation biology is the maintenance of the genetic diversity, so that IUCN recognizes the need to maintain this variation as one of the three global conservation priorities. Two aspects should be considered: first, due to the continuous process of environmental changes the populations evolve and adapt to these changes if there is the necessary genetic variability to their adaptation to such alterations. Second, in general, there is an association between loss of genetic diversity, inbreeding and general decrease in reproductive and survival indexes (Frankham et al., 2002).

In wildlife populations subjected to reduction of the habitat area, as well as the number of reproductively active individuals, there may be loss of genetic variability. It is manifested both by the effect of genetic drift as a result of consanguinity (Wanjtal et al., 1995). Some authors estimate that an effective population size to avoid inbreeding depression is 50 and

500 individuals if it is wanted to avoid the loss of genetic variability due to genetic drift. However, caution is needed when determining the population sizes for conservation purposes, since once the genetic variability is lost, it will be only recovered very slowly (by mutation or migration), so that, even taking measures to increase the population size of a species, it may continue an endangered species (Solé-Cava, 2001).

Given that the genetic diversity provides the adaptive/evolutionary potential of a species, its maintenance is one of the main focuses of conservation biology. Thus, it is fundamental the knowledge of the genetic composition of a species and how it is organized (structured) in their populations, so that the management can be done, when necessary, and for the conservation of the species. It is also important to understand if the genetic structure found is a natural characteristic of the species studied or if it is the result of the presence of physical barriers caused by man, as in the case of habitat fragmentation (Galetti Jr. et al., 2008).

With the advent, in recent times, of several technologies of Genetics and Molecular Biology, especially the recombinant DNA technology, polymerase chain reaction (PCR) and sequencing, numerous possibilities of genetic markers arose allowing the detection of genetic polymorphism directly in the DNA. Such methodologies have allowed numerous genetic studies, whether they are targeted or not to conservation, such as: evolution studies, inter- and intra-specific genetic diversity, genetic origin (phylogeny), etc (Faleiro, 2007).

2. Molecular markers

Genetic variability allows us to compare individuals, populations or different species. A relevant aspect from the conservation genetics is the fact that different molecular biomarkers can have different rates of substitution/evolution, so that by the judicious choice of these markers, we can study from problems of identification of individuals to identification of cryptic species or formulating phylogenetic hypothesis in supraspecific groups (Solé-Cava, 2001).

The main technologies available for obtaining molecular genetic markers can be isoenzymatic markers or based on DNA fragments. DNA markers can be classified into two groups, according to the methodology used: DNA hybridization or amplification. The markers related to the hybridization methodology are the RFLP markers (Restriction Fragment Length Polymorphism) and the Minisatellites or VNTR loci (Variable Number of Tandem Repeats). The markers revealed by DNA amplification are those of RAPD (Random Amplified Polymorphism DNA), AFLP (Amplified Fragment Length Polymorphism), SSR (Short Sequence Repeat) or Microsatellite, CAPS (Cleaved Amplified Polymorphic Sequence) or PCR-RFLP, SNP (Single Nucleotide Polymorphism) and others. Some characteristics of the main methodologies for the markers analyses are described in Table 1.

The choice of the methodology to be used in the approach of each problem depends on several criteria. Among the criteria, firstly, the adequacy of the variability degree of the molecular marker chosen at the divergence level that one wishes to study. Markers that evolve rapidly are useful for the study of individuals, families and populations, while markers that evolve more slowly are better used in the study of species or supraspecific taxa.

Another important criterion in the choice is the type of material available for the studies. Researches based on izoenzymatic markers, for example, need higher quantities of biological material, which must necessarily be fresh or frozen (the enzymes denature or are digested by proteases when kept at room temperature). This problem does not exist for most of the methods with DNA thanks to the advent of PCR, since it allows the use of very small amounts of tissue including those preserved in alcohol or dehydrated. On the other side, in population studies in which a very high number of individuals should be analyzed, techniques such as the DNA sequencing are very expensive.

Variable	RFLP	PCR-RFLP	RAPD	Micro satellite	SNP	AFLP
Quantity of DNA	10ug	50ng	50ng	50ng	50ng	500ng
Quality of DNA	Excellent	Reasonable	Reasonable	Reasonable	Good	Good
Based on PCR	No	Yes	Yes	Yes	Yes	Yes
Radioactivity	Yes	No	No	Yes/No	No	Yes/No
Multiplex	No	No	No	Yes	Yes	No
Ease of Use	No	Yes	Yes	Yes	Yes	No
Automation	No	Yes	Yes	Yes	Yes- high	Yes
Reproducibility	Good	Good	Low	Good	Good	Medium
Number of alleles per locus	Biallelic	Biallelic	Biallelic	Multiallelic	Biallelic	Biallelic
Gene expression	Codominant	Codominant	Dominant	Codominant	Codominant	Dominant
Abundance throughout the genome	High	High	Low	Medium	High	Medium
Cost by data	High	Low	Low	Low	Very low	Low

Table 1. Comparative analysis of the characteristics and methodology of main molecular markers.

The choice of molecular marker to be used depends, therefore, on several factors. The fundamental is that the problem to be studied is well defined, that there is an adequacy in the polymorphism degree of the marker chosen to the type of evolutionary divergence to be studied, that the assumptions of the data analyses are clearly explicit (Solé-Cava, 2001).

2.1 Nuclear DNA and microsatellite marker

DNA analysis shows us genetic variability and allows the refinement of the genealogies of wild and captive populations (Mace et al., 1996). Microsatellites, due to their abundance in the genome and their high mutation rates, are molecular markers used to detect levels of genetic variability and to plan future breeding strategies. Although several types of molecular markers from the nuclear genome are available, short, tandem, repetitive regions within the nuclear genome known as microsatellites currently are the most popular molecular marker for ecological studies (Avise, 2004). Microsatellites consist of small DNA fragments of about 10-100 base pairs (bp) that contain repetitive elements displaying tandem repeats of 1-6 bp (Figure 3), variation in the number of repeats resulting in these loci having a high polymorphism information content (PIC). The heterozygosity of microsatellite loci was first described in humans, but microsatellites have been found to be abundant, randomly distributed and highly polymorphic in all eukaryotic organisms investigated so far (Weber & May, 1989). Microsatellites are codominant markers (*i.e.* heterozygotes can be distinguished from the homozygotes) which can be amplified by PCR and generally show a level of heterozygosity in excess of 0.7 (Ferreira & Grattapaglia, 1998).

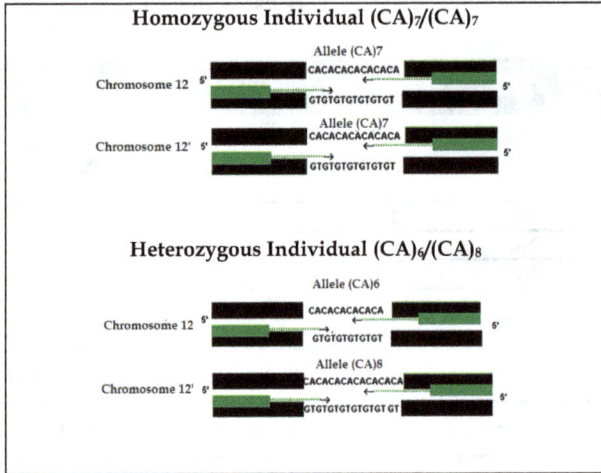

Fig. 3. Representation of the homozygous genotypes (identical alleles) and heterozygous (different alleles) for a genomic region that comprises a microsatellite of (CA)/ (GT) (adapted from Ferreira & Grattapaglia, 1998).

An important characteristic of microsatellites is that primers developed for a specific species can be used for other species of related taxons. Ten highly polymorphic microsatellite loci of dinucleotide repetitions (dC.dA)n/(dG.dT)n developed from the domestic cat genome showed amplified products of the same size in lions, cheetahs, pumas, leopard cat, and Geoffroy's cat (Menotti-Raymond & O'Brien, 1995). A broad range of heterozygosity was observed among the species for a single locus and among loci within a single feline species. This characteristic demonstrates that a microsatellite is an important informative marker (Menotti-Raymond & O'Brien, 1995).

Identifying factors that decrease the potential for inter-population gene flow is of singular importance in the conservation of felids populations. Drift and inbreeding can combine in small populations to reduce fitness in the short term (Keller & Waller, 2002) and theoretically reduce evolutionary potential in the long term (Lacy, 1997). Epps et al. (2005) used a combination of microsatellites and mitochondrial markers to infer greatly reduced gene flow between desert bighorn populations bisected by human-constructed barriers (e.g., major highways, urban development, etc.). While genetic analyses using molecular markers can inform ongoing conservation efforts for established populations felines, such analyses may also contribute to efforts targeted at re-populating vacant feline habitat through reintroductions.

Investigating natural history traits - Some aspects of natural history are difficult or impossible to investigate without using molecular markers. These investigations of wild feline natural history demonstrate the utility of microsatellites: useful not only in population analyses, but also in analyses focusing in scale down to the individual.

2.2 Mitochondrial DNA

A small molecule of circular DNA is present in the mitochondria – the mitochondrial DNA (mtDNA). This material is very abundant in cells in general, since there are many mitochondria per cells; it is inherited maternally in most species and is easily isolated. The restriction enzymes (RFLP), SSCP and sequencing are some of the methods that can be employed in the investigation of genetic diversity in the mtDNA, and, moreover, the availability of several primers for different loci of this DNA allows the amplification of these DNA fragments by PCR and subsequent sequencing of the products (Frankham et al., 2002).

Two interesting features of the mtDNA have led to the constant use of this molecule in phylogenetic and phylogeographic analyses. One is the absence of recombination in this molecule and the other is the fact that the mtDNA has a faster evolutionary rate in relation to the nuclear sequences (there are differences in relation to the molecular evolutionary rates even within the same mtDNA sequence). Based on the purpose of the research, one or another region of the molecule of the mitochondria DNA can be used. In general, the control region of the mtDNA (D-loop region, contains the location of replication origin) is the most employed in population studies, while the NADH genes, ribosomal and cytochrome genes tend to be used in matters related to the species as a whole and its distribution and in the intergeneric systematic (Perez-Sweeney et al., 2003).

Worldwide, researchers involved with the Conservation Genetics area have used mitochondrial markers for different types of analyses in several groups of animals, including the several species of felids. In the case of the representatives of the family Felidae, the phylogenetic and phylogeographic analyses are included in many researches.

3. Applications of molecular markers in the species conservation

Molecular methods have opened the entire biological world to the genetic analysis, as well as the complete spectrum of ecological and evolutionary scales in the genetic differentiation. In a final analysis, biodiversity is genetic diversity, and the magnitudes and patterns of this diversity can now be examined in any organisms (Avise, 1995). With the fast advance of techniques of molecular genetics in the last years, it was noticed that many of them could considerably contribute to the management of endangered species and for the conservation of biodiversity.

There are, essentially, two ways to use genetic markers in the conservation area. The first is based on the assumption that the genetic variability is determinant in the suitability and feasibility of a population. Thus, genetic markers are used to indicate the variability levels and their distribution in populations of a certain species (genetic structure). The second is the measurement of processes, such as migration, which produce measurable effects in the patterns of genetic variability (Neigel, 1996).

Among the molecular markers employed in population studies, evolutionary and/or studies targeted for conservation, stand out: the traditional protein electrophoresis, which initiated great part of the genetic studies in natural population; studies with RFLP and the sequencing of mitochondrial DNA or nuclear genes; DNA fingerprinting using probes for minisatellite regions; RAPD; and polymorphisms in microsatellite regions. Currently, great emphasis is given to studies of microsatellites and mtDNA as markers of genetic diversity (mainly in animal populations), and the latter is extremely useful for investigations of phylogenetic relationships between different taxa and identification of geographic subdivision between population unities (Eizirik, 1996). Microsatellites are markers widely used in genetic studies. The potential use of these markers in small populations, especially in endangered species, is enormous, since these sequences are highly informative and the material required for its analysis can be collected by non-invasive methods (Bruford et al., 1996).

4. Molecular markers in felines

The use of these markers in endangered species is widespread, since these sequences are highly informative. For these studies blood samples, faeces and even hair are used, enabling accurate results from non-invasive methods. Using microsatellites it is possible to evaluate the structure within and between populations, besides the possibility to determine parentage. This fact is crucial for suggesting adequate conservation and management strategies (O'Brien, 1996).

As the genetic progress in recognizing the formulation of conservation strategies is evident that the current information on the genetic structure of natural populations are still insufficient, therefore, becomes necessary to perform reproductive management of the species that are increasingly represented by small populations in forest reserves. It also becomes clear that the initiative to revert the decline of biodiversity should be involved with several areas of knowledge, including molecular genetics. It should be kept in mind that the molecular approach in the study of endangered species is to provide an additional tool that can help to avoid extinction. Based on multidisciplinary approaches, conservation plans have been drawn up with a better knowledge about the real threats to the survival of the species (O'Brien, 1996).

With the increasing destruction of wildlife, captive populations have become an important strategy for the conservation of endangered species. The general discussion on this aspect is how the captive populations and their maintenance programs can be integrated with wild populations and how these programs can maximize the preservation of biodiversity (Foose, 1983).

During the last years, a wide range of studies have been carried out to estimate the genetic variability in feline populations, including the several species represented in this group. O'Brien et al. (1987), using the isoenzymes electrophoresis, analyzed four populations of African lions and Asian lions. They found a moderate variability in the specimens from Africa, while no variations were found in the Asian lions. This fact led to the conclusion that possibly these populations suffered a decrease in the population size followed by inbreeding in the recent past.

To characterize the genetic variability in cheetahs (*Acinonyx jubatus*) studies using alloenzymes, soluble proteins, histocompatibility complex genes, mitochondrial DNA, minisatellites and microsatellites were carried out (O'Brien et al., 1983; Menotti – Raymond

& O'Brien, 1993). Regarding alloenzymes, the variation observed in cheetahs was low, while in mitochondrial DNA it was moderate and high for microsatellites and minisatellites. Perhaps this higher variability in these markers could be explained by the high mutation rate in these sequences. As apparent consequences of this genetic uniformity in cheetahs in relation to isoenzymatic markers, are: great difficulty in captive breeding, high degree of infant mortality in captivity and nature, and high frequency of abnormal spermatozoids in ejaculation.

O'Brien et al. (1990) carried out studies with populations of Florida panther (*Puma concolor coryi*), a highly endangered animal. They reported the existence of two genetically 12 distinct groups using techniques of isoenzymes electrophoresis, 11 mitochondrial DNA and nuclear markers: a group descendant of ancestors of *P. c. coryi*, phylogenetically close to other subspecies of North America, and another group, resembling the cougars of Central America or South America and that must have been introduced in Florida in the recent past. The analysis concluded that, despite the genetic differences presented by the two subpopulations of Florida, they have in their historical separation, a common ancestor sufficiently recent to eliminate the development of mechanisms of reproductive isolation, i.e., to allow the crossbreeding between individuals of two populations. Moreover, according to O' Brien et al. (1990), the formation of hybrids could even enhance the survival chances of Florida panther, which has strongly suffered the consequences of inbreeding.

Menotti-Raymond & O'Brien (1995) isolated, characterized and amplified 10 highly polymorphic loci of microsatellites of dinucleotide repeats (dC.dA)n/(dG.dT)n of the domestic cat (*Felis catus*). They demonstrated that 10 pairs of oligonucleotide primers developed from the domestic cat genome amplified products of similar size in lions, cheetahs, Asian leopard cat and Geoffroy's cat. This fact suggested that the flanking sequences of these microsatellites are conserved in felines. Individual loci showed large heterozygosity in the species of felines studied, thus showing themselves as informative markers. It was observed a mean rate of heterozygosity of 0.77 in *Felis catus*, 0.39 in cheetah, 0.61 in puma and 0.66 in lion, from the 10 oligonucleotide primers analyzed. The polymorphism of the microsatellites loci of felines offers much potential as molecular marker for: i) identification of species or subspecies, ii) accurate measure of the genetic divergence of the population and iii) evaluation of the origin in native populations.

In order to examine the extension of inbreeding in populations of Asian lions and Indian tigers, to identify pure Asian lion from the hybrid and pure Indian tiger from the hybrid and identify lions and tigers with extensive genetic variation for selective inbreeding for their own management, Shankaranarayanan et al. (1997) developed a study using RAPD markers and microsatellites. The RAPD patterns were evaluated in lions and tigers using 30 random primers, of which four produced polymorphic pattern and were used for population studies. It was analyzed a total of 38 individuals and a mean heterozygosity of 25.82% was observed, varying from 16.71% to 34.39% in the individual analysis of the primers. The primers that presented polymorphism in lions did not show any polymorphism in tigers. The analysis of microsatellites was done by 5 loci of CA repetition that are polymorphic in felines. The Asian lions, however, did not show variation for all the 5 loci of microsatellites studied; only 2 of them showed differences between Asian lions and hybrid lions. Analyzing the microsatellites in 30 Indian tigers, Shankaranarayanan et al. (1997) observed mean heterozygosity of 22.65% in three of the five loci tested. The exam of microsatellites in 15 skin samples of museum tigers, aged from 50 to 125 years, showed a mean heterozygosity of 21.01% and, therefore, the difference between present and past populations were not statistically significant. Only the loci that were polymorphic in the actual population showed polymorphism in the individuals of 50 – 125 years ago. These analyses also served to identify the presence of hybrid tigers in the population studied. Moreover, results like this can be used for breeding programs to increase the genetic variability in the population.

Menotti-Raymond et al. (1999) developed a genetic linkage map in felines using 253 microsatellite loci. It was identified and genotyped 235 loci of dinucleotide repetition (dC · dA)n · (dG · dT)n and 18 tetranucleotide, in two families, with 108 members resulting from crossbreeding of lineages of domestic cat (*Felis catus*) and leopard cat (*Prionailurus bengalensis*). Two hundred and twenty-nine loci were linked, identifying 34 linkage groups, and from the 19 pairs of chromosomes, 16 were mapped. According to the authors, the genome measures approximately 2900 cM, and they estimate that the genetic length of the map in 3300 cM.

A study carried out by Eizirik et al. (2001) aimed to investigate the genetic diversity, population structure and demographic history of jaguars throughout their geographic distribution. It was analyzed 715 pb of the control region of mitochondrial DNA and 29 microsatellite loci in approximately 40 individuals sampled from Mexico to southern Brazil. The results showed low to moderate diversity levels for the mtDNA and medium to high for microsatellites and also showed a recent demographic expansion. No well-defined geographic structure was observed but some geographic barriers, such as the Amazon River and the Darien gap between northern South America and Central America, seem to have limited the historical genetic flow in these species, producing measurable genetic diversity. The representatives of *Panthera onca* studied could be divided into four incomplete isolated phylogeographic groups.

The molecular markers have also helped in taxonomic studies in several animal species, including felines. The incorporation of accurate definitions for taxonomic unities in the legislation of wildlife has required a re-evaluation of the taxonomy of threatened or endangered species. Miththapala et al. (1996) carried out a study with 60 leopards (*Panthera pardus*) representing 12 recognized subspecies. Three molecular methods were used: alloenzymes, mtDNA restriction sites and minisatellites feline-specific that showed a considerable genetic variability in the individuals sampled. The mainland populations and subspecies of Africa and Asia presented the highest amount of molecular genetic variation, while in insular populations relatively low amounts of diversity were present. The phylogenetic analysis of molecular data showed the phylogenetic distinction of six groups of leopards geographically isolated: African, central Asian, Hindu, Sri Lankan, Javan and East Asia. Based on the combined molecular analyses and favored by morphological data, the authors suggest a taxonomic revision for eight subspecies of leopards.

Gugolz et al. (2008) used samples of animals preserved in museums of the Alps lynx and other existing populations of representatives of *Lynx lynx* populations in Europe and Asia, in order to evaluate the phylogenetic position of the Alps lynx. The phylogenetic analysis using a region of 345pb of the cytochrome b gene placed the Alps lynx within the lineage of lynxes from Eurasia, while the analysis of a fragment of 300pb of the control regions showed seven different haplotypes but no exclusive haplotype of the Alps lynx. The haplotypes of the extinct population were identical to those previously described in the Scandinavian lynx signifying a recent genetic ancestor with the current European populations.

Patterns of molecular genetic diversity of the largest remaining free-ranging cheetah population were described in a survey of 313 individuals throughout Namibia using 19 polymorphic microsatellite loci. There was limited differentiation among regions, evidence that this is a generally panmictic population. Measures of genetic variation were similar among all regions and were comparable with Eastern African cheetah populations. The long-term maintenance of current patterns of genetic variation in Namibia depends on retaining habitat characteristics that promote natural dispersal and gene flow of cheetahs (Marker et al., 2008).

Driscoll et al. (2009), using DNA samples extracted from specimens held in museums, carried out a study with mtDNA analysis aiming to investigate and interpret the phylogeographic natural history of the Caspian tiger (*Panthera tigris virgata*) in the context of contemporary subspecies of tigers, the Amur tigers (*P. t. altaica*). The data obtained with the mtDNA fragments showed that the Caspian tigers have a large haplotype differing in only one nucleotide of the monomorphic haplotype found in Amur tigers, suggesting, by the phylogeographic analysis, that both species of tigers colonized Central Asia less than 10, 000 years ago. The authors also suggest that, based on their findings, the original habitat of the Caspian tiger in Central Asia is open to reintroductions of genetic stocks of the Amur Tiger, since there is a evolutionary proximity between the two subspecies.

Colocolo (*Leopardus colocolo*) and the Andean mountain cat (*Leopardus jacobita*) are being highly studied in comparison to the South American felids (Redford & Eisenberg, 1992; Nowell & Jackson, 1996). One of these studies included 33 samples of *L.jacobita* and 75 of *L.colocolo* collected in northern Chile and one of the objectives was the determination of general patterns of genetic variation in the two species. Mitochondrial genes were used and, in general, a relatively low genetic variation (2 haplotypes in the mtDNA) was shown for the Andean mountain cat when compared to colocolo (17 haplotypes), which suggests a distinct evolutionary history (Napolitano et al., 2008).

Another example of application of mtDNA is the identification of individuals, either as forensic evidence or in the classification as a certain species. Grahn et al (2011) analyzed a fragment of 402pb of the mtDNA control region for identifying mitotypes of domestic cats (*Felis catus*) of United States and other geographical regions. The total sampling was of 1394 animals (174 previously studied) and the analyses showed that this region has suitable discriminatory power for use in wildlife forensics.

Puma concolor is one of the felines that suffer most from habitat reduction and forest fragmentation and its population is often reduced to few individuals in various regions. A study in the northeastern São Paulo State – Brazil, in two protected areas, used samples of faeces and analysis of a portion of the mitochondrial gene (cytochrome b) to determine the presence of pumas and estimate their minimal population (Miotto et al, 2007). The results indicated that the mtDNA was able to differentiate the faeces of pumas from other felines present in the region, such as ocelot (*Leopardus pardalis*), amplifying in 60% of the samples collected.

Several other species of felines have been recently studied based on mitochondrial genes as markers to assess different aspects in these animals, besides those already mentioned (e.g., population structure, evolutionary history, hybridization rates and introgression): *Panthera leo ssp* (Barnett et al., 2009), *Acinonyx jubatus* (Charrau et al., 2011), *Prionailurus bengalensis* and *Felis chaus* (Mukherjee et al., 2010), *Felis silvestris* (Hertwig et al., 2009; Eckert et al., 2010), *Neofelis diardi* (Wilting et al., 2011), among others.

Trigo et al., (2008) employed mitochondrial DNA (mtDNA) sequences and nine microsatellite loci to identify and characterize a hybrid zone between two Neotropical felids, *Leopardus geoffroyi* and *L. tigrinus*, both of which are well-established species having diverged from each other c. 1 million years ago. They observed that these two felids are mostly allopatric throughout their ranges in South America and present strong evidence for the occurrence of hybridization between these species. Mondol et al. (2009) studying leopards observed 25 of the 29 tested cross-specific microsatellite markers showed positive amplification in 37 wild-caught leopards, these results demonstrated that the selected panel of eight microsatellite loci can conclusively identify leopards from various kinds of biological samples.

There are also less detailed studies, but also of equal importance since they serve as basis for more refined researches in the future. One of them is the study with Amur tigers (*Panthera tigris altaica*) carried out by Russello et al. (2004), who found by the mitochondrial DNA analysis that this population has low rates of genetic diversity, even lower than of the populations in captivity. This study served as a warning to the importance of the integrate management of *in situ* and *ex situ* populations for the conservation of this species. Another study recently performed with two populations of ocelots (*Leopardus pardalis*) in the USA also found low genetic variability and a small and insufficient effective population size for the long-term viability of these populations (Janecka et al., 2008). In Brazil this type of study is still scarce. Most feline genetics researches that are performed in the country are focused on phylogenetic and evolutionary issues. Literature only contains few studies that focus the genetic variability of local populations, of which two of them are with captive animal (Grisolia, et al., 2007; Moreno et al., 2006) and one with two free-living populations of jaguars (Eizirik et al., 2008). This is a worrying situation since the knowledge of the genetic diversity level is one of the main factors to assess the viability of populations, besides bringing crucial information for future decision-making in conservation and management plans.

5. Conclusion

The potential for future contributions of molecular genetics to wild felines conservation and management will only increase as molecular methods become more accessible, cost effective, and practical. Researches about genetic diversity and geographic structure of neotropical felids are basis for further investigations at regional and local levels, including studies of population structure, relatedness between individuals, dispersal patterns, adaptation to different ecosystems, and other ecological and evolutionary aspects that can be addressed using molecular markers. Likewise, we hope that research in this area will contribute to enable the development of efficient strategies for the conservation of these species, their genetic diversity and maintaining ecological and natural processes that influence the continuity of its evolution.

Molecular analysis of genetic structure and integration of ecology, natural history data and feline reproduction could provide a greater comprehension of the factors to be considered in efficacious management plans for these endangered species groups.

6. References

Avise, J.C. (1995). Mitochondrial DNA polymorphism and a connection between genetics and demography of relevance to conservation. *Conservation Biology*, Vol. 9, pp. 686-690

Avise, J.C. (2004). Molecular Markers, Natural History, and Evolution, 2nd Edition. Sinauer, Sunderland, MA, pp. 684

Barnett, R.; Shapiro, B.; Barnes, I.; Ho, S.Y.W.; Burger, J.; Yamaguchi , N.; Higham, T.F.G.; Wheeler, H.T.; Rosendahl, W.; Sher, A.V.; Sotnikova, M.; Kuznetsova, T.; Baryshnikov, G.F.; Martin, L.D.; Harington, C.R.; Burns, J.A. & Cooper,A. (2009). Phylogeography of lions (*Panthera leo* ssp.) reveals three distinct taxa and a late Pleistocene reduction in genetic diversity. *Molecular Ecology*, Vol.18, pp.1668-1677

Bruford, M.; Cheeseman, D.; Coote, T.; Green, H.; Haines, S.; O'ryan, C. & Williams, T. (1996). Microsatellites and their application to conservation genetics. In Smith, T & Wayne, R. Molecular Genetic Approaches in Conservation, *Oxford University Press*, Oxford, pp. 278-297

Charruau, P.; Fernandes, C.; Terwengel, O.; Peters, L.; Hunter, L.; Ziaie, H.; Jourabchian, A.; Jowkar, H.; Schaller, G.; Ostrowski, S.; Vercammen, P.; Grange, T.; Schlötterer, C.; Kotze, A.; Geigl, E.M.; Walzer, C. & Burger, P.A.(2011). Phylogeography, genetic structure and population divergence time of cheetahs in Africa and Asia: evidence for long-term geographic isolates. *Molecular Ecology*, Vol. 20, pp.706-724

Crawshaw Jr., P.G. (1997). Recomendações para um modelo de pesquisa sobre felídeos neotropicais. In: Valladares-Pádua, C.; Bodmer, R.E. Manejo e conservação da vida silvestre no Brasil. Belém, PA: *Sociedade Civil Mamirauá*, pp. 296

De Salle, R. & Amato, G. (2004). The expansion of conservation genetics. *Nature Reviews*, Vol. 5, pp. 702-712.

Driscoll, C.A.; Yamaguchi, N.; Bar-Gal, G.K.; Roca, A.L.; Luo, S.; Macdonald, D.W. & O'Brien, S.J. (2009). Mitochondrial Phylogeography Illuminates the Origin of the Extinct Caspian Tiger and Its Relationship to the Amur Tiger. *Plosone*, Vol. 4, No.1, pp.1-8

Eckert, I.; Suchentrunk, F.; Markov, G. & Hartl, G.B. (2010). Genetic diversity and integrity of German wildcat (*Felis silvestris*) populations as revealed by microsatellites, allozymes, and mitochondrial DNA sequences. *Mammalian Biology*, Vol. 75, pp.160-174

Eizirik, E. (1996). Ecologia molecular, genética da conservação e o conceito de Unidades Evolutivamente Significativas. *Revista Brasileira de Genética*, Vol. 19, pp. 23-29

Eizirik, E.; Kim, J.H.; Menotti-Raymond, M.; Craushaw Jr., P.G.; O'Brien, S.J. & Johdon, W.E. (2001). Conservation Genetics of Jaguars: Phylogeography, population history and conservation genetics of jaguars (*Panthera onca*, Mammalia, Felidae). *Molecular Ecology*, Vol. 10, pp.65-79

Eizirik, E.; Haag, T.; Santos, A.S.; Salzano, F.M.; Silveira, L.; Azevedo, F.C.C. & Furtado, M.M. (2008). Jaguar Conservation Genetics. Cat News Vol. 4, pp. 31-34

Epps, C.W.; Palsbøll, P.J.; Weyhausen, J.D.; Roderick, G.K.; Ramey II, R.R. & McCullough, D.R. (2005). Highways Block Gene Flow and Cause a Rapid Decline in Genetic Diversity of Desert Bighorn Feline. *Ecology Letters*, Vol. 8, pp. 1029-1038

Faleiro, F.G. (2007). Marcadores genético-moleculares aplicados a programas de conservação e uso de recursos genéticos. Planaltina, DF: *Embrapa Cerrados*, pp.102

Ferreira, M.E. & Grattapaglia, D. (1998). Introdução ao Uso de Marcadores Moleculares em Análise Genética. 3. ed. *Embrapa, Brasília*, pp. 220

Foose, T.J. (1983). The relevance of captive populations to the conservation of biotic diversity. In: Genetics and conservation. Benjamin/Cummings Publishing Co., *Menlo Park*, pp. 374-401

Frankham, R.; Ballou, J.D. & Briscoe, D.A. (2002). Introduction to Conservation Genetics. New York: *Cambridige University Press*, pp. 617

Frankham, R.; Ballou, J.D. & Briscoe, D.A. (2008). Fundamentos de Genética da Conservação, *Editora SBG (Sociedade Brasileira de Genética)*, Ribeirão Preto, São Paulo, pp. 259

Galetti Jr., P.M.; Rodrigues, F.P.; Solé-Cava, A.; Miyaki, C.Y.; Carvalho, D.; Eizirik, E.; Veasey, E.A.; Santos, F.R.; Farias, I.P.; Vianna, J.A.; Oliveira, L.R.; Weber, L.I.; Almeida-Toledo, L.F.; Francisco, M.R.; Redondo, R.F.; Siciliano, S.; Del Lama, S.N.; Freitas, T.R.O.; Herbek, T. & Molina, W.F. (2008). Genética da conservação brasileira, pp.244-274. In: Fundamentos de Genética da Conservação. In: Frankham, R.; Ballou, J.D. & Briscoe, D.A. Ribeirão Preto, SP, *Editora SBG*, pp. 290

Grahn, R.A.; Kurushima, J.D.; Billings, N.C.; Grahn, J.C.; Halverson, J.L.; Hammer, E.; Ho, C.K.; Kun, T.J.; Levy, J.K.; Lipinski, M.J.; Mwenda, J.M.; Ozpinar, H.; Schuster, R.K.; Shoorijeh, S.J.; Tarditi, C.R.; Waly, N.E.; Wictum, E.J. & Lyons, L.A. (2011). Feline non-repetitive mitochondrial DNA control region database for forensic evidence. *Forensic Science International: Genetics,* Vol. 5, pp. 33–42

Grisolia, A.B.; Moreno, V.R.; Campagnari, F.; Milazzotto, M.P.; Garcia, J.F.; Adania, C.H. & Souza, E.B. (2007). Genetic diversity of microsatellite loci in *Leopardus pardalis, Leopardus wiedii* and *Leopardus tigrinus. Genetics and molecular research,* Vol. 6, No. 2, pp. 382-389

Gugolz, D.; Bernasconi, M.V.; Breitenmoser-Würsten, C. & Wandeler, P. (2008). Historical DNA reveals the phylogenetic position of the extinct Alpine lynx. *Journal of Zoology,* Vol. 275, pp. 201-208

Hertwig, S.T.; Schweizer, M.; Stepanow, S.; Jungnickel, A.; Böhle, U.R. & Fischer, M.S. (2009). Regionally high rates of hybridization and introgression in German wildcat populations (*Felis silvestris,* Carnivora, Felidae). *Journal Zoology System Evolution Research,* Vol. 47, No. 3, pp. 283-297

IUCN (2001). IUCN Red List Categories and Criteria. Version 3.1. Prepared by the IUCN Species Survival Commission. IUCN, Gland, Switzerland.

Janecka, J.E.; Helgen, K.M.; Lim, N.T.L.; Baba, M.; Izawa, M.; Boeadi, & Murphy, W.J. (2008). Evidence for multiple species of Sunda Colugo. *Current Biology,* pp. 18 - 21

Johson, W.E.; Eizirik, E.; Roelke-Parker, M. & O'Brien, S.J. (2001). Applications of genetic concepts and molecular methods to carnivore conservation. In: Gittleman, J.L.; Funk, S.M.; MacDonald, D.; Wayne, R.K. Carnivore conservation. *Cambridge University Press,* Cambridge, pp. 335-358

Johnson, W.E.; Eizirik, E.; Pecon-Slattery, J.; Murphy, W.J.; Antunes, A.; Teeling, E. & O'Brien, S.J. (2006). The late Miocene radiation of modern Felidae: a genetic assessment. *Science,* Vol. 311, pp. 73–77

Keller, L.F. & Waller, D.M. (2002). Inbreeding Effects in Wild Populations. *Trends in Ecology and Evolution,* Vol.17, pp.230-241

Lacy, R.C. (1997). Importance of Genetic Variation to the Viability of Mammalian Populations. *Journal of Mammalogy,* Vol. 78, pp. 320-335

Linnæus, C. (1758). Systema naturæ per regna tria naturæ, secundum classes, ordines, genera, species, cum characteribus, differentiis, synonymis, locis. Tomus I. Editio decima, reformata. *Holmiæ,* pp. 1-4

Linnaeus C. (1771). Mantissa Plantarum. Generum editionis VI, et Specierum editionis II, pp. 143-558

Mace, G.M.; Smith, T.B.; Brudford, M.W. & Wayne, R.K. (1996). An overview of issues. In: Molecular genetics approaches in conservation, *Oxford University Press,* New York, pp. 3- 21

Marker, L.; Wilkerson, A. & Sarno, R. et al. (2008). Molecular genetic insights on cheetah (*Acinonyx jubatus*) ecology and conservation in Namibia. *Journal of Heredity*, Vol. 99, pp. 2–13

Mattern, M.Y. & Mclennan, D. (2000). Phylogeny and speciation of felids. *Cladistics*, Vol. 16, No.2, pp.232-253

Menotti-Raymond, M.A. & O'Brien, S.J. (1993). Dating the genetic bottleneck of the African cheetah. Proceedings of the National Academy of Sciences, USA, Vol. 90, pp. 3172–3176

Menotti-Raymond, M.A. & O'Brien, S.J. (1995). Evolutionary conservation of ten microsatellite loci in four species of Felidae. *Journal Hered*, Vol. 86, pp. 319-322

Menotti-Raymond, M.A.; David, V.A. & Lyons, L.A. et al. (1999). A genetic linkage map of microsatellites in the domestic cat (*Felis catus*). *Genomics*, Vol. 57, pp. 9–23

Miotto, R.A.; Rodrigues, F.P.; Ciocheti, G. & Galetti Jr., P.M. (2007). Determination of the Minimum Population Size of Pumas (*Puma concolor*) Through Fecal DNA Analysis in Two Protected Cerrado Areas in the Brazilian Southeast. *Biotropica*, Vol. 39, No. 5, pp. 647–654

Miththapala, S.; Seidensticker, J. & O'Brien, J.S. (1996). Phylogeographic Subspecies Recognition in Leopards (*Panthera pardus*). *Molecular Genetic Variation Conservation Biology*, Vol. 10, No. 4, pp.1115–1132

MMA (Ministério do Meio Ambiente). (2008). Lista Oficial das espécies da Flora Brasileira Ameaçadas de Extinção. Avaliable from http://www.mma.gov.br/sitio

Mondol, S.; Karanth, K.U.; Kumar, N.S.; Gopalaswamy, M.A.; Andheria, A. & Ramakrishnan, U. (2009). Evaluation of non-invasive genetic sampling methods for estimating tiger population size. *Biological Conservation*, Vol. 142, pp. 2350-2360

Moreira, N.; Monteiro-Filho, E.L.A.; Moraes, W.; Swanson, W.F.; Graham, L.H.; Pasquali, O.L.; Gomes, M.L.F.; Morais, R.N.; Wildt, D.E. & Brown, J.L. (2001). Reproductive steroid hormones and ovarian activity in felids of the Leopardus genus. *Zoo Biology*, pp. 103-106

Moreno, V.R.; Grisolia, A.B.; Campagnari, F.; Milazzotto, M.; Adania, C.H.; Garcia, J.F.; Souza, E.B. (2006). Genetic variability of Herpailurus yagouroundi, Puma concolor and Panthera onca (Mammalia, Felidae) studied using Felis catus microsatellites". Genetics and Molecular Biology, Vol. 29, No.2, pp. 290-293

Mukherjee, S.; Krishnan, A.; Tamma, K.; Home, C.; R, N.; Joseph, S.; Das, A. & Ramakrishnan, U. (2010). Ecology Driving Genetic Variation: A Comparative Phylogeography of Jungle Cat (*Felis chaus*) and Leopard Cat (*Prionailurus bengalensis*) in India. *Plosone*, Vol. 5, No. 10

Myers, N. et al. (2000). Biodiversity hotspots for conservation priorities. *Nature*, Vol. 403, pp. 853–58

Napolitano C.; Bennett, M.; Johnson, W.E.; O'brien, S.J.; Marquet, P.A.; Barría, I.; Poulin, E. & Iriarte, A. (2008). Ecological and biogeographical inferences on two sympatric and enigmatic Andean cat species using genetic identification of faecal samples. *Molecular Ecology*, Vol. 17, pp. 678–690td

Neigel, J.E. (1996). Estimation of effective population size and migration parameters from genetic data. In: Molecular genetic approaches in conservation genetics. New York: *Oxford University Press*, pp. 329–346

Nowak, R.M. (1999). Walker's Mammals of the World, 6 ed., Baltimore and London, *The Johns Hopkins University Presser*.

Nowell, K. & Jackson, P. (1996). Status Survey and Conservation Action Plan: Wild Cats. IUCN/SSC Cat Specialist Group, *Gland, Switzerland*.

O'Brien, S.J.; Wildt, D.; Goldman, D.; Merril, C. & Bush, M. (1983). The cheetah is depauperate in genetic variation. *Science*, Vol. 221, pp. 459–462

O'Brien, S.J.; Wildt, D.; Bush, M. et al. (1987). East African cheetahs: evidence for two population bottlenecks? Proceedings of the National Academy of Sciences, USA, Vol. 84, pp. 508-511

O'Briens, J. (1990). Genetic Maps: Locus Maposf Complex Genomes, Ed. 5. Cold Spring Harbor Laboratory, *Cold Spring Harbor*, N.Y.

O'Brien, S.J.; Martenson, J.S.; Miththapala, S.; Janczewski, et al. (1996). Conservation genetics of Felidae. In: Conservation genetics: case histories from nature. *Chapman and Hall*, New York, pp. 50-74

Oliveira, T.G. (1994). Neotropical cats: ecology and conservation, São Luís: EDUFMA. 244p.

Perez-Sweeney, B.M.; Rodrigues, F.P. & Melnick, D. (2003). Metodologias moleculares utilizadas em genética da conservação. In: Métodos de estudos em Biologia da Conservação e Manejo da vida silvestre. Cullen Jr., L.; Rudran, R.; Valladares-Padua, C.; Santos, A.J. et al. – Curitiba: Ed. da UFPR; *Fundação O Boticário de Proteção à Natureza*, pp. 667

Primack, R.B. & Rodrigues, E. (2001). Biologia da Conservação, 1. ed, *Midiograf*, Londrina, pp. 328

Redford, K.H. & Eisenberg, J.F. (1992). Mammals of the Neotropics, Vol. 2: the Southern Cone: Chile, Argentina, Uruguay, Paraguay. *University of Chicago Press*, Chicago

Russello, M.A.; Gladyshev, E.; Miquelle, D. & Caccone, A. (2004). Potential genetic consequences of a recent bottleneck in the Amur tiger of the Russian far east. *Conservation Genetic*, Vol. 5, pp. 707–713

Schipper, J.; Hoffmann, M.; Duckworth, J.W. & Conroy, J. (2008). The 2008 IUCN red listings of the world's small carnivores. *Small Carnivore Conservation*, Vol. 39, pp. 29-34

Shankaranarayanan, P.; Banerjee, M.; Kacker, R.K.; Aggarwal, R.K. & Singh, L. (1997). Genetic variation in Asiatic lions and Indian tigers. *Electrophoresis*, Vol. 18, pp. 1693-1700

Solé–Cava, A.M.(2001). Biodiversidade molecular e genética da conservação. In: Matioli, S.R. (2001). Biologia Molecular e Evolução. Ribeirão Preto: *Holos Editora*, pp. 171–192

Trigo, T.C.; Freitas, T.R.O.; Kunzler, G.; Cardoso, L.; Silva, J.C.R.; Johnson, W.E.; O'Brien, S.J.; Bonatto, S.L. & Eizirik E. (2008). Inter-species hybridization among Neotropical cats of the genus Leopardus, and evidence for an introgressive hybridzone between *L. geoffroyi* and *L. tigrinus* in southern Brazil. *Molecular Ecology*, Vol. 17, pp. 4317–4333

Wajntal, A.; Miyaki, C.Y. & Menck, C.F.M. (1995). Usando técnicas de DNA para preservar aves em extinção. *Ciência Hoje*, Vol. 19, No. 111, pp. 30–38

Weber, J.L.; May, P.E. (1989). Abundant class of human DNA polymorphisms which can be typed using the polymerase chain reaction. *Am J Hum Genet*, Vol. 44, pp. 388-396

Wilson, D.E. & Reeder, D.A.M. (1993). Mammalia species of the world: a taxonomy and geographic reference. 2. ed. Washington DC: *Smithsonian Institution Press*, pp. 546

Wilting, A.; Christiansen, P.; Kitchener, A.C.; Kemp, Y.J.M.; Ambuf, L. & Fickel, J. (2011). Geographical variation in and evolutionary history of the Sunda clouded leopard (*Neofelis diardi*) (Mammalia: Carnivora: Felidae) with the description of a new subspecies from Borneo. *Molecular Phylogenetics and Evolution*, Vol. 58, pp. 317–328

7

Genetic Diversity and Evolution of Marine Animals Isolated in Marine Lakes

Naoto Hanzawa[1], Ryo O. Gotoh[1], Hidekatsu Sekimoto[1], Tadasuke V. Goto[1],
Satoru N. Chiba[2], Kaoru Kuriiwa[3] and Hidetoshi B. Tamate[1]
[1]Department of Biology, Faculty of Science, Yamagata University, Yamagata
[2]Center for Molecular Biodiversity Research
National Museum of Nature and Science, Tokyo
[3]Department of Zoology, National Museum of Nature and Science, Tokyo
Japan

1. Introduction

How do marine organisms genetically differentiate and speciate in illimitable oceans? It is considerably difficult to obtain clear answers to this question due to the following reasons. Marine organisms that reproduce by releasing numerous eggs and larvae are able to disperse over large distances and can therefore be distributed over large geographic areas. Such marine organisms have a large population size, gene flow between distant populations occurs frequently, and interspecies hybridization sometimes occurs (Kuriiwa et al., 2007). Even geographically well-separated populations may be connected genetically, because there are few barriers to prevent gene flow in the oceans (Mayr, 1954; Palumbi, 1994). In contrast to the open ocean environment, the marine lakes of Palau (Western Caroline Islands), which are surrounded entirely by land and isolated from the sea, provide unique local environments for genetic differentiation of marine organisms (Dawson and Hamner, 2005; Gotoh et al., 2009; Goto et al., 2011). We have focused on marine lakes as isolated marine environments and have conducted continuous evolutionary studies of marine organisms in the Palau Islands for the past 13 years.

The Palau archipelago, which consists of volcanic islands to the north and a large number of elevated limestone islands to the south (U.S. Geological Survey, 1956; Hamner and Hamner, 1998), is surrounded by multiple hard-coral atolls. The adjacent waters of the Palau Islands are known as a hotspot with the highest level of marine species diversity (Allen, 2008). There are approximately 70 marine lakes on the limestone islands — commonly called the "Rock Islands". It is thought that the marine lakes, which are small bodies of sea water in embayments and depressions on the limestone islands, have been gradually formed by floods resulting from rising sea levels after the Last Glacial Maximum (approx. 18,000 years ago); these waters were isolated from the surrounding barrier-reef lagoons and became sea-level marine lakes (Dawson and Hamner, 2005). Most of these lakes have been avoided by the local population due to the treacherous surface of the surrounding karst, lack of fresh water on the islands, and persistent myths that the marine lakes are haunted (Hamner and Hamner, 1998). Therefore, the ecosystems of most of the marine lakes have not yet been disturbed by human activities. The fauna and flora in most of the marine lakes are quite different from those in the oligotrophic lagoons that have an abundance of hard corals; some diagnostic species of fish, mollusks, jellyfish, sponges, and green algae are observed in the lakes (Dawson and Hamner, 2005).

Thus, the marine lakes of Palau can be regarded as isolated and untouched marine environments, where the inhabitants confined to each lake are likely to develop into genetically distinct populations within a relatively small geographic range. Indeed, unique patterns of speciation and adaptation have been observed for jellyfish (Dawson and Hamner, 2005), sea anemone (Fautin and Fitt, 1991), foraminifera (Lipps and Langer, 1999; Kawagata et al., 2005), and bacteria (Venkateswaran et al., 1993) in the marine lakes; however, little is known about the evolutionary features of the nektonic and benthic animals in the lakes. Therefore, we focused on the fish and bivalve species in the marine lakes and analyzed their genetic diversity, differentiation, and population structures in comparison with those of related species, or different populations of the same species, that inhabit the barrier-reef lagoons.

In this chapter, we firstly introduce the general features of genetic diversity and evolution of the coral fish inhabiting the barrier-reef lagoons. Secondly, we discuss the genetic diversity and differentiation of marine lake populations of fish and bivalves inhabiting different marine lakes in comparison with those of the outer lagoons. Finally, we discuss genetic differentiation and speciation with respect to the marine lake model as an isolated marine environment.

2. Genetic diversity and evolution in coral reefs

Marine organisms that reproduce by releasing numerous eggs and planktonic larvae are able to disperse over considerable distances via large-scale currents; indeed, many adult-stage nektons frequently migrate across oceans. Therefore, marine organisms often show low levels of genetic differentiation even over large geographical ranges (Grant and Bowen, 1998), because ocean currents and/or the apparent lack of physical barriers to movement appear to facilitate extensive gene flow (Avise, 2000; Palumbi, 1994).

However, various groups of marine organisms have speciated and actually show a high level of species diversity, especially on tropical coral reefs. Here, we introduce the genetic diversity and multiple natural hybridizations of rabbitfish (Teleostei: Siganidae) distributed on tropical and subtropical coral reefs (Kuriiwa et al., 2007).

Twenty-two rabbitfish species in the genus *Siganus*, described in the Western Pacific (Woodland, 1990; Randall and Kulbicki, 2005), are easily identified on the basis of species-specific coloration. We conducted phylogenetic analyses among 19 nominal *Siganus* species based on mitochondrial cytochrome b gene and nuclear ribosomal DNA internal spacer 1 (ITS1) sequences to infer their phylogenetic relationships and degree of genetic differentiation among species. We predicted that reproductive isolation is completely established among siganid species because teleost fish with opsin genes, such as RH1, RH-2, LWS, and SWS (Register et al., 1994; Yokoyama and Yokoyama, 1996; Seehausen et al., 2008), can visually discriminate individuals of the same species according to their coloration. As shown in Fig. 1, phylogenetic analyses using 2 different DNA markers indicated that most species were sufficiently genetically differentiated. However, we were surprised to

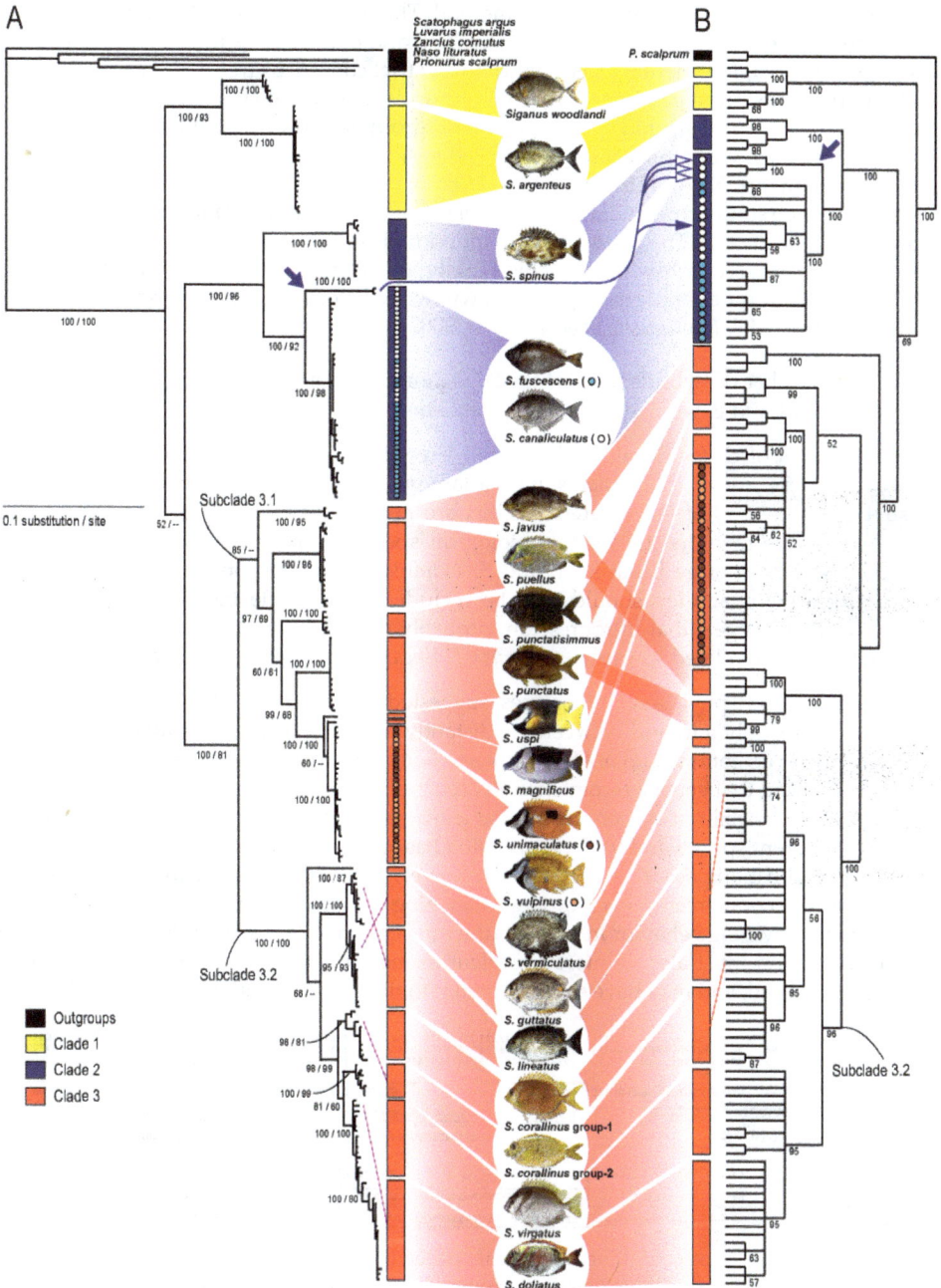

Fig. 1. Phylogenetic trees of siganid fish. (a) A 50% majority rule consensus tree derived from Bayesian analysis of mitochondrial (mt) cytochrome b sequences from 19 siganid species and 5 outgroup species with the GTR + I + Γ model. The topology was basically

identical to that from the neighbor-joining method. The numerals below the nodes indicate the posterior probabilities of the Bayesian trees (left) and bootstrap values (%) from 1,000 replicates in the neighbor-joining tree (right). (b) A strict consensus tree of 9712 trees resulting from maximum parsimony analysis of nuclear ribosomal DNA (internal spacer 1, ITS1) sequences from 19 siganid species and an outgroup species. The numerals below the nodes indicate the bootstrap values (%) from 100 replicates. The broken line indicates individuals with mtDNA or ITS1 sequences different from those expected from their morphology. The solid arrow indicates the Indian-clades of *S. canaliculatus* and the arrowheads indicate the ITS1 types found in individuals of the clade (open arrowheads, Indian-ITS1-types; solid arrowhead, Pacific-ITS1-type). See text for details. Discordance between the phenotypes and genotypes is shown by the broken lines and arrowheads (From K. Kuriiwa et al., *Mol. Phylogenet. Evol.* 45:69-80, 2007).

find a large number of hybrids in 11 of the 20 Siganidae species (2 genetically differentiated species, groups 1 and 2, were included in *S. corallinus*).

The populations of *S. fuscescens*, with a small irregular-dotted skin pattern and found in temperate waters, and *S. canaliculatus*, with a large white-dotted skin pattern and found in the subtropical waters of the Western Pacific, are genetically mixed, indicating that these species should be included in a single biological species. Similarly, the populations of *S. unimaculatus*, with a large black spot, and *S. vulpinus*, without spots, are also genetically mixed, indicating that they should also be included in a single biological species. The analyses further suggest that introgression, probably caused by partial gene flow, occurred between the closely related species *S. guttatus - S. lineatus* and *S. virgatus - S. doliatus*.

On the other hand, phylogenetic analyses suggested that a morphologically unidentified individual is probably a hybrid between the distantly related species *S. corallinus* group 2 *and S. puellus*. However, comparison of ITS-1 sequences among *S. corallinus* group 2, *S. puellus*, and the hybrid showed that genetic recombination occurred in the hybrid in the region (Fig. 2). The data reveal the possibility that genome rearrangements occurred after hybridization between the pair of different species and further speciation has started. Little is known about why such frequent hybridizations occur in rabbitfish; however, it must be related to their sympatric distribution in coral reefs, overlapping spawning season among most species, and simultaneous spawning behavior in which numerous eggs and larvae are released. Such large scale hybridization must also be more prevalent in marine fish that inhabit coral reefs than previously assumed, and may have some relevance to their diversification.

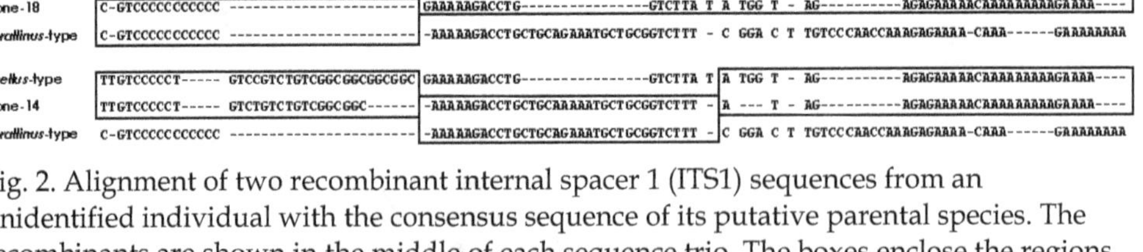

Fig. 2. Alignment of two recombinant internal spacer 1 (ITS1) sequences from an unidentified individual with the consensus sequence of its putative parental species. The recombinants are shown in the middle of each sequence trio. The boxes enclose the regions where the recombinant is identical to one of the parents. Only the variable regions from the alignment are shown (From K. Kuriiwa et al., *Mol. Phylogenet. Evol.* 45:69-80, 2007).

3. Genetic diversity and evolution in the marine lakes of the Palau Islands

The effect of geographical isolation on speciation has been studied in terrestrial and freshwater organisms (Chiba, 1998; Kliman et al., 2000; Caccone et al., 2001; Calsbeek and Smith, 2003; Meyer et al., 1990; Barluenga et al., 2006); however, little is known about what effect this may have on marine organisms. In this section, we introduce the general aspects of "the marine lakes as a model for isolated marine environments" in the Palau Islands and the fauna of the marine lakes, and show the peculiar evolution of some marine animals as inferred from their genetic population structures.

3.1 Geological and limnological aspects of the marine lakes of the Palau Islands

The Palau Islands are located at 7°30' N, 134°30' E in the Western Pacific (Fig. 3) and are composed of approximately 350 small islands (U. S. Geological Survey, 1956). The limestone islands, called the "Rock Islands", are covered with thick tropical forest and are located in the central and southern parts of Palau (Fig. 3). These islands contain more than 70 marine lakes (Fig. 4) (Hamner and Hauri, 1981).

As shown in Fig. 5, it can be geologically inferred that the marine lakes were gradually formed through the following steps. Firstly, the limestone derived from hard corals was shaped and uplifted during the Miocene and Pleistocene periods. Secondly, the islands were eroded by rain and wind, and numerous depressions were formed. Finally, with the rising sea level of the end of the Last Glacial Maximum, seawater flooded the depressions through fissures or tunnels (Hamner and Hauri, 1981; Hamner and Hamner, 1998). Geologically, the marine lakes were estimated to have formed in chronological series, namely, the deeper depressions flooded first (~12,000 years ago) and the shallower depressions flooded later (~5,000 years ago) (Dawson and Hamner, 2005).

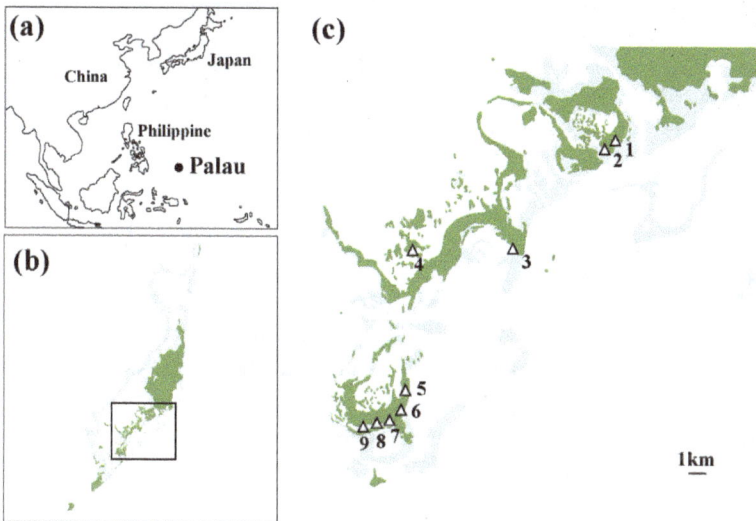

Fig. 3. Location of the Republic of Palau (a) and the surveyed major marine lakes 1–9 in (c) of the Palau Islands (b). The green and light blue portions indicate the islands and coral reefs, respectively.

Fig. 4. (a), (b): Major marine lakes in the Palau Islands; (c), (d): Their shores covered with mangroves.

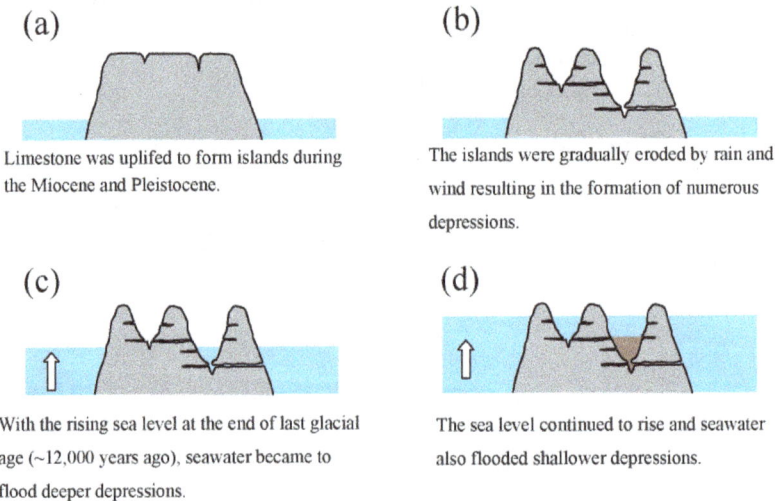

(a)

Limestone was uplifed to form islands during the Miocene and Pleistocene.

(b)

The islands were gradually eroded by rain and wind resulting in the formation of numerous depressions.

(c)

With the rising sea level at the end of last glacial age (~12,000 years ago), seawater became to flood deeper depressions.

(d)

The sea level continued to rise and seawater also flooded shallower depressions.

Fig. 5. (a) – (d). Geologically inferred formation process of the marine lakes. The brown portion indicates an anaerobic layer.

We surveyed more than 25 marine lakes in Palau and show the limnological characteristics of the 9 major marine lakes in Table 1. The major marine lakes of the Palau Islands are limnologically holomictic or meromictic lakes. As shown in Fig. 6, the holomictic marine lakes have retained hydraulic connections to the surrounding lagoons either at the surface or through submarine tunnels and fissures in the fenestrated karst (Hamner et al., 1982), and water exchange and circulation regularly occur by tidal movement. On the other hand, the meromictic marine lakes have no apparent connection to the lagoons (Fig. 6). There are fewer tunnels or fissures where large organisms can pass through, although water exchange occurs through fissures and micropores in the limestone. In the meromictic lakes, thorough water-circulation does not occur because mixing by wind is reduced by the jungle-covered karst ridges. Topographic protection from wind, heavy regular rain throughout the year with precipitation exceeding evaporation, and modest tidal exchange produce stratified brackish waters above permanently anoxic saline hypolimnia (Hamner and Hamner, 1998). Mangrove forests develop on their coasts, and some fish species (e.g., *Apogonidae* spp., *Atherinidae* spp., and *Gobiidae* spp.), jellyfish (e.g., *Aurelia* spp. and *Mastigias* spp.), mussels (*Brachidontes* spp.), many sponges (e.g., *Haliclona* spp.), and sea anemones (e.g., *Entacmaea medusivora*) live in these marine lakes. Even if there were tunnels or fissures in deeper areas, living organisms from the open water could not have colonized the lakes because the deep anoxic layers, which include fatal chemicals such as hydrogen sulfide, would prevent their migration. Therefore, the organisms inhabiting meromictic lakes are thought to have been isolated since the lakes were formed and to have evolved endemically in each lake.

Site	Abbreviation	Depth (m)	Length (m)		Physical structure
			Long axis	Width	
1 Uet era Ngerumeuangel, Koror Is.	NLK	38	270	210	Meromictic
2 Goby Lake, Koror Is.	GLK	15	195	110	Meromictic
3 Shrimp Lake, Ngerktabel Is.	SHN	5	100	55	Meromictic
4 Uet era Ongael, Ongael Is.	OLO	4	150	100	Holomictic
5 North Cassiopea Lake, Mecherchar Is.	NCM	4	135	55	Holomictic
6 Jellyfish Lake, Mecherchar Is.	JFM	32	420	150	Meromictic
7 Big crocodile Lake, Mecherchar Is.	BCM	22	575	290	Meromictic
8 Spooky Lake, Mecherchar Is.	SPM	14	230	65	Meromictic
9 Clear Lake, Mecherchar Is.	CLM	30	320	225	Meromictic

Table 1. Limnological characteristics of the surveyed major marine lakes. Site Nos. 1–9 correspond to those shown in Fig. 3 (c).

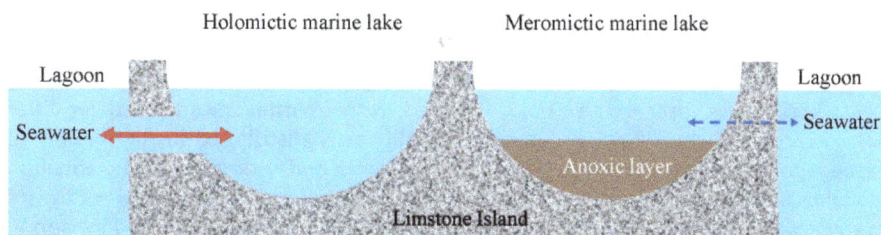

Fig. 6. Schematic figures of the meromictic and holomictic marine lakes in the Palau Islands.

3.2 Fauna in the marine lakes

As mentioned above, species diversity is extremely high in the Palau Islands and numerous fish and benthic animal species inhabit the coral reefs. On the other hand, unique fauna, which is quite different from that of the coral reefs, is maintained in the marine lakes. As shown in Table 2, restricted fish, bivalves, and jellyfish species inhabit the marine lakes and these marine lake-specific diagnostic species are not distributed in all of the marine lakes. The numbers of individuals are extremely high in whichever lakes the species inhabit. In a recent report on copepods, we demonstrated that the species diversity in the marine lakes is clearly lower than in the open sea around the Palau Islands, particularly for copepod species preferring to low-salinity waters, such as *Oithona dissimilis* and *Bestiolina similis*, that inhabit the marine lakes (Saitoh et al., 2011).

Species	Korol Is. NLK	Korol Is. GLK	Ngerktabel Is. SHL	Ongael Is. ONG	Mecherchar Is. NCM	Mecherchar Is. JFM	Mecherchar Is. BCM	Mecherchar Is. SPM	Mecherchar Is. CLM
Cnidaria									
Aurelia sp.4	++			(++)	(++)	++			
Mastigias spp.	++	++		(++)	(++)				++
Cassiopea sp.			+	(++)	+				
Mollusca									
Brachidontes spp.	++	++	++	++	++	++	++	++	++
Chordata									
Atherinomorus endrachtensis				++	++	++	++	+	+
Sphaeramia orbicularis	++				+	++			++
Apogon lateralis				++	++		++	++	++
Acentrogobius janthinopterus	++	++	++		++		++	++	++
Exyrias puntang	++	++		++	++				

++ numerous; + moderate; ( ) indicates large scale fluctuation in the number of individuals

Table 2. Dominant species and their population size in the marine lakes of Palau.

A moon jellyfish, *Aurelia* sp. 4 (Dawson and Jacobs, 2001) only inhabits the marine lakes, while different biological species, *Aurelia* sp. 3 and *Aurelia* sp. 6, only inhabit the surrounding lagoons. Dawson (2005) also described 5 new subspecies of *Mastigias* jellyfish distributed in different marine lakes of Palau. The subspecies are morphologically different, particularly for the shape of their oral arms, from *Mastigias papua*, which is distributed in the open sea of the Pacific. We have also confirmed that each of the 5 subspecies is slightly morphologically different from each other, whereas the morphology of 4 subspecies, except *M. papua etpisoni*, has changed every year. Thus, the 5 subspecies of *Mastigias* jellyfish have been geographically isolated in each marine lake and have evolved, whereas the 4 subspecies are pleiotropic.

Mussels, *Brachidontes* spp., are also only observed in the marine lakes located on different islands. A cardinal fish, *Apogon* sp., is also probably only distributed in the marine lakes. Numerous gobies, such as *Acentrogobius janthinopterus* and *Exyrias puntang*, inhabit the marine lakes, although small numbers of individuals are occasionally observed in the lower reaches of rivers flowing to the lagoons. Several species of sponges (e.g., *Haliclona* spp.), sea anemones (e.g., *Entacmaea medusivora*), sea cucumbers, gastropods, and simple and colonial ascidians are also only found in the marine lakes.

On the other hand, numerous individuals of a cardinal fish, *Sphaeramia orbicularis*, and a silverside, *Atherinomorus endrachtensis*, are found in the marine lakes and outer lagoons.

Furthermore, we found distinctive phenomena indicating that the number of individuals of jellyfish widely fluctuated. First we observed massive deaths of *Aurelia* and *Mastigias* spp. in Jellyfish Lake, Mecherchar Island. A large scale El Niño-Southern Oscillation (ENSO)

occurred during 1997–1998 around Palau, resulting in dry weather without any rain for 6 months, an abnormal rise in the seawater temperature, and the extinction of many hard corals and other animals dependent on the corals (personal communication from Carp Corporation staff). We could not observe any adults of *Aurelia* and *Mastigias* spp. in Jellyfish Lake during our expedition at the beginning of 1999. However, we could find some adults of both species during our expedition at the end of 1999, and the number of adults had explosively increased within a few years after the ENSO.

We also observed similar wide fluctuations in the number of individuals in Ongael Lake. This lake is shallower (~4 m deep) than the other marine lakes and its environmental conditions must, therefore, be more unstable. Three jellyfish species of the genera *Aurelia, Mastigias,* and *Cassiopea* inhabit this lake; however, the number of adult individuals of each species has drastically fluctuated on a cycle of a few years. Namely, adult individuals of any species continue to increase in number for a few years, and suddenly become extinct. But, the number of adults suddenly begins to increase some years later. Thus, fluctuations in the number of adults are common in jellyfish populations. However, the habitats and dynamics of their polyps are unknown. If we can survey them, we may be able to analyze how such short term fluctuations affect the genetic diversity in their populations.

3.3 Genetic diversity and evolution in the marine lake populations

3.3.1 Genetic population structures and divergence of a cardinal fish, *S. orbicularis*

S. orbicularis inhabits meromictic marine lakes and lagoons (Table 3). We determined the complete sequence of the mitochondrial control region (CR: 824 bps) and compared the genetic structure of marine lake and lagoon populations (Gotoh et al., 2009).

CR polymorphisms in *S. orbicularis* A total of 17 haplotypes were detected from 157 individuals collected from 3 meromictic marine lakes and 3 lagoon sites. The base substitutions included 18 transitions and one indel. The base composition of the different haplotypes were 29.4-29.7% for A, 29.9-30.1% for T, 16.9-17.2% for G and 23.2-23.5% for C, and no remarkable deviation in the composition was observed among haplotypes. As for many other teleost fishes, most substitutions were observed in the 5′ terminal side, which is known to be a hyper variable region (Lee et al., 1995). Only 5 of all haplotypes were shared among different populations (Table 3). The haplotype So01 that has the highest frequency among lagoon populations was detected in two individuals from the JFM lake, while none of the other haplotypes were shared between lagoon and marine lake populations.

Genetic differentiation among the marine lake populations in *S. orbicularis* The haplotype diversity (h) and the nucleotide diversity (π) are shown in Table 4. In marine lake populations, the values ranged from 0.067 to 0.465 for h and from 0.008 to 0.113 for π. In lagoon populations, these values ranged from 0.423 to 0.815 and from 0.103 to 0.198, respectively. In pooled lagoon samples, these values were 0.618 and 0.162, respectively.

Lagoon populations share some haplotypes (Table 3). The AMOVA without group design shows a high percentage of variation within populations (90.20%: Table 5) and a significant pairwise Φ_{ST} of 0.098 (P < 1e-03). The AMOVA with two groups calculated with SAMOVA does not show a significant value of F_{CT} (0.159, P = 0.348; Table 5).

Haplotype	3	31	37	132	173	194	221	332	359	478	510	530	568	570	664	Marine lake NLK	Marine lake CLM	Marine lake JFM	PPE	Lagoon JFOS	Lagoon CMR	Accession No.
So01	C	T	C	G	C	A	G	C	G	A	A	A	G	A	–	18 (0.69)		2 (0.08)	18 (0.70)	19 (0.76)	8 (0.33)	AB252837
So02	·	·	·	·	·	G	A	·	·	G	·	G	·	G	·							AB252838
So03	·	·	·	·	·	·	A	·	·	·	·	·	·	G	·		29 (0.97)					AB252839
So04	·	·	·	·	·	G	A	·	·	·	·	·	·	G	·			23 (0.88)				AB252840
So05	·	·	·	A	·	G	A	T	·	·	·	·	·	G	·	7 (0.27)						AB252841
So06	·	·	·	·	·	G	A	·	·	·	·	·	·	G	·	1 (0.04)						AB252842
So07	·	·	·	·	·	·	A	·	A	G	·	·	·	G	·		1 (0.03)					AB252843
So08	·	·	·	·	·	·	A	·	A	·	·	·	·	G	·				1 (0.04)			AB252844
So09	·	·	·	·	·	·	·	·	·	·	·	·	·	G	·				1 (0.04)			AB252845
So10	·	·	·	·	·	·	·	·	A	·	G	·	·	G	·				2 (0.08)		4 (0.17)	AB252846
So11	·	·	·	·	·	·	A	·	·	·	·	·	·	G	·				1 (0.04)	1 (0.04)	2 (0.08)	AB252847
So12	T	·	·	A	·	·	A	·	·	·	·	·	·	G	·				1 (0.04)		1 (0.04)	AB252848
So13	·	·	·	·	·	G	A	·	·	·	·	·	·	G	·				1 (0.04)		4 (0.17)	AB252849
So14	·	·	·	·	·	·	A	·	·	·	·	·	·	G	C			1 (0.04)		2 (0.08)		AB252850
So15	·	·	·	·	·	·	·	·	·	·	·	·	·	G	·				1 (0.04)	2 (0.08)	5 (0.21)	AB252851
So16	·	C	·	·	·	·	·	·	·	·	·	·	·	·	·				1 (0.04)			AB252852
So17	·	·	·	·	·	·	·	·	·	·	·	·	A	·	·					1 (0.04)		AB252853

Table 3. Variable sites of the 17 haplotypes and number of individuals for each haplotype by sampling area. Abbreviations of marine lakes are common with Tabales 1 and 2. PPE: a port of Peleliu Island; JFOS: outside of Jellyfish Lake; CMR: mouth of Comet River (Modified from R. O. Gotoh et al., *Genes Genet. Syst.* 84:287-295, 2009)

Location	Sample size n	H	h	π (%)	Tajima's D D	P	Goodness-of-fit test SSD	P
Marine lakes								
NLK	26	3	0.465±0.086	0.113±0.089	0.447	0.689	0.174	0.100
CLM	30	2	0.067±0.061	0.008±0.019	-1.147	0.038*	0.000	0.250
JFM	26	3	0.219±0.103	0.063±0.061	-1.071	0.180	0.027	0.150
Lagoon								
PPE	26	8	0.526±0.118	0.147±0.108	-0.684	0.293	0.383	0.000*
JFOS	25	5	0.423±0.119	0.103±0.084	-0.543	0.338	0.019	0.550
CMR	24	6	0.815±0.044	0.198±0.135	0.639	0.763	0.020	0.100
LAG[a]	75	10	0.618±0.060	0.162±0.113	-0.501	0.330	0.035	0.180

H: number of haplotypes; h: haplotype diversity; π: nucleotide diversity; D: Tajima's D value; SSD: sum of squqred deviations.
*: P < 0.05.
[a]: LAG consists of pooled samples of all lagoon individuals.

Table 4. Control region sequence diversity, Tajima's D and goodness-of-fit tests (Modified from R. O. Gotoh et al., *Genes Genet. Syst.* 84:287-295, 2009).

The values of pairwise Φ_{ST} ranged from 0.429 to 0.870 among marine lake populations, from -0.008 to 0.181 among lagoon populations, and from 0.531 to 0.848 between marine lake populations and lagoon populations. Almost all of these values were significant at a level of 0.05% (Table 6).

Structure tested	Source of variation	Observed partition % total	Φ statistics	P
1. One gene pool				
	Among populations	9.8	$\Phi_{ST} = 0.098$	0.000
	Within populations	90.2		
2. Two gene pool (PPE, JFOS)(CMR)[a]				
	Among groups	15.9	$\Phi_{CT} = 0.159$	0.348
	Among populations	-1.1	$\Phi_{SC} = -0.013$	0.000
	Within populations	85.2	$\Phi_{ST} = 0.1479$	0.000

[a] By maximizing Φ_{CT}

Table 5. Multiple hierachial analyses of control region in samples of *S. orbicularis* (From R. O. Gotoh et al., *Genes Genet. Syst.* 84:287-295, 2009).

	NLK	CLM	JFM	PPE	JFOS	CMR
NLK		0.000	0.000	0.000	0.000	0.000
CLM	0.844**		0.000	0.000	0.000	0.000
JFM	0.429**	0.870**		0.000	0.000	0.000
PPE	0.713**	0.781**	0.669**		0.448	0.014
JFOS	0.763**	0.848**	0.740**	-0.008		0.002
CMR	0.612**	0.659**	0.531**	0.094*	0.181*	

* P < 0.05, ** P < 0.001

Table 6. Population pairwise Φ_{ST} values for control region (below the diagonal) and P values (above the diagonal) (Modified from R. O. Gotoh et al., *Genes Genet. Syst.* 84:287-295, 2009).

Phylogenetic relationships among haplotypes in *S. orbicularis* We constructed the statistical parsimony network as shown in Fig. 7. For the CLM population, haplotype So07 was derived from the high frequency haplotype So03. In the JFM population, the major haplotype So04 connected to haplotype So14 through one indel and with haplotype So01, which was shared with lagoon populations, through three substitutions. All the haplotypes detected from the NLK population were derived from haplotype So04. Finally, haplotypes recognized in lagoon populations connected to each other through one to four substitutions.

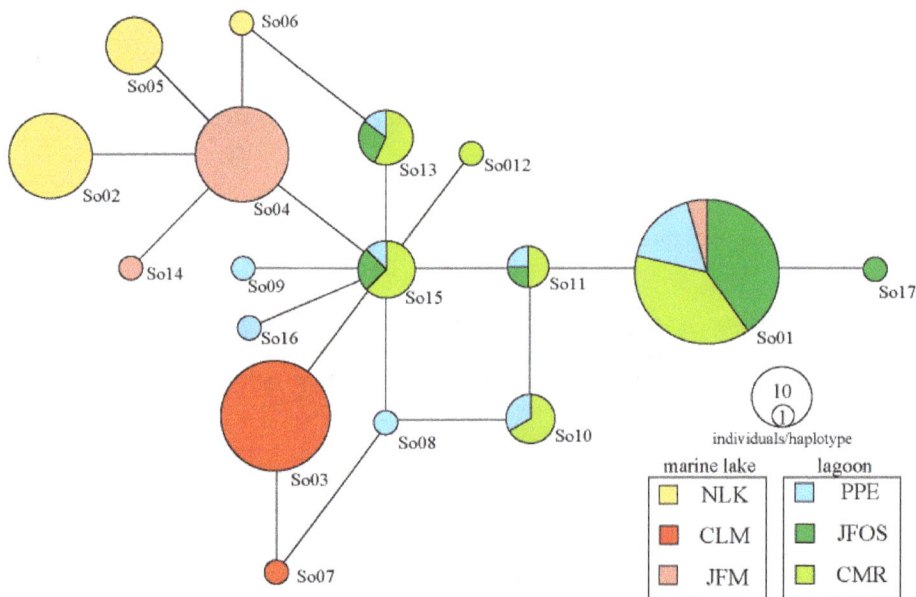

Fig. 7. Statistical parsimony network for haplotypes detected in marine lake and lagoon populations of *S. orbicularis*. Size of circles represents the frequency of each haplotype. Letters beside the circles indicate each labeled haplotype (Modified from R. O. Gotoh et al., *Genes Genet. Syst.* 84:287-295, 2009).

Neutrality and demographic history in populations of *S. orbicularis* Results of Tajima's D and goodness-of-fit tests are shown in Table 4. The values of Tajima's D test were not significant except for the CLM population. In the goodness-of-fit test, only the value of the PPE population was significant. In the mismatch distribution analysis (Fig. 8), the CLM and JFM populations showed similar results where the peak of frequency distribution in the number of nucleotide substitution is near zero. The NLK population showed two distinctive peaks. The pooled lagoon population showed a similar pattern with a slightly moderate curve compared with marine lake populations.

Historical changes of lagoon populations in *S. orbicularis* The results of the AMOVA could not sufficiently clarify the genetic structure of the lagoon populations. However, the pairwise Φ_{ST} values between CMR and PPE, and CMR and JFOS indicate that some genetic divergence has occurred. This divergence is likely due to mouth breeding of *S. orbicularis*, a particular behavior well known among apogonids. In these species, males brood the eggs in their mouth for approximately eight days before releasing larvae (Myers, 1999). After a short pelagic stage, the juveniles settle down in dark habitats among mangrove roots, limestone rocks, or in shallow piers along the shoreline. Therefore, large-scale dispersal by tidal currents is not likely to occur widely. Indeed, previous studies on two fish species, Banggai cardinalfish, *Pterapogon kauderni* and black surfperch, *Embiotoca jacksoni*, lacking a pelagic larval phase revealed low levels of gene flow and strong phylogeographic breaks within their 100 km to 1,000 km geographic ranges (Bernardi and Vagelli, 2004; Bernardi, 2000).

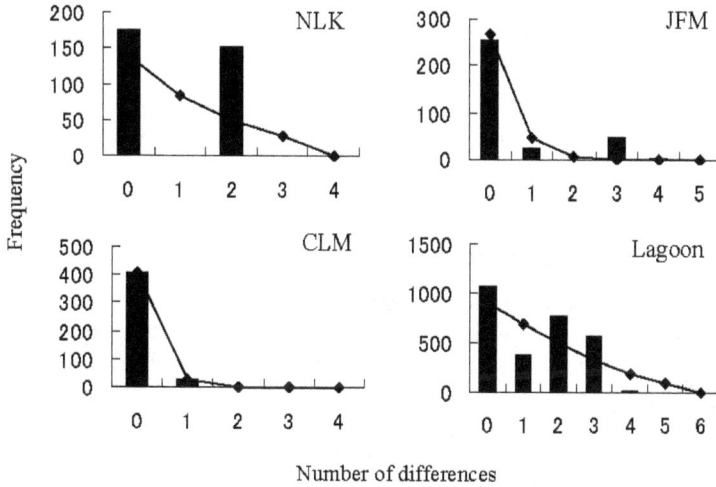

Fig. 8. Mismatch distribution analysis based on haplotypes of control region in each population of *S. orbicularis*. The vertical bars are the observed distribution of mismatches and the line represents the expected distribution under the sudden-expansion model (Modified from R. O. Gotoh et al., *Genes Genet. Syst.* 84:287-295, 2009).

Lagoon populations showed high haplotype and low nucleotide diversities. Grant and Bowen (1998) have analyzed the genetic diversity in populations of sardines and anchovies based on mtDNA sequences and classified them into four patterns. According to their classification, the lagoon populations in *S. orbicularis* correspond to Category 2, indicating that the population size increased and genetic variations have accumulated after its decrease. In mismatch analysis, bottlenecks yield either a bimodal distribution or a distribution close to zero, depending on whether the bottleneck reduced or completely removed the genetic diversity (Frankham et al., 2002). The mismatch analysis performed here suggests that the lagoon populations of *S. orbicularis* have experienced size fluctuations. This sudden expansion model was further supported by the goodness-of-fit test (Pssd = 0.18).

Two historical scenarios are considered as the factors led to these results. The first is that the recent colonization of a small number of individuals in Palau islands (founder event), and afterwards their population size increased with low level of migration with different haplotypes. The migration of individuals having different haplotypes is rarely occurred after the establishment of lagoon population, because *S. orbicularis* is not likely to extensive disperse, as described above. The second is that the reduction of population size caused by large-scale climate fluctuations (bottleneck event). Fauvelot et al. (2003) compared the genetic structure of several Polynesian species of coral fish populations inhabiting outer and inner lagoons, and reported that the genetic diversity in the inner lagoons was lower than that in the outer ones. From these results, they further speculated that the low level of genetic diversity found in the inner lagoons was due to a strong bottleneck that occurred through a drought derived from the lower sea level in inner lagoons during the Ice Age. Approximately 10,000 years ago, the sea level on the Palau Islands was 20 meters lower than at present (Kayanne et al., 2002). It is thus plausible that the population size and genetic diversity of *S. orbicularis* lagoon populations decreased with the lower sea level and then rapidly increased after the bottleneck event so that mutations could accumulate in the populations.

Historical changes of marine lake populations in *S. orbicularis* In contrast to lagoon populations, marine lake populations show low haplotype and nucleotide diversities (Table 4). Although the level of genetic diversity in the NLK population was close to that in the lagoon populations, the NLK population possessed less haplotypes, similar to the other marine lake populations (Table 3). Therefore, we regard this lower genetic diversity as a common conspicuous feature for all types of marine lake.

All marine lake populations indicating low haplotype and nucleotide diversities correspond to Category 1 of Grant & Bowen (1998) classification. This category suggests that a population has experienced founder and/or bottleneck events during the last thousand to tens of thousand years. Moreover, results for the mismatch distribution and Tajima's D value for JFM and CLM populations reveal that their population sizes have increased after a decrease (Tajima's D: CLM D = -1.147, P = 0.038; JFM D = -1.071, P = 0.180). The sudden expansion model was further supported for each population (goodness-of-fit test: CLM Pssd = 0.25; JFM Pssd = 0.15).

The founder events when the marine lakes were formed must have affected the present genetic diversity of marine lake populations. Furthermore, large-scale climate fluctuations, such as the El Niño-Southern Oscillation (ENSO) also possibly affected their population size decrease (bottleneck event). The strongest ENSO actually took place in 1997 - 1998 (McPhaden, 1999) and the precipitations on the Yap Islands (located 100 km north of the Palau Islands) were much lower than those of an ordinary year (Kimura et al., 2001). Furthermore, it did not rain for half-a-year on the Palau Islands and the surface temperature of the outer sea rose by1 to 4 °C, resulting in a large salinity increase (Iijima et al., 2005). It was reported that a large number of marine organisms such as corals and jellyfishes died at the time (Bruno et al., 2001). We also observed that adults of golden and moon jellyfishes inhabiting the JFM lake disappeared in 1999. Because ENSO has taken place every 10 to 15 years (Iijima et al., 2005), such environmental changes must have occurred repeatedly since the marine lakes were formed approximately 10,000 years ago. It is likely that the marine lake species have frequently suffered the heavy bottleneck effect under the short intervals of ENSO and kept low level of genetic diversity in the populations. The number of individuals might have rapidly increased under the stable climate, as we actually observe in jellyfish species, because the marine lake species can reproduce frequently in a year and grow in the tropical eutrophic marine lakes.

It is also possible that the reproduction and the survival of *S. orbicularis* isolated in marine lakes have been affected by ENSO and that their population size decreased, although we could not estimate their decrease exactly at the time. Because genetic drifts have strong impacts on small populations, the level of genetic diversity of marine lake populations decreased to very low levels and then genetic structures changed. Ultimately, each marine lake population developed a distinctive genetic structure (pairwise Φ_{ST} among marine lake populations range from 0.429 to 0.870) (Table 6).

The NLK population shows a slightly different network pattern (Fig. 7), where two major and one minor haprotypes were detected. The effect of increased genetic drift after the founder and the bottleneck events usually leads to the fixation of a single haplotype. However, it is probable that a few haplotypes have been maintained in a population when the population size exponentially increased before the haplotype fixation and then has been stable. We hypothesized that the different pattern between the NLK and the other marine lake populations may be related with areas and depth to anoxic layer of each marine lake because the population size and its stability are dependent on such spatial factors. However, the areas and depth of the marine lakes we studied are very similar (area: JFM = 50,000 m²; NLK = 43,000 m²; CLM = 39,000 m², depth to anoxic layer: JFM = 15 m; NLK = 10 m; CLM = 15 m; Dawson & Hamner, 2005; Hamner & Hamner, 1998). In this study, we cannot conclude why only the NLK population demonstrates the different pattern and whether the pattern was caused by chance or not. To address these questions, we will need to survey more detailed information about each marine lake and conduct population analyses by using nuclear DNA markers.

Evolutionary features of marine lake populations in *S. orbicularis* There was no common haplotype among marine lake populations and all haplotypes were unique to each marine lake except haplotype So01 observed in the JFM lake (Fig. 7). We consider that haplotype So01 did not occur in the JFM lake independently, because haplotype So01 is distant from the major haplotype So04 in the JFM population (Fig. 7). Although two fissures were found in this lake (Hamner & Hamner, 1998), there is no observation that marine organisms moved in or out through these fissures. We think the following two possibilities about this finding. The first is that the haplotype So01 originally possessed in the founder population has remained through severe genetic drifts. The second is that the individuals having the haplotype So01 have been artificially introduced from the lagoon. Many tourists visit the Jellyfish Lake because it is a well-known snorkeling spot where they can swim with numerous jellyfishes and observe them. In fact, the non-indigenous, invasive sea anemone *Aiptasia* sp. and the sponge *Haliclona* sp. were first observed at the foot of the dock where visitors enter the lake and then extend their distribution (Marino et al., 2008). However, we have no measure to confirm which is true for the present.

Pairwise Φ_{ST} among marine lake and lagoon populations ranged from 0.531 to 0.848 (Table 6), indicating a high genetic differentiation between these populations. The peripatric differentiation between marine lake and lagoon populations was caused by a small number of individuals colonizing the lakes from the lagoons (founder event) followed by repetitive bottleneck events, such as those generated by ENSO. So far, such higher genetic divergences in extremely short geographical ranges (approximately 150 - 250 m) have scarcely been reported for marine organisms. The marine lake is thus a model that could clarify the process of evolution by geographical isolation in these organisms.

It is well known that approximately 300 species of cichlids have rapidly speciated in Lake Victoria, Africa, during the last 12,000 years (Johnson et al., 1996). Although most of the species are not divergent both in mitochondrial and nuclear gene sequences, they are morphologically and ethologically differentiated and reproductive isolation is established (Meyer et al., 1990; Verheyen et al., 2003; Nagl et al., 1998; Terai et al., 2004). In a preliminary study, we compared some morphological characters, such as the number of spinous and soft fin ray counts and the scalation among populations in *S. orbicularis* but apparent differences were not found (data not shown). However, we recognized that feeding and escaping behaviors are apparently different between marine lake and lagoon populations. Because individuals in marine lakes bait actively, sometimes even feeding on wounded individual of the same species, and seldom escape without wariness, we can easily collect them by angling. Conversely, individuals in lagoons always hide carefully in dark places, and often escape quickly, making them very difficult to collect. Such behavioral difference between marine lake and lagoon populations may be due to a difference in the number of predators (such as carnivorous fishes) between these two habitats, the lakes having fewer predators than the lagoons. We will further survey quantitative characters such as the length of the caudal fin and the body depth in the future study, because these characters are possibly influenced by behavior. Vamosi (2003) has reported that a change in predation pressure could have promoted speciation in the threespine stickleback *Gasterosteus aculeatus* where a release from predation pressure would have led them to speciation. Such ecological factors may also promote rapid adaptive evolution in completely isolated marine lake populations of *S. orbicularis*.

3.3.2 Genetic population structures and speciation of mussels, *Brachidontes* spp.

Marine lake mussels inhabit many meromictic marine lakes, although we have been never found in holomictic marine lakes and lagoons. We collected the mussels from 9 meromictic marine lakes in Palau and conducted morphological, phylogenetic and population genetic characterization (Goto et al., 2011).

Morphological differences between three morphs of marine lake mussels General morphological characters were similar among the marine lake mussels, and agreed with those of the family Mytilidae. We further found some morphological differences in the maximum shell length, thickness, ratio of shell height to width (height/width ratio), color and clearness of radiating ribs among individuals inhabiting different marine lakes. According to these differences, the mussels were sorted into three morphs: NS, ON and MC (Fig. 9). Only a single morph was found in each marine lake.

Phylogenetic position of the three morphs of mussels We conducted phylogenetic analyses of 3 morphs of mussels based on 18S ribosomal (r) RNA genes to infer their phylogenetic positions. We found no sequence variation in the 18S rRNA genes among the individuals. The phylogenetic trees (Bayesian, ML and MP trees) constructed from the data set consisting of the marine lake mussels and other genera in Mytilidae generally agreed well with each other (Fig. 10; ML and MP trees are not shown). The clade consisting of *Brachidontes* and *Hormomya* species (*Brachidontes–Hormomya* complex) was supported by high posterior probabilities for Bayesian analysis (1.00) and high bootstrap values for ML analysis (100%). The trees clearly indicated that the marine lake mussels are positioned in a single clade within the *Brachidontes–Hormomya* complex. Of the species included in our analysis, the marine lake mussels were most closely related to *H. mutabilis* collected from Okinawa, Japan. The phylogenetic trees also suggested that *G. demissa* is the sister taxon to the complex, although this was not supported statistically. The *Brachidontes–Hormomya* complex and the marine lake mussels formed an unresolved polytomy in these analyses.

NS-morph ON-morph MC-morph

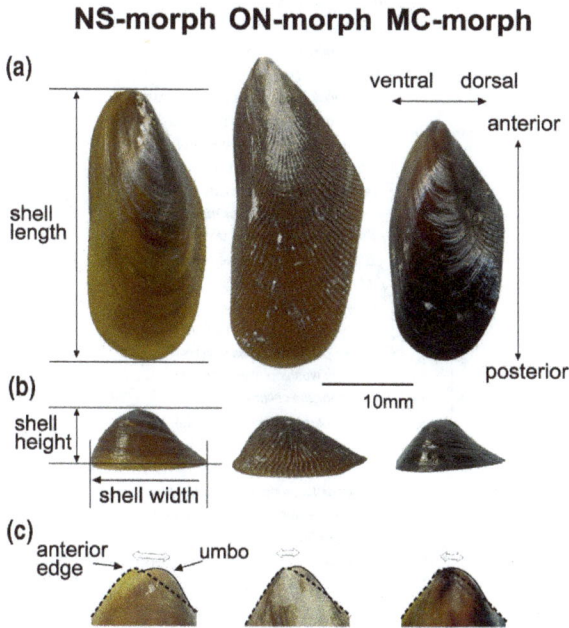

Fig. 9. Three morphs of marine lake mussels. (a) The surface of left valves. Strong divaricate radial ribs are observed over the whole surface of the ON-morph. Weak ribs are present on the whole surface of the NS-morph and dorsal sides of the MC-morph. (b) Left valves viewed from the rear. The ratio of shell height to width of the NS-morph is higher than that of the other morphs. (c) Enlargements of the anterior edge. The umbo of the NS-morph is relatively distant from the anterior edge, whereas those of the ON-and MC-morphs are close to the abterior edges (From T. V. Goto et al., *Zool. Sci.* 28:568-579, 2011).

Phylogenetic relationships among three morphs of marine lake mussels To infer phylogenetic relationships among three morphs of marine lake mussels, we further detected their mitochondrial cytochrome c oxidase subunit 1 (CO I) gene sequences and carried out phylogenetic and population genetic analyses. However, mussels have a unique mode of mitochondrial DNA (mtDNA) inheritance: doubly uniparental inheritance (DUI) (Zouros et al., 1994a, b) or gender-associated inheritance (Skibinski et al., 1994a, b) that two types of mitochondrial genome, the Female (F)- and Male (M)-types, are transmitted to the offspring. Females transmit F-type mtDNA to both female and male offspring through the eggs, whereas males only transmit M-type mtDNA to male offspring through the sperm. In adult males, F-type mtDNA prevails in the somatic tissue, while M-type mtDNA occurs predominantly in the gonads (Stewart et al., 1995; Sutherland et al., 1998). The F- and M-types of mtDNA evolved independently and two highly diverged mitochondrial genomes are retained in all male individuals. F-type mtDNA is more easily used as a genetic marker to study the phylogeny of mussels than M-type mtDNA, because it exists in both sexes, and reference sequences for *Brachidontes* mussels are available (Lee and Foighil, 2004, 2005; Terranova et al., 2007; Samadi et al., 2007). Therefore, in the present study, maternal and paternal CO I gene sequences were first distinguished, and then the F-type CO I gene was used for the phylogenetic analysis.

- Bathymodiolus tangaroa (AY649820)
- Gigantidas gladius (AY649821)
- Bathymodiolus thermophilus (AF221638)
- Tamu fisheri (AF221642)
- Bathymodiolus puteoserpentis (AF221640)
- Bathymodiolus azoricus (AY649822)
- Bathymodiolus brevior (AY649824)
- Bathymodiolus mauritanicus (AY649828)
- Bathymodiolus marisindicus (AY649818)
- Adipicola longissima (DQ340798)
- Bathymodiolus brooksi (AY649825)
- Bathymodiolus heckerae (AY649830)

1.00 91
1.00 100   81
97
0.82 100
1.00 95
91

- Idas arcuatilis (AF221643)
- Myrina pacifica (AF221646)
- Idas macddonaldi (AF221647)
- Adipicola arcuatilis (AF221644)
- Idas washingtonia (AF221645)
- Bathymodiolus childressi (AF221641)

97 ★

- Benthomodiolus lignocola (AF221648)
- Modiolus modiolus (AF124210)
- Modiolus americanus (AF229624)
- Modiolus auriculatus (AF117735)

0.92 ★

1.00 95 Modiolus philippinarum (AB201232)
94 Septifer bilocularis (AF229622)

- Mytilus trossulus (L24490)
- Mytilus edulis (AY527062)
- Mytilus galloprovincialis (L33451)
- Mytilus californianus (L33449)

1.00 98
97
1.00 100
99
1.00 95
90
87

- Perna viridis (EF613234)
- Musculus senhousei (AF124207)
- Musculus discors (AF124206)
- Musculus lateralis (AF229625)

1.00 100
99
1.00 99
90
1.00 100
100

0.91 ★

- Hormomya mutabilis (AB201233)
- marine lake mussels
- Brachidontes variabilis (AJ389643)
- Hormomya exustus (AF229623)
- Hormomya domingensis (AF117736)
- Geukensia demissa (L33450)
- Lithophaga lithophaga (AF120530)
- Lithophaga antillarum (AB201234)
- Lithophaga nigra (AF124209)
- Leiosolenus obesus (AB201237)
- Leiosolenus curtus (AB201235)
- Leiosolenus lithurus (AB201236)
- Mytilus coruscus (EF613242)
- Striarca lactea (AF120531)
- Pinna muricata (AJ389636)
- Atrina pectinata (EF613241)

1.00 95
0.95 ★
1.00 99
93
1.00 99
98
1.00 99
100
1.00 80
85
1.00 82
1.00 100
100

0.2 substitutions/site

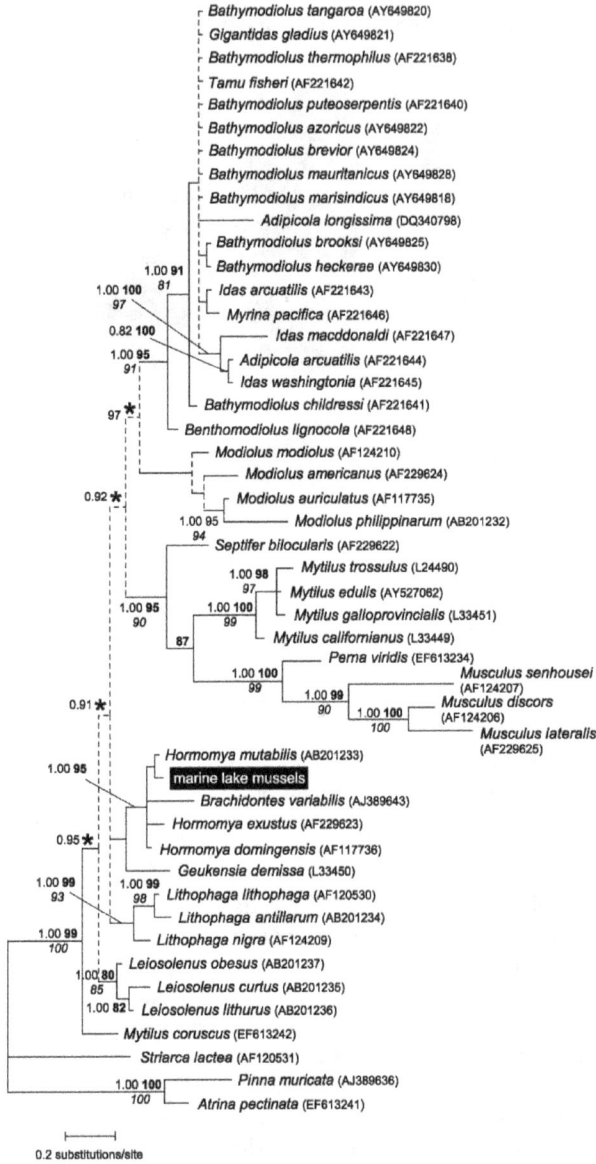

Fig. 10. 18S rRNA gene-based Bayesian tree showing the phylogenetic position of marine lake mussels in the family Mytilidae. The inferred clades of the Bayesian, ML and MP trees (solid line) agreed with each other, but partial topologies (broken line) were not in agreement among the three trees. Asterisks indicate clades of the Bayesian and ML trees that agreed with each other. Bayesian clade posterior probabilities (>0.80, noramal) are indicated with ML (>80, italic) ; bootstrap support values at each branch. The sequence of *Brachidontes exustus* was registered as *Hormomya exustus* in the DNA databases and the registered name was used in this phylogenetic tree (From T. V. Goto et al., *Zool. Sci.* 28:568-579, 2011).

We analyzed F-type COI gene sequences together with M-type COI gene sequences obtained from three specimens to further determine the phylogenetic relationship between the three morphs. No indels were found in 555 bp of the F- and M-type COI gene sequences. A total of 19 haplotypes were detected from the F-type COI genes of the three morphs, and none of these were shared between the different morphs. We found two M-type haplotypes from three ON-morph specimens collected in OLO. The average genetic distance between the F- and M-type sequences was 0.186 (p-distance); the level of genetic divergence was close to that reported among Mytilus species (Mizi et al., 2005).

The phylogenetic relationships between the three morphs and 14 reference sequences retrieved from DDBJ were inferred using Bayesian, ML and MP approaches for these F- and M-type COI gene sequences (only the Bayesian tree is shown in Fig. 11). The F-type COI gene sequences of the marine lake mussels, including 96 polymorphic sites with 75 transitions and 25 transversions, were clustered into two distinct lineages, A and B, with an average genetic distance of 0.149 (p-distance). Monophyly of each of the lineages A and B was strongly supported by posterior probabilities (0.99 and 1.00, respectively) and bootstrap values (lineage A, 95% in ML and 100% in MP; lineage B, 99% in ML, 100% in MP).

Lineage A consisted of all haplotypes of the NS-morph and haplotype A09 of the ON-morph collected from NCM. There were two predominant haplotypes in lineage A, A01 and A06. Haplotypes A01 and A06 were common to NLK and GLK, and GLK and SHN populations, respectively, whereas the minor haplotypes were not shared by different populations. Haplotype A09, detected in the ON-morph from NCM, was highly diverged from A01–A08, being connected to the A06 haplotype with 18 substitutions.

Lineage B contained all haplotypes of the MC- and ON-morphs except for A09, and the haplotype of B. variabilis collected from Hong Kong. Haplotypes in this lineage were further separated into two groups, one consisting of haplotypes of the ON-morph from OLO (B08, B09 and B10), and the other consisting of haplotypes of the MC-morph (B01, B02, B03, B05, B06 and B07) and haplotype B04 of the ON-morph from NCM (Fig. 11). There were two major haplotypes in lineage B, namely, B01 of the MC-morphs and B08 of the ON-morphs; these were diverged by three substitutions. Haplotype B01 was detected in all MC-morph populations. Derivative singleton haplotypes of B01 were only detected in single marine lake populations, with the exception of haplotype B03, which was detected in two populations. The haplotypes detected in the MC-morphs were closely related to each other.

The sequences of B. variabilis were nested within lineage B. The analysis also clearly showed that the ON-morph was not monophyletic. The ON-morph haplotypes detected from OLO and NCM were quite different. A minor haplotype, B10, in the OLO population was diverged from the B08 major haplotype by five substitutions. On the other hand, the haplotypes detected in the NCM population were more diverged. Of the two haplotypes found in NCM, haplotype B04 belonged to lineage B, whereas haplotype A09 was clearly derived from lineage A and was remarkably differentiated (18–23 substitutions) from the other haplotypes of lineage A. Haplotype B04 was more closely related to the haplotypes observed in MC-morphs, as opposed to those observed in OLO.

The other reference sequences were highly diverged from the haplotypes of the three morphs (Fig. 11). The sequence of M. minimus was clearly nested within the Brachidontes–Hormomya complex clade. Some nodes of the reference sequences were strongly supported statistically by posterior probabilities, although most of the basal nodes of the reference sequences were not supported.

The haplotypes of M-type genes formed a highly distinctive clade, which was the sister group to a clade consisting of three reference species, *M. minimus* (Italy, Mediterranian), *B. exustus* (Panama, Atlantic) and *B. adamsianus* (Mexico, Pacific) (Fig. 11). The clade consisting of the M-type haplotypes and the three reference species was strongly supported in Bayesian and ML analysis, but not in MP analysis.

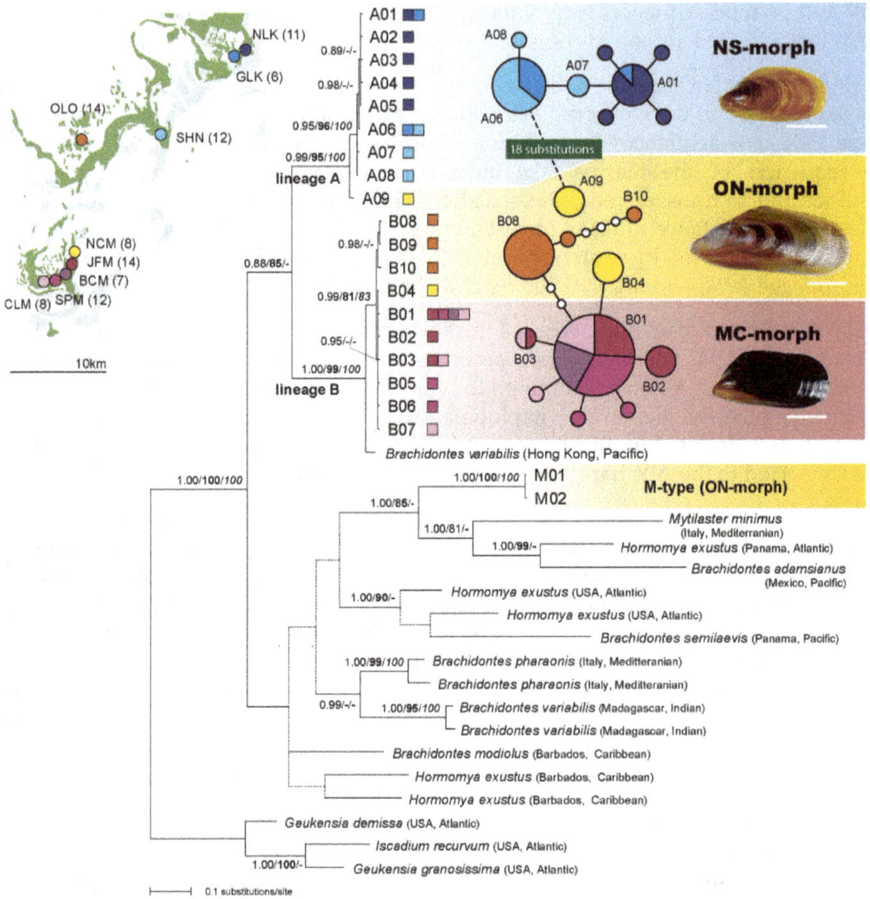

Fig. 11. Phylogenetic relationships among three morphs of marine lake mussels and *Brachidontes-Hormomya* complex. The Bayesian tree using the best-fit model for the COI dataset (GTR I+G), partitioned into the three codon positions is shown. Bayesian clade posterior probabilities (> 0.80, regular) are indicated with ML (> 80, bold) and MP (> 80, italic) bootstrap values at each branch. Squares beside the haplotype symbols indicate the sampling site where the haplotype was found. Topologies of the three trees did not agree with each other in parts within clades consisting of closely related haplotypes. Statistical parsimony network was constructed based on the F-type COI gene haplotypes. Circle size corresponds to the comparative frequency of the haplotypes. Small white circles show missing haplotypes that were not detected. The sequences of *Brachidontes exustus* were registered as *Hormomya exustus* in the DNA databases and the registered name was used here (From T. V. Goto et al., *Zool. Sci.* 28:568-579, 2011).

Genetic diversity of marine lake mussels Haplotype diversity (h) and nucleotide diversity (π) in each marine lake population are shown in Table 7. Haplotype diversity in each population ranged from 0.000 in the BCM population to 0.618 in the NLK population. Nucleotide diversity in all populations, except for NCM, was quite low (0.000–0.001). The nucleotide diversity was highest (0.086) in the NCM population because of its highly diverged haplotypes.

We further calculated pairwise Φ_{ST} as an index of genetic differentiation between the marine lake populations (Table 8). The Φ_{ST} values were statistically significant (p < 0.05), except for between the four populations on Mecherchar Island (JFM, BCM, SPM and CLM). The other exception was for GLK and SHN, which are located on separate islands, approximately 10 km apart. The values between populations of different morphs were higher than those between populations of the same morph. Values for both Tajima's D and Fu's F_s were significantly negative only in the NLK population at the 95% level (p = 0.020 and 0.001, respectively), suggesting a recent rapid demographic expansion. Similarly, in the SPM and CLM populations, negative values of Fu's F_s (p = 0.025 and 0.043, respectively) suggest rapid expansion, but Tajima's D was not significant at the 95% level (p = 0.063 and 0.085, respectively).

Morph	NS-morph				ON-morph			MC-morph				
Sampling sites	NLK	GLK	SHN	Whole	OLO	NCM	Whole	JFM	BCM	SPM	CLM	Whole
Depth of marine lake (m)	38	15	5		4	4		30	22	14	30	
Sample size for genetic analysis	11	6	12	29	14	8	22	14	7	12	8	41
Number of F-type haplotypes	5	2	3	8	3	2	5	3	1	3	3	6
Haplotype diversity (h)	0.618	0.333	0.439	0.704	0.275	0.571	0.662	0.539	0.000	0.318	0.464	0.387
Nucleotide diversity (π) (10^{-3})	1.310	1.201	0.846	2.459	1.247	86.486	48.633	1.049	0.000	0.601	0.901	0.760
Number of Ti/Tv	4/0	2/0	2/0	7/0	4/0	60/24	63/24	2/0	0/0	2/0	2/0	5/0
Tajima's D	-1.712	-1.132	-0.850	-0.698	-1.481	2.615	0.633	-0.201	0.000	-1.451	-1.310	-1.617
p-value	0.020	0.142	0.222	0.297	0.048	1.000	0.804	0.418	1.000	0.063	0.085	0.024
Fu's F_s	-2.908	0.952	-0.725	-2.751	0.117	19.128	19.090	-0.207	0.000	-1.325	-0.999	-4.123
p-value	0.001	0.592	0.097	0.053	0.424	1.000	1.000	0.308	N.A.	0.025	0.043	0.001

N. A. Not applicable.

Table 7. Number of specimens and estimated genetic diversities between populations of the three morphs of marine lake mussels based on COI genes. h: haplotype diversity, π: nucleotide diversity, number of transitions (Ti) and transversions (Tv), values of Tajima's D and Fu's F_S (Modified from T. V. Goto et al., *Zool. Sci.* 28:568-579, 2011).

Genetic divergence among three morphs of mussels F-type COI gene-based phylogenetic analysis clearly showed high genetic divergence between the NS- and MC-morphs of the marine lake mussels (Fig. 11). The genetic distances between NS- and MC-morphs (e.g. p-distance = 0.146 between A01 and B01) were comparable to those reported between species (Terranova et al., 2007) or within cryptic species (Lee and Foighil, 2004, 2005). Lee and Foighil (2004, 2005) estimated the evolutionary rates of the third-codon position for the F-type COI gene in *B. exustus* as 18.3–24.4% per million years per lineage. Applying this mutation rate to the genetic distances between the NS- and MC-morphs (p-distance = 0.378 at the third codon position between A01 and B01), we estimated the time of the NS/MC split to be approximately 0.78–1.03 million years ago. This divergence time is much older than the date of formation of the marine lakes after the Last Glacial Maximum.

Despite the remarkable morphological differences between ON- and MC-morphs, we found lower levels of genetic differentiation between these two morphs. The genetic distance between ON- and MC-morphs (average p-distance at the third codon position = 0.0053; 0.022–0.029 million years ago) was as small as the level within a species or among closely related taxa in the *M. edulis* species complex (Gérard et al., 2008). This may indicate that a rapid change can occur in shell morphology of mussels in the marine lakes. Indeed, it is known that shell characters, such as size, thickness and growth rates, of bivalves readily change according to environmental conditions (Seed and Richardson, 1990; Beadman et al., 2003; Schöne et al., 2003). The limnological conditions of OLO and NCM, which the ON-morph mussels inhabit, are different from the other seven lakes. For example, there is no anaerobic bottom water in OLO and NCM. Such differences in the local environment might have shaped the unique morphology of the mussels via some adaptation in the lakes over a relatively short period.

	NLK	GLK	SHN	NCM	OLO	JFM	BCM	SPM	CLM
NLK		0.009	0.000	0.000	0.000	0.000	0.000	0.000	0.000
GLK	0.653*		0.739	0.027	0.000	0.000	0.000	0.000	0.000
SHN	0.740*	-0.063		0.000	0.000	0.000	0.000	0.000	0.000
NCM	0.542*	0.433*	0.558*		0.000	0.000	0.027	0.000	0.036
OLO	0.991*	0.992*	0.993*	0.538*		0.000	0.000	0.000	0.000
JFM	0.992*	0.993*	0.994*	0.529*	0.822*		0.234	0.090	0.171
BCM	0.994*	0.996*	0.996*	0.404*	0.857*	0.010		0.991	0.991
SPM	0.994*	0.995*	0.995*	0.501*	0.844*	0.121	-0.051		0.622
CLM	0.992*	0.993*	0.994*	0.423*	0.821*	0.076	-0.018	0.009	

Below: Φ_{ST}, The asterisks indicate significant values ($P < 0.05$).
Above: P values.

Table 8. Estimated pairwise Φ_{ST} between marine lake populations based on COI genes (From T. V. Goto et al., *Zool. Sci.* 28:568-579, 2011).

mtDNA lineages of marine lake mussels No COI haplotype was common to all three morphs, suggesting an absence of recent gene flow between the morphs (Fig. 11). Population structures of both NS- and MC-morphs consisted of one or two major and several minor COI haplotypes, and exhibited a star-like pattern of the haplotype network, indicating a founder effect and/or bottlenecking. ON-morphs, on the other hand, showed a more complex pattern of haplotype variations: 1) no haplotype was shared between the two ON-morph populations (NCM and OLO); 2) lineage A and B coexisted in NCM, while only lineage B haplotypes were found in OLO; and 3) the lineage B haplotype in NCM (B04) was more closely related to haplotypes in populations of MC-morphs (JFM, BCM, CLM and SPM) than to the other haplotypes of ON-morphs. These features suggest that the ON-morph may have evolved through a more complicated series of historical events than the NS- and MC- morphs.

In general, the coexistence of multiple lineages of mtDNA haplotypes in a single population can be explained by either ancestral polymorphism or introgression. If the effective population size has been large and constant (or expanding) and the intervals between population splits short, ancestral polymorphisms are likely to be retained in the single population (Pamilo and Nei, 1988; Takahata, 1989). However, this may not be the case in the marine lake mussels because a small effective population size and/or past bottlenecking were suggested by the low level of nucleotide diversity. It is rather likely that mtDNA introgression took place in NCM — the haplotypes of lineage A were possibly introduced into an ancestral population that originated from lineage B. There is another possible scenario for the multiple lineages of mtDNA in ON-morph. The OLO and NCM populations of ON-morphs do not form a monophyletic group in the mtDNA phylogeny, and are separated geographically on different islands, yet the shell morphology of mussels from the two populations is indistinguishable. As mentioned above, the limnological conditions of OLO and NCM might have shaped the unique morphology of ON-morphs through convergence. Alternatively, it is also possible that ON-morph has been formed through hybridization between the morphs, although further genetic analysis using nuclear DNA markers will be necessary to clarify this issue.

In this study, we were unable to obtain M-type sequences from NS- and MC-morphs, and the NCM population of ON-morphs. This could be attributable to either technical difficulties, or the lack of an M-type genome, such as in some Mytilus males (Hoeh et al., 1997; Quesada et al., 1999, 2003). The previous studies suggested that some M-Type genes were newly appeared male-transmitted mtDNA: hypothetical switch of transmitting route from an M-type to F-type. Hence, the discovery of M-type from NS- and MC-types may suggest that such evolutionary event occurred in mitochondrial genomes of the morphs.

Evolutionary aspects of marine lake mussels Genetic diversity of the marine lake mussels, inferred from the haplotype diversity (h = 0.000–0.662) and nucleotide diversity (π = 0.000–0.002), tended to be lower than that reported for other mussels inhabiting seashores, such as Floridian *Brachidontes* (h = 0.546–0.987) (Lee and Foighil, 2004, 2005), Mediterranean *Brachidontes* (h = 0.733–1.000; π = 0.005–0.042) (Terranova et al., 2006, 2007) and *Mytilus* collected from the northern and southern hemispheres (h = 0.600–0.950; π = 0.002–0.021) (Gérard et al., 2008). This indicates a small effective population size and/or past bottlenecking of the marine lake populations. Because the inhabitable area of the marine lake mussels is limited, the effective population size would likely be relatively small, potentially leading to bottlenecking.

Since a single major haplotype was common to all populations of MC-morphs, these populations appear to have been founded and diverged recently. The time of divergence corresponds to the presumptive formation time of the marine lakes, after the Last Glacial Maximum, approximately 18,000 years ago (Hamner and Hamner, 1998; Dawson and Hamner, 2005). In contrast, no haplotype was common to NLK and SHN populations of NS-morphs, and the Φ_{ST} values were statistically significant (p <0.05) between them. Therefore, founder populations of NLK and SHN may have slightly differentiated from each other before colonization of the lakes. This prior genetic differentiation may have been the result of geographic separation of the ancestors. It should be noted, however, that the major haplotype found in GLK was also present in SHN but not NLK, despite GLK being geographically closer to NLK than SHN. Therefore, the level of genetic differentiation between populations does not necessarily correspond to geographical distance separating the marine lakes.

In conclusion, our data clearly indicate that the morphs of the marine lake mussels are differentiated in both morphology and mtDNA lineages. This is especially the case for NS- and MC-morphs, which are differentiated from each other to the level of species. In contrast, the relatively low level of genetic divergence between ON- and MC-morphs suggests that these morphs have rapidly acquired different morphological characters. The present study provides empirical evidence of diversification of mussels in isolated marine environments. Analysis of more sensitive nuclear genetic markers and the identification of additional variant M-type haplotypes of mtDNA markers will assist in better clarifying the evolutionary history of the marine lake mussels.

3.4 Perspective on evolution in the marine lakes

As shown by the examples of a cardinal fish and mussels in the marine lakes, the marine animals in each isolated marine lake have undergone a peculiar evolutionary process, even during a short evolutionary period of 5,000–12,000 years. On the basis of the data for the genetic diversity and differentiation of the animals in the marine lakes, we can predict their future evolutionary patterns after isolation in each marine lake. Fig. 12 (a) shows that individuals of the ancestral lagoon populations are isolated in some meromictic marine lakes and are driven to extinction by severe environmental changes and pathogenic infections. As shown in Table 2, the fact that some diagnostic species do not inhabit some meromictic lakes strongly supports this case. Fig. 12 (b) shows that genetic differentiation does not occur among the marine lake populations, and this is actually observed in most of the holomictic marine lakes. Fig. 12 (c) shows the following situation. Populations isolated in meromictic marine lakes are genetically differentiated during some extent of evolutionary time; however, the meromictic lakes turn into holomictic lakes following the collapse of limestone, individuals from the marine lake populations make secondary contact with those of the lagoon populations, and genetic mixing occurs between the marine lake and lagoon populations. In this case, the genetic diversity of the mixed population must be much higher than that of the marine lake population. Fig. 12 (d) shows that populations that have been isolated in meromictic lakes for a long evolutionary time and are largely genetically differentiated, undergo changes in their characteristics and speciate. The 5 *Mastigias* subspecies inhabiting different marine lakes, described by Dawson (2005), must fit in this case. Recently, a living fossil eel belonging to a new family in the order Anguilliformes was described from an undersea cave in Palau (Johnson et al., 2011). The divergence time between the new species and other eels was surprisingly estimated at 200 million years ago from phylogenetic analysis based on mitochondrial genome sequences, indicating the possibility that this species has been isolated in the undersea cave and speciated. It is known that there are many undersea limestone caves in Palau and some of the caves are probably, geologically, very old (personal communication from Carp Corporation staff). Atolls composed of limestone have probably uplifted and depressed repeatedly over geological time, and marine lakes and undersea caves have been also formed and collapsed repeatedly. Various marine organisms may have speciated in such habitats and some of them may have undergone extinction.

Thus, the "marine lake model as an isolated marine environment" we present here is readily available to study the evolution of various marine species. We have continuously conducted evolutionary studies of other marine organisms including algae and microbes using genetic analyses in the marine lakes of Palau. Further studies on the adaptive evolution of marine organisms in the marine lakes are in progress.

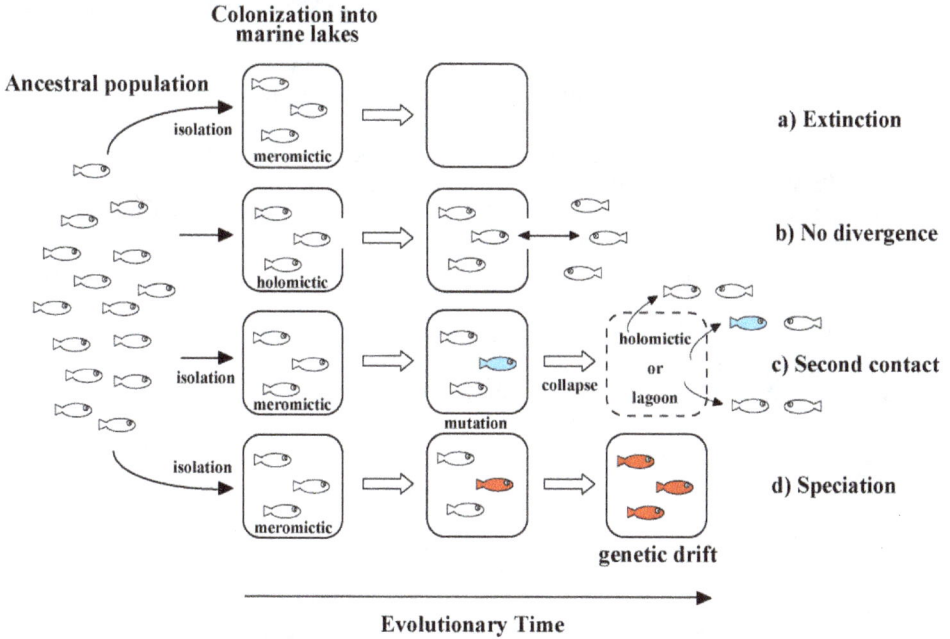

Fig. 12. Schematic figures of the evolutionary patterns in the marine lake populations

4. Acknowledgments

We thank Drs. Y. Hara, H. Kudo, S. Saitoh (Yamagata University) and K. Okuizumi (Kamo Aquarium, Tsuruoka, Yamagata) for support with sampling, helpful advice and discussions. We also thank the Ministry of Resources and Development, Republic of Palau for permitting us to collect specimens, as well as Marino, Urui, Vitk, Baste, I. Kishigawa and other members of the Carp Corporation for supporting our sampling in Palau. This work was supported in part by Grants-in-Aid for Science Research (B), Japan Society for the Promotion of Science (No. 16405012 and 18405015).

5. References

Allen, G. R. (2008). Conservation hotspots of biodiversity and endemism for Indo-Pacific coral reef fishes. *Aquatic Conserv. Mar. Freshw. Ecosyst.* 18: 541-556.

Avise, J. C. (2000). *Phylogeography-The history and formation of species*, Harvard Univ. Press, Cambridge and London.

Barluenga, M., Stolting, K. N., Salzburger, W., Muschick, M. & Meyer, A. (2006). Sympatric speciation in Nicaraguan crater lake cichlid fish. *Nature* 439: 719-723.

Beadman, H. A., Caldow, R. W. G., Kaiser, M. J. & Willows, R. I. (2003). How to toughen up your mussels: using mussel shell morphological plasticity to reduce predation losses. *Marine Biolology* 142: 487-494.

Bernardi, G. (2000). Barriers to gene flow in Embiotoca jacksoni, a marine fish lacking a pelagic larval stage. *Evolution* 54: 226-237.

Bernardi, G. and Vangelli, A. (2004). Population structure in Banggai cardinalfish, *Pterapogon kauderni*, a coral reef species lacking a pelagic larval phase. *Marine Biolology* 145, 803-810.

Bruno, J. F., Siddon, C. E., Witman, J. D. & Colin, P. L. (2001). El Niño related coral bleaching in Palau, Western Caroline Islands. *Coral Reefs* 20: 127-136.

Caccone, A., Gentile, G., Gibbs, J. P., Fritts, T. H., Snell, H. L., Betts, J. & Powell, J. R. (2001). Phylogeography and history of giant Galapagos tortoises. *Evolution* 56: 2052-2066.

Calsbeek, R. & Smith, T. B. (2003). Ocean currents mediate evolution in island lizards. *Nature* 426: 552-555.

Chiba, S. (1998). Synchronized evolution in lineages of land snails in oceanic islands. *Paleobiology* 24: 99-108.

Dawson, M. N. & Jacobs, D. K. (2001). Molecular evidence fro cryptic species of *Aurelia aurita* (Cnidaria, Scyphozoa). *Biological Bulletin* 200: 92-96.

Dawson, M. N. (2005) Five new subspecies of *Mastigias* (Scyphozoa: Rhizostomeae: Mastigiidae) from marine lakes, Palau, Micronesia. *J. Mar. Biol. Ass. UK* 85: 679-694.

Dawson, M. N. & Hamner, W. M. (2005). Rapid evolutionary radiation of marine zooplankton in peripheral environments. *Proc. Natl. Acad. Sci. USA* 102: 9235-9240.

Fautin, D. G. & Fitt, W. K. (1991). A jellyfish-eating sea anemone (Cnidaria, Actiniaria) from Palau: *Entacmaea medusivora* sp. nov. *Hydrobiologia* 216/217: 453-461.

Fauvelot, C., Bernardi, G. & Planes, S. (2003). Reductions in the mitochondrial DNA diversity of coral reef fish provide evidence of population bottlenecks resulting from Holocene sea-level change. *Evolution* 57: 1571-1583.

Frankham, R., Ballou, J. D. & Briscoe, D. A. (2002). *Introduction to conservation genetics*, Cambridge University Press, Cambridge.

Gérard, K., Nicolas, B., Borsa, P., Chenuil, A. & Féral, J.-P. (2008). Pleistocene separation of mitochondrial lineages of *Mytilus* spp. mussels from Northern and Southern Hemispheres and strong genetic differentiation among southern populations. *Mol. Phyl. Evol.* 49: 84-91.

Grant, W. S. & Bowen, B. W. (1998). Shallow population histories in deep evolutionary lineages of marine fishes: insights from sardines and anchovies and lessons of conservation. *J. Heredity* 89: 415-426.

Goto, T. V., Tamate, H. B. & Hanzawa, N. (2011). Phylogenetic characterization of three morphs of mussels (Bivalvia, Mytilidae) inhabiting isolated marine environments in Palau Islands. *Zoological Science* 28: 568-579

Gotoh, R. O., Sekimoto, H., Chiba, S. N. & Hanzawa, N. (2009). Peripatric differentiation among adjacent marine lake and lagoon populations of a coastal fish, *Sphaeramia orbicularis* (Apogonidae, Perciformes, Teleostei). *Genes & Genetic Systems* 84: 287-295.

Hamner, W. M. & Hauri, I. R. (1981). Long-distance horizontal migrations of zooplankton (Scyphomedusae: *Mastigias*). *Limnology and Oceanography* 26: 414-423.

Hamner, W. M., Gilmer, R. W. & Hamner, P. P. (1982). The physical, chemical, and biological characteristics of a stratified, saline, sulfide lake in Palau. *Limnology and Oceanography* 27: 101-118.

Hamner, W. M. & Hamner, P. P. (1998). Stratified marine lakes of Palau (Western Caroline Islands). *Physical Geography* 19: 175-220.

Hoeh, W. R., Stewart, D. T., Saavedra, C., Sutherland, B. W. & Zouros, E. (1997). Phylogenetic evidence for role-reversals of gender-associated mitochondrial DNA in *Mytilus* (Bivalvia: Mytilidae). *Mol. Biol. Evol.* 14: 959-967.

Iijima, H., Kayanne, H., Morimoto, M. & Abe, O. (2005). Interannual sea surface salinity changes in the western Pacific from 1954 to 2000 based on coral isotope analysis. *Geophys. Res. Lett.* 32: 1-4.

Johnson, D. G., Ida, H., Sakaue, J., Sado, T., Asahida, T. & Miya, Masaki. (2011). A 'living fossil' eel (Anguilliformes: Protoanguillidae, fam. nov.) from an undersea cave in Palau. *Proc. R. Soc. B* : doi: 10.1098/rspb.2011.1289.

Johnson, T. C., Scholz, C. A., Talbot, M. R., Kelts, K., Ricketts, K. K., Ngobi, G., Beuning, K., Ssemmanda, I. & McGill, J. W. (1996). Late Pleistocene desiccation of Lake Victoria and rapid evolution of cichlid fishes. *Science* 23: 1091-1093.

Kawagata, S., Yamasaki, M., Genka, R. & Jordan, R. W. (2005). Shallow-water benthic foraminifers from Mecherchar Jellyfish Lake (Ongerui Tketau Uet), Palau. *Micronesica* 37(2): 215-233.

Kayanne, H., Yamamoto, H. & Randall, R. H. (2002). Holocene sea-level changes and barrier reef formation on an oceanic island, Palau Islands, western Pacific. *Sediment. Geol.* 150: 47-60.

Kimura, S., Inoue, T. & Sugimoto, T. (2001). Fluctuation in the distribution of low-salinity water in the North Equatorial Current and its effect on the larval transport of the Japanese eel. *Fish. Oceanogr.* 10: 51-60.

Kliman, R. M., Andolfatto, P., Coyne, J. A., Depaulis, F., Kreitman, M., Berry, A. J., McCarter, J., Wakeley, J. & Hey, J. (2000). The population genetics of the origin and divergence of the *Drosophila simulanus* complex species. *Genetics* 156: 1913-1931.

Kuriiwa, K., Hanzawa, N., Yoshino, T., Kimura, S. & Nishida, M. (2007) Phylogenetic relationships and natural hybridization in rabbitfishes (Teleostei: Siganidae) inferred from mitochondrial and nuclear DNA analyses. *Molecular Phylogenetics & Evolution* 45: 69-80.

Lee, T. & Foighil, D. O. (2004). Hidden Floridian biodiversity: mitochondrial and nuclear gene trees reveal four cryptic species within the scorched mussel, *Brachidontes exustus*, species complex. *Molecular Ecology* 13: 3527-3542.

Lee, T. & Foighil, D. O. (2005). Placing the floridian marine genetic disjunction into a reigional evolutionary context using the scorched mussel, *Brachidontes exustus* species complex. *Evolution* 59: 2139-2158.

Lee, W.-J., Conroy, J., Howell, W. H. & Kocher, T. D. (1995). Structure and evolution of teleost mitochondrial control regions. *J. Molecular Evolution* 41: 54-66.

Lipps, J. H. & Langer, M. R. (1999). Benthic foraminifera from the meromictic Mecherchar Jellyfish Lake, Palau (western Pacific). *Micropaleontology* 45: 278-284.

Marino, S., Bauman, A., Miles, J., Kitalong, A., Bukurou, A., Mersai1, C., Verheij, E., Olkeriil, I., Basilius, K., Colin, P., Patris, S., Victor, S., Andrew, W., Miles, J. & Golbuu, Y.

(2008). *The State of Coral Reef Ecosystems of Palau*, Waddell, J. E. & Clarke, A. M. (ed.), *The State of Coral Reef Ecosystems of the United States and Pacific Freely Associated States: 2008*, NOAA Technical Memorandum NOS NCCOS 73. NOAA/NCCOS Center for Coastal Monitoring and Assessment's Biogeography Team. Silver Spring, MD, pp. 511-539.

McPhaden, M. J. (1999). Genesis and evolution of the 1997-98 El Niño. *Science* 283: 950-954.

Mayr, E. (1954). Geographic speciation in tropical echinoids. *Evolution* 8: 1-18.

Meyer, A., Kocher, T. D., Basasibwaki, P. & Wilson, A. C. (1990). Monophyletic origin of Lake Victoria cichlid fishes suggested by mitochondrial DNA sequences. *Nature* 347: 550-553.

Mizi, A., Zouros, E., Moschonas, N. & Rodakis, G. C. (2005). The complete maternal and paternal mitochondrial genomes of the Mediterranean mussel *Mytilus galloprovincialis*: Implications for thedoubly uniparental inheritance mode of mtDNA. *Mol. Biol. Evol.* 22: 952-967.

Myers, R. F. (1999). *Micronesian reef fish*, A Coral Graphics Production, Guam.

Nagl, S., Tichy, H., Mayer, W. E., Takahata, N. & Klein, J. (1998). Persistence of neutral polymorphisms in Lake Victoria cichlid fish. *Proc. Natl. Acad. Sci. USA* 95: 14238-14243.

Palumbi, S. R. (1994). Genetic divergence, reproductive isolation, and marine speciation. *Annu. Rev. Evol. Syst.* 25: 547-572.

Pamilo, P. & Nei, M. (1988). Relationships between gene trees and species trees. *Mol. Biol. Evol.* 5: 568-583.

Quesada, H., Wenne, R. & Skibinski, D. O. F. (1999). Interspecies transfer of female mitochondrial DNA is coupled with role-reversals and departure from neutrality in the mussel *Mytilus trossulus*. *Mol. Biol. Evol.* 16: 655-665.

Quesada, H., Stuckas, H. & Skibinski, D. O. F. (2003). Heteroplasmy suggests paternal co-transmission of multiple genomes and pervasive reversion of maternally into paternally transmitted genomes of mussel (*Mytilus*) mitochondrial DNA. *J. Molecular Evolution* 57: S138-147.

Randall, J. E. & Kulbicki, M. (2005). Siganus woodlandi, new species of rabbitfish (Siganidae) from New Caledonia. *Cybium* 29(2): 185-189.

Register, E. A., Yokoyama R. & Yokoyama, S. (1994). Multiple origins of the green-sensitive opsin genes in fish. *J. Molecular Evolution* 39: 268-273.

Saitoh, S., Suzuki, H., Hanzawa, N. & Tamate, H. B. (2011). Species diversity and community structure of pelagic copepods in the marine lakes of Palau. Hydrobiologia 666: 85-97.

Samadi, S., Quémeré, E., Lorion, J., Tillier, A., Cosel, R., Lopez, P., Cruaud, C., Couloux, A. & Boisselier-Dubayle M-C. (2007). Molecular phylogeny in mytilids supports the wooden steps to deep-sea vents hypothesis. *C. R. Biol.* 330: 446-456.

Seed, R. & Richardson, C. A. (1990). *Mytilus* growth and its environmental responsiveness. in Stefano, G. B. (ed.), *Neurobiology of Mytilus edulis (Studies in Neuroscience)*, Manchester University Press, Manchester, pp. 12-37.

Seehausen, O., Terai, Y., Magalhaes, I. S., Carleton, K. L., Mrosso, D. J., Miyagi, R., van der Sluijs, I., Schneider, M. V., Maan, M. E., Tachida, H., Imai, H. & Okada, N. (2008). Speciation through sensory drive in cichlid fish. *Nature* 455: 620-626.

Schöne, B. R., Tanabe, K., Dettman, D. L. & Sato, S. (2003). Environmental controls on shell growth rates and $\delta^{18}O$ of the shallow-marine bivalve mollusk *Phacosoma japonicum* in Japan. *Marine Biology* 142: 473-485.

Skibinski, D. O. F., Gallagher, C. & Beynon, C. M. (1994a). Mitocondrial DNA inherince. *Nature* 368: 817-818.

Skibinski, D. O. F., Gallagher, C. & Beynon, C. M. (1994b). Sex-limited mitochondrial DNA transmission in the marine mussel *Mytilus edulis*. *Genetics* 138: 801-809.

Stewart, D. T., Saavedra, C., Stanwood, R. R., Ball, A. O. & Zouros, E. (1995). Male and female mitochondrial DNA lineages in the blue mussel (*Mytilus edulis*) species group. *Mol. Biol. Evol.* 12: 735-747.

Sutherland, B., Stewart, D., Kenchington, E. R. & Zouros, E. (1998). The fate of paternal mitochondrial DNA in developing female mussels, *Mytilus edulis*: Implications for the mechanism of Doubly Uniparental Inheritance of mitochondrial DNA. *Genetics* 148: 341-347.

Takahata, N. (1989). Gene genealogy in three related populations: consistency probability between gene and population trees. *Genetics* 122: 957-966.

Terai, Y., Takezaki, N., Mayer, W. E., Tichy, H., Takehata, N., Klein, J. & Okada, N. (2004). Phylogenetic relationships among East African Haplochromine fish as revealed by short interspersed elements (SINEs). *J. Molecular Evolution* 58: 64-78.

Terranova, M. S., Brutto, S. L., Arculeo, M. & Mitton, J. B. (2006). Population structure of *Brachidontes pharaonis* (P. Fisher, 1870) (Bivalvia, Mytilidae) in the Mediterranean Sea, and evolution of a novel mtDNA polymorphism. *Marine Biology* 150: 89-101.

Terranova, M. S., Brutto, S. L., Arculeo, M. & Mitton, J. B. (2007). A mitochondrial phylogeography of *Brachidontes variabilis* (Bivalvia: Mytilidae) reveals three cryptic species. *J. Zool. Syst. Evol. Res.* 45: 289-298.

U. S. Geological Survey. (1956). *Military geology of Palau Islands, Caroline Islands.* Intelligence division, Office of the Engineer, Headquaters U. S. Army Forces Far East and Eight U. S. Army with Personnel of U. S. G. S.

Venkateswaran, K., Shimada, A., Maruyama, A., Higashihara, T., Sakou, H. & Maruyama, T. (1993). Microbial characteristics of Palau Jellyfish Lake. *Canadian J. Microbiology* 39: 506-512.

Verheyen, E., Salzburger, W., Snoeks, J. & Meyer, A. (2003). Origin of the superflock of cichlid fishes from Lake Victoria, East Africa. *Science* 300: 325-329.

Vamosi, S. M. (2003). The presence of other fish species affects speciation in threespine sticklebacks. *Evol. Ecol. Res.* 5: 717-730.

Woodland, D. J. (1990). Revision of the fish family Siganidae with descriptions of two new species and comments on distribution and biology. *Indo-Pacific Fishes* 19: 1-136.

Yokoyama, S. & Yokoyama, R. (1996). Adaptive evolution of photosensitive receptors and visual pigments in vertebrate. *Ann. Rev. Ecol. Syst.* 27: 543-567.

Zouros, E., Ball, A. O., Saavedra, C. & Freeman, K. R. (1994a). Mitochondrial DNA inheritance. *Nature* 368: 818.

Zouros, E., Ball, A. O., Saavedra, C. & Freeman, K. R. (1994b). An unusual type of mitochondrial DNA inheritance in the blue mussel Mytilus. *Proc. Natl. Acad. Sci. USA* 91: 7463-7467.

8

Genetic Diversity and Genetic Heterogeneity of Bigfin Reef Squid *"Sepioteuthis lessoniana"* Species Complex in Northwestern Pacific Ocean

Hideyuki Imai and Misuzu Aoki
University of the Ryukyus
Nara Women's University
Japan

1. Introduction

The bigfin reef squid *Sepioteuthis lessoniana* Férussac, 1831 in Lesson (1830–1831) is widely distributed in the Indo-Pacific, where it is a very valuable fishery resource (Dunning, 1998). Thus, a lot of ecological research of this species were reported (e.g. Ikeda, 1933; Choe & Ohshima, 1961; Segawa, 1987; Ueta, 2003; Ikeda *et al.*, 2009). Segawa *et al.* (1993a; 1993b) showed that within *Sepioteuthis lessoninana* have diferrences of egg chracteristics and reproductive trait in Ishigakijima Island. Izuka *et al.* (1994) reported an allozyme analysis found so-called *S. lessoniana* around Ishigakijima in Okinawa Prefecture, Japan, includes at least three biological species (Figure 1 & 2). Local fishers call the three species *"aka-ika,"* which has a red body, *"shiro-ika"* or *"aori-ika,"* which has a white body, and *"kua-ika,"* which is smaller than the other two. Of these, the range of *"shiro-ika"* extends to the coast of the main Japanese islands. This is the extent of its taxonomic classification thus far. This is due in part to the limited number of distinguishing morphological characters but also because the type specimens is no longer available and type locality has not been disignated (Lu *et al.*, 1995; Jereb & Roper, 2006). This makes it difficult to determain whether genetically recognized species are undescraibed species or one of 13 known synonymies (Young, 2002). In this study, we treated *"aka-ika"* as *Sepioteuthis* sp. 1, *"shiro-ika"* as *Sepioteuthis* sp. 2, and *"kua-ika"* as *Sepioteuthis* sp. 3.

A previous population genetics study found significant differences in the genetic heterogeneity of *Sepioteuthis* sp. 2 between Pacific Ocean and Japan Sea populations using allozyme analysis (Yokogawa & Ueta, 2000). Yokogawa and Ueta (2000) did not include the Okinawan *Sepioteuthis* sp. 2 population in their study. In addition, Pratoomchat *et al.* (2001) found no significant genetic heterogeneity between Japanese and Thai *Sepioteuthis* sp. 2 populations, while our present study tried significant differences in the genetic heterogeneity of the Japanese and Vietnumese *Sepioteuthis* sp. 2 populations. Recently, Aoki *et al.* (2008a) reported significant genetic heterogeneity between Japanese and Vietnumese populations of *Sepioteuthis* sp. 2 using DNA sequencing analysis of the mitochondrial noncoding region.

Therefore, this study examined the genetic diversity (*i.e.*, the average heterogeneity) and gene flow among *Sepioteuthis* sp. 2 populations using allozyme analysis and among populations of *Sepioteuthis* sp. 1 and *Sepioteuthis* sp. 3 using mitochondrial DNA noncoding region sequencing of populations from Japanese, Taiwanese, and Vietnamese waters.

Fig. 1. A: *Sepioteuthis* sp. 1, B: *Sepioteuthis* sp. 2, C: *Sepioteuhis* sp. 3, D: *Sepioteuthis* sp. 1 egg capsules with 5-13 (mean = 9) per capsule, E: *Sepioteuthis* sp. 2 egg capsules with 3-8 eggs (mode = 6) per capsule and F: *Sepioteuthis* sp. 3 egg capsules with consistently two eggs per capsule laid under dead table coral in shallow waters. Black bar indicated 50mm in length.

2. Materials & methods

2.1 Allozyme analysis of *Sepioteuthis* sp. 2

We collected 327 adults between September 1998 and June 2006 from Noto, Ishikawa, Japan (83 individuals), Mugi, Tokushima, Japan (51), the Goto Islands, Nagasaki, Japan (58), Nakagusuku, Okinawa (52), Keelung, Taiwan (23), and the Gulf of Tonkin, Vietnam (60). All

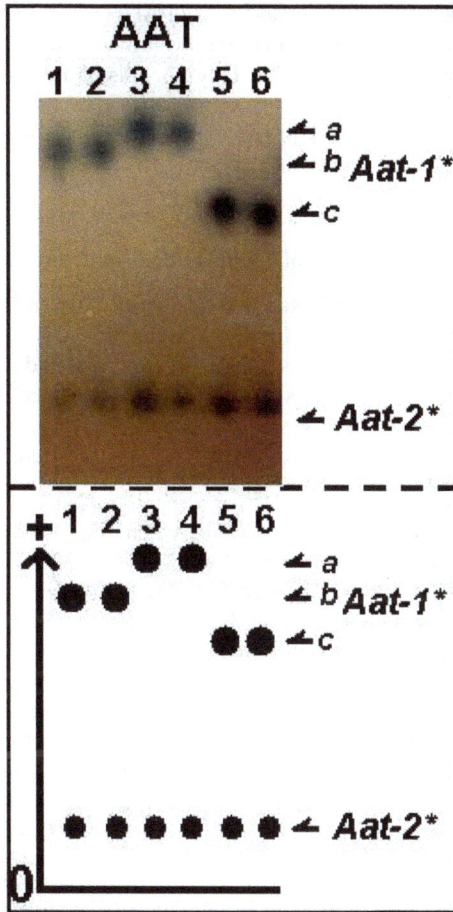

Fig. 2. Electrophoretic patterns of asparate aminotransferase (AAT) of *Sepioteuthis lessoniana* complex. Lane 1-2: *Sepiteuthis* sp. 3, Lane 3-4: *Sepioteuthis* sp. 2 and Lane 5-6: *Sepioteuthis* sp. 1. These three species are clearly identified by the *Aat-1** marker (Izuka *et al.* 1994).

specimens were fresh and immediately sent to a refrigerator in the laboratory. The buccal bulb muscle was removed and kept frozen at –40°C until the allozyme analysis. Small pieces of liver and skeletal muscle were dissected from selected specimens and minced individually in an equal volume of distilled water on ice. Electrophoresis was conducted in a glass box with ice on top of it. The box was in a refrigerator at a constant voltage (250 V) until the Amido Black 10B marker moved seven cm from the origin. The allozymes were tested using 12.5% horizontal starch–gel electrophoresis and the two buffer systems described by Clayton and Tretiak (1972) and modified by Numachi (1989): citric acid N-(3-aminopropyl) diethanolamine (CAEA, pH 7) and citric acid N-(3-aminopropyl) morpholine (CAPM, pH 6). Each gel was sliced into six 1-mm-thick sheets with a wire gel cutter (Numachi, 1981) and stained for the enzymes aspartate aminotransferase (AAT), isocitrate dehydrogenase (IDHP), lactate dehydrogenase (LDH), phosphoglucomutase

(PGM), and phosphogluconate dehydrogenase (6PGD) according to Shaw and Prasad (1970), Numachi (1970a, b), and Taniguchi and Numachi (1978). The locus and gene nomenclature followed Shaklee *et al.* (1990). Polymorphisms involving several alleles with frequencies of more than 5% were tested at a significance level of 0.05 to determine whether they were consistent with Hardy–Weinberg equilibrium. The average heterozygosity H (Nei, 1978) was calculated as a measure of genetic diversity. The χ^2 homogeneity test of allele frequency among samples was also performed.

2.2 Mitochondrial non-coding region of *Sepioteuthis* sp. 1 and sp. 3

In total, 116 *Sepioteuthis* sp. 1 were collected between April 2005 and September 2006 at three localities: Itoman, Okinawajima (49 individuals), Ishigakijima (38), and Keeling, Taiwan (29). An arm or part of the mantle muscles was kept in 90% ethanol and DNA was extracted with TNES 8M-Urea buffer. For *Sepioteuthis* sp. 3, 60 samples were collected between October 2005 and July 2006 from Nago, Okinawajima (30), and Ishigakijima (30). Crude DNA was extracted by TNES 8M Urea buffer and proteinase K digestion followed by a phenol-chloroform isoamyl method described Imai *et al.* (2004).

We analyzed the noncoding region 2 (NC2) between the Ala and Trp transfer RNAs (tRNAs). The original primers SL-Ala (5'-GGTAACCCTTTCTGTATGATTGC-3') and SL-Trp (5'-AAAGACCTTGAAAGTCTTCAG-3'), which target a portion of tRNA-Ala and tRNA-Trp, respectively, were used with the polymerase chain reaction (PCR) to amplify NC2 (Aoki *et al.*, 2008a). The PCR reactions were performed using BIOTAQ (Bioline, UK). A GeneAmp 9700 (Applied Biosystems, USA) thermal cycler was used with the following setting: 94°C for 120 s, followed by 30 cycles at 94°C for 30 s, 60°C for 30 s, and 72°C for 45 s. The PCR products were purified using a PCR Product Pre-sequencing Kit (USB, USA). The nucleotide sequences were determined using ABI 3700 (Applied Biosystems, USA) genetic analyzers. All sequences were initially aligned using ClustalX ver. 1.83.1 (Thompson *et al.*, 1997) and then edited manually using MacClade4 ver. 4.08 (Maddison and Maddison, 2005).

The haplotype diversity h (Nei, 1987) and nucleotide diversity π (Tajima, 1983) within populations were calculated using Arlequin ver. 2.000 (Schneider *et al.*, 2000). An analysis of molecular variance (AMOVA; Excoffier *et al.*, 1992) was used to test the population structure within species for *Sepioteuthis* sp. 1 using Arlequin.

For *Sepioteuthis* sp. 3, AMOVA could not be performed because there were fewer than three localities. Therefore, homogeneity was tested using the chi-square randomization method (Monte Carlo simulation) with 100,000 randomizations of the data (Roff and Benzen, 1989). Significance thresholds were Bonferroni-corrected for multiple pairwise comparisons. Relationships of haplotypes were assessed using a minimum spanning tree created via the Minspanet algorithm in Arlequin and drawn by hand.

3. Results and discussion

3.1 Allozyme analysis of *Sepioteuthis* sp. 2

Regarding the eight loci for the six enzymes analyzed, the five loci *Aat-1**, *Idhp-1**, *Ldh-1**, *Mdh-1**, and *Mdh-3** showed no differences among and within localities, and no genetic

Genetic Diversity and Genetic Heterogeneity of Bigfin Reef Squid "Sepioteuthis lessoniana" Species Complex in Northwestern Pacific Ocean

155

polymorphism was recognized. Two Mdh-2^* heterozygotes were found in Vietnam, although the frequency was 0.017; therefore, it was not considered a polymorphic allozyme locus (Table 1). Genetic polymorphism was detected within a locality for Pgm^* and $6pgd^*$. The polymorphic allozyme loci were in Hardy–Weinberg equilibrium at the localities. Most of the alleles linked to a locus were monomorphic. A marked excess of homogeneity was found. The average observed heterozygosity H=0.005–0.052 was similar to the values of H=0.037 reported by Izuka et al. (1996) and 0.052–0.070 by Yokogawa and Ueta (2000). Other loliginid species have similar heterozygosity values: *Loligo pealeii*, H=0.006; *Lolliguncula brevis*, H=0; *L. plei*, H=0 (Garthwaite et al., 1989); *Ommastrephes bartrami*, H=0.004; *Sthenoteuthis oualaniensis*, H=0.011; *Todarodes pacificus*, H=0.043; *Loliolus japonica*, H=0.030 (Fujio & Kawada, 1989); *L. vulgaris reynaudii*, H=0.030; *L. gahi*, H=0.059 (Carvalho & Loney, 1989); *L. bleekeri*, H=0.003 (Suzuki et al., 1993); and *L. chinensis*, H=0.006–0.009 (Yeatman and Benzie, 1993). The family Loliginidae appears to be characterized by low genetic diversity.

Locus	allele	Ishikawa	Tokushima	Nagasaki	Okinawajima	Taiwan	Vietnam
Aat-1	100 A	1.000	1.000	1.000	1.000	1.000	1.000
Idh-1	100 A	1.000	1.000	1.000	1.000	1.000	1.000
Ldh-1	100 A	1.000	1.000	1.000	1.000	1.000	1.000
Mdh-1	100 A	1.000	1.000	1.000	1.000	1.000	1.000
Mdh-2	110 A	0.000	0.000	0.000	0.000	0.000	0.017
	100 B	1.000	1.000	1.000	1.000	1.000	0.983
Mdh-3	100 A	1.000	1.000	1.000	1.000	1.000	1.000
Pgm	- 60 A	0.006	0.020	0.009	0.019	0.000	0.000
	-100 B	0.946	0.853	0.931	0.952	0.978	1.000
	-140 C	0.048	0.128	0.060	0.029	0.022	0.000
6Pgd	150 A	0.078	0.029	0.043	0.039	0.000	0.033
	100 B	0.892	0.941	0.888	0.933	1.000	0.908
	60 C	0.030	0.029	0.069	0.029	0.000	0.058
Observed	Ho	0.041	0.052	0.045	0.026	0.005	0.027
Expected	He	0.038	0.046	0.042	0.028	0.005	0.025
	Ho / He	1.080	1.120	1.084	0.957	1.000	1.067

Table 1. Allele frequencies at eight loci and indices of genetic heterozygosities within six localities of *Sepioteuthis* sp. 2.

No allele frequency gap was observed among different localities for the polymorphic allozyme loci $6pgd^*$ allele frequency. In contrast, a significant difference was detected between Pgm^* in the Japanese and Vietnamese localities (Table 2). This result differed greatly from that of Pratoomchat et al. (2001), who found the same gene pool in Thailand and Nagasaki, while our result supported the result of Aoki et al. (2008). Pratoomchat et al. (2001) used Pgm^* and $6pgd^*$, but the results might have been influenced by differences in the electrophoresis buffer. No difference in allele frequency was observed among localities in Japan. Yokogawa and Ueta (2000) showed replacement of Ldh-4^* between the main island Japan Sea and Pacific sides, although the allozyme band pattern shown in that paper may have been manipulated, and we find the results suspect. Therefore, we examined Ldh-4^* with a fresh sample following the advice of Dr. Yokogawa, and we did not find it. Pratoomchat et al. (2001) cited Yokogawa and Ueta (2000), but did not detect Ldh-4^*. When Ldh-4^* was eliminated, no difference existed between the Japan Sea and Pacific sides. Aoki et *al.* (2008a) could not show a difference between the Japan Sea and the main island Pacific side, even on analyzing the mitochondrial noncoding region sequence. If *Ldh-4** of Yokogawa and Ueta (2000) is repeatable, the difference in a highly polymorphic marker among populations would not always be detected. For example, regarding *Theragra chalcogramma*, a restriction fragment length polymorphism (RFLP) analysis of mitochondrial DNA and microsatellite DNA could not identify the difference among populations seen in the allele frequency of the superoxide dismutase (SOD) allozyme marker (Iwata, 1975; Mulligan *et al.*, 1992; Bailey *et al.*, 1999; Chow, 2001). Our finding of gene flow between Taiwanese and Japanese localities detected in the allozyme analysis of *Sepioteuthis* sp. 2 was not consistent with the sequencing analysis of the mitochondrial noncoding region by Aoki *et al.* (2008a), perhaps because of the low level of polymorphic loci for the allozyme analysis.

In addition, noncoding regions such as the mitochondrial control region accumulate more variation than allozyme markers. The relative smallness of the effective population size (female) with nuclear DNA made it easier to detection interpopulation genetic differentiation by genetic drift (Williams *et al.*, 2002). Therefore, Aoki *et al.* (2008a) used the mitochondrial noncoding region and found low genetic diversity in Japanese waters. Furthermore, Aoki *et al.* (2008a) revealed the independence of gene flow within the populations in Japanese waters from others.

	Ishikawa	Tokushima	Nagasaki	Okinawajima	Taiwan	Vietnam
Ishikawa	****	0.026	0.591	0.296	0.784	0.009
Tokushima	0.0011	****	0.131	0.241	0.113	0.000*
Nagasaki	0.0000	0.0008	****	0.666	0.543	0.003*
Okinawajima	0.0001	0.0010	0.0001	****	0.368	0.005*
Taiwan	0.0012	0.0017	0.0012	0.0003	****	0.021
Vietnam	0.0004	0.0024	0.0004	0.0002	0.0008	****

Table 2. *P*-value in allele frequencies (*Pgm**; above diagonal) and Nei's genetic distance (below diagonal) among six localities of *Sepioteuthis* sp2. Bonferroni correction *P*<0.05.

3.2 Mitochondrial non-coding region of *Sepioteuthis* sp. 1

We sequenced 552 base pairs (bp) of the NC2 sequence for 116 *Sepioteuthis* sp. 2 specimens from three localities. From a total of 35 haplotypes, 23 variable sites were identified (Table 3). One haplotype was shared among three localities, and the remaining 31 haplotypes were each specific to a single locality. Among the populations, 23.3% of the samples belonged to haplotype no. 1, which was the major haplotype in all Japanese localities. In contrast, haplotype no. 3 was the major haplotype in Taiwan (Figure 3).

The haplotype diversity (h) ranged from 0.7994 for Ishigakijima to 0.8665 for Okinawajima, and the nucleotide diversity (π) varied from 0.0035 for Ishigakijima to 0.0052 for Okinawajima (Table 4). Among the three *Sepioteuthis* sp. 1 localities, the level of genetic diversity did not differ much. The genetic diversity of *Sepioteuthis* sp. 1 was similar to that of *Sepioteuthis* sp. 2: h=0.8972, π=0.0124 in Taiwan and h=0.6828, π=0.0077 in Vietnam. The Japanese values of h=0.2583, π=0.0024 indicate that Japan has three times more genetic diversity than haplotype diversity.

Haplotype	14	19	21	137	199	213	216	228	231	235	236	237	240	242	276	293	294	301	320	330	393	508	536	OK	IG	TA
1	G	A	C	A	T	T	G	C	A	C	A	G	C	A	C	T	C	T	G	T	G	T	A	13	14	0
2	A	·	·	·	·	·	·	·	·	·	·	·	·	·	·	·	·	·	·	·	·	·	·	12	7	0
3	A	·	·	·	·	·	·	·	T	·	G	·	·	·	·	·	·	·	·	·	·	·	·	0	0	11
4	·	·	·	·	·	·	·	·	·	·	G	·	·	·	·	·	·	·	A	·	·	·	·	0	8	0
5	A	·	·	·	·	·	·	·	T	·	G	·	·	·	·	·	·	·	A	·	·	·	·	0	0	7
6	A	·	·	·	·	·	·	·	·	·	·	·	·	·	·	·	·	·	A	·	·	·	·	3	0	1
7	·	·	·	·	·	·	·	·	·	·	G	·	·	·	·	·	·	·	·	·	·	·	·	2	1	0
8	·	·	·	·	·	·	T	·	·	·	G	·	T	·	·	·	·	·	A	C	·	·	·	4	0	0
9	A	·	·	·	·	·	·	·	T	·	G	·	·	·	·	·	·	·	A	C	·	·	·	2	0	0
10	·	·	·	·	·	·	·	G	·	G	·	·	T	·	·	·	·	·	·	·	·	·	·	0	2	0
11	A	·	G	C	·	·	·	·	T	·	·	·	·	·	·	·	·	·	A	·	·	·	·	0	0	2
12	A	·	·	·	·	·	·	·	T	·	G	·	·	·	·	·	·	C	A	·	·	·	·	0	0	2
13	·	·	·	·	·	·	·	·	T	·	G	·	·	·	·	·	·	·	A	C	·	G	·	1	0	0
14	·	·	·	·	·	·	·	·	T	·	G	·	·	·	·	·	·	·	A	·	·	·	·	2	0	0
15	·	·	·	·	C	·	·	·	·	·	·	·	·	·	·	·	·	·	·	·	·	·	·	1	0	0
16	·	·	·	G	·	·	·	·	·	·	·	·	·	·	·	T	·	·	·	·	·	·	·	1	0	0
17	·	·	·	·	·	·	·	·	·	·	G	·	T	G	·	·	·	·	·	·	·	·	·	1	0	0
18	A	·	·	·	·	·	·	-	T	T	G	·	·	·	·	·	·	·	A	·	·	·	·	1	0	0
19	·	·	·	·	·	·	A	·	·	·	·	·	·	·	·	·	·	·	A	·	·	·	·	0	1	0
20	A	·	·	G	·	·	·	·	T	·	G	·	·	·	·	·	·	·	A	·	·	·	·	0	1	0
21	A	·	·	·	·	·	A	·	G	·	·	·	·	·	·	·	·	·	·	·	·	·	·	0	1	0
22	A	·	·	·	·	·	·	·	G	·	·	·	·	·	·	·	·	·	·	·	·	·	·	0	1	0
23	A	·	·	·	·	·	·	·	·	·	·	A	·	·	·	·	·	·	·	·	·	·	·	0	1	0
24	·	·	·	·	·	·	A	·	·	·	·	·	·	·	·	·	·	·	·	·	·	·	·	0	1	0
25	A	·	T	·	·	·	·	·	T	·	G	·	·	·	·	·	·	·	·	·	·	·	·	0	0	1
26	A	·	·	G	C	·	·	·	T	·	·	·	·	·	·	·	·	·	A	·	T	·	·	0	0	1
27	A	·	·	·	·	·	·	·	·	·	G	·	·	·	·	·	·	·	·	·	·	·	·	0	0	1
28	A	·	·	·	·	·	·	·	T	·	G	·	·	·	T	·	·	C	A	·	·	·	·	0	0	1
29	A	·	·	·	·	·	·	·	·	T	G	A	·	·	·	·	T	·	·	·	·	·	·	0	0	1
30	A	·	·	·	·	·	·	·	·	·	G	·	·	·	C	T	·	·	·	·	·	·	·	0	0	1
31	A	·	·	·	C	·	·	·	T	·	·	·	·	·	·	·	·	·	A	·	·	·	·	1	0	0
32	A	G	·	·	·	·	·	·	·	·	·	·	·	·	·	·	·	·	·	·	·	·	·	2	0	0
33	G	·	·	·	·	C	·	·	·	·	·	T	·	·	·	·	·	·	·	·	·	·	·	1	0	0
34	G	·	·	·	·	·	·	·	T	·	G	·	·	·	·	·	·	·	A	C	·	·	·	1	0	0
35	·	·	·	·	·	·	·	·	T	·	G	·	·	·	·	·	·	·	A	C	·	C	·	1	0	0
Total																								49	38	29

Table 3. Haplotype distribution and variable sites of mitochondrial NC2 region of *Sepioteuthis* sp. 1 among three localities.

	Okinawajima	Ishigakijima	Taiwan
Haplotype diversity (h)	0.8665	0.7994	0.8079
Nucleotide diversity (π)	0.0052	0.0035	0.0043
Number of individuals (n)	49	38	29
Number of haplotypes	17	11	11

Table 4. Haplotype diversity and nucleotide diversity of *Sepioteuthis* sp. 1 among three localities.

The AMOVA indicated that the genetic variation over all of the Japanese localities was 34.87%, whereas the within-locality variation was 65.13% ($p<0.01$ [Table 5]). The estimated pairwise Fst values for the three pairs of three localities ranged from 0.0538 to 0.5329. All combinations of locality samples had significant pairwise Fst values ($p<0.05$ [Table 6]). Therefore, each locality had an independent population with restricted gene flow, concurring with Aoki *et al.* (2008a). *Sepioteuthis* sp. 2 had gene flow within the territorial waters of Japan and showed genetic homogeneity. The relationships among the haplotypes

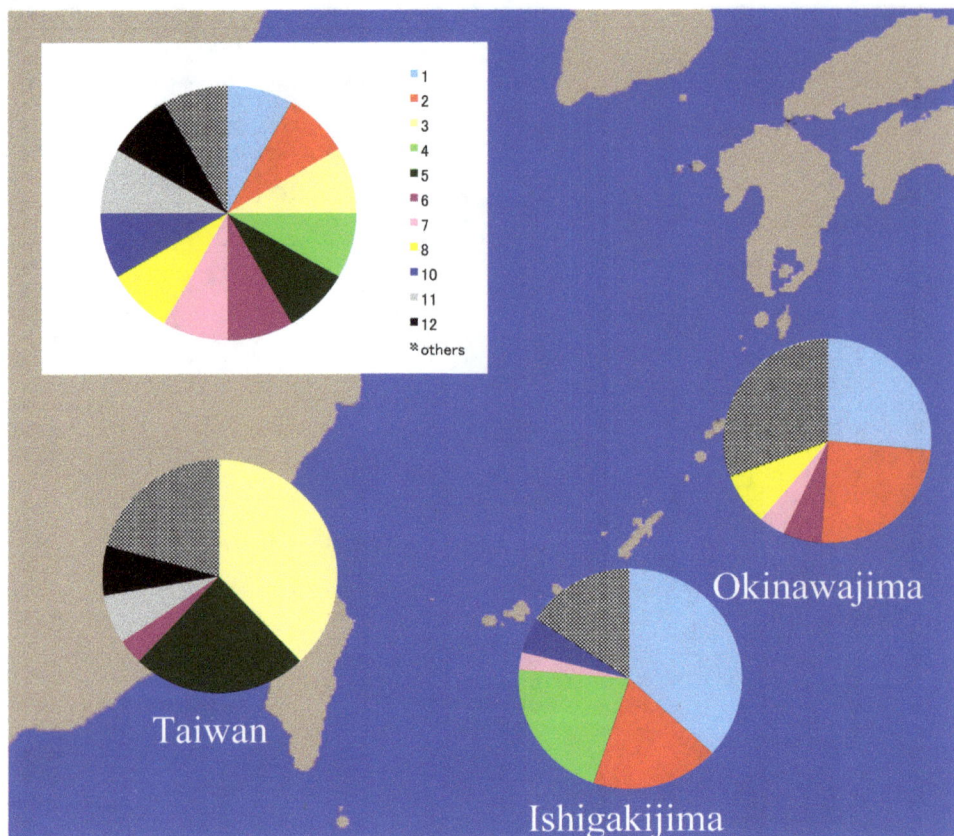

Fig. 3. Pie chart representation of the haplotype frequencies of *Sepioteuthis* sp. 1 of the three localities.

	d.f.	% variation	F-statistics	P
Among populations within groups	2	34.87	$F_{ST} = 0.3487$	< 0.01
Within populations	113	65.13		

Table 5. Analysis of Molecular Variance on pairwaise differences and *P*-value of *Sepioteuthis* sp. 1 among three localities is the probability of a more extreme variance component.

	Okinawajima	Ishigakijima	Taiwan
Okinawajima	****	0.0090*	0.0000*
Ishigakijima	0.0538	****	0.0000*
Taiwan	0.4237	0.5329	****

Table 6. Pairwise *F*st and associated probability (*P*) of *Sepioteuthis* sp. 1 among three localities. *F*st values are below the diagonal and corresponding *P* values are above the diagonal. Bonfferroni correction *P*<0.05.

were represented on a minimum spanning tree, and the shape indicated that the population had long-term stability (Figure 4).

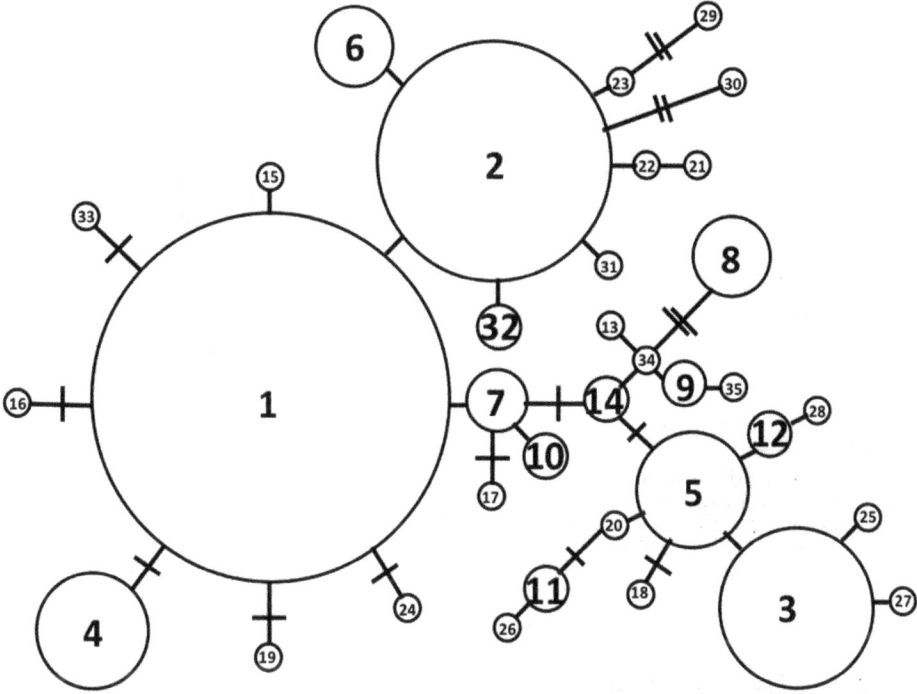

Fig. 4. Minimum spanning tree among 35 hapolotypes of *Sepioteuthis* sp. 1.

3.3 Mitochondrial non-coding region of *Sepioteuthis* sp. 3

We sequenced 557 bp NC2 sequences for 60 *Sepioteuthis* sp. 3 specimens from two localities. From a total of 15 haplotypes, 13 variable sites were identified (Table 7). Seven haplotypes were shared between the two localities, and the remaining eight were specific to a single locality. Overall, 31.6% of the samples belonged to haplotype no. 1, which was not the major haplotype in both localities. Haplotype no. 2 was the major haplotype in Okinawajima (Figure 5). The haplotype diversity (h) ranged from 0.7103 for Ishigakijima to 0.8828 for Okinawajima, and the nucleotide diversity (π) varied from 0.0037 for Ishigakijima to 0.0044 for Okinawajima (Table 8). The level of genetic diversity was similar to that of *Sepioteuthis* sp. 1.

Significant heterogeneity was observed between the Okinawajima and Ishigakijima populations (χ^2=23.89, p<0.01). Therefore, these two populations could be distinguished by the haplotype frequency. Izuka et al. (1996) reported that each population was genetically independent based on the allozyme analysis for Ishigakijima and the Ogasawara Islands. These results suggest that *Sepioteuthis* sp. 3 does not experience larval dispersal, but completes its life history within coral reefs. The relationships among the haplotypes were represented on a minimum spanning tree, and the shape indicated that haplotypes could not be divided into clusters (Figure 6).

Haplotype	91	155	209	235	236	244	254	270	300	321	348	510	542	OK	IG
1	A	A	C	A	C	A	C	T	T	C	T	T	G	3	16
2	·	·	·	·	·	·	·	·	·	·	·	·	A	9	3
3	·	C	·	·	·	·	·	·	·	·	·	·	A	4	2
4	·	C	·	·	·	·	·	·	·	·	A	·	·	3	1
5	·	C	·	·	·	·	·	·	·	·	·	·	·	2	2
6	·	·	T	·	·	G	·	C	·	·	·	C	·	1	2
7	·	·	T	·	T	·	·	C	·	·	·	·	·	2	0
8	·	C	·	·	·	·	·	·	·	·	A	·	A	2	0
9	·	·	T	·	T	·	·	C	·	·	·	·	A	1	1
10	·	·	T	·	·	·	·	C	·	·	·	·	A	1	0
11	·	·	T	·	·	·	·	C	·	T	·	·	A	1	0
12	·	·	T	·	·	·	T	C	·	·	·	·	A	1	0
13	C	·	T	·	·	G	·	C	·	·	·	C	·	0	1
14	·	·	·	G	·	·	·	·	·	·	·	·	A	0	1
15	·	·	T	·	·	G	·	C	C	·	·	C	·	0	1
Total														30	30

Table 7. Haplotype distribution and variable sites of mitochondrial NC2 region of *Sepioteuthis* sp. 3 between Okinawajima and Ishigakijima Island.

	Okinawajima	Ishigakijima
Haplotype diversity (h)	0.8828	0.7103
Nucleotide diversity (π)	0.0044	0.0037
Number of individuals (n)	30	30
Number of haplotypes	12	10

Table 8. Haplotype diversity and nucleotide diversity of *Sepioteuthis* sp. 3 between Okinawajima and Ishigakijima Island.

4. General discussion

In this research showed that genetic differentiation between Okinawajima and Ishigakijima population was identified for *Sepioteuthis* sp. 1 and *Sepioteuthis* sp. 3. The result showed that *Sepioteuthis* sp. 1 and sp. 3 prefer coast lines as habitat that limit periodic dispersal of larva and adult among islands. The result showed that there is no gene flow of *Sepioteuthis* sp1 between Ishigakijima and Taiwan, as Aoki *et al.* (2008a) showed for *Sepioteuthis* sp. 2. Geographical distance of these two areas is 300 km, which has no difference of the geographical distance between Okinawajima and Ishigakijima Island. However, genetic structure differentiation between Ishigakijima and Taiwan is bigger than that of Okinawajima and Ishigakijima. Thus, gene flow between Ishigakijima and Taiwan was disturbed for long period. Kuroshio Current possible is possibly disturbing the gene flow Kuroshio Current is warm current with the surface speed of 2m per second, strong flow that moves more than 50 million tons of water. The current axis starts from north equatorial countercurrent, go up towards north between Taiwan and Yonagunijima, through Tokara strait and flows into southern coast of main island Japan (Figure 7). The width between Taiwan and Yonagunijima is small. Kuroshio Current go up to the north along with

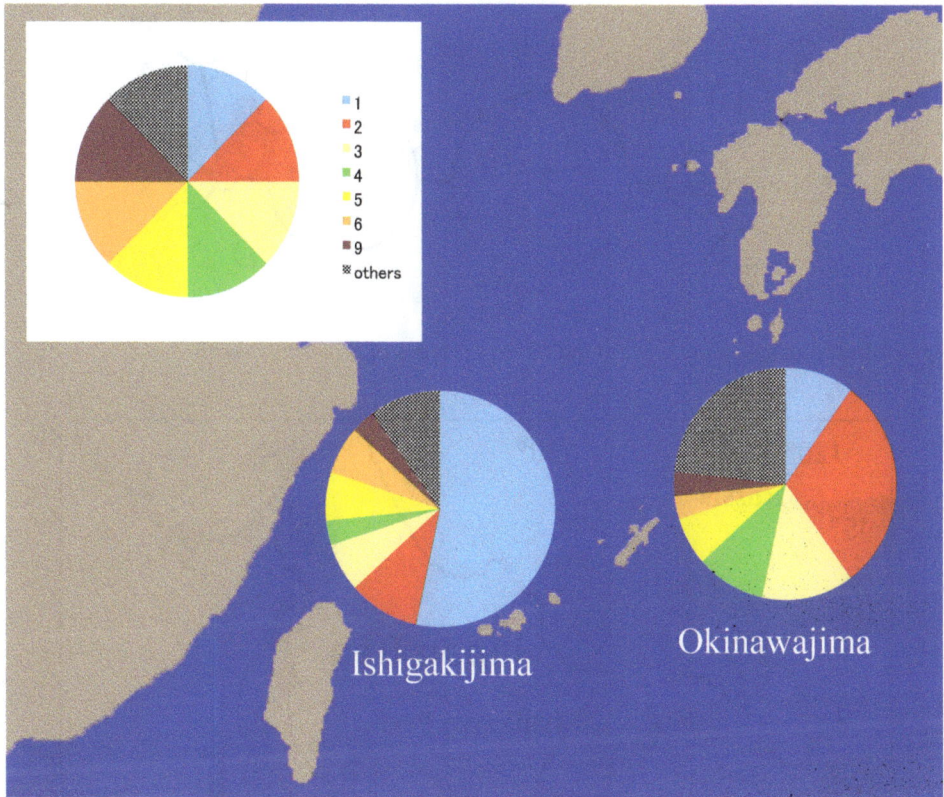

Fig. 5. Pie chart representation of the haplotype frequencies of *Sepioteuthis* sp. 3 of the two localities.

continental shelf. The current split around Kyushu to Tsushima Current along with Japan Sea and main current that goes along with Pacific coast of main island, Japan. The current disperse when it goes to north, the flow of Kuroshio Currnet may take them to very northern part, that has lower temperature of ocean water. Geographical cal distribution of *Panulirus longipes* is another example in which Kuroshio Current is a barrier current between Taiwan and Ryukyu islands (Sekiguchi & Inoue, 2010). Accordingly, several marine organisms, *Uca arcuata* and *Siganus guttatus* shows different genetic structure between Ryukyu Archipelago and Taiwanese population showing no gene flow (Aoki *et al.*, 2008b; Iwamoto *et al.*, 2012). These genetic sturucture pattern may suggest the influence of Kuroshio Current. Especially squids has short longevity, some of the species has only one spawning season in a life. Drastic environmental change and some other accidental events may destroy population. In order to raise the fitness of squids, water temperature and appropriate environment of growth phase are essential (O'Dor & Coelho, 1993). Kuroshio Current may supply appropriate temperature and abundant feed resources for *Sepioteuthis* spp habitat. The comparative study of *Sepioteuthis* sp. 2 reported by Aoki *et al.* (2008a) and *Sepioteuthis* sp. 1 MST shape shows that these two species can belong to each clade of Japan and Taiwan.

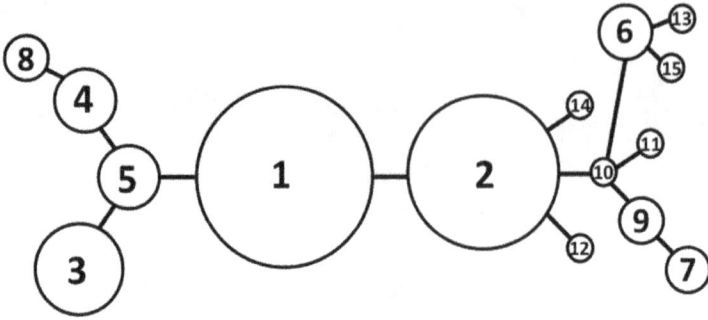

Fig. 6. Minimum spanning tree among 15 hapolotypes of *Sepioteuthis* sp. 3.

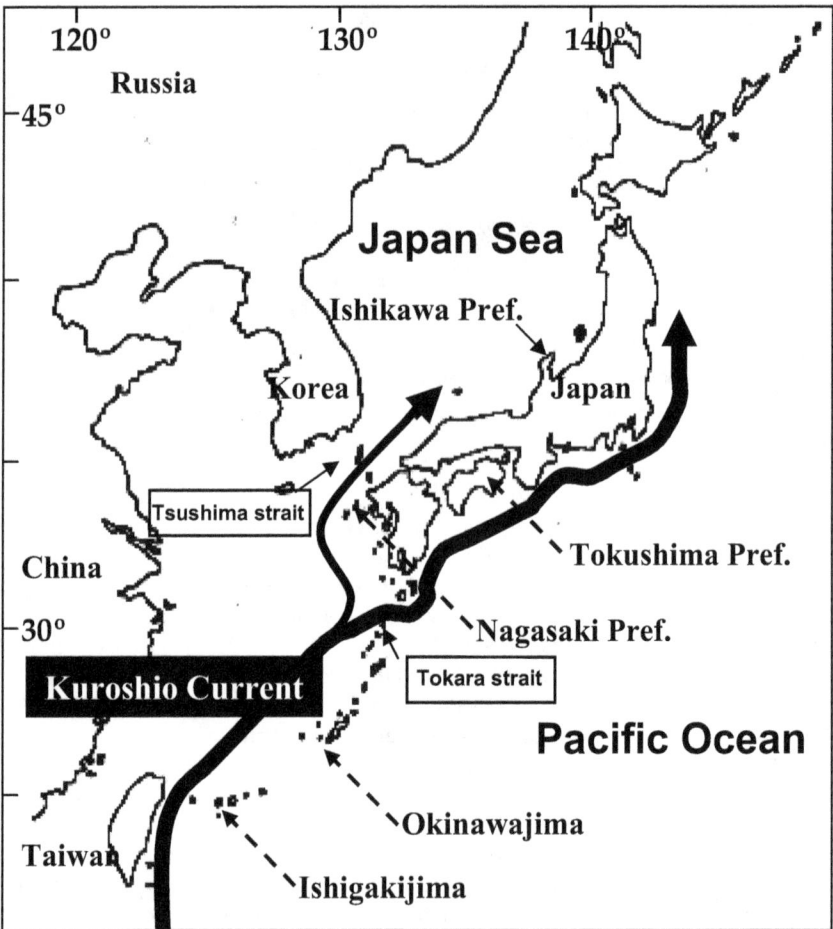

Fig. 7. The location of the main pathway of Kuroshio Current and sampling sites.

5. Conclusion

We analyzed the genetic heterogeneity of populations of *Sepioteuthis* sp. 1 and *Sepioteuthis* sp. 3 in Okinawan waters using DNA analysis. The genetic diversity of *Sepioteuthis* sp. 1 and *Sepioteuthis* sp. 3 was higher than that of *Sepioteuthis* sp. 2 from Japanese waters. Moreover, the genetic heterogeneity of the populations differed significantly. The difference in genetic heterogeneity means that *Sepioteuthis* sp. 1 and *Sepioteuthis* sp. 3 do not have a large gene pool. We postulate that the reason for the genetic differentiation is that these two species prefer coastal habitats. Our results indicate that the Japanese populations of *Sepioteuthis* sp. 2 have very low genetic diversity compared to those of Taiwan and Vietnam. The minimum spanning tree showed that the Japanese populations were of the radiation type, implying that the Japanese populations had experienced founder effects.

The genetic heterogeneity in the Japanese populations was slightly different using AMOVA. This suggests that the mitochondrial noncoding region of the Japanese population lacks sufficient genetic diversity to assess the genetic heterogeneity in the Japanese populations. Moreover, our results suggest that only limited gene flow has occurred between the Ishigakijima and Taiwan populations of *Sepioteuthis* sp. 1 and *Sepioteuthis* sp. 2, implying the presence of barriers to gene flow. Kuroshio Current, a prominent current in this area, which moves at a rate of nearly 50 million m³/s, may prevent dispersal from Taiwan to Ishigakijima.

Lastly, it should be noted that Prof. Segawa's contribute to ecological research for *Sepioteuthis lessoniana* complex. Further study should focus on resolute species complex as soon as possible to develop ecological research of *Sepioteuthis* spp. It is necessary to identify species identification marker, however, frequently used allozyme marker needs fresh samples. Finding DNA marker that can be used for ethanol sample will be useful. DNA marker development should use allozyme marker of Izuka *et al.* (1994) and Triantafillos and Adams (2005) as the standard specimens. Allozyme analysis is the method to detect nuclear DNA polymorphism by the detection of enzyme molecule polymorphism. It cannot show nucleotide base-substitution mutation when it does not have amino-acid sequence variation. The different condition of electrophoresis buffer may produce different results, even though it is worth noting that it is a reliable tool to find cryptic species (Imai, 2006). The first author of this paper, Imai, H is currently working on development of species identification using DNA marker with some other researchers.

6. Acknowledgment

We thank Prof. Y. Ikeda of the Faculty of Science, University of the Ryukyus, Dr. Y. Ueta of the Fisheries Research Institute, Tokushima Agriculture, Forestry, and Fisheries Technology Support Centre; Prof. T. Y. Chan and Dr. M. Mitsuhashi of the Institute of Marine Biology, National Taiwan Ocean University; Dr. B. K. K. Chan of the Research Centre for Biodiversity, Academia Sinica; Mr. T. Higa of the Nago Fishery Cooperative Association; Mr. Y. Yonamine of the Yaeyama Fishery Cooperative Association; and Mr. S. Nagata of Ryukyu-Taiyo Inc. for sample collection. We also   thank Dr. K. Yokogawa Kagawa Prefecture residing for kindly advice of Ldh-4 detection and Ms. K. Hirouchi for checking English the manuscript.

7. References

Aoki, M.; Imai, H. & Ikeda, Y. (2008a). Low genetic diversity of oval squid, *Sepioteuthis* cf. *lessoniana* (Cephalopoda: Loliginidae), in Japanese waters inferred from a mitochondrial DNA non-coding region. *Pacific Science*, Vol. 62, pp. 403-411.

Aoki, M.; Naruse, T.; Cheng, J; Suzuki, Y. & Imai, H. (2008b). Low genetic variability in an endangered population of fiddler crab *Uca arcuata* on Okinawajima Island: analysis of mitochondrial DNA. *Fisheries Science*, Vol. 74, pp. 330-340.

Bailey, K.; Quinn, M.; Benzie, P. & Grant, W. S. (1999). Population structure and dynamics of walleye pollock, *Theragra chalcogramma*. *Advance in Marine Biology*, Vol. 38, pp. 177-255.

Carvalho, G. R. & Loney, K. H. (1989). Biochemical genetic studies on the Patagonian squid *Loligo gahi* d'Orbrigny. 1. Electrophoretic survey genetic variability. *Journal of Experimental Marine Biology and Ecology*, Vol. 126, pp. 231-241.

Choe, S. & Ohshima, Y. (1961). On the embryonal development and growth of the squid, *Sepioteuthis lessoniana* Lesson. *Venus (Japanese Journal of Malaclogy)*, Vol. 21, pp. 462-476.

Chow, S. (2001). Stock identification, *in* Tanaka, S, *et al.* (Ed.), *Textbook for stock analysis method*, 18-27, Japan   Fisheries Resource Conservation Association, Tokyo [in Japanese].

Clayton, J.W. & Tretiak, D.N. (1972). Amine-citrate buffers for pH control in starch gel electrophoresis. *Journal of the Fisheries Research Board of Canada*, Vol. 29, pp.1169-1172.

Excoffier, L.; Smouse, P. E. & Quattro, J. M. (1992). Analysis of molecular variance inferred from metric distance among DNA haplotypes: Application to human mitochondrial DNA restriction data. *Genetics*, Vol. 131, pp. 479–491.

Dunning, M. C. (1998). Loliginidae, *in* Carpenter, K. E. & Niem, V. H. (Ed.), *The living marine resources of the western central Pacific*, 764-780, FAO species identification guide for fishery purposes, Food and Agriculture Organization of the United Nations, Rome.

Fujio, Y. & Kawada, G. (1989). Genetic differentiation and variability in squids, *in* Report on the genetic assessment project (Ed.), *Population analysis of marine organisms by isozyme analysis*, 508–523, Japan Fisheries Resource Conservation Association, Tokyo [in Japanese].

Garthwaite, R. L.; Berg Jr., C. J. & Harrigan, J. (1989). Population genetics of the common squid *Loligo pealei* LeSueur, 1821 from Cape Cod to Cape Hatteras. *Biological Bulletin*, Vol. 177, pp. 287-294.

Ikeda, H. (1933). Sex-correlated marking in *Sepioteuthis lessoniana* Férussac. *Venus(Japanese Journal of Malaclogy)*, Vol. 3, pp. 324-329.

Ikeda, Y.; Imai, H.; Sugimoto, C. & Oshima, Y. (2009). Egg case containing three ova of oval squid *Sepioteuthis lessoniana* in Okinawa Island of the Ryukyu Archipelago. *Aquaculture Science*, Vol. 57, pp. 631-634.

Imai, H. (2006). Animal diversity of bay and tidal in Ryukyu Archipelago, *in Biodiversity of coral reef and island ecosystems of the Ryuykus*, 21st Centry COE Program, University ot the Ryukyu (Ed.), 35-47,. Tokai University Press, Tokyo [in Japanese].

Imai, H.; Cheng, J. H.; Hamasaki, K. & Numachi, K. (2004). Identification of mud crabs (genus *Scylla*) using ITS-1 and 16S rDNA markers. *Aquatic Living Resources*, Vol. 17, pp. 31–34.

Iwamoto, K.; Chang, C.; Takemura, A. & Imai, H. (2012). Genetically structured population and demographic history of the goldlined spinefoot *Siganus guttatus* in northwestern Pacific. Fisheries Science, Vol. 78, in press [doi: 10.1007/s 12562-011-0455-3].

Iwata, M. (1975). Genetic identification of walleye pollock (*Theragra chalcogramma*) populations on the basis of tetrazolium oxidase polymorphism. *Comparative Biochemical Physiology*, Vol. 50B, pp. 197-201.

Izuka, T.; Segawa, S. & Okutani, T. (1996). Biochemical study of the population heterogeneity and distribution of the oval squid *Sepioteuthis lessoniana* complex in southwestern Japan. *American Malacological Bulletin*, Vol. 12, pp. 129-135.

Izuka, T.; Segawa, S.; Okutani, T. & Numachi, K. (1994). Evidence on the existence of three species in the oval squid *Sepioteuthis lessoniana* complex in southwestern Japan. *Venus:(Japanese Journal of Malaclogy)*, Vol. 53, pp. 217-228.

Jereb, P. & Roper, C. F. E. (2006). Cephalopods of the Indian Ocean. A review. Part I. Inshore squids (Loliginidae) collected during the International Indian Ocean Expedition. *Proceedings ot the Biological Society of Washington*, Vol. 119, pp. 91-136.

Lu, C. C.; Boucher-Rodoni, R. & Tillier, A. (1995). Catalogue of types of recent Cephalopoda in the Muséum National d'Histoire Naturelle (France). *Bulletin du Muséum national d'Histoire naturelle*, Paris, 4e sétie, Vol. 17, pp. 307-343.

Maddison, D. R. & Maddison, W. P. (2005). *MacClade: A computer program for phylogenetic analysis. 4.08 OSX*. Sinauer Associates, Inc. Sunderland, Massachusetts.

Mulligan, T. J.; Chapman, R. W. & Brown, B. L. (1992). Mitochondrial DNA analysis of walleye polluck, *Theragra chalcogramma*, from the eastern Bering Sea and Shelikof Strait, Gulf of Alaska. *Canadian Journal of Fisheries and Aquatic Sciences*, Vol. 49, pp. 319-326.

Nei, M. (1978). Estimation of average heterozygosity and genetic distance from a small number of individuals. *Genetics*, Vol. 89, pp. 583–590.

Nei, M. (1987). *Molecular evolutionary genetics*, Columbia University Press, New York.

Numachi, K. (1970a). Lactate and malate dehydrogenase isozyme pattern in fish and marine mammals. *Bulletin of the Japanese Society of Scientific Fisheries*, Vol. 36, pp. 1067-1077.

Numachi, K. (1970b). Polymorphism of malate dehydrogenase and genetic structure of juvenile population in saury *Cololabis saira*. *Bulletin of the Japanese Society of Scientific Fisheries*, Vol. 36, pp. 1235-1241.

Numachi, K. (1981). Simple method for preservation and scanning of starch gels. *Biochemical Genetics*, Vol. 19, pp. 233-236.

Numachi, K. (1989). Identification of population of marne organisums by isozyme analysis, *in* Report on the genetic assessment project (Ed.), *Population analysis of fish and shellfish by isozyme analysis*, 42–63, Japan Fisheries Resource Conservation Association, Tokyo [in Japanese].

O'Dor, R. K. & Coelho, M. L. (1993). Big squid, big currents and big fisheries, *in* Okutani, T.; O'Dor, R. K. & Kubodera, T. (Ed.), *Recent Advance in Fisheries Biology*, 531-535, Tokai University Press, Tokyo.

Pratoomchat, B.; Natsukari, Y.; Maki, I. & Chalermwat, K. (2001). Allozyme determination of genetic diversity in Japanese and Thai populations of oval squid (*Sepioteuthis lessoniana* Lesson, 1930). *Lamar* (Tokyo), Vol. 39, pp. 133-139.

Roff, D. A. & Benzen, P. (1989). The statistical analysis of mitochondrial DNA polymorphisms: X^2 and the problem of small samples. *Molecular Biology and Evolution*, Vol. 6, pp. 539-545.

Segawa, S. (1987). Life history of the oval squid, *Sepioteuthis lessoniana* in Kominato and adjacent waters, central Honshu, Japan. Journal of Tokyo University of Fisheries, Vol. 74, pp. 67-47.

Segawa, S.; Hirayama, S. & Okutani, T. (1993a). Is *Sepioteuthis lessoniana* in Okinawa a single species?, *in* Okutani, T.; O'Dor, R. K. & Kubodera, T. (Ed.), *Recent Advance in Fisheries Biology*, 531-535, Tokai University Press, Tokyo.

Segawa, S.; Izuka, T.; Tamashiro, T. & Okutani, T. (1993b). A note on mating and egg deposition by *Sepioteuthis lessoniana* in Ishigaki Island, Okinawa, southwestern Japan. *Venus (Japanese Journal of Maralacology)*, Vol. 52, pp. 101-108.

Sekiguchi, H. & Inoue, N. (2010). Laval recruitment and fisheries of the spiny lobster *Panulirus japonicus* couplong with the Kuroshio subgyre circulation in the western North Pacific: A review. *Journal of Marine Bilogical Association of India*, Vol. 52, pp. 195-207.

Shaklee, J. B.; Allendorf, F. W.; Morizot, D. C. & Whitt, G. S. (1990). Gene nomenclature for protein-coding loci in fish. *Transactions American Fisheries Society*, Vol. 119, pp. 2–15.

Shaw, C. R. & Prasad, R. (1970). Starch gel electrophoresis of enzyme- A compilation of recipes. *Biochemical Genetics*, Vol. 4, pp. 297-320.

Suzuki, H.; Ichikawa, M. & Matsumoto, G. (1993). Genetic approach for elucidation of squid family, *in* Okutani, T.; O'Dor, R. K. & Kubodera, T. (Ed.), *Recent Advance in Fisheries Biology*, 531-535, Tokai University Press, Tokyo.

Tajima, F. (1983). Evolutionary relationship of DNA sequences in finite populations. *Genetics*, Vol. 105, pp. 437–460.

Taniguchi, N. & Numachi, K. (1978). Genetic variation of 6-phosphogluconate dehydrogenase, isocitrate dehydrogenase, and glutamic-oxaloacetic transaminase in the liver of Japanese eel. *Bulletin of the Japanese Society of Scientific Fisheries*, Vol. 44, pp. 1351-1355.

Thompson, J. D.; Gibson, T. J.; Plewniak, F.; Jeanmougin, F. & Higgins, D. G. (1997). The Clustal-X windows interface: Flexible strategies for multiple sequence alignment aided by quality analysis tools. *Nucleic Acids Reseaech*, Vol. 25, pp. 4876–4882.

Triantafillos, L & Adams, M. (2005). Genetic evidence that the northern calamary, *Sepioteuthis lessoniana*, is a species complex in Australian waters. *ICES Journal of Marine Science*, Vol. 62, pp. 1665-1670

Ueta, Y. (2003). Ecology and stock management of oval squid, *Sepioteuthis lessoniana*. Japan Fisheries Resource Conservation Association, Tokyo [in Japanese].

Schneider, S.; Roessli, D. & Excoffier, L. (2000). *Arlequin: A software for population genetics data analysis. Version 2.000.* Genetics and Biometry Laboratory, University of Geneva, Switzerland.

Williams S.T., Jara, J. A., Gomez, E. & Knowlton, N. (2002) The marine Indo-West Pacific break: contrasting the resolving power of mitochondrial and nuclear genes *Integrative and Comparative Biology*, Vol. 42 pp. 941-952 .

Yokogawa, K. & Ueta, Y. (2000). Genetic analysis of oval squid (*Sepioteuthis lessoniana*) around Japan. *Venus (Japanese Journal of Malacology)*, Vol. 59, pp. 45-55.

Young, R. E. (2002). Taxa associated with the family Loliginidae Lesueur, 1821. Version June 2002. http://tolweb.org/accessory/ Loliginidae_Taxa?acc_id=2326 in The Tree of Life Web Project, http://tolweb.org/.

Yeatman, J. M. & Benzie, J. A. H. (1993). Crytic specification in Loligo from Northern Australia, *in* Okutani, T.; O'Dor, R. K. & Kubodera, T. (Ed.), *Recent Advance in Fisheries Biology*, 641-652, Tokai University Press, Tokyo.

9

Genetic Variation of Host Immune Response Genes and Their Effect on Hepatitis C Infection and Treatment Outcome

Pooja Deshpande[1], Michaela Lucas[2,3] and Silvana Gaudieri[1,2]
[1]School of Anatomy and Human Biology
University of Western Australia, Western Australia
[2]Institute of Immunology and Infectious Diseases, Murdoch University, Western Australia
[3]Department of Health, Western Australia
Australia

1. Introduction

Effective control of pathogens is typically achieved by the timely interplay between the innate and adaptive immune response of an individual. The first-line of defence of the host immune system is our skin and mucosal membranes, which form a physical and chemical barrier. If a pathogen gains access to the body, for example when there is skin damage, cells of our innate immune system recognise pathogens by means of receptors (acting as sentinels) binding common motifs on pathogen surfaces and mount an immediate non-specific effector response. Often this response leads to the destruction of the pathogens, but also the production of secreted proteins such as interferons (IFNs) or cytokines, which help the interaction between the various cells involved in the immune response. This initial "inflammation" of the infected tissue commonly leads to the recruitment of lymphocytes to the site and triggers a more pathogen-specific response known as the adaptive immune response. In addition, local inflammation triggers activation of the complement system, a group of plasma proteins, which can recruit further immune cells to the site or be directly proteolytic to bacterial surfaces (Janeway 2008).

Not surprisingly, there are multiple genes that code for the molecules of the innate and adaptive host immune response. Between individuals, these genes can show variations within the coding sequence itself and/or within associated regulatory elements such as promoters. Furthermore, many of these genes are part of multicopy gene families resulting from previous gene duplication events (Kelley and Trowsdale 2005). Gene duplication can lead to new gene copies developing complementary and/or overlapping functions allowing for redundancy in the immune system and the ability to cope with a number of different pathogens (Nei, Gu et al. 1997).

Some of the genetic characteristics of the immune system predate mammalian divergence while others reflect more recent primate and human evolution. However, all living species, including humans, have been under constant bombardment from pathogens over thousands or millions of generations and it is likely that this host/pathogen interaction has to a large extent shaped the genetics of both host and pathogen populations.

Remarkably, for some highly mutable and therefore rapidly changing pathogens such as HIV and the Hepatitis C virus (HCV), two small RNA viruses, viral adaptation to the host's immune response results in characteristic variations in the virus that allow the virus to effectively escape a particular immune response. In contrast, larger, complex DNA viruses, such as Herpes viruses, with a more rigid genome have the coding capacity for additional genes, which interfere specifically with the immune response of the host (Lucas, Karrer et al. 2001). The multitude of escape strategies presented to our immune system, highlights the diversity of modifications needed to successfully combat the pathogens of our modern times. Furthermore, the result of these selective forces on the host over thousands of generations is the extensive host genetic diversity of immune response genes we observe today in all populations.

2. Examples of genetic diversity of the immune response genes

Two genetic systems that are commonly associated with variable infection outcomes are the killer-immunoglobulin like receptors (KIRs) present on NK cells that can determine the activatory or inhibitory propensity of the NK cell as part of the innate immune response and human leucocyte antigens (HLA) that present processed antigens to T lymphocytes, which are part of the adaptive immune response. These two genetic systems are also functionally connected; HLA molecules are ligands for the KIRs and their particular combinations add another layer of complexity to the differential immune response exhibited by individuals to particular pathogens.

2.1 KIR genetic diversity

The region of chromosome 19q13.4 containing the KIR genes is characterised by polymorphism, gene duplication, and linkage disequilibrium leading to KIR haplotypes spanning several hundred kilobases (Martin, Kulski et al. 2004). Genetic variation between haplotypes presents as allelic variation at specific loci and variation in gene content (Martin, Kulski et al. 2004; Pyo, Guethlein et al. 2010).

The KIRs can be separated into KIR2D and KIR3D genes based on the number of extracellular immunoglobulin domains. Further differentiation occurs based on the length of their cytoplasmic domain (short – S and long – L). Those KIR genes with a short cytoplasmic tail that lack an ITIM motif confer an activatory phenotype to the KIR2D and 3D genes while longer cytoplasmic domains with an ITIM motif confers an inhibitory phenotype. Sequencing of the KIR region and of individual genes shows extensive allelic diversity for both activatory and inhibitory genes (Table 1).

The effects of KIRs on host immune responses are mediated by specific cognate interactions between these receptors and their natural HLA Class I ligands on the surface of target cells, which may result in either activation or inhibition of NK cell cytotoxicity (Lanier 1998). The presence of particular HLA-B/-C alleles will determine the repertoire of inhibitory and activatory ligands that can be utilised by KIRs to modulate NK cell mediated responses. Several KIR and HLA combinations have been associated with both infectious disease outcome (Khakoo, Thio et al. 2004) and autoimmune diseases (Martin, Nelson et al. 2002). A functional model for the interaction between KIR and their HLA ligands has been proposed to account for a suspected hierarchy of inhibitory NK cell responses that may reflect the disease associations (Kulkarni, Martin et al. 2008). In this model, individuals with KIR haplotypes carrying a greater number of activatory genes, and with the appropriate HLA Class I ligands, may exhibit lower inhibition and higher activation (activatory phenotype), which may be beneficial in viral infections but may also correlate to a greater risk of developing autoimmunity.

2.2 HLA genetic diversity

The HLA genetic system within the Major Histocompatibility Complex (MHC) on the short arm of chromosome 6 is the most polymorphic in humans. Similar to the KIR region, the MHC region is characterised by extensive intra- and inter-genic polymorphism (Gaudieri, Dawkins et al. 2000), segment duplications (containing more than a single gene or gene fragment) (Gaudieri, Kulski et al. 1999) and linkage disequilibrium leading to haplotypes containing HLA and other immune-related genes that stretch hundreds of kilobases and even megabases. One major difference between KIR and MHC haplotypes is that within the MHC, for the most part, all individuals have the same number of HLA Class I and II genes but heterozygosity values at each locus is high.

The HLA family comprises class I and class II molecules that can present processed antigen to CD8+ and CD4+ T lymphocytes, respectively. The HLA Class I molecules including HLA-A, -B and –C are expressed on all nucleated cells and share significant similarity reflecting largely overlapping functions but with some important distinctions, including their function as NK cell receptor ligands. The HLA Class II molecules including HLA-DR, -DQ and –DP are expressed on antigen-presenting cells. Orthologues of the HLA Class I and II genes are present in other primates and most vertebrates.

HLA molecules were initially identified via their involvement in self-nonself discrimination in transplantation. Serology-based assays using anti-HLA antibodies were used to match donor and recipient pairs prior to transplantation. These initial HLA typing assays were able to differentiate dozens of HLA Class I and II proteins (Graw, Goldstein et al. 1970). However, it was not until sequence-based typing was developed in the 1990s that the extent of genetic variation within these genes became apparent. Several hundred to thousands of alleles have now been described for many of the HLA Class I and II loci (Table 1).

Much of the variation within HLA genes exists in the peptide binding domains of the molecule. Individuals with different HLA types can therefore present different parts of the pathogen to T lymphocytes and it is thought that the variation observed within the HLA peptide binding domains are driven by positive selection pressures that favour the maintenance of polymorphism in this system (Hughes, Ota et al. 1990). Support for this is provided by several studies that show evidence for heterozygote advantage at the HLA loci following HIV and HCV infection (Carrington, Nelson et al. 1999; Hraber, Kuiken et al. 2007); with heterozygotes presenting a greater variety of peptides to T lymphocytes. Furthermore, studies examining pathogen-load in different human populations suggest pathogen-selection has been one of the driving forces in shaping the extensive variation we observe today in the HLA system (Prugnolle, Manica et al. 2005). However, variation outside the peptide binding domain can also be important as has been shown for HIV (Fellay, Shianna et al. 2007). In some cases, these variations may alter the effect of microRNAs on transcribed species and ultimately the expression level of HLA on the cell surface (Kulkarni, Savan et al. 2011).

Family	Gene	Alleles
HLA Class I	A	1,698
	B	2,271
	C	884
HLA Class II	DRB1	975
	DQB1	158
	DPB1	149
MIC	MICA	77
	MICB	33
KIR2D	2DL1	43
	2DL2	28
	2DL3	34
	2DL4	46
	2DL5	41
	2DS1	15
	2DS2	22
	2DS3	14
	2DS4	30
	2DS5	16
	2DP1	22
KIR3D	3DL1	73
	3DL2	84
	3DL3	107
	3DS1	16
	3DP1	23

*Data from International Immunogenetics Project (www.ebi.ac.uk) release date April 2011 for KIR and July 2011 for MIC and HLA.

Table 1. Number of alleles of variable host immune-related genes*

Given the central role of HLA in the adaptive and innate immune response, it is not surprising that the outcome of hundreds of diseases have been associated with certain HLA alleles and more generally with genes within the MHC region. However, the intrinsic properties of the region, particularly the extensive linkage disequilibrium, have made the delineation of specific disease genes or alleles difficult (Dawkins, Leelayuwat et al. 1999).

2.3 Diversity of other genes involved in host immune response

Other genes within the MHC that also exhibit allelic variation include the MIC genes (MICA and MICB) that are ligands for the activatory NK cell receptor NKG2D. This gene family also has multiple gene copies in the genome with extensive diversity (Table 1). Interestingly, the HLA-B and -C genes and MICA (and to a lesser extent MICB) are within a region of approximately 200kb in the MHC and are in strong linkage disequilibrium such that specific combinations of alleles for these genes are commonly inherited together. Again, this represents a problem when geneticists try to disentangle disease associations with these genes. On the other hand, the genetic proximity of these genes is likely to have functional consequences for NK cell and T lymphocyte function. This highlights the importance of assessing clusters of related genes as opposed to studying genes in isolation. In-vitro and in-vivo studies that can better understand the immunological relevance of specific allelic combinations of these genes are therefore needed.

Other immune-related genes within the MHC are positioned between the HLA Class I and II genes (sometimes referred to as the HLA Class III region). Genes within this region include the central mediating cytokine Tumour Necrosis Factor alpha gene (TNF-α), which exhibits variations within the regulatory portion of the gene; this may lead to different expression and secretion levels of the cytokine. Furthermore, variations within the TNF-α gene and surrounding region also form haplotypes (Allcock, Windsor et al. 2004) reinforcing the need to functionally assess the relevance of combinations of variations within a cluster rather than a single polymorphism.

The IFN genes involved in antiviral immune responses are also an example of a complex gene family including IFN-α, -β, -γ and -λ each with several subtypes and in some cases allelic variation. The IFN-λ family includes the genes *IL28A*, *IL28B* and *IL29* on chromosome 19. Genetic variation of the *IL28B* gene (encoding IFN-λ3) has become of recent interest in the field of Hepatitis C due to the identification of variations upstream of this gene associated with HCV infection and treatment outcome in recent genome-wide association studies (GWAS) (discussed later). Additional variation is also observed for toll-like receptors (TLRs), all of which are associated with the host's immune response.

3. Hepatitis C infection: A case example demonstrating the effect of host and viral genetic diversity on infection outcome

Hepatitis C is a global health issue with more than 170 million people worldwide suspected of carrying the virus. Variations in the global distribution of HCV reflect different exposure risks. In developed nations emergence of new cases is predominantly associated with the use of intravenous drugs while in developing nations the use of unscreened blood or blood products and increasing use of intravenous drugs are the leading causes of new cases (Shepard, Finelli et al. 2005) (Lavanchy 2011) (Dore, Law et al. 2003).

HCV is a single-stranded RNA virus and following infection about 30% of individuals spontaneously clear the virus within the first six months of infection. The majority of individuals develop chronic infection of which about 20% develop cirrhosis and 3% develop hepatocellular carcinoma (Venook, Papandreou et al. 2010) (NIH 2002). In developed countries, HCV infection is a major indicator for liver transplantation. Current standard of care for HCV involves the use of pegylated-interferon alpha in combination with ribavirin (pegIFN-α/RBV) but this treatment regime is effective in only about 50% of cases and is typically associated with a plethora of side-effects. There is no protective or therapeutic vaccine available for HCV.

Like other complex diseases, HCV infection and treatment outcome is the result of interactions between various host and viral determinants. The classical approach of heritability estimates using family and twin studies indicate a significant genetic contribution to HCV infection outcomes (Goncales, Fernando et al. 2000; Fried, Kroner et al. 2006). In a study including 3993 haemophilia individuals infected with HCV, the concordance rate between 257 sibling pairs was two-fold higher than randomly paired subjects for spontaneous and treatment-related clearance of HCV. The heritability estimate from the study was approximately 0.3 (heritability of 1 indicates total phenotype variation due to genetic differences). These findings indicate that genetic factors play an important role in HCV disease outcome. This review will focus on genetic variability in host and viral determinants to provide an insight into the possible mechanisms underlying HCV infection and treatment outcome.

3.1 Host genetic diversity and natural history of HCV infection

Upon infection, HCV triggers the host innate and adaptive immune responses. Accordingly, many studies have shown that polymorphisms in genes involved in the host's innate and adaptive immune response are associated with HCV infection outcome (reviewed in (Rauch, Gaudieri et al. 2009) (Schmidt, Thimme et al. 2011) ; Figure 1 and Table 2). Nevertheless, we still lack clear understanding of the underlying mechanisms by which a person develops chronic infection or progresses to end-stage liver disease.

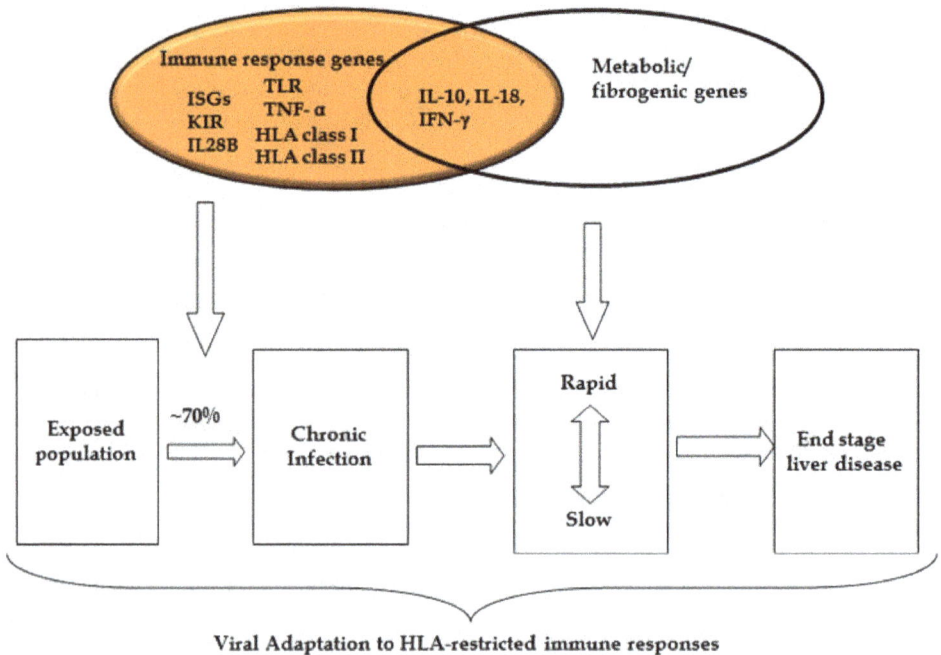

Fig. 1. Host genetic factors and viral adaptation associated with natural history of HCV infection outcome. The metabolic and fibrogenic genes are part of a seven gene signature used to predict liver fibrosis (Huang, Shiffman et al. 2007).

Gene family	Gene/Allele	Outcome
HLA class I &II	*HLA*-A*03,-A*02 , -A*11,**-B*57,-B*27**,-Cw*05, Cw*01 ***HLA* –DRB1*01**,-DQB1*05,**-DQB1*03**,**-DRB1*04, -DRB1*15,-DRB1*11**, -DRB1*08	Spontaneous clearance
	HLA-A*23,-A*01, -B*53,-B*08, -B*61, -B*38, Cw*04, -Cw*07, Cw*03 *HLA*-DRB1*03, -DRB1*07, -DQB1*02,-DRB1*08,-DQB1*06	Persistence
	HLA-A*24,-B*40	SVR
ISGs	*PKR* -168 CT genotype *MxA* -88T	Spontaneous clearance
	OAS-1 polymorphism in 3'UTR- GG genotype	Persistence
	MxA -88T	SVR
	OAS-1 – rs3213545 T, rs1169279 A and rs2859398 C alleles	SVR
	SOCS3 -4874 AA genotype	Treatment
KIR	*KIR2DL3-HLA-C* ligand	Spontaneous Clearance
	KIR2DL2/2DL3-HLA-C1C1	Treatment
IFNs	*IL-28B* - rs12979860 CC genotype *IL-28B* – rs8099917 TT genotype *IFN-γ* -764G	Spontaneous Clearance and SVR
TLR	*TLR-7* variants	Persistence
Cytokines	*IL-10*-592AA genotype *TNFa* -863C/-308G haplotype (black subjects), -863C *IL-18* –607A, -137C	Spontaneous Clearance
	IL-10 -1082 GG genotype	Persistence
	IL-10 - ACC promoter haplotype *IL-10* -592AA,-819TT genotypes *IL-12B* – 3'UTR 1188C *IL-18* – 607A	SVR
	IL-6 -572C IL-18 -607C/*, IL-18 -137G/*, IFN-γ 874T/*, IL-10 -1082A/A IL-10 -824T, -1087A, -1087/-824 haplotypes AT and AC	Progression to liver fibrosis

SOCS3 = suppressor of cytokine signalling 3. Underlined HLA alleles appear in more than one study. References: (Thio, Gao et al. 2002; McKiernan, Hagan et al. 2004; Wang, Zheng et al. 2009), (Thio, Thomas et al. 2001; Yee 2004; Yoon, Han et al. 2005; Ksiaa, Ayed-Jendoubi et al. 2007; Harris, Sugimoto et al. 2008), (Ishida, Ikebuchi et al. 2011), (Falleti, Fabris et al. 2010), (Haas, Weiß et al. 2009), (Mueller, Mas-Marques et al. 2004), (Yee, Tang et al. 2001) (Morgan, Lambrecht et al. 2008), (Thio, Goedert et al. 2004; Paladino, Fainboim et al. 2006), (Knapp, Hennig et al. 2003; Lio, Caruso et al. 2003; Kimura, Saito et al. 2006; An, Thio et al. 2008), (Schott, Witt et al. 2008), (Huang, Yang et al. 2007), (Ge, Fellay et al. 2009; Mangia, Thompson et al. 2010), (Tanaka, Nishida et al. 2009; Rauch, Kutalik et al. 2010), (Vejbaesya, Nonnoi et al. 2011), (Khakoo, Thio et al. 2004; Montes-Cano, Caro-Oleas et al. 2005), (Persico, Capasso et al. 2008), (Su, Yee et al. 2008), (Suzuki, Arase et al. 2004), (Knapp, Yee et al. 2003), (Dai, Chuang et al. 2010).

Table 2. Immune response gene polymorphisms associated with HCV infection and treatment outcome

3.1.1 Candidate gene studies

3.1.1.1 Immune response genes associated with HCV outcome

The innate immune system has a critical role in detecting pathogens and triggering the immune response. TLRs activate the innate immune system by detecting pathogen associated molecular patterns (PAMPs) such as viral RNA. TLR7 and TLR8 are receptors for single-stranded RNA and are localised on the endosomal membrane. Ligation of these TLRs can activate intracellular pathways resulting in the production of IFNs and subsequently several anti-viral IFN stimulated genes (ISGs). Variations in TLR7 and some ISGs have been shown to be associated with HCV infection outcome (reviewed in (Rauch, Gaudieri et al. 2009). One such study has shown the association between variations within *TLR 7* and HCV infection outcome (Schott, Witt et al. 2008). In this study, variants at position 32 and 2403 of TLR7 were associated with viral persistence and IFN-α treatment outcome. Other genetic variations within the ISGs MxA, OAS-1 and PKR have been associated with HCV infection outcome (Knapp, Yee et al. 2003) (Table 2).

Not surprisingly, polymorphisms in genes encoding cytokines and chemokines have been associated with HCV outcome. A recent meta-analysis on IL-10 gene polymorphisms and HCV infection outcome found the IL-10 (-1082) GG genotype was more frequent in subjects with persistent HCV infection and correlated with higher IL-10 serum levels (Zhang, Zhang et al. 2010). IL-12 plays an important role as a pro-inflammatory cytokine in T helper 1 lymphocyte differentiation and a study on 123 chronically infected and 72 spontaneous resolvers showed that viral persistence was associated with homozygosity for a 3'UTR variant of IL-12 and correlated with lower production of IL-12 (Houldsworth, Metzner et al. 2005).

3.1.1.2 NK cell KIRs and HLA ligands

Several groups have investigated the level at which the genetically polymorphic NK receptors and ligands influence HCV infection outcome. Two studies have confirmed that homozygosity for the NK cell inhibitory receptor KIR2DL3 and its ligand, HLA-C alleles belonging to the C1-group (defined by asparagine at residue 80 of the extracellular domain), results in a higher probability of resolution of HCV infection (Khakoo, Thio et al. 2004; Vidal-Castineira, Lopez-Vazquez et al. 2010). It has been hypothesised that this combined genotype results in relatively weak inhibition of NK cells and therefore protects against HCV by rendering NK cells more easily activated than in other subjects. This would be consistent with the interaction between KIR2DL3 and C1-group HLA-C alleles being relatively weak compared to the alternative combinations KIR2DL2 + C1 and KIR2D1 + C2 (Moesta, Norman et al. 2008). Furthermore, other HLA-C and KIR as well as HLA-B and the 3D KIR gene interactions are also likely to be involved in HCV resolution as has been shown for HIV infection (Gaudieri, DeSantis et al. 2005).

3.1.1.3 HLA

Differences in the strength and breadth of CD4+ and CD8+ HCV-specific T cell immune responses is an important correlate of HCV infection outcome (Lechner, Wong et al. 2000; Lauer, Lucas et al. 2005). Host HLA molecules govern the cellular T lymphocyte response to HCV and accordingly HLA class I and class II alleles have been associated with HCV outcome. Genetic association studies have shown that HLA class I alleles HLA-B*27, -A*11, -

Cw*01, -A*03 and -B*57 are associated with viral clearance, while HLA-B*38, -Cw*07, -B*08 and -A*01 are associated with viral persistence or chronic infection. HLA class II alleles, which present antigen to CD4+ T cells, are also shown to be associated with viral clearance or persistence. HLA-DRB1*04, -DRB1*11, -DQB1*03 and -DRB1*01 are associated with viral clearance, while HLA-DRB4*01 and -DRB1*07 are associated with chronic infection (reviewed in (Rauch, Gaudieri et al. 2009; Schmidt, Thimme et al. 2011) (Table 2).

Studies of HCV single source outbreaks provides the opportunity to examine the influence of HLA by removing or reducing confounding effects such as gender and age. One study on individuals from an Irish HCV single source outbreak including 227 women (141 chronic and 86 cleared) that were exposed to HCV contaminated immunoglobulin showed the association of HLA-B*27 with spontaneous resolution (McKiernan, Hagan et al. 2004). This study highlights the contribution of the host's immune CD8+T cell response and HLA genes to HCV outcome. However, the relationship between HLA and HCV outcome should be considered in the context of viral variation. It is likely that viral immune escape mutations present in the incoming virus will affect HCV infection outcome as has been shown using single source outbreak cohorts (Merani, Petrovic et al. 2011) (Salloum, Oniangue-Ndza et al. 2008). Accordingly, viral adaptation to the host's HLA-restricted immune response will be an important correlate of infection outcome (Figure 1).

3.1.2 GWAS

The initial association between polymorphisms upstream of *IL28B* and HCV infection outcome were identified using GWAS. GWAS are used to identify single nucleotide polymorphisms (SNPs) associated with disease outcome without a priori knowledge. The method screens a large number of host variants in hundreds or thousands of case/control subjects. The SNPs are pre-selected based on how well they "tag" certain areas or genes due to the linkage disequilibrium pattern across the genome, such that not all of the more than three million base differences (on average) between two unrelated individuals need be sampled. Although tagging SNPs may not necessarily be the causative variation they are likely to be near putative causative variations that can be further investigated.

Rauch *et al* utilised a GWAS to identify host genetic factors associated with HCV spontaneous clearance (Rauch, Kutalik et al. 2010) . The SNP rs8099917 upstream of the gene *IL28B* was the strongest predictor for HCV clearance (OR=2.31, CI=1.74-3.04). Similarly, Thomas *et al* also showed a SNP upstream of *IL28B* was associated with spontaneous resolution of HCV in different ethnic groups (European OR=2.6, CI=1.9-3.8 and African OR=3.1, CI=1.7-5.8); in this case the SNP was rs12979860 (Thomas, Thio et al. 2009). Individuals homozygous for the C allele at rs12979860 were more likely to clear HCV infection than individuals with the CT or TT genotype. The protective effect of the CC genotype was seen in subjects belonging to Caucasian as well as African-American ethnicity. Although the two studies identified different SNPs, both flanked the *IL28B* gene and are likely to reflect a protective *IL28B* haplotype.

IL28B encodes for the type III IFN-λ3. IFN-λ stimulates an intracellular cascade that turns on IFN-α/β like anti-viral responses. In addition, IFN-λ plays a vital role in inhibiting HCV manifestations by interfering with virus replication (Robek, Boyd et al. 2005). Furthermore, genetic variations upstream and flanking *IL28B* may correlate to different expression levels and many groups are now trying to understand the role of IFN-λ in HCV infection.

3.2 Host and viral factors influencing treatment response

The current standard of treatment for HCV is pegIFN-α/RBV. However, this treatment regimen is associated with several severe side-effects and sustained virological responses (SVR) is observed in 30-80% subjects. Several factors have been shown to be predictive of SVR, which include HCV genotype (GT) and pre-treatment viral load. Subjects infected with GT 1 have a low SVR rate of less than 50%, while subjects infected with GTs 2 or 3 have higher SVR rates of approximately 80% (Selzner and McGilvray 2008).

Various approaches such as candidate gene studies and GWAS have identified genes associated with different response rates to pegIFN-α alone or combination therapy (Table 2). The most widely studied genes encode for the IFN genes and ISGs. A study on subjects in the Hepatitis C Antiviral Long-Term Treatment Against Cirrhosis (HALT-C) trial demonstrated associations between variants in IFN-α pathway genes and treatment outcome. Specifically, variants in the *Interferon a receptor R1 (IFNAR1)* (-22T>G), *IFNAR2* (-33T>C), *Janus kinase 1 (JAK1)* (+112G>T) were associated with SVR. The *Tyrosine kinase 2 (TYK2)* -2256 A allele was also associated with SVR (Welzel, Morgan et al. 2009). Su et al examined the association in 12 ISGs in 374 treatment naïve HCV subjects. Three SNPs in the IFN induced *2'-5' oligoadenylated synthetase -like gene (OASL)* (rs3213545 T , rs1169279 A and rs2859398 C alleles) were found to be associated with SVR (Su, Yee et al. 2008). The T allele in the promoter (–88) of *MxA* was also associated with SVR (Suzuki, Arase et al. 2004). Although these association studies highlight the relevance of IFN and ISG variants in HCV treatment, additional gene expression studies have been carried out to directly correlated expression levels with treatment outcome. Chen et al identified 18 differentially expressed genes in liver biopsy samples obtained before the commencement of IFN-α treatment. Most differentially expressed genes in treatment responders and treatment non responders were IFN sensitive genes (including OAS2 and 3 and MxA) and two (ISG15/USP18) belonged to the IFN regulatory pathway (Chen, Borozan et al. 2005). Up regulation of ISG15/USP18 was identified as a predictor of IFN therapy outcome. ISG15, a ubiquitin-like protein, and USP18, part of the protease pathway, are thought to be important in host innate defense immunity.

Four independent groups utilised GWAS to evaluate the association between SNPs in the human genome and HCV treatment outcome (Ge, Fellay et al. 2009; Suppiah, Moldovan et al. 2009; Tanaka, Nishida et al. 2009; Rauch, Kutalik et al. 2010). The SNP rs8099917 located 10kb upstream of the *IL28B* gene (discussed above) was identified in all four studies as showing a significant association with HCV treatment outcome. SNP rs12979860 located 3kb upstream of *IL28B* was shown by Ge and colleagues (Ge, Fellay et al. 2009) to also be associated with HCV treatment outcome. The four studies included different ethnic groups but genotyping and analytical methods were similar, although the type of commercial SNP set used by the groups varied and may account for the lack of detection of SNP rs12979860 in some of the studies (Rauch, Rohrbach et al. 2010). The same SNPs upstream of *IL28B* were associated with HCV treatment and infection outcome.

The GWAS performed by Ge *et al* utilised subjects in the IDEAL study; a large randomized control trial comparing the effectiveness of different forms of pegIFN-α (Ge, Fellay et al. 2009). The North American study included 1137 subjects from three ethnics groups Caucasian, African American and Hispanic. Seven SNPs were reported to be associated with treatment outcome with the SNP rs12979860 having the strongest correlation to treatment outcome with the CC genotype showing a two-fold greater rate of SVR than genotype TT in all ethnics groups. The C allele frequency is highest in south East Asian populations and accordingly they achieve the highest SVR with combined therapy than subjects from European or African background. Furthermore, the frequency of the protective C allele is significantly higher in individuals from European ancestry than African American and explains half of the difference in response between the two ethnic groups.

Interferon inducible gamma 10kDa protein (IP-10), which belongs to the chemokine family has also been identified as a predictor of treatment response. The low plasma level of IP-10 was shown to be associated with a decrease in HCV RNA and SVR in early or first phase of treatment (Askarieh, Alsiö et al. 2010). In another study by Lagging et al, they correlated variants in three SNPs related to *IL-28B* with pre-treatment plasma levels of IP-10 (Lagging, Askarieh et al. 2011). These studies suggest assessment of IP-10 levels and testing of *IL-28B* variants can help predict treatment response in a clinical setting.

Cytokines profiles have been studied with respect to viral clearance and persistence during the current treatment regime. Polymorphisms in IL-12, IL-18, IL-10 cytokines have been shown to be associated with HCV treatment outcome (Wan, Kung et al. 2009). The findings of most of these studies are inconsistent reflecting the differences in study design and likely ethnic differences.

Similar to host genetic factors, genetic variability within the viral genome is likely to influence treatment response. Studies have indicated that amino acid substitutions in the core region (70 and 91), which forms the nucleocapsid of HCV, are related to treatment outcome. Similarly, amino acid substitutions in the interferon sensitivity determining region (ISDR) located in the non-structural 5A region (NS5A) of the viral genome is found to be associated with favorable treatment outcome (Maekawa and Enomoto 2009).

Better understanding of the complex interactions between genes and environmental factors may significantly help in predicting the response to current HCV therapy in the future.

3.3 Host and viral factors influencing disease progression

Chronic liver disease is a known morbidity of HCV infection with 20% of chronic HCV cases developing these complications, of which approximately 5% go on to develop hepatocellular carcinoma. Studies have shown demographic as well as epidemiological factors are associated with progression to fibrosis. Some subjects progress rapidly to early stage of fibrosis, while some do not develop fibrosis. Thus genetic factors predisposing to fibrosis are crucial in pathogenesis of hepatic fibrosis.

Pro-inflammatory and anti-inflammatory genes have been shown to be associated with liver disease progression in HCV. The anti-inflammatory cytokine IL-10 is likely to have an immuno-modulatory role in fibrosis. Promoter polymorphisms in *IL-10*, which affect the level of IL-10 production, have been shown to be associated with susceptibility to liver fibrosis. In a study of Japanese chronic HCV subjects, the IL-10 -824 T allele, -1087 A allele and -1087/-824 haplotypes AT and AC were shown to be risk factors for progression of hepatic fibrosis (Ishida, Ikebuchi et al. 2011). Another study investigated the combined effect of SNPs in IL-10, IL-18 and IFN-γ genes on 77 chronic HCV infected patients. Subjects carrying 3-4 high risk genotypes were associated with greater risk of developing liver cirrhosis (Bouzgarrou, Hassen et al. 2011).

Risk factors identified in these studies would help in understanding the molecular mechanisms involved in pathogenesis of chronic liver disease in HCV infected individuals. Such studies may also provide new insights for developing drug targets thus reducing the burden of HCV worldwide.

4. HCV diversity and viral adaptation

As HCV is an obligate intracellular virus with a small RNA genome, it evolves by escaping/adapting to the host's immune response. The mechanisms utilised by HCV to establish persistent infection are not fully understood but involve several strategies (Nolan, Gaudieri et al. 2006).

HCV has a high mutation rate due to the lack of proof-reading activity of its RNA dependent RNA polymerase. This characteristic along with the high rate of replication of HCV results in the circulation of genetically related variants/quasispecies within the infected individuals. In addition, HCV is classified into six major GT and more than 50 different subtypes based on the sequence variability of the virus. GTs 1 and 3 are the most prevalent GTs and also account for the majority of HCV infections worldwide (Chayama and Hayes 2011).

The efficacy of HCV-specific T lymphocyte responses is therefore compromised by the mutability of the viral genome within HLA Class I and II-restricted epitopes and this represents a potent escape strategy of HCV (Timm, Lauer et al. 2004; Thimme, Lohmann et al. 2006) as well as other viruses such as HIV (Goulder and Watkins 2004). These mutations may exist in the infecting virus or arise *in-vivo* can affect HLA-peptide binding and therefore influence the selection of peptides presented by an individual. These mutations may also impair HLA-peptide and T lymphocyte interaction and reduce the efficiency of the T lymphocyte-mediated immune response. The relevance of HCV immune escape mutations in chronic infection was first demonstrated in chimpanzees (Weiner, Erickson et al. 1995; Erickson, Kimura et al. 2001; Grakoui, Shoukry et al. 2003) and subsequently in humans (Cox, Mosbruger et al. 2005; Tester, Smyk-Pearson et al. 2005) (reviewed in (Bowen and Walker 2005). Evidence for T lymphocyte escape and reversion has also been recently demonstrated in acute infection (Timm, Lauer et al. 2004) and 18-22 years after a common-source outbreak (Ray, Fanning et al. 2005). Flanking epitope escape mutations have also been described and alter proteasomal epitope processing and subsequent peptide presentation (Seifert, Liermann et al. 2004). Overall, this emphasizes the complexity of multiple effective mechanisms of T lymphocyte immune escape in HCV.

Studies utilizing population-based genetic approaches, which examine the association between specific viral polymorphisms and the HLA types of individuals within a host population, have demonstrated viral escape (adaptation) to HLA-restricted immune pressure evident at the population level for both HIV and HCV (Gaudieri, Rauch et al. 2006; Timm, Li et al. 2007) · The approach is based on the premise that the host's HLA molecules regulate immune responses by presenting specific viral epitopes to T lymphocytes and viral polymorphisms within or flanking these epitopes that compromise the efficacy of these T lymphocyte responses (i.e. allow viral escape) would be identified as HLA-specific HCV polymorphisms (viral adaptations). Accordingly, HCV can escape host immune responses via mutations that confer a selective advantage resulting in the fixation (or dominance) of specific strains within the host and to some extent drive HCV evolution.

Given the extent of sequence diversity between the GTs (Figure 2), it would be anticipated that the sequence context of epitopes restricted for specific HLA molecules would be altered and potentially disrupted for other GTs. The initial population-based genetic study on HCV GT 1 (Gaudieri, Rauch et al. 2006) was extended to examine HCV adaptation to HLA-restricted immune responses in GT 3 (Rauch, James et al. 2009). The study reported that the immune escape profiles differ between the two main circulating GTs 1 and 3 reflecting the extensive variation between the GTs and the observation that individuals can clear one GT but can be re-infected with other GTs with limited protection.

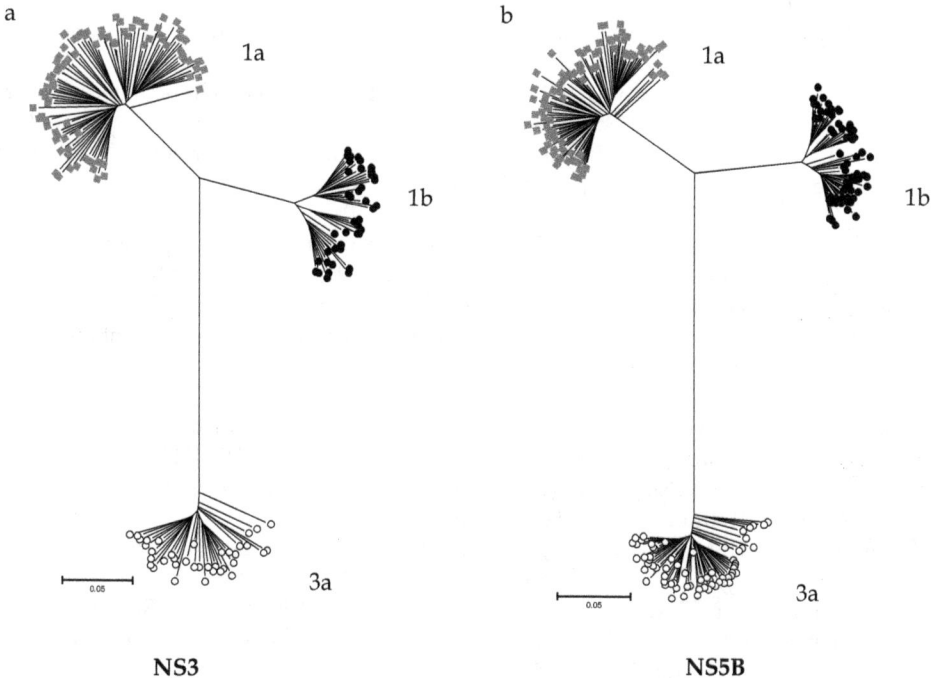

Fig. 2. HCV diversity in the (a) NS3 and (b) NS5B regions for Caucasian chronic HCV populations.

5. Summary

Host and viral genetic determinants within genes related to the host immune response are important in determining infection outcome, including HCV outcome. Studies accounting for variability in both host and pathogen are likely to provide a more complete understanding of the complex host/pathogen interaction and its variable outcomes. Specifically, an understanding of how viral adaptation to host HLA-restricted immune responses (and perhaps to NK cells given recent evidence of "footprints" in HIV genome due to KIR genotypes; (Alter, Heckerman et al. 2011)) affects infection outcome will provide an insight into vaccine design.

Given our current knowledge, the HCV virus would be particularly amenable to a T lymphocyte vaccine strategy given that immunity in those that control the virus consists of broad T lymphocyte responses and viral structural constraints that prevent escape from all mounted T lymphocyte responses. Identifying the viral T lymphocyte targets of a successful host immune response is crucial to vaccine design and should consider the genetic diversity of the virus as well as the HLA genetic system of the host population for which the vaccine is designed.

6. References

Allcock, R. J., L. Windsor, et al. (2004). "High-Density SNP genotyping defines 17 distinct haplotypes of the TNF block in the Caucasian population: implications for haplotype tagging." *Hum Mutat* 24(6): 517-25.

Alter, G., D. Heckerman, et al. (2011). "HIV-1 adaptation to NK-cell-mediated immune pressure." *Nature* 476(7358): 96-100.

An, P., C. L. Thio, et al. (2008). "Regulatory Polymorphisms in the Interleukin-18 Promoter Are Associated with Hepatitis C Virus Clearance." *Journal of Infectious Diseases* 198(8): 1159-1165.

Askarieh, G., Å. Alsiö, et al. (2010). "Systemic and intrahepatic interferon-gamma-inducible protein 10 kDa predicts the first-phase decline in hepatitis C virus RNA and overall viral response to therapy in chronic hepatitis C." *Hepatology* 51(5): 1523-1530.

Bouzgarrou, N., E. Hassen, et al. (2011). "Combined effect of pro- and anti-inflammatory cytokine gene polymorphisms on susceptibility to liver cirrhosis in Tunisian HCV-infected patients." *Hepatol Int.* 5(2): 681-687.

Bowen, D. G. and C. M. Walker (2005). "Mutational escape from CD8+ T cell immunity: HCV evolution, from chimpanzees to man." *J Exp Med* 201(11): 1709-14.

Carrington, M., G. W. Nelson, et al. (1999). "HLA and HIV-1: heterozygote advantage and B*35-Cw*04 disadvantage." *Science* 283(5408): 1748-1752.

Chayama, K. and C. N. Hayes (2011). "Hepatitis C virus: How genetic variability affects pathobiology of disease." *J Gastroenterol Hepatol* 26(suppl 1): 83-95.

Chen, L., I. Borozan, et al. (2005). "Hepatic Gene Expression Discriminates Responders and Nonresponders in Treatment of Chronic Hepatitis C Viral Infection." *Gastroenterology* 128(5): 1437-1444.

Cox, A. L., T. Mosbruger, et al. (2005). "Cellular immune selection with hepatitis C virus persistence in humans." *J Exp Med* 201(11): 1741-52.

Dai, C., W. Chuang, et al. (2010). "Human leukocyte antigen alleles and the response to pegylated interferon/ribavirin therapy in chronic hepatitis C patients." *Antiviral Research* 85(2): 396-402.

Dawkins, R., C. Leelayuwat, et al. (1999). "Genomics of the major histocompatibility complex: haplotypes, duplication, retroviruses and disease." *Immunol Rev.* 167: 275-304.

Dore, G. J., M. Law, et al. (2003). "Epidemiology of hepatitis C virus infection in Australia." *Journal of clinical virology : the official publication of the Pan American Society for Clinical Virology* 26(2): 171-184.

Erickson, A. L., Y. Kimura, et al. (2001). "The outcome of hepatitis C virus infection is predicted by escape mutations in epitopes targeted by cytotoxic T lymphocytes." *Immunity* 15(6): 883-95.

Falleti, E., C. Fabris, et al. (2010). "Genetic polymorphisms of interleukin-6 modulate fibrosis progression in mild chronic hepatitis C." *Human Immunology* 71(10): 999-1004.

Fellay, J., K. V. Shianna, et al. (2007). "A whole-genome association study of major determinants for host control of HIV-1." *Science* 317(5840): 944-7.

Fried, M. W., B. L. Kroner, et al. (2006). "Hemophilic Siblings With Chronic Hepatitis C: Familial Aggregation of Spontaneous and Treatment-Related Viral Clearance." *Gastroenterology* 131(3): 757-764.

Gaudieri, S., Dawkins, RL, Habara, K, Kulski, JK, Gojobori, T. (2000). "SNP profile within the human major histocompatibility complex reveals an extreme and interrupted level of nucleotide diversity." *Genome Research* 10(10): 1579-1586.

Gaudieri, S., D. DeSantis, et al. (2005). "Killer immunoglobulin-like receptors and HLA act both independently and synergistically to modify HIV disease progression." *Genes Immun* 6(8): 683-690.

Gaudieri, S., J. K. Kulski, et al. (1999). "Different evolutionary histories in two subgenomic regions of the major histocompatibility complex." *Genome Research* 9(6): 541-549.

Gaudieri, S., A. Rauch, et al. (2006). "Evidence of viral adaptation to HLA class I-restricted immune pressure in chronic hepatitis C virus infection." *J Virol* 80(22): 11094-12004.

Ge, D., J. Fellay, et al. (2009). "Genetic variation in IL28B predicts hepatitis C treatment-induced viral clearance." *Nature* 461(7262): 399-401.

Goncales, J. R., L. Fernando, et al. (2000). "Hepatitis C virus in monozygotic twins." *Revista do Instituto de Medicina Tropical de São Paulo* 42: 163-165.

Goulder, P. J. and D. I. Watkins (2004). "HIV and SIV CTL escape: implications for vaccine design." *Nat Rev Immunol* 4(8): 630-40.

Grakoui, A., N. H. Shoukry, et al. (2003). "HCV persistence and immune evasion in the absence of memory T cell help." *Science* 302(5645): 659-62.

Graw, R. G. J., I. M. Goldstein, et al. (1970). "Histocompatibility testing for leucocyte transfusion." *Lancet* 1970(1): 7663.

Haas, S., C. Weiß, et al. (2009). "Interleukin 18 Promoter Variants (−137G>C and −607C>A) in Patients with Chronic Hepatitis C: Association with Treatment Response." *Journal of Clinical Immunology* 29(5): 620-628.

Harris, R. A., K. Sugimoto, et al. (2008). "Human leukocyte antigen class II associations with hepatitis C virus clearance and virus-specific CD4 T cell response among Caucasians and African Americans." *Hepatology* 48(1): 70-79.

Houldsworth, A., M. Metzner, et al. (2005). "Polymorphisms in the IL-12B Gene and Outcome of HCV Infection." *Journal of Interferon & Cytokine Research* 25(5): 271-276.

Hraber, P., Kuiken, C, Yusim, K. (2007). "Evidence for human leukocyte antigen heterozygote advantage against hepatitis C virus infection." *Hepatology* 46(6): 1713-1721.

Huang, H., M. L. Shiffman, et al. (2007). "A 7 gene signature identifies the risk of developing cirrhosis in patients with chronic hepatitis C." *Hepatology* 46(2): 297-306.

Huang, Y., H. Yang, et al. (2007). "A functional SNP of interferon-γ gene is important for interferon-α-induced and spontaneous recovery from hepatitis C virus infection." *Proceedings of the National Academy of Sciences* 104(3): 985-990.

Hughes, A. L., T. Ota, et al. (1990). "Positive Darwinian selection promotes charge profile diversity in the antigen-binding cleft of class I major-histocompatibility-complex molecules." *Mol Biol Evol* 7(6): 515-524.

Ishida, C., Y. Ikebuchi, et al. (2011). "Functional Gene Polymorphisms of Interleukin-10 are Associated with Liver Disease Progression in Japanese Patients with Hepatitis C Virus Infection." *Internal Medicine* 50(7): 659-666.

Janeway (2008). *Immunobiology*. New York, Garland Sciences, Taylor and Francis Group.

Kelley, J. and J. Trowsdale (2005). "Features of MHC and NK gene clusters." *Transplant Immunology* 14(3-4): 129-134.

Khakoo, S. I., C. L. Thio, et al. (2004). "HLA and NK Cell Inhibitory Receptor Genes in Resolving Hepatitis C Virus Infection." *Science* 305(5685): 872-874.

Kimura, T., T. Saito, et al. (2006). "Association of Transforming Growth Factor–β1 Functional Polymorphisms with Natural Clearance of Hepatitis C Virus." *Journal of Infectious Diseases* 193(10): 1371-1374.

Knapp, S., B. W. Hennig, et al. (2003). "Interleukin-10 promoter polymorphisms and the outcome of hepatitis C virus infection." *Immunogenetics* 55(6): 362-369.

Knapp, S., L. J. Yee, et al. (2003). "Polymorphisms in interferon-induced genes and the outcome of hepatitis C virus infection: roles of MxA, OAS-1 and PKR." *Genes Immun* 4(6): 411-419.

Ksiaa, L., S. Ayed-Jendoubi, et al. (2007). "Clearance and Persistence of Hepatitis C Virus in a Tunisian Population: Association with HLA Class I and Class II." *Viral Immunology* 20(2): 312-319.

Kulkarni, S., M. P. Martin, et al. (2008). "The Yin and Yang of HLA and KIR in human disease." *Seminars in Immunology* 20(6): 343-352.

Kulkarni, S., Savan, R, Qi, Y, Gao, X, Yuki, Y, Bass, SE, Martin, MP, Hunt, P, Deeks, SG, Telenti, A, Pereyra, F, Goldstein, D, Wolinsky, S, Walker, B, Young, HA, Carrington, M. (2011). "Differential microRNA regulation of HLA-C expression and its association with HIV control." *Nature* 472(7344): 495-498.

Lagging, M., G. Askarieh, et al. (2011). "Response Prediction in Chronic Hepatitis C by Assessment of IP-10 and IL28B-Related Single Nucleotide Polymorphisms." *PLoS ONE* 6(2): e17232.

Lanier, L. L. (1998). "NK Cell Receptors." *Annual Review of Immunology* 16(1): 359-393.

Lauer, G., Lucas, M, Timm, J, Ouchi, K, Kim, AY, Day, CL, Schulze, Zur Wiesch, J, Paranhos-Baccala, G, Sheridan, I, Casson, DR, Reiser, M, Gandhi, RT, Li, B, Allen, TM, Chung, RT, Klenerman, P, Walker, BD (2005). " Full-breadth analysis of CD8+ T-cell responses in acute hepatitis C virus infection and early therapy." *J Virol* 79(20): 12979-12988.

Lavanchy, D. (2011). "Evolving epidemiology of hepatitis C virus." *Clinical Microbiology and Infection* 17(2): 107-115.

Lechner, F., Wong, DK,Dunbar, PR,Chapman, R,Chung, RT,Dohrenwend, P,Robbins, G,Phillips, R,Klenerman, P,Walker, BD (2000). "Analysis of successful immune responses in persons infected with hepatitis C virus." *J Exp Med* 191(9): 1499-1512.

Lio, D., C. Caruso, et al. (2003). "IL-10 and TNF-α polymorphisms and the recovery from HCV infection." *Human Immunology* 64(7): 674-680.

Lucas, M., U. Karrer, et al. (2001). "Viral escape mechanisms--escapology taught by viruses." *Int J Exp Pathol* 82(5): 269-86.

Maekawa, S. and N. Enomoto (2009). "Viral factors influencing the response to the combination therapy of peginterferon plus ribavirin in chronic hepatitis C." *J Gastroenterol* 44(10): 1009-1015.

Mangia, A., A. J. Thompson, et al. (2010). "An IL28B Polymorphism Determines Treatment Response of Hepatitis C Virus Genotype 2 or 3 Patients Who Do Not Achieve a Rapid Virologic Response." *Gastroenterology* 139(3): 821-827.e1.

Martin, A. M., J. K. Kulski, et al. (2004). "Comparative genomic analysis, diversity and evolution of two KIR haplotypes A and B." *Gene* 335(0): 121-131.

Martin, M. P., G. Nelson, et al. (2002). "Cutting Edge: Susceptibility to Psoriatic Arthritis: Influence of Activating Killer Ig-Like Receptor Genes in the Absence of Specific HLA-C Alleles." *The Journal of Immunology* 169(6): 2818-2822.

McKiernan, S. M., R. Hagan, et al. (2004). "Distinct MHC class I and II alleles are associated with hepatitis C viral clearance, originating from a single source." *Hepatology* 40(1): 108-114.

Merani, S., D. Petrovic, et al. (2011). "Effect of immune pressure on hepatitis C virus evolution: Insights from a single-source outbreak." *Hepatology* 53(2): 396-405.

Moesta, A. K., P. J. Norman, et al. (2008). "Synergistic polymorphism at two positions distal to the ligand-binding site makes KIR2DL2 a stronger receptor for HLA-C than KIR2DL3." *J Immunol* 180(6): 3969-79.

Montes-Cano, M. A., J. L. Caro-Oleas, et al. (2005). "HLA-C and KIR Genes in Hepatitis C Virus Infection." *Human Immunology* 66(11): 1106-1109.

Morgan, T. R., R. W. Lambrecht, et al. (2008). "DNA polymorphisms and response to treatment in patients with chronic hepatitis C: Results from the HALT-C trial." *Journal of hepatology* 49(4): 548-556.

Mueller, T., A. Mas-Marques, et al. (2004). "Influence of interleukin 12B (IL12B) polymorphisms on spontaneous and treatment-induced recovery from hepatitis C virus infection." *Journal of Hepatology* 41(4): 652-658.

Nei, M., X. Gu, et al. (1997). "Evolution by the birth-and-death process in multigene families of the vertebrate immune system." *Proceedings of the National Academy of Sciences* 94(15): 7799-7806.

NIH (2002). "NIH Consensus Statement on Management of Hepatitis C: 2002." *NIH Consens State Sci Statement* 19(13): 1-46.

Nolan, D., S. Gaudieri, et al. (2006). "Host genetics and viral infections: immunology taught by viruses, virology taught by the immune system." *Current Opinion in Immunology* 18(4): 413-421.

Paladino, N., H. Fainboim, et al. (2006). "Gender Susceptibility to Chronic Hepatitis C Virus Infection Associated with Interleukin 10 Promoter Polymorphism." *J. Virol.* 80(18): 9144-9150.

Persico, M., M. Capasso, et al. (2008). "Elevated expression and polymorphisms of SOCS3 influence patient response to antiviral therapy in chronic hepatitis C." *Gut* 57(4): 507-515.

Prugnolle, F., A. Manica, et al. (2005). "Pathogen-driven selection and worldwide HLA class I diversity." *Curr Biol* 15(11): 1022-7.

Pyo, C.-W., L. A. Guethlein, et al. (2010). "Different Patterns of Evolution in the Centromeric and Telomeric Regions of Group A and B Haplotypes of the Human Killer Cell Ig-Like Receptor Locus." *PLoS ONE* 5(12): e15115.

Rauch, A., S. Gaudieri, et al. (2009). "Host genetic determinants of spontaneous hepatitis C clearance." *Pharmacogenomics* 10(11): 1819-1837.

Rauch, A., I. James, et al. (2009). "Divergent adaptation of hepatitis C virus genotypes 1 and 3 to human leukocyte antigen-restricted immune pressure." *Hepatology* 50(4): 1017-1029.

Rauch, A., Z. Kutalik, et al. (2010). "Genetic Variation in IL28B Is Associated With Chronic Hepatitis C and Treatment Failure: A Genome-Wide Association Study." *Gastroenterology* 138(4): 1338-1345.e7.

Rauch, A., J. Rohrbach, et al. (2010). "The recent breakthroughs in the understanding of host genomics in hepatitis C." *European Journal of Clinical Investigation* 40(10): 950-959.

Ray, S. C., L. Fanning, et al. (2005). "Divergent and convergent evolution after a common-source outbreak of hepatitis C virus." *J Exp Med* 201(11): 1753-9.

Robek, M. D., B. S. Boyd, et al. (2005). "Lambda interferon inhibits hepatitis B and C virus replication." *J Virol* 79(6): 3851-4.

Salloum, S., C. Oniangue-Ndza, et al. (2008). "Escape from HLA-B*08-restricted CD8 T cells by hepatitis C virus is associated with fitness costs." *J Virol* 82(23): 11803-12.

Schmidt, J., R. Thimme, et al. (2011). "Host genetics in immune-mediated hepatitis C virus clearance." *Biomark Med* 5(2): 155-69.

Schott, E., H. Witt, et al. (2008). "Association of TLR7 single nucleotide polymorphisms with chronic HCV-infection and response to interferon-a-based therapy." *Journal of Viral Hepatitis* 15(1): 71-78.

Seifert, U., H. Liermann, et al. (2004). "Hepatitis C virus mutation affects proteasomal epitope processing." *J Clin Invest* 114(2): 250-9.

Selzner, N. and I. D. McGilvray (2008). "Can genetic variations predict HCV treatment outcomes?" *J Hepatol* 49(4): 494-497.

Shepard, C. W., L. Finelli, et al. (2005). "Global epidemiology of hepatitis C virus infection." *The Lancet Infectious Diseases* 5(9): 558-567.

Su, X., L. J. Yee, et al. (2008). "Association of single nucleotide polymorphisms in interferon signaling pathway genes and interferon-stimulated genes with the response to interferon therapy for chronic hepatitis C." *Journal of Hepatology* 49(2): 184-191.

Suppiah, V., M. Moldovan, et al. (2009). "IL28B is associated with response to chronic hepatitis C interferon-[alpha] and ribavirin therapy." *Nat Genet* 41(10): 1100-1104.

Suzuki, F., Y. Arase, et al. (2004). "Single nucleotide polymorphism of the MxA gene promoter influences the response to interferon monotherapy in patients with hepatitis C viral infection." *Journal of Viral Hepatitis* 11(3): 271-276.

Tanaka, Y., N. Nishida, et al. (2009). "Genome-wide association of IL28B with response to pegylated interferon-[alpha] and ribavirin therapy for chronic hepatitis C." *Nat Genet* 41(10): 1105-1109.

Tester, I., S. Smyk-Pearson, et al. (2005). "Immune evasion versus recovery after acute hepatitis C virus infection from a shared source." *J Exp Med* 201(11): 1725-31.

Thimme, R., V. Lohmann, et al. (2006). "A target on the move: innate and adaptive immune escape strategies of hepatitis C virus." *Antiviral Res* 69(3): 129-41.

Thio, C. L., X. Gao, et al. (2002). "HLA-Cw*04 and Hepatitis C Virus Persistence." *J. Virol.* 76(10): 4792-4797.

Thio, C. L., J. J. Goedert, et al. (2004). "An analysis of tumor necrosis factor [alpha] gene polymorphisms and haplotypes with natural clearance of hepatitis C virus infection." *Genes Immun* 5(4): 294-300.

Thio, C. L., D. L. Thomas, et al. (2001). "Racial Differences in HLA Class II Associations with Hepatitis C Virus Outcomes." *Journal of Infectious Diseases* 184(1): 16-21.

Thomas, D. L., C. L. Thio, et al. (2009). "Genetic variation in IL28B and spontaneous clearance of hepatitis C virus." *Nature* 461(7265): 798-801.

Timm, J., G. M. Lauer, et al. (2004). "CD8 epitope escape and reversion in acute HCV infection." *J Exp Med* 200(12): 1593-604.

Timm, J., B. Li, et al. (2007). "Human leukocyte antigen-associated sequence polymorphisms in hepatitis C virus reveal reproducible immune responses and constraints on viral evolution." *Hepatology* 46(2): 339-49.

Vejbaesya, S., Y. Nonnoi, et al. (2011). "Killer cell immunoglobulin-like receptors and response to antiviral treatment in Thai patients with chronic hepatitis C virus genotype 3a." *Journal of Medical Virology* 83(10): 1733-1737.

Venook, A. P., C. Papandreou, et al. (2010). "The Incidence and Epidemiology of Hepatocellular Carcinoma: A Global and Regional Perspective." *The Oncologist* 15(suppl 4): 5-13.

Vidal-Castineira, J. R., A. Lopez-Vazquez, et al. (2010). "Effect of Killer Immunoglobulin-Like Receptors in the Response to Combined Treatment in Patients with Chronic Hepatitis C Virus Infection." *J. Virol.* 84(1): 475-481.

Wan, L., Y. J. Kung, et al. (2009). "Th1 and Th2 cytokines are elevated in HCV-infected SVR (-) patients treated with interferon-alpha." *Biochem Biophys Res Commun* 379(4): 855-860.

Wang, J., X. Zheng, et al. (2009). "Ethnic and geographical differences in HLA associations with the outcome of hepatitis C virus infection." *Virology Journal* 6(1): 46.

Weiner, A., A. L. Erickson, et al. (1995). "Persistent hepatitis C virus infection in a chimpanzee is associated with emergence of a cytotoxic T lymphocyte escape variant." *Proc Natl Acad Sci U S A* 92(7): 2755-9.

Welzel, T. M., T. R. Morgan, et al. (2009). "Variants in interferon-alpha pathway genes and response to pegylated interferon-Alpha2a plus ribavirin for treatment of chronic hepatitis C virus infection in the hepatitis C antiviral long-term treatment against cirrhosis trial." *Hepatology* 49(6): 1847-58.

Yee, L. J. (2004). "Host genetic determinants in hepatitis C virus infection." *Genes Immun* 5(4): 237-245.

Yee, L. J., J. Tang, et al. (2001). "Interleukin 10 polymorphisms as predictors of sustained response in antiviral therapy for chronic hepatitis C infection." *Hepatology* 33(3): 708-712.

Yoon, S. K., J. Y. Han, et al. (2005). "Association between human leukocytes antigen alleles and chronic hepatitis C virus infection in the Korean population." *Liver International* 25(6): 1122-1127.

Zhang, L.-Z., T.-C. Zhang, et al. (2010). "Interleukin-10 gene polymorphisms in association with susceptibility to chronic hepatitis C virus infection: a meta-analysis study." *Archives of Virology* 155(11): 1839-1842.

Part 2

Conservation of Genetic Resources

10

Managing the Genetic Resources in the Intensive Stock Enhancement Program Carried out on Black Sea Bream in Hiroshima Bay, Japan

Enrique Blanco Gonzalez and Tetsuya Umino
Graduate School of Biosphere Science, Hiroshima University
Japan

1. Introduction

The establishment of sustainable fishery management strategies requires detailed characterization of the stocks, including their genetic diversity and structure (Allendorf and Ryman, 1987; Ryman, 1981). Traditionally, the large population size, wide distribution and the potential high mobility during the pelagic eggs and larval phase was presumed to explain the lack of genetic differentiation and population structure found in marine species (Hauser & Carvalho, 2008; Ward et al., 1994). Natural selection and high gene flow were considered the main evolutionary forces affecting genetic structure of marine organisms. However, recent genetic studies on marine fish have evidenced population structure at different geographical and temporal scales (Knutsen et al., 2003; Umino et al., 2009; Watts et al., 2010). This information is especially relevant to preserve the genetic identity of wild stocks and minimize the negative genetic interaction between wild and hatchery specimens from escapees and stock enhancement programs (Blanco Gonzalez & Umino, 2009; Glover at al., 2010). Imprecision or lack of genetic information may not only exacerbate problems that affect yields of fisheries but also erode the gene pool and the potential adaptive response of the stock in an irretrievable manner (Laikre et al., 2010; Reiss et al., 2009; Ward, 2006). Despite its importance for conservation and management, it has not been until recently that marine stock enhancement programs started integrating genetic analysis and monitoring data prior to, during and after release (Bert et al., 2001; Blanco Gonzalez et al., 2008a). Meanwhile, our knowledge about the genetic resources of commercially-farmed stocks for their identification in case of escapees is still very scarce (Glover et al., 2010; Svåsand et al., 2007).

Microsatellite DNA markers generally exhibit high levels of genetic polymorphism and are a priori presumed to behave as neutral markers, i.e. the effects of selection are neglected. Consequently, genetic changes among stocks will be explained by their origins and population demography processes; mainly by gene flow and genetic drift, and mutation to some extent (Luikart et al., 2003). Microsatellites have been the most common genetic marker employed in stock enhancement programs, being extensively used for delineating the genetic diversity and population structure of the species prior to and after the release

(Blanco Gonzalez & Umino, 2009; Jeong et al., 2003; Perez-Enriquez et al., 2001), assigning individuals to baseline hatchery stocks (Blanco Gonzalez et al., 2008a; Ortega-Villaizán et al., 2005) or tracing the pedigree to manage the captive broodstock (Blanco Gonzalez et al., 2010; Jeong et al., 2007; Perez-Enriquez et al., 1999).

Black sea bream *Acanthopagrus schlegelii* is an important commercial and sport fishing species in Japan and represent one of the most intensively stocked fish in the country with millions of juveniles released annually. The largest releases have been conducted in Hiroshima Bay (Fig. 1) which is also the primary fishing area in the country (Blanco Gonzalez et al., 2008b). The stock enhancement program in this bay started in the early 1980s after a drastic reduction of landings during the previous two decades, when fishing yields dropped from about 500 mt in 1960s to 150 by the end of 1970s. Juveniles for stocking were usually originated from fertilized eggs produced in one night during the spawning peak and collected by tank overflowing. They were reared in indoor tanks and released at 3 cm in total length. Initially, released juveniles were identified marking their otoliths with alizarin-complexon and by ventral fin-clipping (Nakagawa et al., 2000; Umino et al., 1999; Yamashita et al., 1997). Using non-genetic markers, it was possible to investigate the acclimation process (Nakagawa et al., 2000; Yamashita et al., 1997), migration (Anonymous, 1987) and optimum size-at-release (Umino et al., 1999). Later on, the development of microsatellite DNA markers (Jeong et al., 2003, 2007) has contributed to address fundamental questions regarding the effects of the releases on the genetic resources of the natural stock.

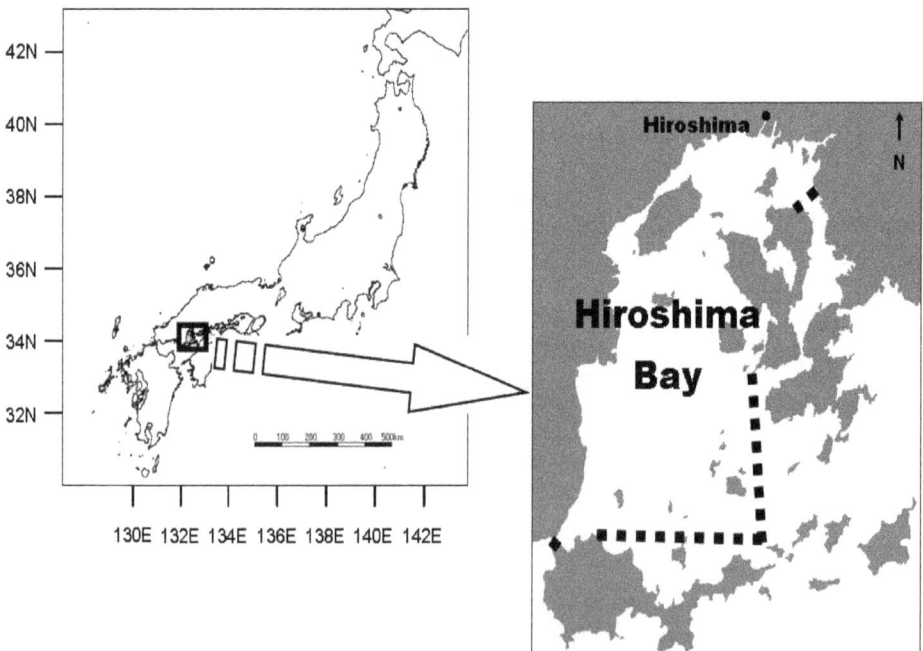

Fig. 1. Map of Hiroshima Bay.

In addition to providing a better estimation of the contribution to the fishing yield; microsatellite genotyping helped to characterize the genetic architecture of black sea bream and minimize the potential deleterious effects of fish releases by improving broodstock management. This paper reports on the progress of using microsatellite DNA markers to gain understanding of the genetic implications of the stock enhancement program carried out on black sea bream in Hiroshima Bay.

2. Genetic risks associated to releases

Stock enhancement programs have been implemented as a mean to recover depleted populations from many taxonomic groups worldwide (Blanco Gonzalez et al., 2008b; Laikre et al., 2010). However, the limited number of breeders reared to produce the offspring for release and the differential population origins have raised awareness about harmful loss of genetic diversity and changes in allele composition that large-scale releases may cause on the native stocks (Allendorf and Ryman, 1987; Laikre et al., 2010; Ryman, N. 1981). The relative large population size and higher dispersion reported in marine fish compared to anadromous and freshwater species favors gene flow and reduces genetic differentiation between stocks (DeWoody & Avise, 2000; Ward et al., 1994). Hence, freshwater and anadromous species are more vulnerable to the harmful genetic effects of large-scale releases (Cross, 1999; Hauser & Carvalho, 2008). Nevertheless, genetic risks on marine species should not be underestimated.

2.1 Genetic diversity among-populations

The first step to be accomplished before implementing a stock enhancement program is to delineate the genetic architecture of the recipient population (Allendorf & Ryman, 1987; Ryman 1981). Whenever possible, the broodstock should be of local origin. Non-local breeders may carry alleles that were previously absent in the wild. Interbreeding of their offspring with wild specimens can replace the original gene pool by genotypes that are locally non-adapted; thus, affecting survival, growth or disease resistance, and ultimately compromising the viability and productivity of the stock (Cooke & Philipp, 2006; Laikre et al., 2010). Sometimes, interbreeding of genetically divergent stocks may result in hybrid vigor in the F1 generation (Shikano & Taniguchi, 2003) that may evolve into outbreeding depression and breakdown of co-adapted gene complexes after the second generation (Cross, 1999).

2.2 Genetic diversity within populations

The broodstock should comprise enough specimens to accurately represent the genetic identity of the stock to be enhanced (Allendorf & Ryman, 1987; Taniguchi, 2003). This number is crucial because the gene pool present in the broodstock will determine the maximum genetic diversity that may be inherited by the offspring (Allendorf & Ryman, 1987). Therefore, efforts should be directed towards avoiding loss of genetic diversity or changes in the genetic composition caused by genetic drift, selection due to domestication and inbreeding depression.

2.2.1 Genetic drift

Genetic drift is a stochastic process that changes allele frequencies in the next generation. The loss on genetic variation may be reflected in the heterozygosity, proportion of polymorphic loci and number of alleles per locus (Allendorf & Phelps, 1980) and will increase exponentially to the reduction on the population size (Nei et al., 1975). Ideally, in order to maximize the transference of the gene pool to the offspring and minimize a potential reduction on fitness, broodstock management should aim at keeping the sex ratio between male and female at 1:1 and ensuring that all breeders kept in the hatchery contribute equally to the offspring (Allendorf & Ryman, 1987). Under this ideal situation, the effective number of breeders (Nb) will double the broodstock size, while skewed sex ratio or variance in family sizes will reduce Nb. Allendorf & Ryman (1987) reported that keeping Nb at 50 will lead to the loss of approximately 10 % of the genetic variation after ten generations, a fact that may have important deleterious effects on survival and growth (Falconer, 1981). In hermaphroditic species, including some sea breams, collecting mature breeders will prevent from posterior undesirable sex changes that may skew the proportion of males and females (Cross, 1999).

2.2.2 Domestication

Domestication selection tends to favor genetic profiles better adapted to the hatchery conditions. Good broodstock management practices contribute to maximize the transference of genetic information to the offspring; however, domestication selection may change or reduce the genetic variability at some loci and erode the adaptive potential of the stock (Taniguchi, 2004). In stock enhancement programs, it is noteworthy to consider that the direction and intensity of local selective forces between hatchery and wild environments are likely to differ; hence, the longer the fish are reared in captivity the greater effects of domestication selection will be (Araki & Schmid, 2010; Bekkevold et al., 2006; Milles & Kapauscki, 2003). Good survival and fitness under hatchery conditions may evolve into poor performance once the juveniles are released into the wild. In this regard, most of studies suggest that unintentional selection domestication produces negative effects on fitness-related traits including survival, morphology, behavior, response to predation or disease resistance, and that ultimately can compromise the reproductive success and the viability of the stock (see reviews by Araki & Schmid, 2010; Fraser, 2008; Reisenbichler & Rubin, 1999; Thorstad et al., 2008).

2.2.3 Inbreeding depression

Inbreeding depression represents the reduction on fitness produced by breeding related specimens. Initially, the ratio of homozygous genotypes will be augmented. Consequently, harmful recessive alleles that were unexpressed under heterozygosity will be exposed to selective forces that will reduce fitness. As relatives are more likely to carry the same rare deleterious alleles, inbreeding may increase the occurrence of harmful effects on fitness (Lynch, 1991). Several empirical studies on fish species have suggested inbreeding to be associated with morphological abnormalities, slow growth or low reproductive success (Araki et al., 2009; Kincaid, 1983; Shikano & Taniguchi, 2003). Chances of mating relatives will increase proportionally to the population size and scale of the releases. Hence, stock enhancement programs should keep a large wild-born broodstock and foster mating schemes that maximize Nb. Once released into the wild, in addition to contributing to the fishing yield, a portion of the offspring is expected to interbreed with their wild counterparts (Bell et al., 2008). Ryman & Laikre (1991) warned that in a successful stock enhancement program, the increment in the portion of hatchery-reared offspring may reduce the total effective population (wild and hatchery) and favor inbreeding. However, in their review of marine fish stocking programs, Kitada et al. (2009) found no evidence of long-term negative effects of large-scale releases on fitness in the wild population. Furthermore, the minimal kinship approach (Doyle et al., 2001), collecting several batches of eggs over the spawning season (Nugroho & Taniguchi, 2004) or at different time intervals over a single night (Blanco Gonzalez et al., 2010) have proven promising results to increase Nb and minimize the loss of genetic diversity.

3. Genetic resources of Black sea bream in Hiroshima Bay

The success of any stock enhancement program greatly depends on the accurate identification and characterization of the genetic architecture in the population to be managed (Allendorf & Ryman, 1987; Ryman, 1981). Unfortunately, by the time the stock enhancement program in Hiroshima Bay started, information on the genetic resources of black sea bream in Japan was very scarce and limited to other regions (Sumantadinata & Taniguchi 1982; Taniguchi et al., 1982, 1983). In fact, it was not until 1997, when almost twenty million juveniles had been released already (Fig 2.), that the first genetic analysis of samples collected from Hiroshima Bay was carried out by minisatellites (Jeong et al., 2002). Consequently, due to the lack of genetic information prior to the commencement of the stock enhancement program and the large number of juveniles released, the evaluation of the genetic impact of the releases has focused on the stock inhabiting Hiroshima Bay at the time of the research rather than on the original native stock.

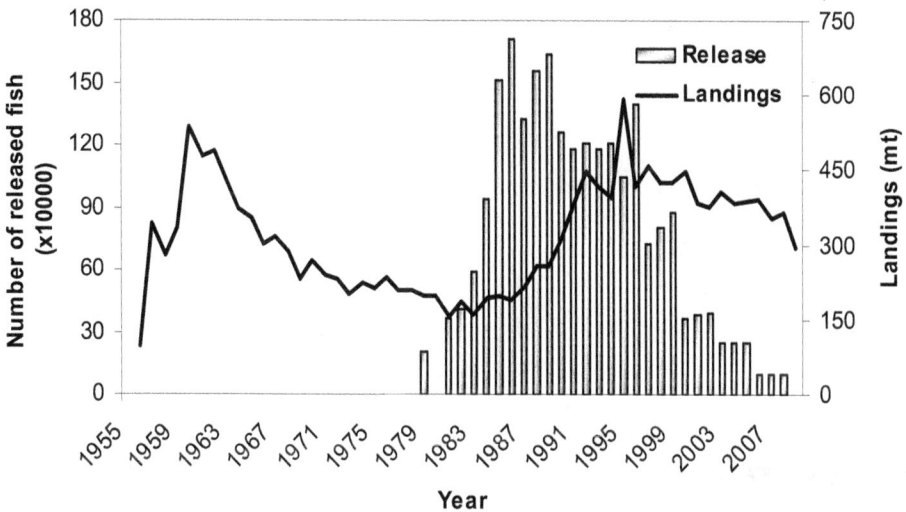

Fig. 2. Annual fluctuation in released and landed black sea bream in Hiroshima Bay.

In contrast to minisatellites, microsatellite markers estimate allele frequencies at a given locus (Estoup & Agners, 1998) hence are very useful for studies on population genetics. In order to characterize the gene pool of black sea bream in Hiroshima Bay and evaluate the effectiveness and genetic implications of the stock enhancement program carried out in the bay, eight highly polymorphic microsatellite markers were developed in our laboratory (Jeong et al. 2003, 2007). This set of microsatellites has been used at different stages of the stock enhancement program (Table 1).

3.1 Genetic diversity and population structure

Initially, four microsatellites were genotyped to assess the genetic divergence and population structure of black sea bream collected at six locations in Western Japan and Korea (Jeong et al., 2003). Despite the large stocking histories, the genetic diversity of the natural stock from Hiroshima Bay, expressed either as number of alleles per locus (6-20) or observed heterozygosity (0.65-0.96), was very high and similar to the wild stocks from other locations (Jeong et al., 2003).

Between 2000 and 2002, the offspring released by the Hiroshima City Marine Products Promotion Association (HCMPPA) were produced by only 51 breeders (Blanco Gonzalez et al., 2008a; Jeong et al., 2007). In spite of its small size, levels of genetic variation in the broodstock were similar to the natural stock from Hiroshima Bay (Table 1). In contrast, rare alleles at the loci presenting the highest polymorphism were missing in the offspring prior to the release (Jeong et al. 2003, 2007); warning about genetic drift associated to the differential contribution among parental fish, as reported during the seed production of other species commonly used in stock enhancement programs (Nugroho et al. 2000; Sekino et al., 2003; Sugama et al., 1988).

Sample origin	Hatchery (%)	Number of alleles (loci)	Observed Heterozygosity	Reference
Natural 2000	–	6-20 (4)	0.65-0.96	Jeong et al., 2003
Broodstock 2000-2002		8-18 (7)	0.74-0.92	Jeong et al., 2007; Blanco Gonzalez et al., 2008a
Pre-release 2000	100	6-16 (4)	0.57-1.0	Jeong et al., 2003
Pre-release 2000	100	7-16 (4)	0.76-0.92	Jeong et al., 2007
Pre-release 2001	100	7-14 (7)	0.74-0.92	Jeong et al., 2007
Post-release 2001	50*	7-17 (7)	052-1.0	Jeong et al., 2003
Post-release 2003	12.5	8-20 (6)	0.77-0.90	Blanco Gonzalez et al., 2008a
Post-release 2004	13.5	7-17 (6)	0.84-0.90	Blanco Gonzalez et al., 2008a
Post-release 2006	–	7-24 (6)	0.61-0.98	Blanco Gonzalez & Umino, 2009

* Proportion assumed based on tag-recapture data

Table 1. Studies on the genetic diversity of black sea bream in Hiroshima Bay

Once released, stock enhancement programs aim at maximizing fitness performance and minimizing negative genetic or ecological interactions between wild and hatchery fish. Several experiments using non-genetic markers indicated high survival and fast acclimation of black sea bream juveniles released in Hiroshima Bay (Ji et al., 2003; Nakagawa et al., 2000; Umino et al., 1999; Yamashita et al., 1997). However, limitation of food resources related to the stock enhancement program was suggested to explain the reduction on size-at-age recorded on the adults collected in 2000 compared to 1983 (Blanco Gonzalez et al., 2009). Genetically, levels of diversity 10 days after the release were similar to those observed in the natural stock (Jeong et al., 2003). Meanwhile, given the small sample sizes (n = 10 and 14 specimens in 2003 and 2004, respectively), the lower number of alleles scored in the fish collected three and four years after the release were not conclusive about any erosion due to fish stocking (Blanco Gonzalez et al., 2008a). In order to deduce the putative origin of the samples, a population assignment test was performed using WHICHRUN (Banks & Eichert, 2000) with a jackknife procedure; choosing the first natural stock collected in Hiroshima Bay whose genotype was characterized (Jeong et al., 2003) and the broodstock reared at the HCMPPA between 2000-2002 (Blanco Gonzalez et al., 2008a) as baseline populations (Fig. 3). Most of the fish collected in 2003 (82.5%) and 2004 (84.6%) were assigned to the broodstock; nevertheless, about 60% of them remained within the two dashed lines indicating deviation from equality by a factor of 10. Consequently, although these results supports the above mentioned idea of high survival and contribution of juvenile releases, the small differences in the genotype probability of the fish between the baseline populations requires a careful interpretation of the conclusions.

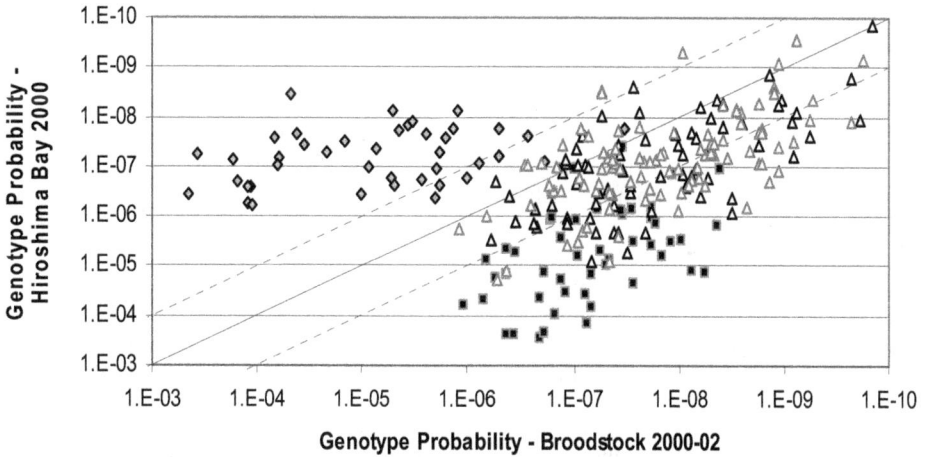

◇ Hiroshima Bay 2000 ■ Broodstock 2000-02 ▲ Hiroshima Bay 2003 △ Hiroshima Bay 2004

Fig. 3. Genotype assignment test to putative origin baselines (Hiroshima Bay 2000 and Broodstock 2000-02). The value of the X and Y coordinates represents the genotype probability of the fish collected in Hiroshima Bay in 2003 and 2004 for each baseline plotted on a log scale. The solid line represents equal probabilities of belonging to each baseline populations and separates the region of the graph where the probability is assigned to Hiroshima Bay 2000 (above) and Broodstock 2000-02 (below). The dashed lines deviate from equality by a factor of 10.

The genetic integrity of black sea bream within Hiroshima Bay and the existence of certain population structuring related to the stock enhancement program was evaluated genotyping samples collected from five locations at six microsatellites (Blanco Gonzalez & Umino, 2009). At Ninoshima, the location where stocking was most intense, the lowest number of alleles was scored. Moreover, an initial evaluation of the pairwise F_{ST} value suggested genetic differences between black sea bream from the western and eastern part of Hiroshima Bay. However, the differences disappeared once the analysis was performed standardizing the age class composition of the samples and black sea bream evidenced high genetic homogeneity among all locations, suggesting that the effects of the stock enhancement program were extensive over Hiroshima Bay.

3.2 Individual assignment and parental contribution

The polymorphism of the genetic markers determines their ability to correctly assign each offspring to a parental couple and facilitates the posterior identification once recaptured. The microsatellite DNA markers developed for black sea bream were highly polymorphic (Table 1) and exhibited high exclusion power and good performance for parentage analysis and pedigree reconstruction (Blanco Gonzalez et al., 2008a; Jeong et al., 2007).

The relevance of the origin and size of the broodstock to ensure a high Nb and low rate of inbreeding to preserve the native gene pools, contrasts with the high costs and space requirements of rearing a large number of breeders. In Hiroshima Bay, the broodstock kept at the HCMPPA never comprised more than 100 fish. Moreover, part of the fish was usually of hatchery origin and/or maintained to produce offspring for several years. During the period 2000-2002, the broodstock comprised 29 dams and 22 sires originated from wild captive and hatchery strains (Jeong et al., 2007). Genotyping seven microsatellites, the pedigree analysis conducted on the juveniles produced in 2000 and 2001 revealed that the proportion of breeders contributing to the offspring was 59% and 63%, respectively (Table 2), increasing to 76.5% combining both years (Jeong et al., 2007). This proportion is much larger than previously reported with isozymes, where only 15.7-25.5% of the breeders contributed to the first two generations of offspring (Taniguchi et al., 1983), likely related to the improvement in the markers performance and the broodstock management techniques. Jeong et al. (2007) found that the sex ratio among contributors maintained 1:1, however, the variance in family size due to the differential contribution among breeders reduced Nb to 20 in 2000 and 9 in 2001, and resulted in a high inbreeding coefficient, F (Table 2).

Management efforts to preserve the genetic integrity and maximize fitness performance of black sea bream should offset the potentially deleterious environmental and genetic effects (Allendorf and Ryman, 1987). Genetic drift has drastically reduced the genetic resources prior to the release (Jeong et al., 2007), eroding the adaptive potential of the juveniles. However, the large number of hatchery-reared fish identified two months (Jeong et al., 2007) and four years after the release (Blanco Gonzalez et al., 2008a) suggested no harmful effects on adaptation to the natural conditions. Moreover, rates of inbreeding of the latter group were similar to those observed before the release (Table 2) and the contribution of additional contributors was also detected (Blanco Gonzalez et al., 2008a). Although promising results were achieved, it is noteworthy drawing conclusions carefully because the small sample sizes analyzed may mask some underlying effects.

Origin	Size	Number of contributors (%)	Nb	F (%)	Reference
Pre-release 2000	70	32 (63)	20	2.5	Jeong et al. 2007
Pre-release 2001	110	30 (59)	9	5.6	Jeong et al. 2007
Post-release 2003	10	13 (26)	17	3.0	Blanco Gonzalez et al. 2008a
Post-release 2004	14	14 (28)	16	3.3	Blanco Gonzalez et al. 2008a

Table 2. Summary of parentage analysis conducted on black sea bream in Hiroshima Bay. Sample origin and size, loss of alleles, number of contributors, Nb, F and reference.

4. Conclusions

Microsatellite-based studies have provided insights into key aspects to preserve the natural gene pool and minimize any potential harmful genetic or ecological effects derived from the stock enhancement program carried out on black sea bream in Hiroshima Bay. The large number of juveniles released and their good acclimation to the natural environment have partly contributed to increase the fishing yields in the bay (Fig. 2); nevertheless, limitation for food was suggested to be responsible for the reduction in size-at-age observed two decades after the commencement of the stock enhancement program (Blanco Gonzalez et al., 2009). Interestingly, levels of genetic diversity in Hiroshima Bay were similar to other locations where stocking activities had never been conducted (Jeong et al., 2003) and black sea bream was suggested to comprise a large panmictic stock in western Japan. Reviewing the only two large-scale marine stock enhancement programs conducted worldwide over multiple generations and where data on the catches and genetic diversity of wild and hatchery-released fish have been monitored, Kitada et al. (2009) emphasized the importance of replacing the broodstock annually. The panmictic genetic structure, large population size and gene flow of red sea bream (*Pagrus major*) inhabiting Kagoshima Bay, Japan, was suggested to contribute to attenuate the genetic differentiation resulted from producing the hatchery-released offspring from the same small broodstock over several years. On the other hand, on Pacific herring (*Clupea pallasii*), they found no evidence of genetic erosion attributable to the stock enhancement programs, using a native broodstock with annual replacement. None of the programs showed any sign of fitness decline.

Genetic drift due to the limited number of breeders kept in the hatchery was identified as a major constraint to preserve the genetic resources of black sea bream in Hiroshima Bay (Blanco Gonzalez et al., 2008a; Jeong et al., 2007). The small broodstock reared at the HCMPPA provided a relatively good representation of the genetic diversity of the natural stock inhabiting Hiroshima Bay (Jeong et al., 2003); however, their differential contribution resulted in a very small Nb and high rates of inbreeding, warning about the risks of genetic erosion and the need to improve broodstock management practices (Jeong et al., 2007). In this regard, Blanco Gonzalez et al. (2010) demonstrated that a larger number of breeders and the collection of eggs at two-hourly intervals contributed to increase Nb and reduce the rate of inbreeding. The minimal kinship approach (Doyle et al., 2001) and the collection of several batches of eggs at different days over the spawning season have also proven good results to preserve the genetic resources of stock enhancement programs on red sea bream (Nugroho & Taniguchi, 2004).

Parentage analysis confirmed that the released juveniles had good survival and growth (Jeong et al., 2007), and some of them reached maturity (Blanco Gonzalez et al., 2008a), suggesting that hatchery conditions had no negative effects on the posterior fitness performance in the natural environment. Moreover, despite the loss of genetic diversity observed before the release, black sea bream showed high genetic homogeneity within Hiroshima Bay and no sign of population structuring (Blanco Gonzalez & Umino, 2009). Given the large number of juveniles produced by a small number of parental fish and their high survival, interbreeding between natural and hatchery fish is likely to have taken place (Blanco Gonzalez et al., 2008a); hence, broodstock management strategies and genetic monitoring should aim at avoiding genetic swamping of the natural population (Ryman & Laikre, 1991).

The stock enhancement program conducted on black sea bream in Hiroshima Bay is expected to continue providing first-hand empirical information about the genetic and ecological implications of fish releases in marine ecosystems. In 2008, due to several socio-economical problems (Umino et al., 2011), the program was stopped. Therefore, future studies will deepen our understanding of the evolutionary processes underlying large-scale releases and how to preserve the genetic resources. Indeed, genetic diversity at selective markers should insight into the genetic component involved in juvenile survival and local adaptation.

5. Acknowledgements

We would like to thank Dr. D.S. Jeong and the students of our laboratory for their assistance during the sampling and data collection. We are also very grateful to the staff working at HCMPPA for providing hatchery specimens. The present study was partly supported by a Grant-in-Aid for Scientific Research (C) from the Japan Society for the Promotion of Science (JSPS) to T.U. (No.14560152&19580205).

6. References

Allendorf, F.W. & Phelps, S. (1980). Loss of genetic variation in a hatchery stock of cutthroat trout. *Transactions of the American Fisheries Society*, Vol.109, No.5, (September 1980), pp. 537-543, ISSN 0002-8487.
Allendorf, F.W. & Ryman, N. (1987). Genetic management of hatchery stocks. In: *Population Genetics and Fishery Management*, Ryman, N. & Utter, F., (Eds.), pp. 141-159, University of Washington Press, ISBN 0295964367, Seattle, WA.
Anonymous. (1987). Results in Hiroshima Prefecture. In: *Work on the Development of Releasing Techniques for the Stock Enhancement for Fiscal Year 1986 - Synthetic Report of the Group for Black Sea Bream*, Nansei Regional Fisheries Research Laboratory, (Ed.), pp. 1-26, Ohno, Japan (in Japanese).
Araki, H., Cooper, B. & Blouin, M.S. (2009). Carry-over effect of captive breeding reduces reproductive fitness of wild-born descendants in the wild. *Biology Letters*, Vol.5, No.5, (October 2009), pp. 621-624, ISSN 1744-957X.
Araki, H. & Schmid, C. (2010). Genetic effects of long-term stock enhancement programs. *Aquaculture*, Vol.308, Supp.1, (October 2010), pp. S2-S11, ISSN 0044-8486.

Banks, M.S. & Eichert, M. (2000). WHICHRUN (Version 3.2): a computer program for population assignment of individuals based on multilocus genotype data. *Journal of Heredity*, Vo.91, No.1, (January-February 2000), pp. 87-89, ISSN 1465-7333.

Bekkevold, D., Hansen, M.M. & Nielsen, E.E. (2006). Genetic impact of gadoid culture on wild fish populations: predictions, lessons from salmonids, and possibilities for minimizing adverse effects. *ICES Journal of Marine Science*, Vol.63, No.2, pp. 198-208, ISSN 1095-9289.

Bell, J.D., Leber, K.M., Blankenship, H.L., Loneragan, N.R. & Masuda, R. (2008). A new era for restocking, stock enhancement and sea ranching of coastal fisheries resources. *Reviews in Fisheries Science*, Vol.16, No.1-3, pp. 1-9, ISSN 1064, 1262.

Bert, T.M., Leber, K.M., Mcmichael, R.H., Neidig, C.L., Cody, R.P., Forstchen, A.B., Halstead, W.G., Tringali, M.D., Winner, B.L. & Kennedy, F.S. (2001). Evaluating stock enhancement strategies: a multi-disciplinary approach. *UJNR Technical Report* Vol.30, (December 2001), pp. 105–126.

Blanco Gonzalez, E., Murakami, T., Yoneji, T., Nagasawa, K. & Umino, T. (2009). Reduction in size-at-age of black sea bream (*Acanthopagrus schlegelii*) following intensive releases of cultured juveniles in Hiroshima Bay, Japan. *Fisheries Research*, Vol. 99, No.2, (August 2009), pp. 130–133, ISSN 0165-7836.

Blanco Gonzalez, E., Nagasawa, K. & Umino, T. (2008a). Stock enhancement program for black sea bream (*Acanthopagrus schlegelii*) in Hiroshima Bay: monitoring the genetic effects. *Aquaculture*, Vol.276, No.1-4, (April 2008), pp. 36–43, ISSN 0044-8486.

Blanco Gonzalez, E., Taniguchi, N. & Umino, T. (2010). Can ordinary single-day egg collection increase the effective population size in broodstock management programs? Breeder-offspring assignment in black sea bream (*Acanthopagrus schlegelii*) through two-hourly intervals. *Aquaculture*, Vol.308, Supp.1, (October 2010), pp. S2-S11. ISSN 0044-8486.

Blanco Gonzalez, E. & Umino, T. (2009). Fine-scale genetic structure derived from stocking black sea bream, *Acanthopagrus schlegelii* (Bleeker, 1854), in Hiroshima Bay, Japan. *Journal of Applied Ichthyology*, Vol.25, No.4, (August 2009), pp.407–410, ISSN 0175-8659.

Blanco Gonzalez, E., Umino, T. & Nagasawa, K. (2008b). Stock enhancement program for black sea bream (*Acanthopagrus schlegelii*) in Hiroshima Bay, Japan: a review. *Aquaculture Research*, Vol.39, No.12, (September 2008), pp. 1307–1315, ISSN 1365-2109.

Cooke, S.J. & Philipp, D.P. (2006). Hybridization among divergent stocks of largemouth bass (*Micropterus salmoides*) results in altered cardiovascular performance: the influence of genetic and geographic distance. *Physiological and Biochemical Zoology*, Vol.79, No.2, (March-April 2006), pp. 400–410, ISSN 1522-2152.

Cross, T.F. (1999). Genetic considerations in enhancement and ranching of marine and anadromous species. In: *Stock Enhancement and Sea Ranching*, Howell, B.R., Moksness, E. & Svåsand, T., (Eds.), pp. 37-48, Blackwell Science Lt., ISBN 0-85238-246-4, Oxford, UK.

DeWoody, J.A. & Avise, J.C. (2000). Microsatellite variation in marine, freshwater and anadromous fish compared with other animals. *Journal of Fish Biology*, Vol.56, No.3, (March 2000), pp. 461-473, ISSN 0022-1112.

Doyle, R.W., Perez-Enriquez, R., Takagi, M. & Taniguchi, N. (2001). Selective recovery of founder genetic diversity in aquacultural broodstocks and captive, endangered fish populations. *Genetica*, Vol.111, No.1-3, (January 2001), pp. 291-304, ISSN 0016-6707.

Estoup, A. & Angers, B. (1998). Microsatellites and minisatellites for molecular ecology: theoretical and empirical considerations, In: *Advances in Molecular Ecology*, Carvalho, G. (Ed.), pp. 55-86, NATO Science Series A, Vol.306, IOS Press, ISBN 90 5199 440 0, Amsterdam, The Netherlands.

Falconer, D.S. (1981). *Introduction to quantitative genetics* (2nd edition), Longman, ISBN 978-0582441958, London. UK.

Fraser, D.J. (2008). How well can captive breeding programs conserve biodiversity? A review of salmonids. *Evolutionary Applications*, Vol.1, No.4, (November 2008), pp. 535-585, ISSN 1752-4571.

Glover, K.A., Dahle, G., Westgaard, J.I., Johansen, T., Knutsen, H. & Jørstad, K.E. (2010). Genetic diversity within and among Atlantic cod (*Gadus morhua*) farmed in marine cages: a proof-of-concept study for the identification of escapees. *Animal genetics*, Vol.41, No.5, (October 2010), pp. 515-522, ISSN 1365-2052.

Hauser, L. & Carvalho, G.R. (2008). Paradigm shifts in marine fisheries genetics: ugly hypotheses slain by beautiful facts. *Fish and Fisheries*, Vol.9, No.4, (December 2008), pp.333-362, ISSN 1467-2979.

Jeong, D.S., Blanco Gonzalez, E., Morishima, K., Arai, K. & Umino, T. (2007). Parentage assignment of stocked black sea bream, *Acanthopagrus schlegelii* in Hiroshima Bay using microsatellite DNA markers. *Fisheries Science*, Vol.73, No.4, (July 2007), pp. 823-830, ISSN 0919-9268.

Jeong, D.S., Hayashi, M., Umino, T., Nakagawa, H., Morishima, K., Arai, K. (2002). Comparison of genetic variability between wild and hatchery-stocked black sea bream *Acanthopagrus schlegeli* using DNA fingerprinting. *Fish Genetics and Breeding Science*, Vol.32, No.2, (December 2002), pp.135-139, ISSN 13437917.

Jeong, D.S., Umino, T., Kuroda, K., Hayashi, M., Nakagawa, H., Kang, J.C., Morishima, K. & Arai, K. (2003). Genetic divergence and population structure of black sea bream *Acanthopagrus schlegeli* inferred from microsatellite analysis. *Fisheries Science*, Vol.69, No.5, (October 2003), pp. 896-902, ISSN 0919-9268.

Ji H., Om A.D., Umino T., Nakagawa H., Sasaki T., Okada K., Asano M.E. & Nakagawa A. (2003). Effect of dietary ascorbate fortification on lipolysis activity of juvenile black sea bream *Acanthopagrus schlegeli*. *Fisheries Science*, Vol.69, No.1, (February 2003), pp. 66-73, ISSN 0919-9268.

Kincaid, H.L. (1983). Inbreeding in fish populations used for aquaculture. *Aquaculture*, Vol.33, No.1-4, (June 1983), pp. 215-227, ISSN 0044-8486.

Knutsen, H., Jorde, P.E., André, C. & Stenseth, N.C. (2003). Fine-scaled geographical population structuring in a highly mobile marine species: the Atlantic cod. *Molecular ecology*, Vol.12, No.2, (February 2003), pp. 385-394, ISSN 0962-1083.

Kitada, S., Shishidou, H., Sugaya, T., Kitakado, T., Hamasaki, K. & Kishino, H. (2009). Genetic effects of long-term stock enhancement programs. *Aquaculture*, Vol.290, No.1-2, (May2009), pp. 69-79, ISSN 0044-8486.

Laikre, L., Schwartz, M.K., Waples, R.S., Ryman, N. & The GeM Working Group. (2010). Compromising genetic diversity in the wild: unmonitored large-scale release of plants and animals. *Trends in Ecology & Evolution*, Vol.25, No.9, (August 2010), pp. 520-529, ISSN 0169-5347.

Luikart, G., England, P.R., Tallmon, D., Jordan, S. & Taberlet, P. (2003). The power and promise of population genomics: from genotyping to genome typing. *Nature Reviews Genetics*, Vol.4, No.12, (December 2003), pp. 981-994, ISSN 1471-0056.

Lynch, M. (1991). The genetic interpretation of inbreeding depression and outbreeding depression. *Evolution*, Vol.45, No.3, (May 1991), pp. 662-629, ISSN 1558-5646.

Miller, L.M. & Kapuscinski, A.R. (2003). Genetic guidelines for hatchery supplementation programs, In: *Population Genetics: Principles and Applications for Fisheries Scientists*,

Hallerman, E.M., (Ed.), pp. 329-355, American Fisheries Society, ISBN 978-1888569278, Bethesda, MD.

Nakagawa, H., Umino, T., Hayashi, M., Sasaki, T. & Okada K. (2000). Changes in biochemical composition of black sea bream released at 20 mm size in Daio Bay, Hiroshima. *Suisanzoshoku*, Vol.48, No.4, (December 2000), pp. 643-648, ISSN 0371-4217.

Nei, M., Maruyama, T. & Chakraborty, R. (1975). The bottleneck effect and genetic variability in populations. *Evolution*, Vol.29, No.1, (March 1975), pp. 1-10, ISSN 1558-5646.

Nugroho, E. & Taniguchi, N. (2004). Daily change of genetic variability in hatchery offspring of red sea bream during spawning season. *Fisheries Science*, Vol.70, No.4, (August 2004), pp. 638-644, ISSN 0919-9268.

Nugroho, E., Taniguchi, N., Miyashita, M. & Kato, K. (2000). Genetic difference among seed populations of greater amberjack used as seed fish in aquaculture of Japan. *Suisanzoshoku*, Vol.48, No.4, (December 2000), pp. 665-674, ISSN 0371-4217.

Ortega-Villaizan Romo, M.M., Suzuki, S., Ikeda, M., Nakajima, M. & Taniguchi, N. (2005). Monitoring of the genetic variability of the hatchery and recaptured fish in the stock enhancement program of the rare species barfin flounder *Verasper moseri*. *Fisheries Science*, Vol.71, No.5, (September 2005), pp.1120-1130, ISSN 0919-9268.

Perez-Enriquez, R, Takagi, M. & Taniguchi, N. (1999). Genetic variability and pedigree tracing of a hatchery-reared stock of red sea bream (*Pagrus major*) used for stock enhancement, based on microsatellite DNA markers. *Aquaculture*, Vol.173, No.1-4, (March 1999), pp. 413-423, ISSN 0044-8486.

Perez-Enriquez, R, Takemura, M., Tabata, K. & Taniguchi, N. (2001). Genetic diversity of red sea bream *Pagrus major* in western Japan in relation to stock enhancement. *Fisheries Science*, Vol.67, No.1, (February 2001), pp. 71-78, ISSN 0919-9268.

Reisenbichler, R.R. & Rubin, S.P. (1999). Genetic changes from artificial propagation of Pacific salmon affect the productivity and viability of supplemented populations. *ICES Journal of Marine Science*, Vol.56, No.4, (August 1999), pp. 459–466, ISSN 1095-9289.

Reiss, H., Hoarau, G., Dickey-Collas, M. & Wolff, W.J. (2009). Genetic population structure of marine fish: mismatch between biological and fisheries management units. *Fish and Fisheries*, Vol.10, No.4, (December 2009), pp.361-395, ISSN 1467-2979.

Ryman, N. (1981). Conservation of genetic resources: experiences from the brown trout (*Salmo trutta*). In: *Fish Gene Pools: Preservation of Genetic Resources in Relation to Wild Fish Stocks*, Ryman, N., (Ed.), pp. 61-74, Ecological Bulletin Vol. 34, ISBN 978-9154602995, Stockholm, Sweden.

Ryman, N. & Laikre, L. (1991). Effects of supportive breeding on the genetically effective population size. *Conservation Bioloy*, Vol.5, No.3, September 1991), pp. 325-329, ISSN 0888-8892.

Sekino, M., Saitoh, K., Yamada, T., Kumagai, A., Hara, M. & Yamashita, Y. (2003). Microsatellite-based pedigree tracing in a Japanese flounder *Paralichthys olivaceus* hatchery strain: implications for hatchery management related to stock enhancement program. *Aquaculture*, Vol.221, No.1-3, (May 2003), pp. 255-263, ISSN 0044-8486.

Shikano, T. & Taniguchi, N. (2003). DNA markers for estimation of inbreeding depression and heterosis in the guppy *Poecilia reticulate*. *Aquaculture Research*, Vol.34, No.11, (September 2003), pp. 905-911, ISSN 1365-2109.

Sugama, K., Taniguchi, N. & Umeda, S. (1988). An experimental study on genetic drift in hatchery population of red sea bream. *Nippon Suisan Gakkaishi*, Vol.54, No.5, pp. 739–744, ISSN 0021-5392.

Sumantadinata, K. & Taniguchi, N. (1982). Biochemical genetic variations in black sea bream. *Nippon Suisan Gakkaishi*, Vol.48, No.2, (Febrero 1982), pp. 143-149, ISSN 0021-5392.

Svåsand T., Crosetti D., García-Vázquez E. & Verspoor E. (2007). Genetic impact of aquaculture activities on native populations. Genimpact final scientific report (EU contract n. RICA-CT-2005-022802), Available from http://genimpact.imr.no/

Taniguchi, N. (2003). Genetic factors in broodstock management for seed production. *Reviews in Fish Biology and Fisheries*, Vol.13, No.2, (June 2003), pp. 177-185, ISSN 0960-3166.

Taniguchi, N. (2004). Broodstock management for stock enhancement programs of marine fish with assistance of DNA marker (a review), In: *Stock Enhancement and Sea Ranching: Developments, Pitfalls and Opportunities, Second Edition*, Leber, K.M., Kitada, S., Blankenship, H.L., Svåsand, T., (Eds.), pp.329-338, Blackwell Publishing Ltd., ISBN 978-1-4051-1119-5, Oxford, UK.

Taniguchi, N., Sumantadinata, K. & Iyama, S. (1983). Genetic change in the first and second generations of hatchery stock of black seabream. *Aquaculture*, Vol.35, pp. 309-320, ISSN 0044-8486.

Taniguchi, N., Sumantadinata, K., Suzuki, A. & Yamada, J. (1982). Genetic variation in isoelectric focusing pattern of sarcoplasmic protein of black seabream. *Nippon Suisan Gakkaishi*, Vol.48, No.2, (February 1982), pp. 139-141, 0021-5392.

Thorstad, E.B., Fleming, I.A., McGinnity, P., Soto, D., Wennevik, V. & Whoriskey, F. (2008). *Incidence and impacts of escaped farmed Atlantic salmon Salmo salar in nature*. NINA Special Report 36, ISBN 978-82-426-1966-2, Norway.

Umino, T., Blanco Gonzalez, E., Saito, H. & Nakagawa, H. (2011). Problems associated with the recovery on landings of black sea bream (*Acanthopagrus schlegelii*) intensively released in Hiroshima Bay, Japan, In: *Global Change: Mankind-Marine Environment Interactions*, Ceccaldi, H.-J., Dekeyser, I., Girault, M. & Stora, G., (Eds.), pp. 37-40, Springer Science+Bussiness Media B.V., ISBN 978-90-481-8629-7, Dordrecht, The Netherlands.

Umino, T., Hayashi, M., Miyatake, J., Nakayama, K., Sasaki, T., Okada, K. & Nakagawa, H. (1999). Significance of release of black sea bream at 20-mm size on stock enhancement in Daiô Bay, Hiroshima. *Suisanzoshoku*, Vol.47, No.3, (May 1999), pp. 337-342, ISSN 0371-4217.

Umino, T., Kajihara, T., Shiozaki, H., Ohkawa, T., Jeong, D.-S. & Ohara, K. (2009). Wild stock structure of *Girella punctata* in Japan revealed shallow genetic differentiation but subtle substructure in subsidiary distributions. *Fisheries Science*, Vol.75, No.4, (July 2009), pp. 909-919, ISSN 0919-9268.

Ward, R.D. (2006). The importance of identifying spatial population structure in restocking and stock enhancement programmes. *Fisheries Research*, Vol.80, No.1, (August 2006), pp. 9-18, ISSN 0165-7836.

Ward, R.D., Woodwark, M. & Skibinski, D.O.F. (1994). A comparison of genetic diversity levels in marine, freshwater and anadromous fish. *Journal of Fish Biology*, Vol.44, No.2, (February 1994), pp. 213–232, ISSN 1095-8649.

Watts, P.C., Kay, S.M., Wolfenden, D., Fox, C.J., Geffen, A.J., Kemp, S.J. & Nash, R.D.M. (2010). Temporal patterns of spatial genetic structure and effective population size in European plaice (*Pleuronectes platessa*) along the west coast of Scotland and in the Irish Sea. *ICES Journal of Marine Science*, Vol.67, No.4, (May 2010), pp. 607-616, ISNN 1095-9289.

Yamashita, H., Umino, T., Nakahara, S., Okada, K. & Nakagawa, H. (1997). Changes in some properties of black sea bream released into the Daio Bay. *Fisheries Science*, Vol.63, No.2, pp. 267-271, (February 1997), ISSN 0919-9268.

11

Landscape Genomics in Livestock

Lorraine Pariset[1], Stephane Joost[2],
Maria Gargani[1] and Alessio Valentini[1]
*[1]Department for Innovation in Biological, Agro-Food and Forest
Systems (DIBAF), University of Tuscia, Viterbo*
*[2]Laboratory of Geographic Information Systems (LASIG), School of Architecture
Civil and Environmental Engineering (ENAC)
Ecole Polytechnique Fédérale de Lausanne (EPFL), Lausanne*
[1]Italy
[2]Switzerland

1. Introduction

Landscape genomics correlates genetic variation patterns with geographic variables to investigate how geographical and environmental characteristics affect the genetic structure of populations (Luikart et al., 2003; Joost et al., 2007; Holderegger & Wagner, 2008; Pariset et al., 2009; Shwartz et al., 2009). A field combining molecular markers, genetics and landscape structure was first described by Manel et al. (2003) and its definition evolved and changed in the following years (Storfer et al., 2007; Holderegger & Wagner, 2006).

Landscape genomics requires the recording of the exact location of the sampling and the assessment of a number of molecular markers on a representative number of individuals in a population, obtaining the allelic frequency at these loci (Joost et al., 2008). Markers can vary from mtDNA to Y chromosome, microsatellites, Single Nucleotide Polymorphisms (SNP) (Manel et al., 2003; Wang, 2011). The nature of the markers used for the analysis can affect the detection of geographical structuring of populations, as suggested by Naderi et al. (2007). The simultaneous use of both nuclear and single parent transmitted markers (mitochondrial and Y chromosome) is likely to provide more significant results respect to the use of either marker alone (Hewitt, 2004; Gonçalves et al., 2010; Wang et al., 2011; Pariset et al., 2011).

Recently, the availability of high density SNP devices for a few species has given new possibilities of analysis. High throughput genomics is providing new DNA sequences suitable for gene discovery and for the study of genetic variation at different levels. Most important, the quick expansion of molecular genetic technologies not only is providing a huge amount of genomic data and suitable markers: it is offering data at affordable and constantly declining costs. Therefore the genomic information available for most species, including livestock, is rapidly increasing (Luikart et al., 2003; Marnis et al., 2007; Segelbacher et al., 2010; Helyar et al., 2010) making possible the setup of high throughput SNP devices for livestock (Box 7).

Box 1. Molecular Markers for livestock landscape genomics

To analyse the interaction between geographic variables and genetic patterns, landscape genomics requires a high number of molecular markers for providing enough power of resolution. The main marker systems utilized in livestock studies are AFLPs, microsatellites or STR, SNPs and CNV.

AFLP

The AFLP technique consists of two steps: a DNA digestion and a PCR amplification. Specifically, the fragments obtained from digestion are linked to adapters and amplified using specific primers (complementary to the adapters). This technology generates a mixture of fragments that are separated and identified by polyacrylamide gel electrophoresis or by sequencing devices. These markers are assumed to be dominant and neutral (Vos et al., 1995) and were widely applied to livestock (Negrini et al., 2007; Ajmone-Marsan et al., 2008, 2011), but they can be used to find genes under selection with a population genomics method (Luikart et al., 2003). The advantage of AFLPs is that they allow analyzing several loci in a single experiment and that they can be applied to species of which genome sequences are unknown. However, their mostly dominant behaviour makes it difficult to assess allelic frequencies when Hardy Weinberg equilibrium is not assured. They also require substantial hand work.

Microsatellites

Microsatellites are codominant markers located in the nuclear DNA. They are short tandem repeats of two to eight or more nucleotides which occur as interspersed repetitive elements in all eukaryotic genomes (Tautz & Renz, 1984). The microsatellite polymorphism consists in the number variation of the tandem repeated units; the alleles at the same locus differ in their length. After PCR amplification performed using primers flanking the repeated motif, the different alleles are analyzed by gel electrophoresis or by automated sequencers. The microsatellites are not very suitable for massively parallel analysis because traditional multiplexing cannot be expanded to more than a few loci (far less than 20); too many DNA amplifications are required and microsatellite are now being replaced by more efficient systems. However, recently a STR profiling method was developed using the Roche Genome Sequencer FLX to sequence multiple microsatellite loci (Fordyce et al., 2011).

Mithocondrial DNA

Mitochondrial DNA (mtDNA) is a uniparental marker as its inheritance is clonal (maternal) and it is not subject to recombination (Galtier et al., 2009). mtDNA is present in hundreds of copies per cell and evolves 10 times faster than nuclear DNA. The evolution of mitochondrial genes varies from region to region: the ribosomal DNA genes are highly conserved; the region known as non-coding control region (containing the displacement loop or D-loop and the origin of replication) is much more variable since it is deemed to be not being subject to selective pressure. Mitochondrial markers are amplified by PCR and then sequenced or, today more rarely, assessed by restriction enzymes. Sequencing can be performed by Sanger or by parallel devices (Galtier et al., 2009). The multiplicity of copy number of mtDNA in a cell makes it suitable for the analysis of ancient specimens. Sequencing data obtained are then aligned and analyzed with appropriate bioinformatics programs.

Single Nucleotide Polymorphisms

SNPs (Single Nucleotide Polymorphisms) are the most common form of polymorphism among individuals, arising approximately every 200 base pairs in livestock (Williams, 2005). A SNP marker is a single base change in a DNA sequence (Vignal, 2002). The main methods for the identification of new single nucleotide variations are direct sequencing of regions of interest and *in silico* research that allows the identification of SNPs through alignment and comparison of sequences using public sequence databases (Guryev et al., 2005). Expressed sequence tags (ESTs) including numerous copies of the same gene sequence can be used to identify SNPs *in silico* (Primmer, 2008). ESTs are becoming publicly available with increasing frequency; they represent a cheap and effective tool for identifying gene-linked markers and are available also for species not yet fully sequenced (Pariset et al., 2009c, 2010b). Today, millions of SNPs are discovered by resequencing different individuals by parallel sequencers. After sequencing, SNPs can be selected and a chip that allows the diagnosis of the alleles for a very low cost (well below 1/1000 of USD each, depending on market and species) can be devised. Y chromosome SNPs constitute another important source of uniparental markers (Nijman et al., 2008). Due to their abundance and to the recent availability of high throughput analysis technologies, SNP markers are being used more and more and have begun the most suitable markers for landscape genomics.

CNV

Copy Number Variation (CNV) refers to genomic structural variations that involve DNA segments ranging from 1 kb to 5 Mb (Feuk et al., 2006). The quantitative variants comprising insertions and deletions, as well as inversions and translocations, are relative to a reference genome sequence (Scherer et al., 2007). CNVs represent an important source of genetic variation among individuals and cover more nucleotide sequences per genome compared to SNPs markers (Conrad et al., 2010). Some studies have demonstrated that CNVs can influence gene expression through position effects, and can be related to human diseases (Feuk et al., 2006; Zhang et al., 2009). The common methods for CNVs detection and analysis are SNP arrays and array comparative genome hybridization microarrays (CGH). The advantages of using the arrays are their low cost and high-density which make them ideal for large population screening (Perkel, 2011). About 29,000 CNVs have been identified in humans (Conrad et al., 2009), and it was estimated that two individual genomes have differences in CNVs on order of 500 to 1000 (Perkel, 2011). CNVs dataset have been identified in cattle using both high-density CGH and SNP array data. For example, CGH arrays have been used to identify 25 germline CNVs in three Holstein bulls (Liu et al., 2008); Liu et al. (2010) discovered over 200 candidate CNV regions some of which contribute to the breed formation and adaptation. Fadista et al. (2010) identified 304 CNV regions in 20 animals belonging to four cattle breeds. SNP data from Bovine HapMap Consortium samples were used to identify 682 candidate CNV regions in a diverse panel of 521 animals from 21 different breeds (Hou et al., 2011). Using the BovineSNP50 genotyping data 368 CNV regions from 265 Korean Hanwoo cattle and 682 candidate CNV regions in a diverse panel of 521 animals from 21 different breeds have been identified (Bae et al., 2010).

Landscape Genomics has proven ability to detect statistical signals that associate loci with environmental parameters in very different species, like *Hylobius abietis* and *Ovis aries* (Joost et al., 2007). The use of a high number of markers allows the identification of loci that may be under selection. In fact, loci under selection may be non-optimal for calculating population parameters while they can be useful in assessing local speciation, adaptation and, in the case of livestock, the effects of human selection (Storz, 2005; Joost et al., 2008; Pariset et al., 2009b; Manel et al., 2010). This approach presents differences when analyzing livestock or wild species (Bruford, 2004). Gene flow in natural populations depends on ecological characteristics and global or local environment, while for livestock it is influenced mainly by human activities (Manel et al., 2003; Storfer et al., 2007; Berthouly et al., 2009; Anderson et al., 2010). Present day livestock breeds are the result of years of human selection, adaptation to different environments and demographic effects as domestication, migration, selection of the more desired individuals, all contributing to the actual patterns of genetic diversity (Bruford et al., 2003).

Threats to biodiversity, in terms of extinction rate, destruction of ecosystems and habitat, or loss of genetic diversity, are increasing within the species utilized in agriculture. Livestock plays a fundamental role in human society, as source of both food and nitrogen and greenhouse gas contributing to environmental pollution and climate change (Joost et al., 2009). Livestock sector is losing genetic diversity as large-scale production expands (Taberlet et al., 2008; Joost et al., 2010; Anderson et al., 2010). After domestication process, livestock sector has changed remarkably because of the intense anthropogenic selection (Taberlet et al., 2008; Joost et al., 2011). As a consequence, farmers progressively substituted less productive local breeds with highly productive cosmopolitan breeds and progressively a significant number of native breeds disappeared (Simianer et al., 2003). It is more strategically important than ever to preserve as much of the livestock diversity as possible, to ensure a prompt and proper response to the needs of future generations. In this context, approaches based on the combination of genomics and spatial analysis is of great help.

2. Landscape analysis methods

Analysis techniques in landscape genetic employ various statistical approaches that can be applied using several statistical software, some of which are listed in Box 3. Statistical procedures for estimating genetic subdivision as AMOVA or F statistics (Wright, 1951; Excoffier et al., 1992) calculate divergence among populations. Statistical methods suitable for landscape analysis rely mostly on IBD, multivariate analysis and clustering models. They relate genetic variations to demographic data that can include aspects of the landscape.

2.1 IBD (isolation by distance)

The genetic structure of natural populations is influenced by the limited gene flow occurring when geographical distances increase between them. The non random mating is a result of the preferentially reproduction between geographically close individuals; this means that the genetic distance between individuals or populations is proportional to their spatial distance (isolation by distance). The IBD models are used to study demographic, migratory reproductive aspects of populations (Loiseau et al., 2009; Petit et al., 2001; Prugnolle et al.,

---

Box 2. GIS and GIScience

Geographical Information Systems (GIS) are specific systems designed to capture, store, manipulate, manage, analyze, and present digital geographically referenced data. In short, GIS constitute the merging of cartography, statistical analysis, and database technology. They belong to a rich set of methods, approaches and techniques, gathered together within a science (GIScience) to investigate the fundamental issues arising from the use of geographic information (Longley et al., 2001). GIScience consists of a two-sided discipline made up of its own technology driven research and development closely related to computer science (software, databases, formats, etc.), and of a collection of theories and statistical modelling approaches that explicitly use the spatial referencing of data (Goodchild & Haining, 2004).

2005). The presence of an IBD pattern is usually inferred using Mantel test (1967) which is a regression typically used to test for non-random associations between genetic differentiation (between pairs of individuals) and their geographical distance matrices (Manel et al., 2003; Guillot et al., 2009). A partial Mantel test is used to compare three or more variables allowing to identify, among a set of landscape variables, those that are associated with a significant levels of the genetic distance among individuals (Manel et al., 2003; Epps et al., 2005; Guillot et al., 2009).

Box 3. Software for Landscape genetics analysis	
BAPS (Corander et al., 2008)	BAPS is a program for Bayesian inference of the genetic structure in a population. BAPS assigns individuals to genetic clusters by either mixture or admixture models assuming HWE within cluster. The analyses can be done using a non-spatial, and spatial model for genetic discontinuities in populations. The spatial model requires that coordinate data is available for the clustered units (groups or individuals). (http://web.abo.fi/fak/mnf/mate/jc/software/baps.html)
Genclass (Piry et al., 2004)	Genclass uses Bayesian and likelihood approaches to detect migrants and to assign individuals to populations. Assumes HWE and calculates if a genotype can be excluded from a given population. (http://www.montpellier.inra.fr/URLB/index.html)
Structure (Pritchard et al., 2000)	Structure uses a Bayesan approach to investigate the population genetic structure using multi-locus genotype. Assuming HWE it infers the presence of distinct populations, detects new migrant and admixed individuals, assigns individuals to populations and studies hybrid zones. (http://pritch.bsd.uchicago.edu/software/structure2_1.html)
Geneland (Guillot et al., 2005)	Geneland is a R package that processes individual multilocus genetic data to detect population structure, assuming HWE and linkage equilibrium. Genland integrates spatial contiguity of individuals with a Bayesian genetic assignment. As a result, individuals are assigned to the genetic cluster not only on the basis of their genotype, but also of their geographic locations. The program provides a graphic with a spatial distribution of the subdivision. (http://www2.imm.dtu.dk/~gigu/Geneland/#)

Fdist (Beaumont & Nichols, 1996)	Fdist is a program for the identification of loci that might be under selection in structured populations. Fdist detect loci that show unusually low or high levels of genetic differentiation using the statistic of *Fst*. A plot of *Fst* and heterozigosity, using a coalescent model, identifies outlying *Fst* values. The program assumes an infinite or finite model of migration. (http://www.rubic.rdg.ac.uk/~mab/software.html)
Lamarc (Kuhner, 2006)	Lamarc is a program based on likelihood analysis to calculate effective population sizes assuming constant mutation rates among loci, a recombination rate, population exponential growth rates, and past migration rates assuming a stable migration structure. (http://evolution.gs.washington.edu/lamarc/lamarc_prog.html)
SAM (Joost et al., 2008)	SAM (Spatial Analysis Method) is an approach that gives the possibility to identify loci likely to be under natural selection. SAM analyzes the association between the allelic frequencies at molecular markers and data from various environmental variables. To this end SAM uses one or more environmental variable describing the sampling location and a molecular marker matrix. Using a logistic regression, this method associates the frequency of molecular markers with the environmental parameters at each site and highlights the potential markers linked to genomic regions involved in adaptation. (http://www.econogene.eu/software/sam/)
BayesAss (Wilson & Rannala, 2003)	BayesAss uses a Markov chain Monte Carlo (MCMC) method to estimates recent migration rates between populations. It also estimates each individual's immigrant ancestry, and inbreeding within populations. Loci are assumed to be in linkage equilibrium. (http://www.rannala.org/labpages/software.html)
BayeScan (Foll & Gagiotti, 2008)	BayeScan is a program that identifies candidate loci under natural selection from genetic data, using differences in allele frequencies between populations and it is based on the multinomial-Dirichlet model. BayeScan accepts different types of data: *(i)* codominant data (as SNPs or microsatellites), *(ii)* dominant binary data (as AFLPs) and *(iii)* AFLP amplification intensity, which are neither considered as dominant nor codominant. (http://cmpg.unibe.ch/software/bayescan/)
Allele in space (AIS). (Miller, 2005)	AIS is a program that combines information from genetic and spatial data. It performs some spatial analyses with genetic data: Mantel Tests, Spatial Autocorrelation Analyses, Allelic Aggregation Index Analyses (AAIA), Mommonier's Algorithm, and "Genetic Landscape Shape" interpolations. (http://www.marksgeneticsoftware.net/AISInfo.htm)
BATWING (Wilson et al., 2003)	BATWING uses a Markov chain Monte Carlo (MCMC) method to assess the past demography of populations based on multilocus genotypes. It estimates mutation rates, effective population sizes and growth rates, and times of population splitting events. (http://www.maths.abdn.ac.uk/~ijw)
NewHybrids (Anderson &	NewHybrids is a program for computing the posterior distribution that individuals fall into different hybrid categories. It uses a Bayesian

Thompson, 2002).	approach and assumes the HWE within parental populations. (http://ib.berkeley.edu/labs/slatkin/eriq/software/software.htm)
Migrate (Beerli, 2008)	The program estimates the effective population size and migration rates in different populations using maximum likelihood and Bayesian inference. It assumes constant migration rates. (http://popgen.sc.fsu.edu/)
IM (Hey & Nielsen, 2004)	IM program uses Bayesian inference to estimate the divergence time and migration occurred in the ancestry of two populations. The program assumes no linkage and recombination between loci. (http://genfaculty.rutgers.edu/hey/software)
Hickory (Holsinger et al., 2002)	Hickory is a program that estimates F population statistics from subdivided subpopulations. It uses a Bayesian approach and reports the posterior distribution of inbreeding coefficients and F_{ST}. (http://darwin.eeb.uconn.edu/hickory/hickory.html)

2.2 Spatial clustering

Spatial clustering models start by the assumptions that loci are in HWE and there is no admixture within clusters. They assign individuals assuming that some allele frequencies are cluster-specific. The early and still widely used program based on clustering is STRUCTURE (Pritchard et al., 2000). This program describes the genetic structure of populations using multilocus genotype data, assuming that there are K clusters, defined by allele frequencies at each locus. Modifications of the original model, recently reviewed by François & Durand (2010), include presence of genetic linkage (Falush et al., 2003; Hoggart et al., 2004; Corander & Tang, 2007), inbreeding (François et al., 2006), migration (Zhang, 2008), mutation (Shringarpure & Xing, 2009) and dominance (Falush et al,. 2007). Recently developed Bayesian clustering models such as those implemented in GENELAND (Guillot et al., 2005) and BAPS5 (Corander et al., 2008) take into account of individual geographic coordinates.

2.3 Multivariate analysis

Principal component analysis (PCA) is mainly used to represent individual or population-specific variations in allele distribution. Firstly used by Cavalli-Sforza and Edwards (1963), the method is again widely used to investigate population structure being computationally efficient and capable of handling wide datasets, where STRUCTURE requires high computational cost. PCA is still largely employed (Cavalli-Sforza et al., 1993, 1994; Novembre & Stephens, 2008) and recently applied to the study of population structure dealing with large datasets (Patterson et al., 2006; Lao et al., 2006; Jakobsson et al., 2008; Li et al., 2008; Novembre et al., 2008; Price et al., 2010), also in comparison with the Bayesian approach implemented in STRUCTURE (Price et al., 2006; Seldin et al., 2006). PCA represents population structure by means of genetic correlations among individuals. Another related method capable of detecting population structure is multidimensional-scaling (MDS) (Purcell et al., 2007; Li & Yu, 2008; Wang et al,. 2009), a method that explains observed genetic distance among individuals by the configuration of points and visually displays the structures hidden in the original data.

Box 4. Georeferenced data

Georeferenced data are geographic coordinates defining the location of investigated objects at the surface of the Earth. They constitute additional descriptors or variables in the data sets, generally X for longitude, Y for latitude and Z for altitude when recorded. Within a GIS, it is possible to simultaneously use different data sets. These datasets constitute separate information layers, whose overlay is possible only if their geographic components (X,Y) use the same projection system. A projection system is a method of representing the surface of a sphere on a plane, necessary for creating maps. Data sets from diverse national origins are produced in diverse projection systems, most often conforming to the geographical specificities of the country where the information is produced. Indeed, the location on the earth and the surface of a country influence the choice of the projection system. Given a frequent heterogeneity of data sets, the use of a common system facilitates the management and use of geodata. Such a universal projection system is longitude-latitude with a standard World Geodetic comprising a standard coordinate frame for the Earth, a standard spheroidal reference surface for raw altitude data, and a gravitational equipotential surface (the geoid) defining the nominal sea level. This system is made of latitude lines (parallels) that run horizontally, and of vertical longitude lines called meridians. Parallels are equidistant from each other, and each degree of latitude is approximately 111 km apart. Degrees of latitude are numbered from 0° to 90° north and south. Zero degrees is the equator, 90° north is the North Pole and 90° south is the South Pole. Meridians, on the other hand, converge at the poles and are widest at the equator (111 km apart). Zero degrees longitude is located at Greenwich, England. The degrees continue 180° east and 180° west, where they meet and form the International Date Line in the Pacific Ocean. To precisely locate points on the earth's surface, degrees longitude and latitude were divided into minutes (') and seconds ("). There are 60 minutes in each degree, and each minute is divided into 60 seconds. Seconds can be further divided into tenths, hundredths, or thousandths. Geographic coordinates can be displayed either in decimal degrees (e.g. 68.135°) or in sexagesimal system (degrees, minutes, and seconds: 68°8'6").

3. How landscape genetics/genomics can be used to infer history and migration of modern breeds

Domestication of many livestock species started about 10,000-5,000 years BP (Bruford et al., 2003). The localisation of the domestic centres can be traced back by simply observing the patterns of genetic diversity (GD) among individuals/populations of the species. GD is higher at the centre of domestication and decreases radiating from it. Landscape genetics is particularly powerful for the identification and illustration of these historical events. The domestication centres for the major livestock species have so far been assessed, e.g. goat (Naderi et al., 2008), sheep (Chessa et al., 2009), cattle (Ajmone- Marsan et al., 2010), pig (Larson et al., 2007), chicken (Kanginakudru et al., 2008), yak (Wiener et al., 2003). From domestication centres, livestock followed human migrations by demic expansion (Cavalli Sforza, 1966) or by active trade (Ajmone-Marsan et al., 2010). For the reconstruction of migration routes landscape genetics has been extensively used to infer also possible alternative routes through land or sea (Pariset et al., 2011). Since molecular markers include sex specific ones, like Y chromosome and mtDNA markers, geographic maps of genetic diversity can be constructed for inferring male or female mediated gene flow (Hanotte et al., 2000). Landscape genetics can also provide the basis for ascertaining co-migration of livestock and humans. Pellecchia et al. (2007) found in Italian cattle haplotypes shared with Turkey breeds, nevertheless distance and discontinuity between the two countries. The data were interpreted as co-migration of Etruscan people along with their cattle around 3000 years BP, therefore corroborating the hypothesis of the middle-East origin of this people. Wild yak are believed to have been domesticated about 5000 years ago in the Qinghai-Tibet Plateau and then dispersed to occupy their current distribution (Zhang, 1989; Wiener et al., 2003). Xuebin et al. (2005) report lack of evidence for recent bottleneck in any of yak populations studied suggesting that the low level of genetic differentiation and the high level of diversity within populations observed today is more likely explained by the recent origin of these populations from a common ancestral population with large effective population size.

4. The importance of landscape genomics in livestock: Main differences with respect to wild animals

Landscape genetics has been mainly applied to the study of the genetic structure in wild populations, where components such as habitat preference can be assessed. In landscape genetics the matrix of habitat, morphological, climatic (etc.) features is considered as a major cause of biological and ecological processes influencing the population structure of wild population (Holderegger & Wagner, 2008). Recently, the field of livestock landscape genomics, where data such as environmental, socio-economic and demographic characteristics are geo-referenced, has boosted during the last decade (Joost et al., 2009).

In livestock the study of adaptation of different breeds to the environment is of crucial importance in order to support production systems based on adapted breeds reducing impact on the environment, and making better goods available to consumers: landscape genomics, studying the animal genome coupled with the description of landscape (including biotic, abiotic, human and market influences), offers a tool to identify the genotypes suitable to a given environment (Joost & Negrini, 2010). Within the landscape genomics studies in livestock, the Econogene project (http://www.econogene.eu) developed a programme with the aim of promoting the sustainable conservation of genetic resources in sheep and goats. Combining a molecular analysis of biodiversity, socio-economics and geostatistical systems the project defined strategies of genetic management and rural development.

Landscape genomics in livestock depends on the topography and on farmer market preferences and social structure. When analysing domesticated populations, characteristics of farms and farmers, as their isolation or farmers' practices, could affect the genetic structure of the animals. For example, the geographical location of farmers may facilitate or reduce animal exchanges influencing the gene flow. When dealing with livestock populations it would then be useful collecting as environmental factors human activities such as human density, roads or hunting activities (Bertouly et al., 2009).

One of the differences between landscape analysis in wild and domesticated species is that, in the latter, anthropic selection plays a relevant role. Therefore many loci reflect the domestication history of the species and are influenced by the human needs for certain characteristics. As a consequence, a high percentage of loci purposely chosen for influencing potentially selected traits could result under selection (Pariset et al., 2009a). This will be discussed in the next paragraph.

Box 5. GPS

The Global Positioning System (GPS) is a worldwide radio-navigation system developed and maintained by the U.S. government, formed from a constellation of 27 satellites (24 in operation and three extras in case one fails) and their ground stations. Initially designed for military applications, a decision directive was signed by President Clinton in 1996 describing GPS as an international information utility.
Each of these satellites circles the globe at about 19'300 km, making two complete rotations of the Earth every day. The orbits are arranged so that at any time and anywhere, there are at least four satellites visible in the sky.
Radio signals are sent from orbiting satellites to Earth. GPS receivers on the ground can collect and convert these radio signals into position, velocity, and time information and calculate positions accurate to a matter of meters.
To this end, a GPS receiver has to locate four or more of these satellites, figure out the distance to each, and use this information to deduce its own location by means of triangulation (or trilateration). To triangulate, a GPS receiver measures distance using the travel time of radio signals. GPS accuracy is affected by a number of factors, including satellite positions, noise in the radio signal, atmospheric conditions, and natural barriers to the signal. These factors can create an error between 1 to 10 meters and may result from interferences caused by a physical obstacle near the receiver or a radio emission on the same frequency. For instance, objects such as mountains or buildings can also error sometimes up to 30 meters. The most accurate determination of position occurs when the satellite and receiver have a clear view of each other and no other objects interfere.

---

5. Detection and consequences of artificial selection in livestock

Landscape genomics needs the simultaneous study of a high number of markers, mainly neutral but including also genes under selection. This combination of loci with different characteristics can aid in understanding the action of evolutionary forces (selection, drift, migration) influencing the genetics of livestock populations. In this matter we are helped by the fact that many innovative tools, such as medium or high density SNP chips, are now available for many domesticated species (cattle, pig, sheep, chicken already available; goat in progress - see Box 7) and custom chip (see Box 7) are sold at relatively low cost.

Recently many authors emphasized the need of accompanying the analysis of neutral markers with those of loci under selection, which may directly reflect environmental change (Kohn et al., 2006; Hoffmann & Willi, 2008; Primmer, 2009). The analysis of population genetic measures like Wright's F statistics (Wright, 1978; Weir & Cockerham, 1984) as continuous distributions across a genome (Storz, 2005; Foll & Gaggiotti, 2008; Excoffier et al., 2009) can help in identifying genomic regions showing significant differentiation among populations, thus regions that have likely been under natural selection (Nielsen et al., 2007;

Box 6. Next Generation Sequencing

Next generation sequencing (NGS) generates hundreds of millions of sequence reads in parallel. New NGS technologies are based on production of 'libraries' obtained by breaking the entire genome into small pieces which are then ligated to designated adapters. The DNA templates are read randomly during DNA synthesis (sequencing-by-synthesis) (Zang et al., 2011). Several NGS platform recently developed allow larger-scale DNA sequencing:

454 sequencing

The 454 system developed by Roche was the first commercial platform. During library construction the DNA is fragmented and ligated to adapters. The fragments are linked to microbeads that have millions of oligomers complementary to the adaptor sequences and then amplified by emulsion PCR (Dressman et al., 2003; Margulies et al., 2005). The 454 Genome Sequencer uses the pyrosequencing technology in which the nucleotide addition leads to a pyrophosphate release that triggers an enzymatic cascade and consequently a light signal (Ronaghi et al., 1998). The 454 platform gives a length reads of 500bp and 400-600 Mb per run.
(http://www.454.com).

Illumina

Illumina – Solexa sequencing technology is a platform based on massively parallel sequencing of millions of fragments using reversible terminator-based sequencing chemistry (modified Sanger). This technology uses a bridge amplification in which the fragmented genomic DNA is arranged on an optically surface and amplified to create a high density sequencing flow cell. The sequencing system uses reversible terminator dideossinucleotides with removable fluorescent dyes. The Illumina HiSeq 2000 Genome Analyzer produces single reads of 2 x 100 bp (pair-end reads), and generates about 200 giga basepair (Gb) of short sequences per run with accuracy of 99%.
(http://www.solexa.com).

ABI solid

The SOLiD platform uses the emulsion PCR in which the amplicons are captured on small magnetic beads (1μm). The sequencing reaction is catalyzed by a DNA ligase and relies on serial ligation of labelled oligonucleotides. SOLiD4 platform produces 80-100 Gbp per run and read length of up to 50 bp with system accuracy greater than 99.94%.
(http://www.solid.appliedbiosystems.com).

Ion Torrent

Ion Torrent technology uses a new approach to sequencing based on the detection of hydrogen ions released by DNA polymerization process. A semiconductor chip captures voltage measurements due to hydrogen ions and directly convert the chemical information to digital sequence information. Ion Torrent offers different sequencing chip densities producing from 10 Mb to more than 1 Gb of sequences. The technology produces a read length of 200bp, and it is expected to get read length of 400bp in 2012.
(http://www.iontorrent.com/).

Bonin, 2008; Akey, 2009) that can be used to identify candidate genes for functional analysis (Akey et al., 2002; Vigouroux et al., 2002; Roberge et al., 2007; Bonin et al., 2009; Bigham et al., 2010).

The methods of Fdist and SAM were used to detect signatures of selection in goats, confirming the usefulness of both methods in outlier loci identification (Pariset et al., 2009b). By adding or removing neutral markers from datasets it could be possible to individuate the effects of the forces acting on a population (Pariset et al., 2009a, 2011). This will work in an efficient way if the number of markers is high, as in the case of landscape genomics.

On a short timescale, migration-drift equilibrium should result in the conservation of genetic differentiation. Differentiation between populations would increase with drift occurrence and decrease in the case of migration.

In the evolution of functional traits, diversification occurs as a result of chance and selective processes. In wild populations, founder effects can result in stochastic evolution acting against adaptive evolution; then founder effects can be assessed by testing for the signature of natural selection. This is not the case in livestock, where human selection acts sometimes against stochastic and natural selection. Particularly during the last century, the livestock sector has undergone striking changes as large-scale production expanded, leading to the formation of well-defined breeds, exposed to intense anthropogenic selection. Selective breeding in fact results in the increase of the phenotypes with desired characteristics (and sometimes with undesired characteristics, as in the case of CVM in Holstein cattle, see Agerholm et al., 2004). This will be better analysed in the next paragraph.

Therefore human selection can have effects similar to those of bottleneck and genetic drift, particularly amplified by the fact that sex ratio is strongly biased and by the progress of management practices, the introduction of artificial insemination and embryo transfer, resulting in reduced allelic diversity and heterozygosity, a non-random sample of the genes in the population and the loss of rare alleles (Nei et al., 1975; Allendorf, 1986; England et al., 2003).

Bottleneck detection is mainly used for the interpretation of historical demography of wild populations and for endangered wild species management (Hundertmark & Daele, 2010). Anyway its detection results of crucial importance also in livestock, where local and typical flocks are represented by small number of animals. In fact, a significant number of cattle, sheep, and goat breeds already disappeared and many are presently endangered (FAO, 2007) because farmers progressively substitute the less productive, locally adapted, native breeds with highly productive cosmopolitan breeds and progressively abandon marginal areas (Taberlet et al., 2008).

6. Landscape analysis and regions associated with adaptation and disease resistance

Landscape genetics can be useful at identifying environmental and landscape components in the spreading of diseases, for example tracking hosts to assess aetiological agents spread (Archie et al., 2009), providing data relevant for health, studying epidemiology of zoonoses, understanding the spread of disease, designing optimal surveillance and control programs and identifying interactions affecting spatial patterns of disease incidence (Guillot et al.,

2009). SAM could assist the discovery of genomic regions linked to quantitative trait loci implicated in selection and adaptation (e.g. for disease resistance). One striking example in Scottish Blackface sheep is the identification of an allele at locus DYMS1 (from the major histocompatibility complex), associated with the number of wet days using SAM (Joost et al., 2007). In a previous study Buitkamp and collaborators found this locus linked to parasite resistance in the same sheep breed (Buitkamp et al., 1996). After *Ostertagia circumcincta* infection the faecal egg counts were associated with the major histocompatibility complex alleles (Buitkamp et al. 1996). By using both SAM and Fdist methods, Joost et al. (2007) detected an outlier allele at locus OARJMP29 that has been showed to be implicated in a disease resistance. The SAM method can also contribute to monitor and control the infectious disease processes (Biek & Real, 2010). The spatially explicit Bayesian clustering methods were used to analyse the genetic structure of European wild boar affected by classical swine fever in order to identify geographical barriers for disease management units. The results showed an overestimation of genetic structure when using Bayesian clustering methods in data sets characterized by isolation by distance. This bias could lead to the erroneous delimitation of management or conservation units (Frantz et al., 2009).

Cringoli et al. (2007) used the landscape genomics approach to understand the role of sheep in the cystic echinococcosis disease transmission to cattle and buffalo. The authors found a higher incidence of the disease in cattle and buffaloes farms that showed a close proximity with the sheep farms in the studied area. Moreover the higher prevalence found in cattle compared to water buffalo farms is explained by the lower distance between the sheep and cattle farms than those between the sheep and water buffalo.

An example of how landscape genomics can provide analytical tools in the mapping of diseases is reported by Tum et al. (2007). The authors compared the maps produced using geographic information systems and field measurements to predict the levels of risk of fasciolosis due to *Fasciola gigantica* in Cambodia cattle and buffalo. They found a good correlation between the two methods indicating the power of using GIS (Box 2).

7. An overview of geographical patterns of livestock genetic diversity

The loss of diversity of livestock breeds is affected by genetic drift, inbreeding, introgression, natural and artificial selection (Bruford, 2004). The application of landscape genomics is suitable to develop our understanding of the mechanisms leading to livestock genetic change (Luikart et al., 2003).

Unlike cattle or water buffaloes, sheep and goats are raised by almost all ethnic groups, representing suitable systems to study the effect of farmer connectivity on livestock genetic structure. Berthouly et al. (2009) reported the effects of gene flow due to the spatial distribution of ethnic groups, farmer ethnicity and husbandry practices on goat spatial pattern. Good examples of application of spatial genomics to sheep and goats was performed by the Econogene project (http://www.econogene.eu) (Pariset et al. 2006; Bertaglia et al., 2007; Peter et al., 2007; Cañon et al., 2007; Pariset et al., 2011). Within the same project Joost et al. (2007) analysed European sheep breeds and found 40 alleles significantly associated with at least one environmental parameter; Pariset et al. (2009b) using different landscape genomics approaches identified adaptive variation in goat, which is characterized by a large range of climatic conditions in the rearing areas and by a history of intense trade.

Box 7. High-throughput SNP genotyping in livestock

New DNA sequencing technologies have recently made feasible the discovery of SNPs virtually in all species. Millions of SNPs have been discovered in recent years and deposited in the NCBI (National Center for Biotechnology Information) database dbSNPs (http://www.ncbi.nlm.nih.gov/projects/SNP/).

Technological progresses now make available tools for typing hundreds of thousands of SNPs in the same individual. Some of the techniques described in Box 1 are suitable for high-throughput SNP genotyping; however one of the most efficient systems to analyse many SNPs simultaneously is microarray/chip analysis (Perkel, 2008). Hundreds of probes synthesized on a single chip allow the analysis of many SNPs simultaneously.

Affymetrix and Illumina systems offer the densest platforms for SNP genotyping in livestock. Using data from the bovine HapMap project, Affymetrix has produced a 25K SNP panel. This chip can analyze approximately 25000 SNPs discovered by sequencing the bovine genome. The Illumina technology allows the analysis of thousands of SNPs and recently became the most popular platform. The Illumina iSelect BeadChip are thus more functional for large amounts of SNP genotyping assays and achieve high densities that include tens of thousands of point mutations distributed throughout the genome. Illumina has developed several panels for the SNP genotyping including the GoldenGate Bovine3K BeadChip that provides 2,900 SNPs, the BovineSNP50 BeadChip consisting of 54,609 informative SNP probes that uniformly span the entire bovine genome, the BovineHD BeadChip which is the most comprehensive genome-wide genotyping array with more than 777,000 informative SNP across bovine genome; the PorcineSNP60 BeadChip with 62,000 SNP validated in seven economically important pig breeds; the 60k SNP BeadChip in chicken; the OvineSNP50 BeadChip with 52,000 SNP partially discovered by sequencing 60 animals from 15 breeds and partially derived from the ovine draft genome and the GoatSNP50 chip that will be released by December 2011. Low density SNP chips are also available from a selection of markers from the largest devices. Imputation methods allow to reconstruct the more dense genotype with accuracy higher than 95% (Nothnagel et al., 2009).

Currently commercial microarray platforms allow the development of custom genotyping array. Illumina provides the platform Golden Gate to identify 384 to 3,072 SNPs per sample and Affymetrix provides MyGeneChip Custom Arrays.

Sequenom MassARRAY platform combines a primer extension chemistry with the MALDI-TOF mass spectrometry to characterize genotypes with the highest levels of reproducibility. It is possible to multiplex up to 40 SNPs in a single well and process up to 384 samples in parallel.

Fluidigm performs TaqMan SNP genotyping and offer some benefits including the low cost per genotype and high-sample throughput for low- to mid-multiplex SNP genotyping.

Kijas et al. (2009) used a SNP panel to analyse sheep nuclear genome, providing the indication that breeds cluster into large groups based on geographic origin, and that SNPs can successfully identify population substructures within individual breeds. Sheep generally show a moderate geographic structure and a high genetic variability within breeds (Kijas et al., 2009) compared to cattle (Achilli et al., 2009). This can be explained by the easiness of transportation of sheep compared to cattle. Anyway, using a different dataset, Pariset et al. (2011) show a good correspondence of breeds to geographical locations.

The geographical patterns in sheep breeds has been also analysed by Tapio et al. (2010) who performed a Bayesian clustering analysis on 52 sheep breeds from the Eurasian subcontinent using 20 microsatellite markers. They found three genetic clusters: Nordic, Composite and Fat-tailed. The differentiation of the Fat-tailed cluster from the others indicates restricted gene flow between steppe or mountain environments in central Eurasia and cooler and moister northern areas of the continent.

Most of the information about history of the species have been gathered using mtDNA. In a recent study Meadows et al. (2011) analyzed complete mtDNA sequences from each haplogroup previously identified in domestic sheep, and from a sample of their wild relatives. Bayesian, maximum likelihood revealed that among various mtDNA components the control region is the more suitable to detect the true relationship between sheep.

A recent study on retrovirus integrations (Chessa et al., 2009) has provided information on the introduction of sheep into Europe, indicating an early arrival of the primitive sheep populations (European mouflons, North-Atlantic Island breeds) and a subsequent advent of wool producing sheep.

Other examples of geographical patterns in livestock concern goat populations studies. The phylogenetic history and population structure of domestic goats was assessed using both mtDNA and nuclear markers. Cañón et al. (2006) analysed thirty microsatellites in 1426 goats from 45 traditional or rare breeds in 15 European and Middle Eastern countries. They found at least four discrete clusters using Bayesian-based clustering analysis and multivariate analysis. About 41% of the genetic variability among the breeds could be explained by their geographical origin. The analysis of mtDNA polymorphism in the domestic goat revealed six different haplogroups that have been found also in its wild ancestor (Naderi et al., 2008), suggesting that the domestication process occurred over a very large area encompassing eastern Anatolia and North-West Iran (Taberlet et al., 2011).

Among livestock, buffalo plays a fundamental role in the agricultural economy. The study of geographical pattern in buffalo population can contribute to breeding management. Gargani et al. (2009) analyzed six Turkish water buffalo populations using a set of 26 heterologous (bovine) microsatellite markers. Principal component and Bayesian cluster approach revealed three genetically distinct groups with a good correspondence of population to geographical locations. The analysis of mtDNA from different Indian breeds revealed that the river buffalo was domesticated in the Western region of the Indian subcontinent and that different maternal lineages contributed to the domestication process (Kumar et al., 2007).

The genetic diversity and the geographical patterns in cattle has been compared using different molecular markers. Mitochondrial DNA and microsatellite loci, for example, showed that taurine and zebu cattle were domesticated independently (Bradley et al., 1996). The selective breeding and genetic isolation of taurine cattle leads to the formation of many specialized dairy and beef breeds, with a complex spatial pattern of genetic differentiation.

Pariset et al. (2010a) assessed the relationship among some Podolic breeds and verified whether their genetic state reflects their history using SNP polymorphisms. The Bayesian inference assignment confirmed that the set of chosen SNPs is able to distinguish among the breeds and that the breeds are genetically distinct. Gautier et al. (2010) investigated the genetic diversity of cattle breeds analysing 47 populations from different parts of the world with 44,706 autosomal SNPs markers. The differentiation of the African taurine, the European taurine and zebus, indicated a support for three distinct domestication centres. Spatial principal component (sPCA) analysis and spatial metric multidimensional scaling (sMDS) was applied to 101 cattle breeds using microsatellite markers (Laloë et al., 2010). The results showed a strong geographic structure along a southeast to northwest cline, corresponding to a gradient from Indian zebu to European taurine cattle. The diversity and differentiation of the African Ankole Longhorn cattle breed have been analysed on the basis of genotypic and spatial distance data by Ndumu et al. (2008). Using analyses on distance-based and model-based methods they found an isolated sub-population that it is well differentiated from the others.

In yak, cattle microsatellites are commonly used for the study of genetic diversity (Ritz et al., 2000; Dorji et al., 2002; Xuebin et al., 2005; Qi, 2004, Nguyen et al., 2005). Population structure of nine Chinese yak breeds were analyzed by means of 16 microsatellite markers and the Neighbor-Joining phylogenetic tree constructed based on Nei's standard genetic distances revealed a separation in 3 clusters (Zhang et al., 2008). Recently, a study of mitochondrial DNA haplotypes identified taurine cattle mtDNA in two samples of Tibetan yak and yak maiwë populations (Lai et al., 2007). Qi et al. (2008), using three different methods, show the results of admixture in populations of yak-cattle across the range of the species geographical distribution using mtDNA haplotypes and 17 autosomal microsatellites. Cattle bulls are commonly used to hybridize with yak cows at relatively high altitudes, while reciprocal crossing is adopted at a lower altitude of their distribution range (Davaa, 1996; Tshering et al., 1996). Some factors influence the variation of frequency of cattle introgression in yak between and among regions. For example, a generally low frequency of cattle introgression was observed at relatively high altitudes (about 3500 meters) (Wiener et al., 2003; Qi et al., 2008). Today, the yak geographical distribution extends from the southern slopes of the Himalayas to the Altai mountains Hangai of Mongolia and Russia, and the Pamir and Tianshan mountains and the Qilian Mountains Minshan (Wiener et al., 2003). The Mongolian and Russian yak populations may have originated from a large effective population size and frequent gene flow between two populations that live close to each other. Xuebin et al. (2005) report lack of evidence for recent bottleneck in any of yak populations studied suggesting that the low level of genetic differentiation and the high level of diversity within populations observed today is more likely explained by the recent origin of these populations from a common ancestral population with large effective population size.

8. Livestock conservation

Landscape genetic investigations are particularly useful in the management and conservation of species (Bruggeman et al., 2009; Segelbacher et al., 2010), even if the examples of landscape approaches to practical conservation management of species are still a few and mainly focused on wild species (Epps et al., 2005; Segelbacher et al., 2008).

Threats to biodiversity in terms of extinction rate, destruction of ecosystems and habitat, or loss of genetic diversity are increasing within the species utilized in agriculture. Since mid-

1800s, the modern breed concept and its application to breeding and husbandry practices led to the formation of well-defined breeds, exposed to intense anthropogenic selection and during the last century the livestock sector has undergone striking changes as large-scale production expanded. This has led farmers to progressively substitute the less productive, locally adapted, native breeds with highly productive cosmopolitan breeds and to progressively abandon marginal areas. Therefore a significant number of cattle, sheep, and goat breeds already disappeared and many are presently endangered (Taberlet et al., 2008). According to the Food and Agriculture Organization of the United Nations (FAO), a total of 1491 livestock breeds world-wide are classified as being either critically endangered, critical-maintained, endangered, or endangered-maintained; it is likely that a high number of breeds are being, and will be lost in the near future, suffering of the effects of rapid climate change, increasing market demand and human demographic expansion. It is then strategically important to preserve as much the farm animal diversity as possible, and this will be better accomplished with the aid of landscape genomic studies. Landscape genomics, by combining geo-referencing of breed distributions, spatial genetic diversity, climatic, ecological, epidemiological and production system information (Hanotte & Jianlin, 2005) will help in formulate priority decisions for in situ breed conservation. It could help to understand the genetic basis of animal adaptation to the environment, and the co-evolution of livestock and their production systems (Joost & Negrini, 2010).

Breeding systems on genetic management of threatened species needs to be fully evaluated (Frankham, 2010). SNP markers (see Box 1 and 7) open up new perspectives to livestock genomics, in particular for the investigation of genome diversity within and among individuals and populations, population structure, search of causative genes, and for the identification of signatures left by selection. They can also fulfil what has been envisaged in the recent past by Bertaglia et al. (2007) when only a reduced set of markers were known, i.e. the possibility to geographically map the socio-economy of rural areas and the genetic variation patterns of livestock, in order to match policies of intervention with the capability of the system to be driven to a more sustainable status.

9. References

Achilli, A., Bonfiglio, S., Olivieri, A., Malusà, A., Pala, M., Kashani, B.H., Perego, U.A., Ajmone-Marsan, P., Lotta, L., Semino, O., Bandelt, H.J., Ferretti, L., & Torroni, A. (2009). The Multifaceted Origin of Taurine Cattle Reflected by the Mitochondrial Genome. *PLoS ONE*, vol. 4, No. 6, e5753.

Adachi, A., & Kawamoto, Y. (1992). Hybridization of yak and cattle among the Sherpas in Solu and Khumbu, Nepal. *Report of the Society for Researches on Native Livestock*, vol. 14, pp. 79–87.

Agerholm, J.S., Andersen, O., Almskou, M.B., Bendixen, C., Arnbjerg, J., & Aamand, G.P. (2004). Evaluation of the inheritance of the complex vertebral malformation syndrome by breeding studies. *Acta Veterinaria Scandinavica*, vol. 45, pp. 33-137.

Ajmone-Marsan, P., Gorni, C., Milanesi, E., Mazza, R., van Eijk, M.J., Peleman, J.D., & Williams, J.L. (2008). Assessment of AFLP marker behaviour in enriching STS radiation hybrid maps. *Animal Genetics*, vol. 39, pp. 383-394.

Ajmone-Marsan, P., Garcia, J.F., & Lenstra, J.A. (2010). On the origin of cattle: How aurochs became cattle and colonized the world. *Evolutionary Anthropology: Issues, News, and Reviews*, vol. 19, No. 4, pp. 148-157.

Ajmone-Marsan, P., Negrini, R., Crepaldi, P., Milanesi, E., Gorni, C., Valentini, A., & Cicogna, M. (2001). Assessing genetic diversity in Italian goat populations using AFLP markers. *Animal Genetics*, vol. 32, pp. 281-288.

Akey, J.M. (2009). Constructing genomic maps of positive selection in humans: where do we go from here? *Genome Research*, vol. 19, pp. 711–722.

Akey, J.M., Zhang, G., Zhang, K., Jin, L., & Shriver, M.D. (2002). Interrogating a high-density SNP map for signatures of natural selection. *Genome Research*, vol. 12, pp. 1805-1814.

Allendorf, F.W., Hohenlohe, P.A., & Luikart, G. (2010). Genomics and the future of conservation genetics. *Genetics*, vol. 11, pp. 697-709.

Allendorf, F.W. (1986). Genetic drift and the loss of alleles versus heterozygosity. *Zoo Biology*, vol. 5, pp. 181–190.

Anderson, E.C., & Thompson, E.A. (2002). A model-based method for identifying species hybrids using multilocus genetic data. *Genetics*, vol. 160, pp. 1217–1229.

Archie, E.A., Luikart, G., & Ezenwa, V.O. (2009). Infecting epidemiology with genetics: a new frontier in disease ecology. *Trends in Ecology & Evolution*, vol. 24, pp. 21–30.

Bae, J.S., Cheong, H.S., Kim, L.H., NamGung, S., Park, T.J., Chun, J.Y., Kim, J.Y., Pasaje, C.F.A., Lee, J.S., & Shin, H.D. (2010). Identification of copy number variations and common deletion polymorphisms in cattle. *BMC Genomics*, vol. 11, p.232.

Beaumont, M.A., & Nichols, R.A. (1996). Evaluating loci for use in the genetic analysis of population structure. *Proceedings of the Royal Society B: Biological Sciences,* vol. 263, pp. 1619–1626.

Beerli, P. (2008). Migrate-N, Version 2.4. Available from: (http://popgen.sc.fsu.edu/).

Bertaglia, M., Joost, S., Roosen, J., & the Econogene Consortium. (2007). Identifying European Marginal Areas in the Context of Local Sheep and Goat Breeds Conservation: A Geographic Information System Approach. *Agricultural Systems*, vol. 94, pp. 657-670.

Berthouly, C., Do Ngoc, D., Thévenon, S., Bouchel, D., Van, T.N., Danes, C., Grosbois, V., Thanh, H.H., Chi, C.V., & Maillard, J.C. (2009). How does farmer connectivity influence livestock genetic structure? A case-study in a Vietnamese goat population. *Molecular Ecology*, vol. 18, pp. 3980-3991.

Bigham, A., Bauchet, M., Pinto, D., Mao, X., Akey, J.M., Mei, R., Scherer, S.W., Julian, C.G., Wilson, M.J., López-Herráez, D., Brutsaert, T., Parra, E.J., Moore, L.G., & Shriver, M.D. (2010). Identifying signatures of natural selection in Tibetan and Andean populations using dense genome scan data. *PLoS Genetics*, vol. 6, e1001116.

Bonin, A. (2008). Population genomics: a new generation of genome scans to bridge the gap with functional genomics. *Molecular Ecology*, vol. 17, pp. 3583–3584.

Bonin, A., Paris, M., Tetreau, G., David, J.P., & Després, L. (2009). Candidate genes revealed by a genome scan for mosquito resistance to a bacterial insecticide: sequence and gene expression variations. *BMC Genomics*, vol. 10, p. 551.

Botstein, D., White, R.L., Skolnick, M., & Davis R.W. (1980). Construction of a genetic linkage map in man using restriction fragment length polymorphisms. *The American Journal of Human Genetics*, vol. 32, pp. 314–331.

Bruford, M.W. (2004). Conservation genetics of UK livestock: from molecules to management, In: *Farm Animal Genetic Resources,* Eds G. Simm, B. Villanueva and S. Townsend, pp. 151-169, Nottingham University Press, Nottingham, UK.

Bruford, M.W., Bradley, D.G., & Luikart, G. (2003). Dna markers reveal the complexity of livestock domestication. *Nature,* vol. 4, pp. 900-910.

Buitkamp, J., Filmether, P., Stear, M.J., & Epplen, J.T. (1996). Class I and class II Major histocompatibility complex alleles are associated with faecal egg counts following natural, predominantly *Ostertargia circumcincta* infection. *Parasitology Research,* vol. 82, pp. 693-696.

Cañón, J., García, D., García-Atance, M.A., Obexer-Ruff, G., Lenstra, J.A., Ajmone-Marsan, P., Dunner, S., & the Econogene Consortium (2006). Geographical partitioning of goat diversity in Europe and the Middle East. *Animal Genetics,* vol. 37, pp. 327-334.

Cavalli-Sforza, L.L., & Edwards, A.W.F. (1963). Analysis of human evolution. *Proceedings of the XI International Congress of Genetics, The Hague, Genetics Today,* vol. 3, pp. 923-933.

Cavalli-Sforza, L.L. (1966). Population Structure and Human Evolution. *Proceedings of the Royal Society of London Series B-Biological Sciences,* vol. 164, pp. 362-379.

Cavalli-Sforza, L.L., Menozzi, P., & Piazza, A. (1993). Demic expansions and human evolution. *Science,* vol. 259, pp. 639-646.

Cavalli-Sforza, L.L., Menozzi, P., & Piazza, A. (1994). The History and Geography of Human Genes. Princeton University Press.

Chessa, B., Pereira, F., Arnaud, F., Amorim, A., Goyache, F., Mainland, I., Kao, R.R., Pemberton, J.M., Beraldi, D., Stear, M., Alberti, A., Pittau, M., Banabazi, M.H., Kazwala, R., Zhang, Y-P., Arranz, J.J., Ali, B.A., Wang, Z., Uzun, M., Dione, M., Olsaker, I., Holm, L-E., Saarma, U., Ahmad, S., Marzanov, N., Eythorsdottir, E., Holland, M., Ajmone-Marsan, P., Bruford, M.W., Kantanen, J., Spencer, T.E., & Palmarini, M. (2009). Revealing the history of sheep domestication using retrovirus integrations. *Science,* vol. 324, pp. 532-536.

Cingoli, G., Rinaldi, L., Musella, V., Veneziano, V., Maurelli, M.P., Di Pietro, F., Frisiello, M., & Di Pietro, S. (2007). Geo-referencing livestock farms as tool for studying cystic echinococcosis epidemiology in cattle and water buffaloes from southern Italy. *Geospatial Health,* vol. 2, No. 1, pp. 105-111.

Conrad, D.F., Pinto, D., Redon, R., Feuk, L., Gokcumen, O., Zhang, Y., Aerts, J., Andrews, T.D., Barnes, C., Campbell, P., Fitzgerald, T., Hu, M., Ihm, C.H., Kristiansson, K., Macarthur, D.G., Macdonald, J.R., Onyiah, I., Pang, A.W., Robson, S., Stirrups, K., Valsesia, A., Walter, K., Wei, J., Wellcome Trust Case Control Consortium, Tyler-Smith, C., Carter, N.P., Lee, C., Scherer, S.W., & Hurles, M.E. (2010). Origins and functional impact of copy number variation in the human genome. *Nature,* vol. 464, pp. 704-712.

Corander, J., & Tang, J. (2007). Bayesian analysis of population structure based on linked molecular information. *Mathematical Biosciences,* vol. 205, pp. 19-31.

Corander, J., Siren J., & Arjas, E. (2008). Bayesian spatial modeling of genetic population structure. *Computational Statistics,* vol. 23, No. 1, pp. 111-129.

Davaa, M. (1996). Conservation and Management of Domestic yak Genetic Diversity in Mongolia. *Proceedings of a Workshop on Conservation and Management of Yak Genetic Diversity,* pp. 41-46, Kathmandu, Nepal, 29-31 October 1996.

Dorji, T., Goddard, M., Perkins, J., Robinson, N., & Roder, W. (2002). Genetic diversity in bhutanese yak (*Bos Grunniens*) populations using microsatellite markers. *Proceedings of the third International Congress on Yak,* pp. 197–201, Lhasa, P.R. China, 4-9 September, 2000.

Dressman, D., Yan, H., Traverso, G., Kinzler, K.W., & Vogelstein, B. (2003). Transforming single DNA molecules into fluorescent magnetic particles for detection and enumeration of genetic variations. *Proceedings of the National Academy of Science of the Unite States of America,* vol. 100, pp. 8817–8822.

England, P.R., Osler, G.H.R., Woodsworth, L.M., Montgomery, M.E., Briscole, D.A., & Frankham, R. (2003). Effects of intense versus diffuse population bottlenecks on microsatellite genetic diversity and evolutionary potential. *Conservation Genetics,* vol. 4, pp. 595–604.

Epps, C.W., Palsboll, P.J., Wehausen, J.D., Roderick, G.K., Ramey, I.R., & McCullough, D.R. (2005). Highways block gene flow and cause a rapid decline in genetic diversity of desert bighorn sheep. *Ecology Letters,* vol. 8, pp. 1029–1038.

Excoffier, L., Smouse, P.E., & Quattro, J.M. (1992). Analysis of molecular variance inferred from metric distances among DNA haplotypes: application to human mitochondrial DNA restriction data. *Genetics,* vol. 131, pp. 479-491.

Excoffier, L., Hofer, T., & Foll, M. (2009). Detecting loci under selection in a hierarchically structured population. *Heredity,* vol. 103, pp. 285–298.

Fadista, J., Thomsen, B., Holm, L.E., & Bendixen, C. (2010). Copy number variation in the bovine genome. *BMC Genomics,* vol. 11, p. 284.

Falush, D., Stephens, M., & Pritchard, J.K. (2003). Inference of population structure using multilocus genotype data: linked loci and correlated allele frequencies. *Genetics,* vol. 164, pp. 1567–1587.

Falush, D., Stephens, M., & Pritchard, J.K. (2007). Inference of population structure using multilocus genotype data: dominant markers and null alleles. *Molecular Ecology Notes,* vol. 7, pp. 574-578.

Feuk, L., Carson, A.R., & Scherer, S.W. (2006). Structural variation in the human genome. *Nature Reviews Genetics,* vol. 7, pp. 85-97.

Foll, M., & Gaggiotti, O. (2008). A genome-scan method to identify selected loci appropriate for both dominant and codominant markers: a Bayesian perspective. *Genetics,* vol. 180, pp. 977–993.

Fordyce, S.L., Ávila-Arcos, M.C., Rockenbauer, E., Børsting, C., Frank-Hansen, R., Petersen, F.T., Willerslev, E., Hansen, A.J., Morling, N., & Gilbert, M.T.P. (2011). High-throughput sequencing of core STR loci for forensic genetic investigations using the Roche Genome Sequencer FLX platform. *Biotechniques,* vol. 51, pp. 127-133.

François, O., & Durand, E. (2010). The state of the art - Spatially explicit Bayesian clustering models in population genetics. *Molecular Ecology Resources,* vol. 10, pp. 773-784.

François, O., Ancelet, S., & Guillot, G. (2006). Bayesian clustering using hidden Markov random fields in spatial population genetics. *Genetics,* vol. 174, pp. 805-816.

Frankham, R. (2010). Where are we in conservation genetics and where do we need to go? *Conservetion Genetics,* vol. 11, pp. 661–663.

Frantz, A.C., Cellina, S., Krier, A., Schley, L., & Burke, T. (2009). Using spatial Bayesian methods to determine the genetic structure of a continuously distributed population: clusters or isolation by distance? *Journal of Applied Ecology*, vol. 46, pp. 493-505.

Galtier, N., Nabholz, B., Glémin, S., & Hurst, G. (2009). Mitochondrial DNA as a marker of molecular diversity: a reappraisal. *Molecular Ecology*, vol. 18, 4541-4550.

Gargani, M., Pariset, L., Soysal, M.I., Ozkan, E., & Valentini, A. (2009). Genetic variation and relationships among Turkish water buffalo populations. *Animal Genetics*, vol. 41, pp. 93-96.

Gonçalves, G.L., Moreira, G.R.P., Freitas, T.R.O., Hepp, D., Passos, D.T., & Weimer, T.A. (2010). Mitochondrial and nuclear DNA analyses reveal population differentiation in Brazilian Creole sheep. *Animal Genetics*, vol. 41, pp. 308-310.

Goodchild, M.F., & Haining, R.P. (2004). GIS and Spatial Data Analysis: Converging Perspectives. *Papers in Regional Science*, vol. 83, pp. 363- 385.

Greger, L., Albarella, U., Dobney, K., Rowley-Conwy, P., Schibler, J., Tresset, A., Vigne, J-D., Edwards, C.J., Schlumbaum, A., Dinu, A., Balaçsescu, A., Dolman, G., Tagliacozzo, A., Manaseryan, N., Miracle, P., Van Wijngaarden-Bakker, L., Masseti, M., Bradley, D.G., & Cooper, A. (2007). Ancient DNA, pig domestication, and the spread of the Neolithic into Europe. *Proceedings of the National Academy of Sciences*, vol. 104, No. 39, pp. 15276 -15281.

Guillot, G., Mortier, F., & Estoup, A. (2005). Geneland: a computer package for landscape genetics. *Molecular Ecology Notes*, vol. 5, No. 3, pp. 712-715.

Guillot, G., Leblois, R., Coulon, A., & Frantz A.C. (2009). Statistical methods in spatial genetics. *Molecular Ecology*, vol. 18, pp. 4734-4756.

Guryev, V., Berezikov, E. & Cuppen, E. (2005). CASCAD: A database of annotated single nucleotide polymorphisms associated with expressed sequences. *BMC Genomics*, vol. 6, p. 10.

Hanotte, O., & Jianlin, H. (2005). Genetic characterization of livestock populations and its use in conservation decision-making. *The role of biotechnology*, pp. 131-136, Villa Gualino, Turin, Italy, 5-7 March 2000.

Hanotte, O., Tawah, C.L., Bradley, D.G., Okomo, M., Verjee, Y., Ochieng, J., & Rege, J.E.O. (2000). Geographic distribution and frequency of a taurine *Bos taurus* and an indicine *Bos indicus* Y specific allele amongst sub-Saharan African cattle breeds. *Molecular Ecology*, vol. 9, pp. 387-396.

Helyar, S.J., Hemmer-Hansen, J., Bekkevold, D., Taylor, M.I., Ogden, R., Limborg, M.T., Cariani, A., Maes, G.E., Diopere, E., Carvalho, G.R. & Nielsen, E.E. (2011). Application of SNPs for population genetics of non-model organisms: new opportunities and challenges. *Molecular Ecology Resources*, vol. 11, pp. 123-136.

Hewitt, G.M. (2004). Genetic consequences of climatic oscillations in the Quaternary. *Philosophical Transactions of the Royal Society B: Biological Sciences*, vol. 359, pp. 183-195.

Hey, J., & Nielsen, R. (2004). Multilocus methods for estimating population sizes, migration rates and divergence time, with applications to the divergence of *Drosophila pseudoobscura* and *D. persimilis*. *Genetics*, vol. 167, pp. 747-760.

Hoffmann, A.A., & Willi, Y. (2008). Detecting genetic responses to environmental change. *Nature Reviews Genetics*, vol. 9, pp. 421-432.

Hoggart, C.J., Shriver, M.D., Kittles, R.A., Clayton, D.G., & McKeigue, P.M. (2004). Design and analysis of admixture mapping studies. *American Journal of Human Genetics*, vol. 74, pp. 965–978.

Holderegger, R., & Wagner, H.H. (2006). A brief guide to landscape genetics. *Landscape Ecology*, vol. 21, pp. 793–796.

Holderegger, R., & Wagner, H.H. (2008). Landcspe genetics. *Bioscience*, vol. 58, pp. 199-207.

Holsinger, K.E., & Lewis, P.O. (2003). Hickory: a package for analysis of population genetic data v1.0. University of Connecticut, Storrs: Distributed by the authors.

Hou, Y., Liu, G.E., Bickhart, D.M., Cardone, M.F., Wang, K., Kim, E., Matukumalli, L.K., Ventura, M., Song, J., VanRaden, P.M., Sonstegard, T.S., & Van Tassell, C.P. (2011). Genomic characteristics of cattle copy number variations. *BMC Genomics*, vol. 12, p. 127.

Hundertmark, K.J., & Van Daele, L.J. (2010). Founder effect and bottleneck signatures in an introduced, insular population of elk. *Conservation Genetics*, vol. 11, pp. 139–147.

Jenkins, S., & Gibson, N. (2002). High-throughput SNP genotyping. *Comparative and Functional Genomics*, vol. 3, pp. 57–66.

Joost, S., & Negrini, R. (2010). Early Stirrings of Landscape Genomics: Awaiting Next-next Generation Sequencing Platforms before Take-off. *Sustainable Improvement of Animal Production and Health*. Eds. N.E. Odongo, M. Garcia & G.J. Viljoen, pp. 185–189, Joint FAO/IAEA Division of Nuclear Techniques in Food and Agriculture, Rome, Italy.

Joost, S., Bonin, A., Bruford, M.W., Després, L., Conord, C., Erhardt, G., & Taberlet, P. (2007). A spatial analysis method (SAM) to detect candidate loci for selection: towards a landscape genomics approach to adaptation. *Molecular Ecology*, vol. 16, pp. 3955–3969.

Joost, S., Kalbermatten, M., & Bonin, A. (2008). Spatial Analysis Method (SAM): a software tool combining molecular and environmental data to identify candidate loci for selection. *Molecular Ecology Resources*, vol. 8, pp. 957–960.

Joost, S., Colli, L., Baret, P.V., Garcia, J.F., Boettcher P.J., Tixier-Boichard, M., Ajmone-Marsan, P., & the Globaldiv Consortium. (2010). Integrating geo-referenced multiscale and multidisciplinary data for the management of biodiversity in livestock genetic resources. *Animal Genetics*, vol. 41, pp. 47–63.

Joost, S., Colli, L., Bonin, A., Biebach, I., Allendorf, F., Hoffmann, I., Hanotte, O., Taberlet, P., Bruford, M., & the Globaldiv Consortium. (2011). Promoting collaboration between livestock and wildlife conservation genetics communities. *Conservation Genetics Resources*, vol. 3, pp. 785–788.

Kanginakudru, S., Metta, M., Jakati, R.D., & Nagaraju, J. (2008). Genetic evidence from Indian red jungle fowl corroborates multiple domestication of modern day chicken. *BMC Evolutionary Biology*, vol. 8, p. 174.

Kijas, J.W., Townley, D., Dalrymple, B.P., Heaton, M.P., Maddox, J.F., McGrath, A., Wilson, P., Ingersoll, R.G., McCulloch, R., McWilliam, S., Tang, D., McEwan, J., Cockett, N., Oddy, V.H., Nicholas, F.W., Raadsma, H., & International Sheep Genomics Consortium. (2009). A genome wide survey of SNP variation reveals the genetic structure of sheep breeds. *PLoS One*, vol. 4, e4668.

Kohn, M.H., Murphy, W.J., Ostrander, E.A., & Wayne, R.K. (2006). Genomics and conservation genetics. *Trends in Ecology and Evolution*, vol. 21, pp. 629–637.

Kokoris, M., Dix, K., Moynihan, K., Mathis, J., Erwin, B., Grass, P., Hines, B., & Duesterhoeft, A. (2000). Highthroughput SNP genotyping with the Masscode system. *Molecular Diagnosis*, vol. 5, pp. 329–340.

Kuhner, M.K. (2006). LAMARC 2.0: maximum likelihood and Bayesian estimation of population parameters. *Bioinformatics applications note*, vol. 22, pp. 768–770.

Kumar, S., Nagarajan, M., Sandhu, J.S., Kumar, N., Behl, V., & Nishanth, G. (2007). Mitochondrial DNA analyses of Indian water buffalo support a distinct genetic origin of river and swamp buffalo. *Animal Genetics*, vol. 38, pp. 227–232.

Larson, G., Albarella, U., Dobney, K., Rowley-Conwy, P., Schibler, J., Tresset, A., Vigne, J.D., Edwards, C.J., Schlumbaum, A., Dinu, A., Bălăçsescu, A., Dolman, G., Tagliacozzo, A., Manaseryan, N., Miracle, P., Wijngaarden-Bakker, L.V., Masseti, M., Bradley, D.G., & Cooper, A. (2007). Ancient DNA, pig domestication, and the spread of the Neolithic into Europe. *Proceedings of the National Academy of Science of the United States of America*, vol. 104, pp. 15276–15281.

Lai, S.J., Chen, S.Y., Liu, Y.P., & Yao, Y.G. (2007). Mitochondrial DNA sequence diversity and origin of Chinese domestic yak. *Animal Genetics*, vol. 38, pp. 77–80.

Li, J.Z., Absher D.M., Tang, H., Southwick, A.M., Casto, A.M., Ramachandran, S., Cann, H.M., Barsh, G.S., Feldman, M. Cavalli-Sforza, L.L., & Myers, R.M. (2008). Worldwide Human Relationships Inferred from Genome-Wide Patterns of Variation. *Science*, vol. 319, No. 5866, pp. 1100-1104.

Li, Q., & Yu, K. (2008). Improved correction for population stratification in genome-wide association studies by identifying hidden population structures. *Genetic Epidemiology*, vol. 32, pp. 215-226.

Liu, G.E., Van Tassell, C.P., Sonstegard, T.S., Li, R.W., Alexander, L.J., Keele, J.W., Matukumalli, L.K., Smith, T.P., & Gasbarre, L.C. (2008). Detection of germline and somatic copy number variations in cattle. *Developments in Biologicals*, vol. 132, pp. 231-237.

Liu, G.E., Hou, Y., Zhu, B., Cardone, M.F., Jiang, L., Cellamare, A., Mitra, A., Alexander, L.J., Coutinho, L.L, Dell'Aquila, M.L., Lacalandra, G., Li, R.W., Matukumalli, L.K., Nonneman, D., Regitano, L.C.A., Smith, T.P.L., Song, J., Sonstegard, T.S., Van Tassell, C.P., Ventura, M., Eichler, E.E., McDaneld, T.G., & Keele, J.W. (2010). Analysis of copy number variations among diverse cattle breeds. *Genome Research*, vol. 20, pp. 693-703.

Loiseau, L., Richard, M., Garnier, S., Chastel, O., Julliard, R., Zoorob, R., & Sorci, G. (2009). Diversifying selection on mhc class I in the house sparrow (*Passer domesticus*). *Molecular Ecology*, vol. 18, pp. 1331-1340.

Longley, P.A., Goodchild, M.F., Maguire, D.J., & Rhind, D.W. (2001). Geographic Information Systems and Science, Chichester, Wiley.

Luikart, G., Allendorf, F.W., Cornuet, J.M., & Sherwin, W.B. (1998). Distortion of allele frequency distributions provides a test for recent population bottlenecks. *Journal of Heredity*, vol. 89, pp. 238–247.

Luikart, G., England, P.R., Tallmon, D., Jordan, S., & Taberlet, P. (2003). The power and promise of population genomics: From genotyping to genome typing. *Nature Reviews Genetics*, vol.4, pp. 981-994.

Manel, S., Joost, S., Epperson, B.K., Holderegger, R., Storfer, A., Rosenberg, M.S., Scribner, K., Bonin, A., & Fortin, M.J. (2010). Perspective on the use of landscape genetics to detect genetic adaptive variation in the field. *Molecular Ecology*, vol. 19, pp. 3760-3772.

Manel, S., Schwartz, K., Luikart, G., & Taberlet, P. (2003). Landscape genetics: Combining landscape ecology and population genetics. *Trends in Ecology and Evolution*, vol. 18, pp. 189–197.

Margulies, M., Egholm, M., Altman, W.E., Attiya, S., Bader, J.S., Bemben, L.A., Berka, J., Braverman, M.S., Chen, Y.J., Chen, Z., Dewell, S.B., Du, L., Fierro, J.M., Gomes X.V., Godwin, B.C., He, W., Helgesen, S., Ho, C.H., Irzyk, G.P., Jando, S.C., Alenquer, M.L., Jarvie, T.P., Jirage, K.B., Kim, J.B., Knight, J.R., Lanza, J.R., Leamon, J.H., Lefkowitz, S.M., Lei, M., Li, J., Lohman, K.L., Lu, H., Makhijani, V.B., McDade, K.E., McKenna, M.P., Myers, E.W., Nickerson, E., Nobile, J.R., Plant, R., Puc, B.P., Ronan, M.T., Roth, G.T., Sarkis, G.J., Simpson, J.W., Srinivasan, M., Tartaro, K.R., Tomasz, A., Vogt, K.A., Volkmer, G.A., Wang, S.H., Wang, Y., Weiner, M.P., Yu, P., Begley, R.F., & Rothberg, J.M. (2005). Genome sequencing in microfabricated high-density picolitre reactors. *Nature*, vol. 437, pp. 376–380.

Meadows, J.R.S., Hiendleder, S., & Kijas, J.W. (2011). Haplogroup relationships between domestic and wild sheep resolved using a mitogenome panel. *Heredity*, vol. 106, pp. 700-706.

Miller, M.P. (2005). Alleles In Space (AIS): Computer Software for the Joint Analysis of Interindividual Spatial and Genetic Information. *Journal of Heredity*, vol. 96, No. 6, pp. 722-724.

Naderi, S., Rezaei, H.R., Pompanon, F., Blum, M.G., Negrini, R., Naghash, H.R., Balkiz, O., Mashkour, M., Gaggiotti, O.E., Ajmone-Marsan, P., Kence, A., Vigne, J.D., & Taberlet, P. (2008). The goat domestication process inferred from large-scale mitochondrial DNA analysis of wild and domestic individuals. *Proceedings of the National Academy of Science of the United States of America*, vol. 105, No. 46, pp. 17659-17664.

Negrini, R., Milanesi, E., Colli, L., Pellecchia, M., Nicoloso, L., Crepaldi, P., Lenstra, J.A., & Ajmone-Marsan, P. (2007). Breed assignment of Italian cattle using biallelic AFLP markers. *Animal Genetics*, vol. 38, pp. 147-153.

Nei, M., Maruyama, T., & Chakraborty, R. (1975). The bottleneck effect and genetic variability in populations. *Evolution*, vol. 29, pp. 1-10.

Nguyen, T.T., Genini, S., Menetrey, F., Malek, M., Vogeli, P., Goe, M.R., & Stranzinger, G. (2005). Application of bovine microsatellite markers for genetic diversity analysis of Swiss yak (*Poephagus grunniens*). *Animal Genetics*, vol. 36, pp. 484–489.

Nielsen, R., Hellmann, I., Hubisz, M., Bustamante, C., & Clark, A.G. (2007). Recent and ongoing selection in the human genome. *Nature Reviews Genetics*, vol. 8, pp. 857–868.

Nijman, I.J., Van Boxtel, D.C.J., Lisette, M., Van Cann, L.M., Marnoch, Y., Cuppen, E., & Lenstra, J.A. (2008). Phylogeny of Y chromosomes from bovine species. *Cladistics*, vol. 24, No. 5, pp. 723-726.

Novembre, J., & Stephens, M. (2008). Interpreting principal component analyses of spatial population genetic variation. *Nature Genetics*, vol. 40, pp. 646–649.

Oleksyk, T.K., Smith, M.W., & O'Brien, S.J. (2010). Genome-wide scans for footprints of natural selection. *Philosophical Transactions of the Royal Society B*, vol. 365, pp. 185-205.

Olivier, M. (2005). The Invader assay for SNP genotyping. *Mutation Research*, vol. 573, pp. 103-110.

Pariset, L., Cappuccio, I., Ajmone Marsan, P., Dunner, S., Luikart, G., Obexer-Ruff, G., Peter, C., Marletta, D., Pilla, F., Valentini A., & the Econogene Consortium (2006). Assessment of population structure by single nucleotide polymorphisms (SNPs) in goat breeds. *Journal of Chromatography B*, vol. 833, pp. 117-120.

Pariset, L., Cuteri, A., Ligda, C., Ajmone Marsan, P., Valentini, A., & the Econogene Consortium (2009a). Geographical patterning of sixteen goat breeds from Italy, Albania and Greece assessed by Single Nucleotide Polymorphisms. *BMC Ecology*, vol. 9, p. 20.

Pariset, L., Joost, S., Ajmone Marsan, P., & Valentini, A. (2009b). Landscape genomics and biased FST approaches reveal Single Nucleotide Polymorphisms under selection in goat breeds of North-East Mediterranean. *BMC Genetics*, vol. 10, p. 7.

Pariset, L., Chillemi, G., Bongiorni, S., Spica, V.R., & Valentini, A. (2009c). Microarrays and high throughput transcriptomic analysis for species with limited knowledge of genomic sequences. *New Biotechnology*, vol. 25, No. 5, pp. 272-279.

Pariset, L., Mariotti, M., Nardone, A., Soysal, M.I., Ozkan, E., Williams, J.L., Dunner, S., Leveziel, H., Maroti-Agots, A., Bodò, I., & Valentini A. (2010a). Relationships of podolic cattle breeds assessed by Single Nucleotide Polymorphisms (SNPs) genotyping. *Journal Animal breeding and Genetics*, vol. 127, pp. 481-488.

Pariset, L., Bueno, S., Bongiorni, S., & Valentini, A. (2010b). Not working on mice or humans? A pipeline to rapidly generate species- specific microarrays from sequence databases. *G.I.T. Laboratory Journal Europe* 1-2/2010, pp. 020-021. http://www.laboratory-journal.com/science/informationstechnologie-it/not-working-mice-or-humans.

Pariset, L., Mariotti, M., Gargani, M., Joost, S., Negrini, R., Perez, T., Bruford, M., Ajmone Marsan, P., Valentini, A., & the Econogene Consortium (2011). Genetic diversity of sheep breeds from Italy, Albania and Greece assessed by mitochondrial DNA and nuclear polymorphisms (SNPs). *TheScientificWorldJOURNAL*, vol. 11, pp. 1641-1659.

Pellecchia, M., Negrini, R., Colli, L., Patrini, M., Milanesi, E., Achilli, A., Bertorelle, G., Cavalli-Sforza, L.L., Piazza, A., & Torroni, A. (2007). The mystery of Etruscan origins: novel clues from *Bos taurus* mitochondrial DNA. *Proceedings of the Royal Society B*, vol. 274, No. 1614, pp. 1175–1179.

Perkel, J.M. (2008). SNP genotyping: six technologies that keyed a revolution. *Nature Methods*, vol. 5, pp. 447-453.

Perkel, J.M. (2011). Copy Number Variants: Mapping the Genome's 'Land Mines'. *BioTechniques*, vol. 51, pp. 21–24.

Peter, C., Bruford, M., Perez, T., Dalamitra, S., Hewitt, G., Erhardt, G., & the Econogene Consortium (2007). Genetic diversity and subdivision of 57 European and Middle-Eastern sheep breeds. *Animal Genetics*, vol. 38, pp. 37–44.

Petit, C., Freville, H., Mignot, A., Colas, B., Riba, M., Imbert, E., Hurterez-Bousses, S., Virevaire, M., & Olivieri, I. (2001). Gene flow and local adaptation in two endemic plant species. *Biological Conservation*, vol. 100, pp. 21-34.

Piry, S., Alapetite, A., Cornuet, J.M., Paetkau, D., Baudouin, L., & Estoup, A. (2004). GeneClass2: A Software for Genetic Assignment and First-Generation Migrant Detection. *Journal of Heredity*, vol. 95, pp. 536-539.

Price, A.L., Patterson, N.J., Plenge, R.M., Weinblatt, M.E., Shadick, N.A., & Reich, D. (2006). Principal components analysis corrects for stratification in genome-wide association studies. *Nature Genetics*, vol. 38, No. 8, pp. 904-909.

Price, A.L., Kryukov, G.V., de Bakker, P.I.W., Purcell, S.M., Staples, J., Wei, J.L., & Sunyaev, S.R. (2010). Pooled Association Tests for Rare Variants in Exon-Resequencing Studies. *The American Journal of Human Genetics*, vol. 86, No. 6, pp. 832-838.

Primmer, C.R. (2009). From conservation genetics to conservation genomics. *Annals of the New York Academy of Science*, vol. 1162, pp. 357-368.

Pritchard, J.K., Stephens, M., & Donnelly, P. (2000). Inference of population structure using multilocus genotype data. *Genetics*, vol. 155, pp. 945–959.

Prugnolle, F., Theron, A., Pointier, J., Jabbour-Zahab, R., Jarne, P., Durand, P., & de Meeûs, T. (2005). Dispersal in a parasitic worm and its two hosts: Consequence for local adaptation. *Evolution*, vol. 59, pp. 296-303.

Purcell, S., Neale, B., Todd-Brown, K., Thomas, L., Ferreira, M.A.R., Bender, D., Maller, J., Sklar, P., de Bakker, P.I.W., Daly, M.J., & Sham, P.C. (2007). PLINK: a toolset for whole-genome association and population-based linkage analyses. *The American Journal of Human Genetics*, vol. 81, pp. 559–575.

Qi, X-B. (2004). Genetic Diversity, Differentiation and Relationship of Domestic yak Populations - a Microsatellite and Mitochondrial DNA Study. Lanzhou, P.R. China: Lanzhou University.

Qi, X-B., Jianlin, H., Wang, G., Rege, J.E.O., & Hanotte, O. (2008). Assessment of cattle genetic introgression into domestic yak populations using mitochondrial and microsatellite DNA markers. *Animal Genetics*, vol. 41, pp. 242–252.

Rasool, G., Khan, B.A., & Jasra, A.W. (2002). Yak Pastoralism in Pakistan. Nairobi, Kenya: International Livestock Research Institute (ILRI). *Proceedings of the third International Congress on Yak*, pp. 95–99, Lhasa, P.R. China, 4-9 September, 2000.

Ritz, L.R., Glowatzki-Mullis, M.L., MacHugh, D.E., & Gaillard, C. (2000). Phylogenetic analysis of the tribe Bovini using microsatellites. *Animal Genetics*, vol. 31, pp.178–185.

Roberge, C., Guderley, H., & Bernatchez, L. (2007) Genomewide identification of genes under directional selection: gene transcription Qst scan in diverging Atlantic salmon subpopulations. *Genetics*, vol. 177, pp. 1011–1022.

Biek, R., & Real, L.A. (2010). The landscape genetics of infectious disease emergence and spread. *Molecular Ecology*, vol. 19, pp. 3515–3531.

Ronaghi, M., Uhlen, M., & Nyren, P. (1998). A sequencing method based on real-time pyrophosphate. *Science*, vol. 281, pp. 363–365.

Schwartz, M.K., Copeland, J.P., Anderson, N.J., Squires, J.R., Inman, R.M., Mckelvey, K.S., Pilgrim, K.L., Waits, L.P., & Cushman, S.A. (2009). Wolverine gene flow across a narrow climatic niche. *Ecology*, vol. 90, No. 11, pp. 3222-3232.

Segelbacher, G., Tomiuk, J., & Manel, S. (2008). Temporal and spatial analyses disclose consequences of habitat fragmentation on the genetic diversity in capercaillie (*Tetrao urogallus*). *Molecular Ecology*, vol. 17, pp. 2356-2367.

Segelbacher, G., Cushman, S., Epperson, B., Fortin, M.J., Francois, O., Hard, O., Holderegger, R., Taberlet, P., Waits, L.P., & Manel, S. (2010.) Landscape Genetics: concepts and Challenges in a Conservation Context. *Conservation Genetics*, vol. 11, pp. 375–385.

Seldin, M.F., Shigeta, R., Villoslada, P., Selmi, C., Tuomilehto, J., Silva, G., Belmont, J.W., Klareskog, L., & Gregersen, P.K. (2006). European population substructure: clustering of Northern and Southern populations. *PLoS Genetics*, vol. 2, pp. 1339-1351.

Simianer, H., Marti, S.B., Gibson, J., Hanotte, O., & Rege J.E.O. (2003). An approach to the optimal allocation of conservation funds to minimize loss of genetic diversity between livestock breeds. *Ecological Economics*, vol. 45, pp. 377-392.

Storfer, A., Murphy, M.A., Evans, J.S., Goldberg, C.S., Robinson, S., Spear, S.F., Dezzani, R., Delmelle, E., Vierling, L., & Waits, L.P. (2007). Putting the "landscape" in landscape genetics. *Heredity*, vol. 98, pp. 128-142.

Storz, J.F. (2005). Using genome scans of DNA polymorphism to infer adaptive population divergence. *Molecular Ecology*, vol. 14, pp. 671-688.

Syvanen, A.C. (1998). Solid-phase minisequencing as a tool to detect DNA polymorphism. *Methods in Molecular Biology*, vol. 98, pp. 291-298.

Taberlet, P., Valentini, A., Rezaei, H.R., Naderi, S., Pompanon, F., Negrini, R., & Ajmone-Marsan, P. (2008). Are cattle, sheep, and goats endangered species? *Molecular Ecology*, vol. 17, pp. 275-284.

Taberlet, P., Coissac, E., Pansu, J., & Pompanon, F. (2011). Conservation genetics of cattle, sheep, and goats. *Comptes Rendus Biologies*, vol. 334, No. 3, pp. 247-254.

Tapio, M., Ozerov, M., Tapio, I., Toro, M.A., Marzanov, N., Cinkulov, M., Goncharenko, G., Kiselyova, T., Murawski, M., & Kantanen, J. (2010). Microsatellite-based genetic diversity and population structure of domestic sheep in Northern Eurasia. *BMC Genetics*, vol. 11, pp. 76-86.

Tautz, D., & Renz, M. (1984). Simple sequences are ubiquitous repetitive components of eukaryotic genomes. *Nucleic Acid Research*, vol. 12, pp. 4127-4138.

Tshering, L., Gyamtsho, P., & Gyeltshen, T. (1996). Yaks in Bhutan. *Proceedings of a Workshop on Conservation and Management of Yak Genetic Diversity*, pp. 13-24, Kathmandu, Nepal: 29-31 October, 1996.

Tum, S., Puotinen, M.L., Skerratt, L.F., Chan, B., & Sothoeun, S. (2007). Validation of a geographic information system model for mapping the risk of fasciolosis in cattle and buffaloes in Cambodia. *Veterinary Parasitology*, vol. 143, pp. 364-367.

Vignal, A., Milan, D., SanCristobal, M., & Eggen, A. (2002). A review on SNP and other types of molecular markers and their use in animal genetics. *Genetic Selection and Evolution*, vol. 34, pp. 275-305.

Vigouroux, Y., McMullen, M., Hittinger, C.T., Houchins, K., Schulz, L., Kresovich, S., Matsuoka, Y., & Doebley, J. (2002). Identifying genes of agronomic importance in maize by screening microsatellites for evidence of selection during domestication. *Proceedings of the National Academy of Sciences*, vol. 99, pp. 9650-9655.

Vos, P., Hogers, R., Bleeker, M., Reijans, M., van de Lee, T., Hornes, M., Frijters, A., Pot, J., Peleman, J., & Zabeau, M. (1995). AFLP: a new technique for DNA fingerprinting. *Nucleic Acid Research*, vol. 23, pp. 4407-4444.

Wang, I.J. (2011). Choosing appropriate genetic markers and analytical methods for testing landscape genetic hypotheses. *Molecular Ecology*, vol. 20, No. 12, pp. 2480-2482.

Wang, Y.H., Yang, K.C., Bridgman, C.L., & Lin, L.K. (2008). Habitat suitability modelling to correlate gene flow with landscape connectivity. *Landscape Ecology*, vol. 23, pp. 989-1000.

Weir, B.S., & Cockerham, C.C. (1984). Estimating F-statistics for the analysis of population structure. *Evolution*, vol. 38, pp. 1358–1370.

Wiener, G., Jianlin, H., & Ruijun, L. (2003). The Yak. The Regional Office for Asia and the Pacific. Bangkok, Thailand: Food and Agriculture Organization of the United Nations.

Williams, J.L. (2005). The use of marker-assisted selection in animal breeding and biotechnology. *Revue Scientifique et Technique*, vol. 24, pp. 379-391.

Wilson, G.A., & Rannala, B. (2003). Bayesian inference of recent migration rates using multilocus genotypes. *Genetics*, vol. 163, pp. 1177–1191.

Wilson, I.J., Weale, M.E., & Balding, D.J. (2003). Inferences from DNA data: population histories, evolutionary processes and forensic match probabilities. *Journal of the Royal Statistical Society: Series A (Statistics in Society)*, vol. 166, pp. 155–188.

Wright, S. (1951). The genetical structure of populations. *Annals of Eugenics*, vol. 15, pp. 323–354.

Wright, S. (1978). Evolution and the genetics of populations. Chicago: University of Chicago Press.

Xuebin, Q., Jianlin, H., Rege, J.E.O., & Hanotte, O. (2002). Y-Chromosome Specific Microsatellite Polymorphisms in Chinese yak. *Proceedings of 7th World Congress on Genetics Applied to Livestock Production*, pp. 509–512.

Xuebin, Q., Jianlin, H., Lkhagva, B., Chekarova, I., Badamdorj, D., Rege, J.E., & Hanotte, O. (2005). Genetic diversity and differentiation of Mongolian and Russian yak populations. *Journal of Animal Breeding and Genetics*, vol. 122, pp. 117–216.

Zhang, F., Gu, W., Hurles, M.E., & Lupski, J.R. (2009). Copy number variation in human health, disease, and evolution. *Annual Review of Genomics and Human Genetics*, vol. 10, pp. 451-481.

Zhang, G., Chen, W., Xue, M., Wang, Z., Chang, H., Han, X., Liao, X., & Wang, D. (2008). Analysis of genetic diversity and population structure of Chinese yak breeds (*Bos grunniens*) using microsatellite markers. *Journal of Genetics and Genomics*, vol. 35, pp. 233-238.

Zhang, J., Chiodini, R., Badr, A. & Zhang, G. (2011). The impact of next-generation sequencing on genomics. *Journal of Genetics and Genomics*, vol. 38, No. 3, pp. 95-109.

12

Loss of Genetic Diversity in Wild Populations

Shawn Larson
Seattle Aquarium
United States

1. Introduction

The importance of genetic diversity within populations has been debated since the study of genetics began. There are two major camps: the classical school (genetic variability is low within species) and the balanced school (genetic variability is high within species; Avise, 1994). This debate was brought to a head when O'Brien et al., 1985 suggested that population viability problems within the cheetah (*Acinonyx jubatus*) could be linked to low genetic diversity. Since then there has been a significant increase in conservation genetics studies within other species focusing on genetic variability and it's importance in conservation, particularly within those populations that have passed through significant population bottlenecks (Coulson et al., 1999; Hoelzel, 1997; O'Brien et al., 1985; O'Brien et al., 1996; Wildt et al., 1987). In addition to the classical and balanced genetic variability school debate, there has been criticism of the use of neutral genetic markers in conjunction with fitness related indices to suggest inbreeding depression (Hedrick & Kalinowski, 2000; Lynch, 1996). The reasoning is that since neutral markers are by definition not affected by natural selection then they cannot be causally linked to inbreeding indices (Avise, 1994). However, since they are not affected by natural selection they may actually indicate a more realistic picture of overall potential genetic diversity not yet affected by selective pressure following population bottlenecks than those regions affected by selection (Amos &Balmford, 2001; Avise, 1994). The premise here is that the total genetic variation lost following such events may be estimated and that any loss of genetic diversity may negatively affect long term population adaptability and viability because future conditions and stochastic events cannot be predicted (Avise, 1994). Due to the rapid increase in the study of genetic variation within wildlife populations there is a growing body of evidence linking fitness variables with both qualitative and quantitative genetic methods, suggesting meaningful trends that may affect population survival and viability (Amos & Balmford, 2001).

Populations that experience bottlenecks are thought to lose genetic diversity through genetic drift and inbreeding (Charlesworth & Charlesworth, 1999; Crnokrak & Roff, 1999; Hedrick & Kalinowski, 2000; Lacy, 1997; Lynch, 1996; Ralls et al., 1988). Small population size may lead to inbreeding where related individuals produce offspring (Eldridge et al., 1999; Lynch, 1996; Slate et al., 2000). Inbreeding may lead to the buildup of deleterious recessive genes, termed inbreeding depression, that may cause decreased fecundity, increased mortality, slowed growth, developmental defects, increased susceptibility to disease, decreased ability to withstand stress, and decreased ability to compete (Lacy, 1997). The cost of inbreeding to fecundity in small populations has been documented by several studies within the Felidae

(O'Brien et al., 1985; O'Brien et al., 1987; O'Brien et al., 1996; and Wildt et al., 1987). These studies revealed significantly decreased genetic variation in isolated wild felid populations and inbreeding depression measured by increased sperm abnormalities, decreased reproductive hormone levels and difficulty breeding in captivity (O'Brien et al., 1985; O'Brien et al., 1996; and Wildt et al., 1987). In addition, other work has linked inbreeding within several wild species with increased mortality, decreased fecundity, and slowed growth (Crnokrak & Roff 1999; Ralls et al., 1988; Slate et al., 2000).

Inbreeding and the potential effects on population viability are important because populations that experience a bottleneck are thought to suffer from decreased genetic variation for hundreds of thousands of years (in the absence of immigration) due to the low locus mutation rate that prevails at most genetic locations or loci (10-8 to 10–5 mutations per year; Driscoll et al., 2002; Lande &Barrowclough, 1987; Lynch, 1996). Recovery of lost genetic variation in the absence of immigration is the inverse of the mutation rate (Driscoll et al., 2002; Nei et al., 1975). Small populations that retain relatively low genetic variability over several generations run the risk of fixation of alleles. Most individuals are assumed to contain unique deleterious genes at several loci, all subject to chance fixation in a small founder population (Lynch, 1996). The probability of fixation of deleterious alleles in founder populations of 30 or fewer individuals is close to the alleles' initial frequencies in the population because genetic drift overwhelms natural selection, and may lead to relatively rapid extinction of the population even in the absence of any stochastic events (Lynch, 1996; Shaffer, 1981).

Genetic studies of small populations that regularly contain fewer than 100 breeding individuals, or effective population size (N_E), suggest that these populations are extremely vulnerable to the loss of genetic variation (Lacy, 1997; Lynch, 1996). Low diversity combined with inbreeding depression increase a small population's vulnerability to extinction from stochastic events (Lacy, 1997; Lynch, 1996). Population sizes of at least 1000 are suggested to protect against the fixation of deleterious genes and a breeding population of 10,000 adults is suggested to protect adaptive genetic variation (Lynch, 1996). Unfortunately, these population sizes are usually not found in endangered species or those that have been fragmented by human activities. The fixation of deleterious alleles, genetic drift, the loss of adaptive genetic variation and inbreeding depression all dramatically increase a small population's vulnerability to extinction from stochastic events (Charlesworth & Charlesworth. 1999; Crnokrak & Roff 1999; Fowler & Whitlock 1999; Hedrick & Kalinowski 2000; Lacy 1997; Lynch 1996; Ralls et al., 1988).

Inbreeding depression is defined in general terms as a reduction in fitness of inbred offspring, but its specific effects are highly variable and depend on how the genotypes interact with the environment (Fowler & Whitlock, 1999; Hedrick & Kalinowski, 2000). Current theory is that inbreeding depression is more severe in stressful environments because of reduced genetic plasticity to adapt to changes or stessors (Dahlgard & Hoffman, 2000; Hedrick & Kalinowski, 2000; Miller, 1994). Inbreeding depression causes an overall decrease in population growth (decreasing N_E) due to either decreased birth rate or increased mortality or a combination of the two.

Documenting the effects of inbreeding can be difficult in wild and captive populations of vertebrates due to long generation intervals and/or the inability to control for varying environmental conditions. Consequently, many studies have employed laboratory species as models to study inbreeding depression under controlled conditions. Studies done using the fruit fly, Drosophila melanogaster, have shown significant relationships between inbreeding, changing environmental conditions, and population viability (Bijlsma et al., 1999, 2000; Dahlgard & Hoffman, 2000; Miller, 1994). Specifically, inbred lines of Drosophila suffered 20–50% higher extinction rates compared to outbred lines when challenged with stressful environmental conditions such as high temperatures or exposure to ethanol (Bijlsma et al., 1999, 2000). These studies suggest inbred populations may not respond as well as outbred lines to stress.

When vertebrate individuals are "stressed" (e.g., exposed to adverse environmental conditions including short-term events such as human disturbance, severe weather, food restrictions, or exposure to contaminants or parasites), the hypothalamic–pituitary–adrenal (HPA) axis between the brain and the adrenal gland mediates the stress response (Norris, 1996). Physical, psychological, chemical, and other stressors trigger the release of corticotrophin-releasing hormone (CRH) in the hypothalamus leading to the release of adrenocorticotropic hormone (ACTH) in the pituitary, which in turn mediates the release of the adrenal glucocorticoids (GCs) from the adrenal glands (Norris, 1996). The major GCs include cortisol and corticosterone, and it is the circulating levels of these hormones that produce the physiological and behavioral response to stress (i.e., fight or flight response). Most animals will secrete primarily one of the GCs (either cortisol or corticosterone) in response to stress. Birds, amphibians, and reptiles generally secrete corticosterone as their major glucocorticoid. Mammals secrete both, with cortisol tending to dominate, while fish generally secrete cortisol (Idler & Truscott, 1980; Norris, 1996).

The cost of elevated GCs is well documented. Observations include a decrease in overall health with reduced individual growth, changes in metabolism, delayed repair of tissues or healing, and immunosuppression resulting in an increased incidence of disease in stressed animals (Astheimer et al., 2000; Gorbman et al., 1983; Johnson et al., 1992; Norris, 1996; Rivier & Rivest, 1991). Chronically high basal GC levels also may negatively affect reproductive potential by depressing reproductive hormone levels (Brann & Mahesh, 1991; Moberg, 1991; Norman, 1993; Kosowska & Zdrojewicz 1996).

Genetic load is the longer term effect of loss of fitness over time due to an increase in detrimental mutations becoming fixed creating an overall loss of genetic variation for future adaptation (Kirkpatrick & Jarne 2000). Genetic load may be documented as low estimates of fitness related variables when compared to other populations that have not experience bottlenecks or have no evidence of loss of fitness associated with loss of genetic diversity.

Genetic rescue is a management technique that has been used to manage critically endangered populations with high genetic load. This strategy uses unrelated individuals from one population that are introduced into the population with apparent low fitness. This introduction on new genetic material acts to reduce genetic load by lowering or eliminating the frequency of detrimental gene variants in the population. This has been used successfully in captive settings and within wild populations (Hedrick & Fredrickson, 2010). There are clearly recognized guidelines for the successful use of genetic rescue as a conservation management strategy (Hedrick & Fredrickson, 2010). They are as follows: Evidence of low fitness; the existence of a closely related donor population from the same species and from similar habitat; experimental data in a captive situation such as successful mating, good survival of progeny and molecular data; a translocation protocol; a detailed monitoring plan and a commitment for long term management (Hedrick & Fredrickson, 2010).

2. Case studies of loss of genetic diversity and subsequent loss of fitness

2.1 Sea otters

Sea otters, *Enhydra lutris*, were once abundant across their range in the north Pacific Rim from northern Japan to the central Pacific Coast of Baja California, Mexico (Lensink, 1962; Kenyon, 1969; Riedman and Estes 1990). During the 18th and 19th centuries, the Pacific maritime fur trade eliminated or greatly reduced sea otter populations throughout this area, eventually resulting in 11 scattered populations after protection in 1911, with a combined population totaling approximately 1% of the original abundance estimated to be at one time approximately 100,000 animals (Lensink, 1962; Kenyon, 1969).

By the 1970's, few sea otter populations had recovered to pre-exploitation levels, with the majority of historic sea otter habitat vacant along the west coast of North America from Prince William Sound, Alaska, southward to California (Estes, 1990; Kenyon, 1969; Riedman & Estes, 1990). In an effort to re-establish sea otter populations throughout their former range, management authorities made several translocations from the 1950's through the 1970's (Jameson et al., 1982). In total, 715 otters were captured at Amchitka Island in the Aleutian chain and Prince William Sound, Alaska, and then released at various unoccupied habitats in Alaska, British Columbia, Washington and Oregon (Jameson et al., 1982). The translocations to Washington, Oregon and the Pribilof Islands included only animals captured at Amchitka, and only the Washington effort was successful (Bodkin et al., 1999; Jameson et al., 1982). The translocations to Southeast Alaska and British Columbia (off the west coast of Vancouver Island) included a mix of Amchitka and Prince William Sound animals, and both were successful (Bodkin et al., 1999; Jameson et al., 1982). In spite of these successful translocation efforts, sea otter populations today remain fragmented with extant populations geographically separated and, in most cases, reproductively isolated (Bodkin et al., 1999).

The remaining sea otter populations constitute three subspecies: Russian (*E.l. lutris*), Northern (*E.l. kenyoni*), and Southern (*E.l. nereis*) based on skull morphometrics (Wilson et al., 1990) spread among remnant and translocated populations throughout the former range. Investigation into the remaining neutral genetic variation post exploitation found only half the original genetic variation left within sea otter populations (Larson et al., 2002b). Genetic diversity estimates for all extant populations were similar, with expected heterozygosities ranging from 40% within California to 47% within Southeast Alaska and the allelic richness (mean number of alleles corrected for sample size) ranging from 3.20 in Prince William Sound to 3.70 in Amichitka and Washington (Larson et al., 2002a).

This low diversity evident throughout sea otter populations make distinguishing significant genetic differences between subspecies difficult because there are so few alleles available for differentiation. Studies of mitochondrial DNA (mtDNA) have found little genetic support for distinguishing the Northern and Southern subspecies (Cheney, 1995; Cronin et al., 1996; Larson et al., 2002a; Sanchez, 1992; Scribner et al., 1997). However, population level comparisons using hypervariable nuclear (microsatellite) genetic markers have indicated some phylogeographic structure among contemporary sea otters. Cronin et al., 2002 found stock differences in Northern sea otters using microsatellite markers: a Southwest stock including the Aleutian Islands and Kodiak Island; a Southcentral stock including Prince William Sound, the Kenai Peninsula and Cordova; and a Southeastern stock including the Alexander Archipelago. In addition, Larson et al. (2002a) found significant genetic differentiation using seven variable microsatellite nuclear markers among three remnant sea otter populations: Amchitka in the western Aleutian Islands of Alaska (a United States Endangered Species Act (ESA) listed stock), Prince William Sound in southcentral Alaska, and Southern sea otters from the single California population (also an ESA listed group).

Do sea otters suffer inbreeding depression because of their loss of genetic diversity and population fragmentation and resuction? Stress as measured by circulating adrenal glucocorticoids (GC's) was found to be significantly negatively correlated with heterozygosity in sea otters (Larson et al., 2009). This significant relationship was found at both the individual and population level for corticosterone, but only at the individual level for Cortisol (Larson et al., 2009).

These results are strong evidence that genetic diversity may be a significant determinant of how individuals and thus populations are able to respond to and handle stress — from exposure to disease agents or contaminants to dealing with various levels of food availability or even short-term severe storm events. To fully evaluate the biological meaning of relative levels of circulating GCs requires an understanding, specific for each species, of the role and the affinity of the binding proteins, the cell surface and nuclear receptors to GCs, and ultimately the role of GCs through differential gene transcription. The relative importance of cortisol vs. corticosterone on gene transcription varies widely within mammals (Gayrard et al. 1996, Tanigawa et al. 2002), and the effect of either GC on sea otter physiology has not been investigated.

If corticosterone levels do relate to low genetic diversity and diminished ability to cope with environmental stress, then these relationships could also suggest a fitness repercussion from a population bottleneck and may serve as an indicator of inbreeding depression. One potential consequence of higher circulating GCs is reduced immune response (Norris 1996). California sea otters historically have had the lowest population growth rate of any sea otter population as well as the highest reported incidence of disease within adult animals, those that make up the breeding population (Gerber et al., 2004; Hanni et al., 2003; Kreuder et al., 2003; Miller et al., 2004; Thomas & Cole, 1996). In addition, a high incidence of infectious diseases among this population suggests a weakened immune system or immunosuppression (Gerber et al., 2004; Hanni et al., 2003; Kreuder et al., 2003; Miller et al., 2004). It seems at least conceivable that the low genetic diversity and corresponding high corticosterone levels in this population have played a role in susceptibility to infectious disease via immunosuppression. The higher incidence of disease eventually leads to elevated mortality of prime-aged animals and low overall population growth rates.

Many extant sea otter populations appear to be in a precarious balance. Because they inhabit coastal areas, they are often in contact with humans and can suffer negative interactions associated with fishing activities and exposure to shoreline and near-shore pollution sources. Add to these difficulties the potential for negative effects associated with the loss of genetic diversity and buildup of deleterious alleles called genetic loading, and we may improve our understanding of why some sea otter populations fail to thrive, like the threatened California population.

2.2 Florida panther

The American mountain lion, cougar or puma (*Puma concolor*) once was widely distributed throughout North, Middle and South America. There are currently seven recognized subspecies based on genetic data: *P. c. cougar*: North America, *P. c. corryi*: Florida, *P. c. costaricensis*: Central America, *P. c. capricornensis*: eastern South America, *P. c. concolor*: northern South America, *P. c. cabrerae*: central South America, and *P. c. puma*: southern South America (Culver et al., 2000). The cougar subspecies in southern Florida swamplands, *P. c. corryi*, is the only subspecies exclusively referred to as "panther" and is the only subspecies listed as critically endangered by the IUCN (2008 IUCN red list) and endangered by the ESA (Endangered Species List 1967). The Florida panther has been restricted to a small area that includes the Big Cypress National Preserve, Everglades National Park, and the Florida Panther National Wildlife Refuge, which is thought to represent only 5% of its historic range. The number of living Florida panthers was estimated to be approximately 100 in 2010 (Johnson et al. 2010). Overhunting was responsible for the initial population bottleneck in Florida panthers which created a small, isolated population that became inbred.

The population was found to have several problems associated with reproduction (cryporchidism in males, high neonate mortality), as well as shared genetic abnormalities such as heart defects and kinked tails (Roelke et al., 1983). This prompted a program to protect and enhance the genetic diversity of the remaining panthers through a program of genetic rescue. In 1995, conservation managers translocated eight unrelated female pumas from Texas to increase depleted genetic diversity, improve population numbers, and reverse indications of inbreeding depression (Hedrick, 1995). The effect of the genetic rescue was dramatic. Since the introduction of those few unrelated females the panther numbers have increased threefold with the panther population growth rate increasing from 0 to 12% annually, genetic heterozygosity has doubled, individual survival and fitness measures have improved, and inbreeding correlates declined significantly. (Hedrick & Fredrickson, 2010; Johnson et al. 2010). The remarkable success of the genetic rescue experiment within Florida panthers makes it a model for other genetic rescue programs. However even with the recovery of the Florida panther after the influx of new genes the population continues to struggle with issues related to habitat loss, and disease (Johnson et al. 2010).

2.3 Mexican wolf

The Mexican wolf (*Canis lupis baileyi*) is a an endangered subspecies of the grey wolf (*Canis lupus*) (Leonard et al, 2005). The population was driven to critically low numbers because of hunting pressure resulting in their extinction in the United States and only a few small isolated populations in Mexico by 1980. These small, isolated populations were thought to have been negatively affected by inbreeding resulting in zero population growth due in large part to reduced reproductive rate and relatively high levels of mortality. Based on the population fragmentation and low growth rate this wolf was listed as endangered by the ESA in 1976 (Hedrick & Fredrickson, 2010). The surviving individuals were eventually caught and placed in captivity. It is thought that all Mexican wolves surviving today are from descended three captive lineages produced by from a total of seven founders caught between 1960 and 1980 (Hedrick et al., 1997). The captive wolves were kept and bred both within founder lineages and between. These captive groups were monitored for several fitness related traits to determine effects of inbreeding and out-breeding. Initial fitness improvements were significantly higher numbers of pups in each litter within the groups with crossed lineages (Hedrick & Fredrickson, 2010). In 1998 a pure lineage group was reintroduced into the wild. Then in 2009 a group from another lineage was introduced resulting in the first genetically mixed progeny in the wild (Hedrick & Fredrickson, 2010). Almost immediately the population rate started growing with the average litter size doubling and a significant decline in the measured inbreeding coefficients (Hedrick & Fredrickson, 2010). Based on these reproductive indices (litter size and inbreeding coefficients in pups) there appeared to be successful genetic rescue in wild Mexican wolves but after only two distinct lineages were allowed to interbreed (Hedrick & Fredrickson, 2010).

3. Case studies of loss of genetic diversity and little subsequent loss of fitness

3.1 Northern elephant seals

There are examples in nature where populations that have experienced bottlenecks and consequently have very little measurable genetic variation that seem to suffer no ill effects from inbreeding depression and grow rapidly. One case in point is the northern elephant seal (*Mirounga angustirostris*) which was exploited for its blubber in the 19th century resulting in only 10-30 remaining individuals (Hoelzel, 1997). The population recovered quickly, numbering approximately 127,000 individuals in 1991 (Hoelzel, 1997). Genetic analysis using a variety of markers revealed little genetic variation (Bonnell & Selander, 1974; Hoelzel, 1997). Furthermore, a study of pre-exploitation northern elephant seal genetic diversity compared to contemporary seals, revealed a significant loss of genetic diversity (55% loss, Weber et al., 2000). Even though the northern elephant seal recovery has been remarkable, recent data suggests that they may be experiencing some ill effects such as decreased reproductive fitness. Paternity success of male northern elephant seals was lower than expected when compared to observed copulations and compared to the average paternity success of the southern elephant seals, *M. leonine*, which did not experience the severe exploitation similar to the Northern elephant seal (Hoelzel, 1997).

3.2 Canadian moose

Another case study of loss of genetic variation within bottlenecked populations and dramatic recovery is that found within the moose, *Alces alces*. In the late 19th and early 20th century, six moose were translocated from mainland Canada to two areas of Newfoundland (Broders et al., 1999). The population in 1999 was estimated to be approximately 150,000 with more than 400,000 harvested since their introduction (Broders et al., 1999). Average genetic diversity as measured by microsatellite heterozygosity over all populations was 33%, with the loss of heterozygosity from the three founder events (two translocations and one natural colonization) ranging from 14%-30% (Broders et al., 1999). The cumulative loss of heterozygosity from a translocated population which then seeded another by natural colonization was 46% (Broders et al., 1999). All the moose populations suffered a significant loss of diversity but have not exhibited inbreeding depression as indicated by population growth rates, although the long term viability of the population and individuals remains unknown (Broders et al. 1999).

4. Conclusion

Genetic studies of small populations that regularly contain fewer than 100 breeding individuals suggest that these populations are extremely vulnerable to the loss of genetic variation (Lacy, 1997; Lynch, 1996). Low diversity combined with inbreeding depression increase a small population's vulnerability to extinction from stochastic events (Lacy, 1997; Lynch, 1996). Population sizes of at least 1000 are suggested to protect against the fixation of deleterious genes and a breeding population of 10,000 adults is suggested to protect adaptive genetic variation (Lynch, 1996). Unfortunately, these population sizes are usually not found in endangered species or those that have been fragmented by human activities such as sea otters, Florida panthers and Mexican wolves. The fixation of deleterious alleles, genetic drift, the loss of adaptive genetic variation and inbreeding depression all dramatically increase a small population's vulnerability to extinction from stochastic events (Charlesworth & Charlesworth 1999; Crnokrak & Roff, 1999; Fowler & Whitlock, 1999; Hedrick & Kalinowski, 2000; Lacy, 1997; Lynch, 1996; Ralls et al., 1988). In an increasingly fragmented world where wildlife individuals have little opportunity to regularly migrate freely between isolated populations to maintain geneflow and thus overall effective population size, it becomes imperative for management policies to encourage genetic diversity. If the ultimate goal is to maintain maximal diversity within wildlife populations to ensure maximal potential to respond to environmental changes then management decisions may need to be based on maintaining genetic diversity rather than maintaining unique populations such as subspecies. Genetic rescue may require the mixing of subspecies and thus the loss of unique subspecific characteristics. Managers must decide what is most important to preserve, the species with maximal genetic diversity for future stochastic changes or a genetically unique race that may be negatively affected by isolation and inbreeding and incapable of responding to future challenges. That ultimately may be the cost of maintaining genetically healthy populations, the loss of genetically unique groups and adaptive variation.

5. References

Amos, W. & Balmford, A. (2001). When does conservation genetics matter? *Heredity* 87: 257-265.

Astheimer, L. B., Buttemer, W. A. , & Wingfield, J. C. (2000). Corticosterone treatment has no effect on reproductive hormones or aggressive behavior in free-living male tree sparrows, Spizella arborea. *Hormones and Behavior* 37:31–39.

Avise, J.C. (1994). *Molecular markers, Natural history and evolution*. Chapman and Hall. New York.

Bijlsma, R., Bundgaard, J., & Van putten, W. F. (1999). Environmental dependence of inbreeding depression and purging in *Drosophila melanogaster*. *Journal of Evolutionary Biology* 12:1125–1137.

Bijlsma, R., Bundgaard, J., & Boerema, A. C. (2000). Does inbreeding affect the extinction risk of small populations?: Predictions from Drosophila. *Journal of Evolutionary Biology* 13:502–514.

Bodkin J.L., Ballachey, B.E., Cronin, M.A., & Scribner, K.T. (1999). Population demographics and genetic diversity in remnant and translocated populations of sea otters. *Conservation Biology* 13:1378-1385.

Bonnell M.L., & Selander, R.K. (1974). Elephant seals: Genetic variation and near extinction. *Science* 184:908-909.

Brann, D. W., & Mahesh, V. B. (1991). Role of corticosteroids in female reproduction. *The FASEB Journal* 5:2691-2697.

Broders, H.G., Mahoney, S.P., Montevecchi, W.A., & Davidson, W.S. (1999). Population genetic structure and the effect of founder events on the genetic variability of moose, *Alces alces*, in Canada. *Molecular Ecology* 8:1309-1315.

Charlesworth, B., & Charlesworth, D. (1999). The genetic basis of inbreeding depression. *Genetics Research* 74: 329-340.

Cheney, L.C. (1995). *An assessment of genetic variation within and between sea otter (Enhydra lutris) populations off Alaska and California.* Masters thesis. Moss Landing Marine Laboratories. California State University, San Jose, California.

Coulson, T., Albon, S., Slate, J. & Pemberton, J. (1999). Microsatellite loci reveal sex-dependent responses to inbreeding and outbreeding in red deer calves. *Evolution* 53(6): 1951-1960.

Crnokrak, P., & Roff, D.A. (1999). Inbreeding depression in the wild. *Heredity* 83: 260-270.

Cronin, M.A., Bodkin, J.L., Ballachey, B.E., Estes, J.A., & Patton, J.C. (1996). Mitochondrial DNA variation among subspecies and populations of sea otters (*Enhydra lutris*). *Journal of Mammalogy* 77: 546-557.

Culver, M., Johnson, W.E., Pecon-Slatterym, J., & O'Brien, S.J. (2000). Genetic ancestry of the American Puma (*Puma concolor*). *Journal of Heredity* 91:186-197.

Dahlgard, J, & Hoffman, A. A. (2000). Stress resistance and environmental dependency of inbreeding depression in Drosophila melanogaster. *Conservation Biology* 14:1187–1193.

Driscoll, C.A., Raymond, M.M., Nelson, G., Goldstein, D., & O'Brien, S.J. (2002). Genomic Microsatellites as Evolutionary Chronometers: A Test in Wild Cats. *Genome Research* 12: 414-423.

Ellegren H., Primmer, C.R., & Sheldon, B.C. (1995). Microsatellite evolution: directionality or bias in locus selection. *Nature Genetics* 11:360-362.

Estes, J.A. (1990). Growth and equilibrium in sea otter populations. *Journal of Animal Ecology* 59:385-401.

Fowler, K. & Whitlock, M.C. (1999). The distribution of phenotypic variance with inbreeding. *Evolution* 53: 1143-1156.

Fowler, K., & Whitlock, M. C. (1999). The variance in inbreeding depression and the recovery of fitness in bottlenecked populations. *The Proceedings of the Royal Society, Series B* 266:2061-2066.

Gerber, L. R., Tinker, T. M., Doak, D. F., Estes, J. A., & D.A. Jessup. 2004. Mortality sensitivity in life-stage simulation analysis: A case study of southern sea otters. *Ecological Applications* 14:1554–1565.

Gorbman, A., Dickhoff, W. W., Vigna, S. R., Clark, N. B., & Ralph, C. L. (1983). *Comparative endocrinology.* 2nd edition. JohnWiley and Sons, Inc., New York, NY.

Hanni, K. D., Mazet, J. A. K., Gulland, F. M. D., Estes, J., Staedler, M., Murray, M.J., Miller, M., & Jessup, D.A. (2003). Clinical pathology and assessment of pathogen exposure in southern and Alaskan sea otters. *Journal of Wildlife Diseases* 39:837–850.

Hedrick P.W., & S.T. Kalinowski. (2000). Inbreeding depression in conservation biology. *Annual Review of Ecology, Evolution, and Systematics* 31:139-162.

Hedrick, P.W., & Fredrickson, R. (2010). Genetic rescue guidelines with examples from Mexican wolves and Florida panthers. *Conservation Genetics* 11:615-626.

Hedrick, P.W., Miller, P.S., Geffen, E., & Wayne, R. (1997). Genetic evaluation of the three captive Mexican wolf lineages. *Zoo Biology* 16:47-69.

Hedrick, P.W. (1995). Gene flow and genetic restoration: the Florida panther as a case study. *Conservation Biology* 9:996-1007.

Hoelzel, A.R. (1997). Molecular ecology of pinnipeds. In: *Molecular genetics of marine mammals*. Dizon, A.E., Chivers, S.J., & Perrin, W.F., eds. Pp. 147-157 Special Publication, 3. Society for Marine Mammalogy.

Idler, D. R., & Truscott, B. (1980). Phylogeny of vertebrate adrenal corticosteroids. In: *Evolution of vertebrate endocrine systems*. Pang, P. K. T. & Epple, A. eds. Pp.357–372. Texas Tech University Press, Lubbock, TX.

Jameson, R.J., Kenyon, K.W., Johnson, A.M., & Wright, H.M. (1982). History and status of translocated sea otter populations in North America. *Wildlife Society Bulletin* 10:100-107.

Johnson, E. O., Kamilaris, t. C. , Chrousos, g. P., Gold, & P. W. (1992). Mechanisms of stress: A dynamic overview of hormonal and behavioral homeostasis. *Neuroscience and Behavioral Reviews* 16:115–130.

Johnson, W.E., Onorato, D. P., Roelke, M.E., Land, E. D., Cunningham, M., Belden, R.C, McBride, R., Jansen, D., Lotz, M., Shindle, D., Howard, J., Wildt, D.E., Penfold, L.M. , Hostetler, J.A. , Oli, M.K., & O'Brien, S.J. (2010). Genetic Restoration of the Florida Panther. *Science* 49:329.

Kenyon, K.W. (1969). *The sea otter in the eastern Pacific Ocean*. North American Fauna 68. U.S. Department of the Interior, Washington, D.C.

Kirkpatrick, M., & Jarne, P. (2000). The effects of a bottleneck on inbreeding depression and the genetic load. *American Naturalist* 155:154-167.

Kreuder, C., Miller, M.A., Jessup, D.A., Lowenstine, L.J., Harris, M.D. , Ames, J.A. , Carpenter, T.E., Conrad, P.A., & Mazet, J.A.K. (2003). Patterns of mortality in southern sea otters (*Enhydra lutris nereis*) from 1998–2001. *Journal of Wildlife Diseases*. 39:495-509.

Kosowska, B., & Zdrojewicz, Z. (1996). Relationship between inbreeding and sex hormone concentration in rats under stress. II. The influence of various inbreeding levels on steroid-sex-hormone concentrations in two types of stress. *Journal of Animal Breeding and Genetics* 113:135–143.

Lacy, R.C. (1997). Importance of genetic variation to the viability of mammalian populations. *Journal of Mammalogy* 78(2):320-335.

Lande, R., & Barrowclough, G.F. (1987). Effective population size, genetic variation, and their use in population management. In: *Viable populations for conservation*. Soule, M.E., ed. Pp. 87-123 Cambridge University Press, New York, NY.

Larson, S.E., Jameson, R.J., Bodkin, J.L., Staedler, M., & Bentzen, P. (2002). Microsatellite DNA and mtDNA variation within and among remnant and translocated sea otter, *Enhydra lutris*, populations. *Journal of Mammalogy* 83 (3): 893-906.

Larson, S.E., Jameson, R.J., Etnier, M., Fleming, M., & Bentzen, P. (2002). Loss of genetic diversity in sea otters (*Enhydra lutris*) associated with the fur trade of the 18th and 19th centuries. *Molecular Ecology* 11: 1899-1903.

Larson, S.E., Monson, D. Ballachey, B., Jameson, R., Wasser, S.K. (2009). Stress-related hormones and genetic diversity in sea otters (*Enhydra lutris*). *Marine Mammal Science* 25(2): 351–372.

Lensink, C.J. (1962). *The history and status of sea otters in Alaska*. PhD dissertation. Purdue University, New York.

Leonard, J.A., Vila, C., & Wayne, R.K. (2005). Legacy lost: genetic variability and population size of extirpated US grey wolves (*Canis lupus*). *Molecular Ecology* 14:9-17.

Lynch, M. (1996). A quantitative-genetic perspective on conservation issues. In: *Conservation genetics. Case histories from nature* Avise, J. C., & Hamrick, J. L., eds. Pp. 471-501. Chapman & Hall. New York.

Miller, M. A., Grigg, M. E., Kreuder, C., James, E. R., Melli, A. C., Crosbie, P. R., Jessup, D.A., Boothroyd, J. C., Brownstein, D., & Conrad, P. A. (2004). An unusualgenotype of Toxoplasma gondii is common in California sea otters (*Enhydra lutris nereis*) and is a cause of mortality. *International Journal for Parasitology* 34:275-284.

Miller, P. S. (1994). Is inbreeding depression more severe in a stressful environment? *Zoo Biology* 13:195-208.

Moberg, G. P. (1991). How behavioral stress disrupts the endocrine control of reproduction in domestic animals. *Journal of Dairy Science* 74:304-311.

Nei, M., Maruyama, T., & Chakraborty, R. (1975). The bottleneck effect and genetic variability in populations. *Evolution* 29:1-10.

Norman, R. L. (1993). Effects of corticotropin-releasing hormone on leutenizing hormone, testosterone, and cortisol secretion in intact male rhesus macaques. *Biology of Reproduction* 49:148-153.

Norris, D. O. (1996). Vertebrate endocrinology. 3rd edition. Academic Press, San Diego, CA.

O'Brien, S.J., Martenson, J.S., Packer, C., Herbst, L., De Vos, V., Joslin, P., Ott-Joslin, J., Wildt, D.E., & Bush, M. (1987). Biochemical genetic variation in geographic isolates of African and Asiatic lions. *National Geographic Research* 3:114-124.

O'Brien, S.J., Mortenson, J.S., Miththapala, S., Janczewski, D., Pecon-Slattery, J., Johnson, W., Gilbert, D.A., Roelke, M., Packer, C., Bush, M., & Wildt, D.E. (1996). Conservation genetics of the felidae. In: *Conservation Genetics. Case Histories from Nature* Avise, J.C., & Hamrick, J.L., eds. Pp. 50-74 Chapman and Hall. New York, NY.

O'Brien, S.J., Roelke, M.E., Marker, L., Newman, A., Winkler, C.A., Meltzer, D., Colly, L., Evermann, J.F., Bush, M., & Wildt, D.E. (1985). Genetic basis for species vulnerability in the cheetah. *Science* 227: 1428-1434.

Ralls, K., Ballou, J.D., & Templeton, A. (1988). Estimates of lethal equivalents and the cost of inbreeding in mammals. *Conservation Biology* 2 (2): 185-193.

Riedman, M.L., & Estes, J.A. (1990). The sea otter (*Enhydra lutris*): Behavior, ecology and natural history. U.S. Department of the Interior, Fish and Wildlife Service. *Biological Report 90* (14).

Rivier, C., & Rivest, S. (1991). Effect of stress on the activity of the hypothalamic-pituitarygonadal axis: Peripheral and central mechanisms. *Biology of Reproduction* 45:523-532.

Roelke, M.E., Martenson, J.S., & O'Brien, S.J. (1993). The consequence of demographic reduction in the endangered Florida panther. *Current Biology* 3:304-350.

Sanchez, M.S. 1992. Differentiation and variability of mitochondrial DNA in three sea otter, *Enhydra lutris*, populations. *M.S. thesis* University of California, Santa Cruz, California.

Scribner, K.T., Bodkin, J., Ballachey, B., Fain, S.R., Cronin, M.A., & Sanchez, M. (1997). Population genetic studies of the sea otter (*Enhydra lutris*): A review and interpretation of available data. In: *Molecular Genetics of Marine Mammals*. Dizon, A.E, Chivers, S.J., & Perrin, W.F., eds. Pp. 197-208. Society for Marine Mammalogy Special Publication 3.

Shaffer, M.L. (1981). Minimum population sizes for species conservation. *BioScience* 31:131-134.

Slate, J., Kruuk, L. E. B., Marshall, T. C., Pemberton, J. M., & Clutton-brock, T. H. (2000). Inbreeding depression influences lifetime breeding success in a wild population of red deer (*Cervus elaphus*). *Proceedings of the Royal Society B: Biological Sciences* 267:1657–1662.

Thomas, N. J., & Cole, R. A. (1996). The risk of disease and threats to the wild population. *Endangered Species Update* 13:23–27.

Weber D.S., Stewart, B.S., Garza, J.C., & Lehman, N. (2000). An empirical genetic assessment of the severity of the northern elephant seal population bottleneck. *Current Biology* 10: 1287-1290.

Wildt, D.E., Bush, M., Goodrowe, K.L., Packer, C., Pusey, A.E., Brown, J.L., Joslin, P., & O'Brien, S.J. (1987). Reproductive and genetic consequences of founding isolated lion populations. *Nature* 329:328-331.

Wilson, D.E., Bogan, M.A., Brownell Jr., R.L., Burdin, A.M., & Maminov, M.K. (1991). Geographic variation in sea otters, *Enhydra lutris. Journal of Mammalogy* 72(1):22-36.

13

Low Danube Sturgeon Identification Using DNA Markers

Marieta Costache, Andreea Dudu and Sergiu Emil Georgescu
University of Bucharest, Faculty of Biology
Department of Biochemistry and Molecular Biology
Romania

1. Introduction

Sturgeons represent a class of fish with a distinctive importance, both from a scientific and a commercial point of view. This group of fish appeared 200 million years ago (Bemis *et al.*, 1997) and the early diversification of the species took place in a geographical area currently situated in Asia. Since these are some of the oldest fish species in the world, having survived several mass extinction events, authentic "living fossils" with a slow evolution, sturgeons may constitute a study model for the development of all vertebrates. The commercial importance of these species is due to the value of the products obtained from them, especially caviar, which is considered a gourmet delicacy. Unfortunately, due to their economic importance, these species have been overly exploited through overfishing and poaching. As a result, nowadays they are on the brink of extinction.

General knowledge about sturgeons has been available starting with the last decades of the nineteenth century. In the twentieth century, the interest in sturgeons spread and currently there is a continuous request for new information about the physiology, genetics, phylogeny and evolution history of the species within the *Acipenseriformes* order. The reasoning for this continued interest is the fact that these species are truly endangered, have an increasingly limited distribution area and a continually decreasing number of individuals.

The *Acipenseriformes* order includes two families: *Acipenseridae* (sturgeons proper) and *Polyodontidae* (paddlefish). The distribution of the current 26 species (27, in accordance with Bemis & Kynard, 1997) belonging to the *Acipenseriformes* order is limited to the northern hemisphere (Rochard *et al*, 1991). It has been estimated that approximately half of the sturgeon species are found in Europe (especially in the Pontic-Caspian region), a third of North America and the remainder in eastern Asia and Siberia (Billard & Lecointre, 2001). Sturgeons are distributed on a wide-spread geographical territory, ranging from subtropical to subarctic waters, from North America to Eurasia, in rivers, lakes, coastal waters and inland seas. These species can be found on the European coast of the Atlantic, in the Mediterranean Sea basin, in the Black Sea, the Caspian Sea, the Azov Sea and the rivers in north-western Russia flowing into the Arctic Ocean (Ob, Yenisei, Lena, Kolyma), in Central Asian rivers (Amu Darya and Syr Darya) and in the Baikal Lake. On the Pacific Ocean coast these fish may be encountered in the Amur River, along the Russian-Chinese border, in the hydrographic basin of the Yangtze River and of other rivers in north-eastern China.

Currently, the greatest diversity of *Acipenseridae* species can be found in the Pontic-Caspian region. Some of the sturgeon species from this area possess certain obvious morphologic traits which set them apart from all other sturgeon species, such as the elongated rostrum of the *Acipenser stellatus* species. The Pontic-Caspian region has been extremely unstable during the last 150 million years, a period during which the scientists hypothesise that the *Acipenseridae* diversification had occurred. These changes refer to major variations of the seawater level, the transformation of great water expanses such as the Black Sea from fresh water lakes into enclosed salt-water environments, as well as important modifications regarding river flows along with the formation of mountains. The Black Sea was repeatedly connected to and disconnected from the Caspian and Aral Seas. Thus, the link between the great diversity of *Acipenseridae* in this region and its extremely complex evolutionary history is more than evident.

Most of the sturgeon species are on the brink of extinction. The majority of specialists believe that the changes that have intervened in the habitat of these fish represent the main causes of decline in the sturgeon populations. Among the dramatic anthropogenic factors influencing these fish the most important are overfishing and poaching, pollution and destruction of breeding sites, construction of dams which limit the migration towards spawning sites.

As a consequence of this situation, protection measures of these extremely valuable species have been adopted on an international level. Thus, all these species have been included on the red list of threatened species initiated by IUCN (International Union for Conservation of Nature), and trading with products originating from these fish is controlled by CITES (Convention on International Trade in Endangered Species of Wild Fauna and Flora).

2. General characterization of sturgeons

The classification of sturgeons in accordance with the traditional systematic is the following: regnum *Animalia*, superphylum *Vertebrata*, subphylum *Gnatosthomata*, superclass *Pisces*, class *Osteichtyes* (bony fish), subclass *Actinopteryngi*, superorder *Chondrostei*, order *Acipenseriformes*. The majority of authors agree with the traditional conception that the order *Acipensiformes* consists of two families, *Acipenseridae* and *Polyodontidae*. The *Acipenseridae* family includes two subfamilies *Acipenserinae* and *Scaphirhynchinae* and four genera (*Acipenser, Huso, Scaphirhynchus* and *Pseudoscaphirhynchus*), while the *Polyodontidae* family has two genera (*Polyodon* and *Psephurus*), each with only one species (Rochard *et al.*, 1991; Bemis *et. al.*, 1997).

2.1 Morphological characteristics

The species of the *Acipenseriformes* order are characterized by their large or very large size. Sturgeons have an elongated body with a thicker foreside, large head, small eyes and fins oriented towards the posterior. In this group of fish the scales and lateral line are missing. The dorsal part of the body has a varied colouring, ranging from grey, brown, dark blue to close to black; the colouring becomes less intense on the ventral side. The majority of the species has a light colouring or even white on their ventral side.

Sturgeons possess certain relict characteristics, which set them apart from other fish and prove their extremely old origin. They have a cartilaginous endoskeleton, heterocercal caudal fin, low pectoral fins, and undifferentiated vertebrae, with ganoid scales only in the caudal part of their body. The rest of the body is either naked or covered in five longitudinal rows of bony plates (scutes). The scutes are located one dorsally, two laterally and two ventrally, among these rows small scutes are set irregularly. The skeleton is mostly cartilaginous, with partial ossification only in the case of the cranial arch and maxillary. The cranium is elongated and forms a conical, spatulated rostrum with four ventral sensory barbels used for capturing food, and the inferior mouth is toothless and has thick lips. The endocranium is massive, without an interorbital septum and with few endochondral ossifications. The palatoquadrate cartilages are united on the median line and they are not directly attached to the cranium. The premaxillary is sealed to the maxillaries, which in their turn are sealed to the palatoquadrates and they are not independently mobile. The dermal bones of the cranium are hard. The vertebral bodies are not differentiated; the first vertebrae are attached to the cranium. The mouth, which is situated ventrally, is protrusible. The teeth are reduced or lacking, the branchiostegal rays are also missing. The intestine has a spiral valve and it communicates with the air bladder. The gills are protected by opercula or gill covers, the pectoral fins extend laterally from the body in a generally horizontal orientation and the caudal fin has two asymmetrical lobes. The number of basal elements in the fins is smaller than that of the rays. The operculum consists of a subopercle and a rudimentary preopercle. Some of the characteristics of the *Acipenseriformes* are primitive, while others indicate degeneracy, while others yet are specialized.

2.2 Sturgeon reproduction

Sturgeons migrate for reproduction and feeding. In what regards their life cycle, two migration models have been described (Bemis & Kynard, 1997):

1. **diadromous:** migrating from sea water to fresh water, and in reverse.
   The majority of the *Acipenseriformes* are diadromous species and exhibit two types of migration behaviour:
a. **anadromous** – spend most of their life cycle in their feeding areas in the sea and migrate to fresh water areas for breeding. The two species of the *Huso* genus and the majority of the species of the *Acipenser* genus are anadromous.
b. **amphidromous** – in fresh water – the breeding phase of their life cycle occurs in fresh water, while the feeding and growth phases occur during the migration to sea;
2. **potamodromous:** the fish migrate within the river/lake (McDowall 1988, 1992).
   The moment of reproduction for the *Acipenseridae* varies widely. Sturgeons breed during any of the seasons depending on water temperature and water flow velocity. In addition, the distance covered by sturgeons in their migration from the feeding grounds in the sea to the breeding areas in the rivers varies and is correlated with the fact that the existence of adequate breeding areas is essential for the success of breeding. The sites preferred by females for spawning are areas with a hard substrate consisting in gravel and rocks, with numerous crevices and with moderate currents. The annual success of breeding is difficult to estimate; in fact there may be no success if the water flow velocity is too high, whether it is due to natural phenomena, or the controlled release of water from dams and reservoirs. Water currents which are too strong may totally compromise or significantly reduce the success of breeding.

An interesting trait of sturgeon reproduction is the fact that these fish repeatedly use the same breeding site. The constancy and "fidelity" towards the breeding areas have been emphasised in the sturgeon populations in the Danube. This constancy and "fidelity" have led to the hypothesis that there are several sturgeon subpopulations in this river. For now there is no direct proof to this effect. However, if it is demonstrated that the sturgeons possess this strong instinct to maintain their breeding territories, this may explain the existence of the various types (morphs) within the *Acipenseridae* species.

2.3 Sturgeons species in the Lower Danube

Measuring 2857 km in length, the Danube is the second longest river in Europe. The river is divided into three regions: the Upper Danube (from the river's source in the Black Forest in Germany to Vienna) – 890 km, Middle Danube (from Vienna to the Iron Gates I Dam) – 993 km and Lower Danube (from Iron Gates I to its flow into the Black Sea) – 942 km.

Currently three diadromous species of sturgeons may be found in the Lower Danube basin and in the Black Sea - *H. huso* (beluga sturgeon), *A. stellatus* (stellate sturgeon) and *A. gueldenstaedtii* (Russian sturgeon) and one potamodromous species – *A. ruthenus* (sterlet) (Bacalbaşa-Dobrovici, 1999). The other two species that used to be found in this area, *A. sturio* (common European sturgeon) and *A. nudiventris* (ship sturgeon), are considered extinct, although, by way of exception, ship sturgeons are still being caught in the Danube basin (Kynard *et al.*, 2002).

Huso huso (Linaeus, 1758) (Beluga Sturgeon)

Distribution: in the Black Sea, Azov Sea, Caspian Sea and the rivers tributary to them, sporadically in the Adriatic Sea. In Romania these fish may be found along the Black Sea coastline and in the Danube from the Iron Gates to its inflow (863 km).

Specific characteristics: large transversal mouth, protractile, in the shape of an inverted half-moon; the middle of the lower lip is crescent-shaped. The upper lip is continuous while the lower one has a large interruption. Two pairs of barbels reaching almost to the mouth; they are long, flat laterally, and with fringed ends. Normally the beluga sturgeon measures 200–250 cm in length and weighs 100–150 kg, however, it may grow up to 6 m in length and reach over 1000 kg in weight. The adults may range from 30 to 60 years of age; however the older individuals may exceed 100 years.

Characteristics of habitat and reproduction: For their reproduction cycle the beluga sturgeons migrate upstream, sometimes covering great distances from the inflow mouths. The beluga sturgeon has the longest upstream migration. As a consequence, embankments and damming have had a dramatic impact on the natural breeding of this species. Currently, in the Danube, the beluga sturgeons no longer go beyond the dams at the Iron Gates. Two biological forms have been described for the beluga sturgeon: the spring form and the autumn form. The breeding sites are situated in areas with gravel and crevices on the river bed, and with depths of 4–20 m. The males reach sexual maturity at 12–14 years and the females at 14–16 years, when they have a body length of approximately 2 m. After reproduction the males return to sea, where they remain at depths of up to 100 m. The migration of the spawners to the sea is slower and occurs in July-September.

In the Danube, the beluga sturgeon results in natural hybrids with the sterlet, ship sturgeon and Russian sturgeon. In comparison with the genitor species, the hybrids have the tendency to remain in the fresh water environment for longer periods of time. The artificial hybrid *H. huso* x *A. ruthenus* (bester) is reared as a species of economic interest.

The beluga sturgeon is a predatory fish. Its food depends on the season, the adults prefer to feed on other fishes, and juvenile specimens feed on benthic invertebrates.

Acipenser gueldenstaedtii (Brandt, 1833) (Russian Sturgeon)

Distribution: in the Black Sea and the Azov Sea, as well as the Caspian Sea and their tributaries. Apart from the migratory form, there is a fresh water Russian sturgeon, which does not migrate to sea and which lives in the Volga, Ural and the Danube. In Romania, the Russian sturgeon is present along the Black Sea coastline and in the Danube from the Iron Gates to the inflow mouth.

Specific characteristics: the Russian sturgeon has transversal mouth with the lower lip interrupted. Apart from the typical form (short rostrum and barbels insertion reaching almost to the tip of the rostrum rather than the mouth), in the Danube there have been specimens of Russian sturgeon with longer rostrum, in which case the barbels insertion tends to be further away from the tip of the rostrum. The Russian sturgeon has a wide morphologic variability (Antipa, 1909). Thus, Antipa described three rostrum morphs (*typica*, *longirostris* and *acutirostris*) and two varieties of "integument" (naked and with scutes). Other authors consider that these varieties are actually hybrids (Ene and Suciu, 1996). Normally, the Russian sturgeon measures 100–200 cm in length and weighs 20–30 kg; however, it may reach a maximum of 236 cm and 115 kg. The longevity mentioned by several authors is of 33 years (Otel, 2007).

Characteristics of habitat and reproduction: Depending on the migration model, it is assumed that the Russian sturgeon from the Danube has three different intraspecific biological forms: spring, autumn, as well as a sedentary form which does not migrate to sea (Otel, 2007). Depending on the season, the Russian sturgeons live in water with depths ranging from 2 to 100 m. In rivers they live in depths of 2–30 m and swim upstream for the breeding cycles, sometimes to great distances from the inflow mouths. The sexual maturity occurs at ages ranging from 11–13 years in males and 12–16 years in females, when they have reached a body length exceeding one metre. The Russian sturgeon does not breed in consecutive years, but at intervals ranging from 2 to 5 years. The Russian sturgeon give birth to natural hybrids with the sterlet (*Acipenser ruthenus*), stellate sturgeon (*Acipenser stellatus*), ship sturgeon (*Acipenser nudiventris*), beluga sturgeon (*Huso huso*) and common sturgeon (*Acipenser sturio*). Their food consists in various benthic invertebrates and the adults also feed on certain fish species.

Acipenser stellatus (Linnaeus, 1758) (Stellate Sturgeon)

Distribution: in the Black, Caspian and Azov Seas, from where they migrate upstream to their tributaries. The greatest density may be found in the northern area of the Caspian Sea. In the past it was considered to have spread along the entire length of the Danube River and its tributaries; however, currently it does not pass beyond the dam at the Iron Gates.

Specific characteristics: the stellate sturgeon is characterized by a slightly cleaved upper lip, and a lower lip with an interruption in the centre. The rostrum is elongated, narrow and dorsoventrally flat and its length amounts to 59–65% of the length of the body, which constitutes a particular characteristic, making this species easier to differentiate from the other species of the *Acipenser* genus (Shubina *et al.*, 1989). The body is elongated and fusiform. The mouth is transversal, of medium size. The stellate sturgeon reaches a maximum length of 210 cm and a maximum weight of 68 kg. In the Danube the average weight of the specimens caught is around 6–8 kg. The maximum age mentioned in the literature is of 35 years.

Characteristics of habitat and reproduction: the stellate sturgeon migrates to the sea for feeding and wintering and to the rivers for breeding. The greatest part of the stellate sturgeon's life is spent in the sea. The stellate sturgeon is a species typical for the sea coastal waters and the lower sectors of rivers; however, in contrast with the other sturgeon species, it may also be found in the upper layers of water (Shubina *et al.*, 1989). Adult specimens may be found in pelagic waters, at depths of 10–40 m, and juvenile specimens are to be found closer to the inflow mouths of rivers where there are better places for feeding. For breeding, the stellate sturgeon migrates upstream, sometimes covering great distances from the inflow mouths. The onset of sexual maturity in the Danube is reached at 7–10 years for males and 10–14 years for females, when they reach a body length of 90–120 cm (Manea 1980). The stellate sturgeon does not breed every year. In what regards the breeding migration, there have been mentions of specimens living in the Danube in spring, in the March-April period. There is a secondary period of migration which occurs from August to October, in which case the specimens wintering in the Danube and breed during the April-May period. The preferred sites for breeding are represented by deeper diggings, with depths of 8–10 m, on a rocky substrate, with gravel and ratchel mixed with shell fragments (Shubina *et al.*, 1989). Immediately after breeding the adult individuals return to sea, while the majority of the spawn remains in the river until the autumn. The stellate sturgeon interbreeds with the beluga sturgeon (*Huso huso*), ship sturgeon (*Acipenser nudiventris*), sterlet (*Acipenser ruthenus*) and Russian sturgeon (*Acipenser gueldenstaedti*). In the Black Sea, the spawn feeds on worms, molluscs and crustaceans, while the adults feed on fish and invertebrates (Manea, 1980).

Acipenser ruthenus (Linnaeus, 1758) (Sterlet)

Distribution: the sterlet is an Eurasian species and it is widespread in the rivers that inflow into the Caspian Sea, the Black Sea, the Azov Sea, the Baltic Sea, the White Sea, the Barents Sea and the Kara Sea. This species was introduced through aquaculture in Germany, France, Italy and Belgium. In the Danube the distribution of the species is fragmented.

Specific characteristics: the starlet has a continuous upper lip, and an interrupted lower lip. One of the differences which assist in the differentiation of the sterlet from the other species of the *Acipenser* genus is the fringed barbels (Sokolov and Vasiliev, 1989). The rostrum is elongated, triangular, pointed, slightly raised towards the tip. Antipa (1909) described two varieties of colouring: *alba* (the Latin term for white), with depigmented skin (in which case the eyes may be red) and *erythraea*, in a fire brick red shade. Apart from the typical sterlet shape, with the long and angular rostrum, Antipa (1909) describes a variety with a short rostrum (*Acipenser ruthenus var. brevirostris*). Some authors consider that these forms are in fact distinct genetic groups. Other authors deny the existence of the two forms considering that they are determined by the environment conditions (food) or that they may be the result of hybridization (Sokolov and Vasiliev, 1989). The sterlet may attain a maximum length of 100–125 cm and a maximum weight of 16 kg. The maximum age recorded was of 24 years.

Characteristics of habitat and reproduction: the sterlet is a rheophile fresh water species, preferring the deep areas of great rivers, with strong currents and hard substrate river beds. The sterlet only rarely passes from the Danube to the Black Sea and it is only sporadically found in the Danube Delta. The sterlet is a sedentary species and does not undertake long migrations. The onset of sexual maturity occurs between 3–7 years in males and 5-12 years in females, when the individuals have reached the body length of approximately 40–50 cm. The sterlet interbreeds with the ship sturgeon, beluga sturgeon, stellate sturgeon and most often with the Russian sturgeon. Its food consists of benthic organisms, almost exclusively crustaceans and insect larvae, only occasionally roe and fish spawn.

Hybrids: the interspecific hybridization in vertebrates represents a less frequent phenomenon due to the genetic incompatibility of the parents' genomes (Arnold, 1997) and, generally, these hybrids are not viable. In the case of fish and some amphibians, hybridization occurs more frequently than in the case of other groups of vertebrates. Due to certain characteristics, for instance, polyploidy, sturgeon species hybridize more easily than other fish species (Birstein *et al.*, 1997). It is a well-known fact that sturgeons hybridize in natural conditions, leading, sometimes, to fertile intergeneric or interspecific hybrids. The main cause of hybridization is the overlapping in time and space of the breeding sites. The hybrids are frequently found in the rivers. For instance, it is considered that the sturgeon hybrids from the Volga River represent around 0.02-3.1% of the total number of spawn. In the 1964-1981 period the *A. gueldenstaedtii* x *A. ruthenus* hybrid was the most common (51.3%) among all hybrids encountered in this river.

For an extended period of time, sturgeon hybrids were considered species or subspecies. As a rule, it is considered that the female hybrids from generation F1 obtained through interbreeding of genitor species with the same degree of ploidy are fertile; on the other hand, the hybrids resulted from the interbreeding of genitor species with different degrees of ploidy are sterile (Arefyev, 1997, 1998). Until now the identification of hybrids was carried out based on morphometric characteristics taking into account the fact that they inherit morphological traits characteristic of the species of both genitor species. A relatively high number of morphological descriptions confirm the existence in natural water basins of sturgeon interspecific hybrids (Birstein *et al.*, 1997). Nevertheless, the identification of a certain individual as a hybrid based on morphology is not sufficient and becomes truly complex when three species are involved, as in the case of the hybrid resulted from the interbreeding of the bester (*H. huso* x *A. ruthenus* hybrid) with *A. gueldenstaedtii*. Moreover, the identification of juvenile hybrids based only on morphometric data has been demonstrated to be much more difficult than the identification of adult hybrids. Under these circumstances, both genetic and morphological studies may offer the certain proof of classification as pure species or hybrid.

3. The situation of sturgeon populations in the Lower Danube

From ancient times, sturgeons have had a great economic relevance in the Danube region and they were an important part of the welfare of the local communities from the area. In the old Greek port of Histria the inhabitants were allowed to fish at the Danube inflow mouths and to export salted fish to Greece and Rome, without paying any taxes. Starting with the Middle Age up to the end of the eighteenth century, sturgeons represented an inexhaustible resource of this river (Giurescu 1964). The beluga sturgeon was caught in the entire Romanian sector of the river, as well as in the middle sector of the Danube, and upstream, up to Bavaria. In the period between the twelfth and the fifteenth centuries, sturgeons were exported from the Danube to Poland (Giurescu, 1964). Starting with the Middle Age more complex tools were used for sturgeon fishing in the Danube. Due to overfishing and poaching, starting with the nineteenth century a decline of the *Acipenseridae* populations in this river has been observed. In the twentieth century, the numbers of sturgeons caught in Romania has decreased dramatically; for instance, in 1994 only 11.5 tons were caught in comparison with approximately 200 tons/year in the 1960s (Bacalbaşa-Dobrovici, 1999). A relevant impact on these species – apart from overfishing – was relegated to the construction of hydroelectric dams on the Danube River.

After the fall of the communist regime, the next decade was characterized by intensive and uncontrolled sturgeon fishing in the Lower Danube, due to a lack of legislation which would regulate the situation of these species of fish (Năvodaru, 1999a; Năvodaru, 1999b; Suciu, 2008). In the period between 2002 and 2005 the sturgeon catches in Romania decreased from 37 tons in 2002, to 11 tons in 2005.

A more thorough analysis of the Danube sturgeon species has shown that *H. huso* is considered to be an extinct species in the Upper Danube, critically endangered in the middle sector of the Danube and vulnerable in the Lower Danube (Hensel & Holcick, 1997). In accordance with other authors this species is extinct in the upper and middle sectors and rare in the Lower Danube (Suciu, 2008). *A. gueldenstaedtii* is considered a critically endangered species in all the sectors of the river and *A. stellatus* is deemed an extinct species in the Upper Danube and the first half of the middle sector of the Danube. In the Lower Danube, the stellate sturgeon population has declined dramatically from a numeric perspective, and the Russian sturgeon populations are on the brink of extinction as a result of an „Allee" effect similar to the one that has led to the disappearance of the *A. sturio* species from the rivers in Western Europe (Suciu, 2008). The arguments brought to support this statement are the decline of catches, the discontinuous structure of the age categories, the lack of natural recruitment or low natural recruitment. In the case of sturgeon populations, the low density of the population is correlated with the negative increase and under these circumstances the populations is condemned to disappearance.

3.1 Impact factors for the sturgeon populations in the Lower Danube

The factors that have led to the dramatic reduction of the sturgeon populations are numerous. Among them, the most important are overfishing, the construction of hydroelectric dams, poaching, pollution and the destruction of breeding habitats.

Since these are species of fish with a very slow growth rate, the sturgeon populations have not been able to cope with this intensive exploitation. The effects of pollution coupled with the construction of dams that have hindered the migration during the breeding cycle have resulted in a considerable decrease in the number of sturgeon populations. Thus, in less than one hundred years these species of fish have been endangered despite their existence for millions of years.

The construction and calling into operation in 1972 of the Iron Gates I Hydroelectric Dam has had a devastating impact on all the sturgeon species that breed within the Danube ecosystem. This dam is situated at 862 km upstream from the inflow mouths of the river into the Black Sea and it hinders sturgeon migration towards the breeding sites located upstream from the dam. Several hundred years in the past, the Danube sturgeons would reach upstream to Bratislava, Budapest and Vienna. However, nowadays these fish have had to limit their habitat to sea waters and the lower sector of the river (Kiss, 1997).

The sturgeon species that was most affected was the beluga sturgeon. Due to the large sizes of the adults the beluga sturgeons have been unable to reach their breeding sites upstream, despite the existence of the "falling sluices" and "giants" on the dam meant to facilitate the access of fish to breeding sites. At the same time, the area where the dam and reservoir were built used to be an important habitat for spawning for species such as the Russian sturgeon or the beluga sturgeon, since this sector of the river bed is characterized by gravel, fast currents and it is rather deep.

The flooded plains of the Danube have significantly altered when the dams and embankments were built. In the past, the floodable area of the Danube represented 573000 hectares and the delta of the river amounted to 524000 hectares. The Danube Delta has also been embanked, but to a lower extent and the embankments have been halted after the fall of the communist regime in 1989 due to ecological reasons. Thus, the dams and embankments built before that time determined the modification and alteration of entire habitats and trophic networks of which the sturgeon species were an integral part.

A negative impact can also be relegated to the agricultural irrigations and the excavations for building materials from the river bed. The irrigation pumps lead to the destruction of larvae and juvenile fish.

The water pollution is another factor with a significant impact on all the species in the Danube ecosystem. The heavy metal and pesticide pollution is relatively high at the level of the Lower Danube (Oksiyuk et al., 1992) and it affects the biotope comprehensively (Pringle et al., 1993). The pollution level causes relevant problems in the Black Sea as well, where sturgeon populations unfold their trophic cycle. The studies undertaken in this region (Zaitsev, 1993) have shown that the pollution in the area ranges from tens to hundreds of times higher than the one in the Atlantic and Pacific Oceans or in the Mediterranean Sea. The high concentrations of toxins from oil and various industrial wastes modify the hormonal balance of fish, disturb the metabolism and lead to an increase in the number of hermaphrodite fish. Notwithstanding, the breaking up of the former Soviet Union, the main polluter in the Black Sea region, as well as the decrease of the industrialization in the area (including in Romania) and taking more exacting steps to protect the environment, have led to a lower degree of pollution.

Other causes that have resulted in a dramatic decrease in the sturgeon populations were overfishing and poaching. Due to certain characteristics such as large sizes and predictable migration routes, sturgeons are easy to catch. Sturgeon fishing used for meat and caviar represents a profitable business as a result of the expensive prices of these products.

In order to save the sturgeon species in the Danube basin, various steps have been taken aiming to monitor the natural populations, launching efficient programmes for the repopulation and the long term sustainable exploitation of existing reserves. Globally, CITES (Convention of International Trade in Endangered Species of Wild Fauna and Flora) has included the sturgeon species in Annex II in 1998. This initiative was meant to track the regulation of trade with products originating from these species in order to hinder overexploitation. In order for the sturgeon populations to remain in existence, the decision makers need to bear in mind the fact that the rate of exploitation should not exceed the rate of species regeneration. This ratio is carefully monitored by introducing fishing and exportation quotas. Romania is among the 167 signatories of a convention and together with other countries in the Black Sea region, under the sponsorship of CITES, has founded in 2001 BSSMAG (Black Sea Sturgeon Management Action Group). In 2003, specialists from Romania, Bulgaria, Serbia and Ukraine have established a regional strategy for the protection of these species, focusing on certain aspects such as: maintaining the genetic diversity within the populations, the recovery of the healthy population, able to breed, reconstructing the breeding habitats and developing fisheries to rear sturgeons for consumption as well as roe, leading to market saturation and the elimination of poaching and fishing in the natural environment.

The development of efficient programmes for the management of these extremely valuable natural resources involved various non-governmental organizations. IAD (International Association for Danube Research) adopted an action plan in 2006 for the conservation of *Acipenseridae* species in the Danube River basin („Action Plan for the Conservation of the Sturgeons (*Acipenseridae*) in the Danube River Basin"). In Romania, as a result of the critical situation of the populations, the Ministry of Agriculture, Forestry and Rural Development together with the Ministry of the Environment and Water Husbandry have issued regulations in 2006 meant to strictly forbid the fishing of all sturgeon species on Romanian territory for a period of 10 years. This Order allows the fishing of sturgeons only for aquaculture in order to encourage artificial breeding with the goal to restock the natural habitat with individuals resulted from the crossing of wild genitors.

3.2 European perspectives

Because of the great importance of the sturgeons from both an economical and a scientific point of view there is a growing global interest in studies involving these species. There are many international organizations (World Sturgeon Conservation Society, Wildlife Conservation Society, Natural Resources Defence Council) and research institutes actively involved in the study and conservation of these important natural resources.

In the last years, the release of artificially reproduced fish or the transfer of non-native specimens has become a common practice in conservation programmes intended to reduce the risk of extinction for sturgeon populations. The Elbe, the Weser, the Oder and other German rivers feeding the North Sea and the Baltic Sea once belonged to the habitat of these fish, but the sturgeon population severely declined until extinction. Currently in Germany there is a growing interest for the study of sturgeon species and restocking programs have been initiated in the Baltic Sea. Thus, the beluga sturgeon was introduced in estuaries situated to the east of Rostock city, in the Baltic Sea and more upstream. In these habitats other sturgeon species - *Acipenser gueldenstaedtii* – have also been introduced. Another major concern of biologists in this country is reintroduction of the *Acipenser oxyrhynchus* species in German rivers based on the consideration that the Atlantic sturgeon migrated to the Baltic about 1000 years ago, displacing the European variety, and adopting the river between Germany and Poland as its natural home. Other European countries such as Italy, Spain, Portugal and Hungary are also interested in evaluating the genetic diversity of the existing sturgeon populations and also in saving these populations from extinction. For example, in Italy initiatives have been undertaken to conserve the *Acipenser naccarii* (Adriatic sturgeon) population in the Po River. Once relatively abundant in the Po River and its estuary, the Adriatic sturgeon underwent a dramatic decline in numbers and became rare by the early twentieth century. In the attempt of saving this population, several measures have been taken such us restocking with artificially reproduced individuals. The restocking actions targeted not only the Po River, but also other rivers from Italy. The Standing Committee of the Bern Convention adopted an international action plan for the restoration of the European sturgeon (*Acipenser sturio*) in November 2007. Countries like France and Germany are directly involved in this international plan and they have already started restoration actions for this sturgeon species.

4. Molecular analyses

The correct identification of sturgeon species is very important for various reasons. Thus, the accurate identification of the species from which the various products, especially caviar, originate is required in order to avoid commercial frauds. The main types of caviar are the beluga, ossetra and sevruga, produced by *H. huso*, *A. gueldenstaedtii* and *A. stellatus*, respectively. These types of caviar differ in taste, price and market availability. As a consequence, there are frequent fraud attempts.

Another aim of the identification of sturgeon species is represented by the protection of these endangered species, this being the first step in the progress of the conservation programs. Taking into consideration the dramatic decline of wild sturgeon populations, that comes dangerously close to disappearance, the practice of repopulating rivers with specimens obtained from artificial reproduction and reared in aquaculture conditions has become increasingly frequent. This phenomenon may have a negative impact on native sturgeon species, by drastically reducing the genetic diversity. In addition, an accurate identification of the specimens to be released in the rivers is also required from a genetic point of view in order to avoid the repopulation with hybrids. It is a well-known fact that sturgeons easily hybridize both in their natural environment and in aquaculture conditions, giving birth to viable and sometimes fertile hybrids. In the case of the sturgeon species in the Danube, the reduction and overlapping of breeding territories have led to a considerably increased chance of interspecific hybridization. Currently, apart from sturgeon species from the Danube, local stugeon fisheries also rear exotic species such as *Acipenser baerii*, as well as their hybrids with *Acipenser gueldenstaedtii*. The majority of fisheries are located near the river or its tributaries; therefore there is an increased risk that these species escape to the Danube River and may threaten the genetic diversity of native species. Ludwig *et al.* (2009) have reported the existence in the Danube of the exotic species *A. baerii*, as well as that of its hybrids with the native species *A. ruthenus*.

The characterization of sturgeons, as a pure species or hybrid is difficult to achieve based exclusively on morphometric criteria, especially in the juvenile stage. Taking into account the previously mentioned aspects, it is evident the importance of the characterization of Danube sturgeon species both on morphological and meristic criteria, as well as from a genetic point of view.

The development of DNA markers has had a revolutionary impact on genetics in generally, and on fish genetics implicitly, as well as on its application in aquaculture. The application of molecular markers in fish has enabled the recording of some rapid progresses regarding the studies of genetic variability and inbreeding, determining the paternity, species identification, building genetic linkage maps for aquaculture species, by identifying the loci correlated to certain quantitative traits (QTL - Quantitave Trait Loci) for marker-assisted selection.

In order to perform some molecular analyses, we have collected biological samples from the following Danube sturgeons: *A. stellatus*, *A. gueldenstaedtii*, *A. ruthenus* and *H. huso*. For some of these analyses we have included control samples originating from the aquaculture species: *A. baerii*, *A. gueldenstaedtii* and the interspecific hybrid *A. baerii* X *A. gueldenstaedtii*.

The samples were represented by fin fragments and their collection did not require the death of the sampled individuals. This particular aspect is extremely important in the case of species which are on the brink of extinction and are included in conservation programs. The fin fragments were fixed in 96% ethanol before DNA isolation. The total DNA was isolated by using a specific extraction method with phenol-chloroform-isoamyl alcohol (Taggart et al., 1991) or the the Chelex extraction method (Estoup et al., 1996).

In order to identify the Danube sturgeon species there are various techniques that may be employed.

4.1 PCR-RFLP (Polymerase Chain Reaction - Restriction Fragment Length Polymorphism)

This technique is based on the amplification by PCR of a specific DNA region which comprises one or more polymorphic sites for various restriction enzymes. Along with the increase of the number of "universal" primers available in the literature, different DNA regions which are relatively conserved (for instance, mtDNA) can be targeted. Furthermore, PCR products can be digested with restriction enzymes and visualized by ethidium bromide staining due to the increased quantity of DNA resulted from the amplification reaction. Since the size difference between restriction fragments is usually noticeable, these fragments may be observed relatively easily by electrophoresis.

The literature on sturgeons mentions a relevant number of studies based on RFLP-type markers. These markers were initially used as a focus in two directions: in order to establish the genetic diversity of sturgeon populations with a distinct geographic distribution (Miracle & Campton, 1995; Szalanski et al., 2001; Waldman et al., 2002) and in order to establish the usefulness of these markers for recovering the phylogeny of the *Acipenseridae* class (Brown et al., 1996). In recent years PCR-RFLP was proposed as a molecular method for the identification of sturgeon species and implicitly for the determination of the origin of the caviar on the market. Wolf et al. (1999) identified various species-specific restriction profiles at the level of a 462bp fragment from the mitochondrial gene for cytochrome b in the cases of 10 species from the *Acipenser* and *Huso* genera. Ludwig *et al.* (2002) have extended the study to 22 sturgeon species, by investigating the restriction profiles resulting from the action of five enzymes on a 1121bp fragment representing the entire gene for cytochrome b. Advantages:

i.   The simplicity of the protocol (a single pair of primers is necessary).
ii.   The technique is applicable for the accurate identification of all Danube sturgeon species for breeding, propagation, cryopreservation in order to exclude the possibility of misidentification.

Disadvantages: the main inconvenient of RFLP is the fact that it is possible to identify a relatively low degree of polymorphism. In addition, it requires the knowledge of a particular sequence, which makes difficult to find to find new markers in the species for which the known molecular data is quite reduced. The identification of species through PCR-RFLP is efficient for other vertebrate species, including fish, but it has been proven to be indecisive in the case of *Acipenseridae*.

Thus, it is proven to be very difficult to identify specific substitutions with diagnostic role for species that are very close genetically and evolutionary, as is the case of the *A. gueldenstaedtii/A. persicus* species and of the three species of the *Scaphirhynchus* genus. On the other hand, the analysis of restriction fragments length polymorphisms of mitochondrial DNA has the disadvantage that mtDNA is inherited exclusively through the maternal lineage. This may lead to erroneous interpretations, as in the case of hybrid identification and/or of ancient introgression, since both phenomena may be found in the case of sturgeons and are reported in the literature (Arefyev, 1997; Ludwig et al., 2003; Tranah et al., 2004).

In order to identify the Danube sturgeon species, we have amplified by PCR a 462bp fragment on the level of the $RNA_t{}^{Glu}$/cytochome b mitochondrial region. The fragment was subsequently subjected to restriction endonuclease digestion (in accordance with the method proposed by Wolf *et al.*, 1999), thus obtaining species-specific restriction profiles.

In order to amplify the $RNA_t{}^{Glu}$/cyt b fragment we used a single pair of primers: F (AAAAACCACCGTTGTTATTCAACTA) (Brugner & Hubner, 1998) and R (GCCCCTCAGAATGATATTTGTCCTCA) (Kocher *et al.*, 1989). These primers are annealing to conserved regions of mtDNA and they allow the amplification of the fragment of interest for all analyzed sturgeon species.

In order to identify the sites with a role in the diagnosis and selection of enzymes that allow the interspecific discrimination, the fragments obtained were sequenced by the "dye terminator" method (using *Applied Biosystems* reagents and equipment) and they were aligned with Clustal X 2.0.9 (Larkin *et al.*, 2007) (Figure 1).

Thus, for the restriction reaction we selected the *SspI*, *RsaI* and *Tru9I* enzymes. The digestion of the 462bp fragment with the three restriction enzymes followed by the 3.5% agarose gel electrophoresis resulted in species-specific restriction profiles (Table 1). Fragments which are smaller than 50bp could not be visualized by gel electrophoresis.

The calculated RFLP profiles as a result of the sequence analysis were confirmed experimentally, given that after the digestion with the three enzymes we obtained the species-specific profiles presented in figures 2, 3, 4.

In the case of the hybrids analyzed, it was demonstrated that they present a nucleotide sequence identical with that of the maternal species, *A. baerii*. This proves that such methods are limited in their utility for the identification of pure sturgeon species. In the case of hybrids this method is limited to the identification of their maternal origin. Although

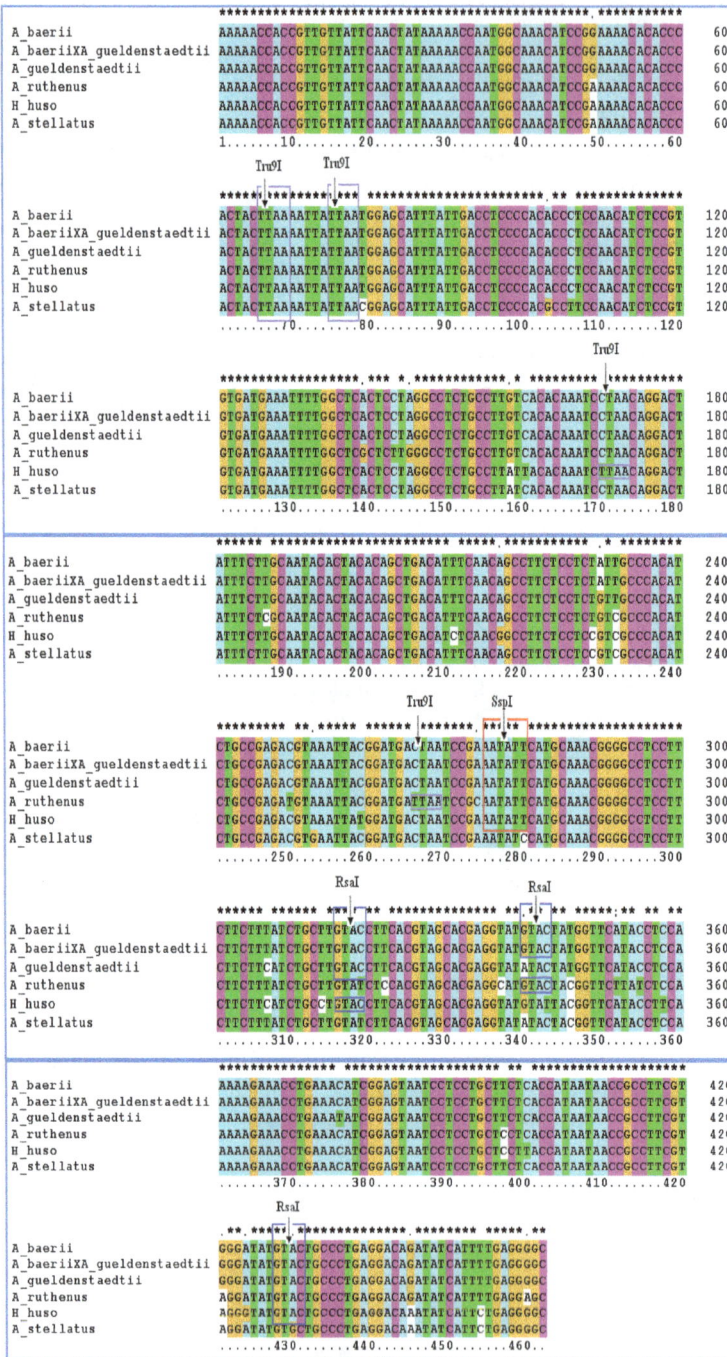

Fig. 1. Alignment of the RNAt^{Glu}/cytochrome b fragment in the analyzed sturgeon species.

mitochondrial markers are frequently used for identifying species, we must take into account the exclusively maternal inheritance of mtDNA, given that this characteristic may lead to erroneous interpretations in the identification of hybrids and/or introgression.

	H. huso	A. stellatus	A. ruthenus	A. gueldenstaedtii	A. baerii
SspI	277;185	462	277;185	277; 185	277;185
RsaI	317;112;33	462	341;88;33	317;112;33	317;88;33;24
Tru9I	292;95;66;9	387;66;9	196;191;66;9	387;66;9	387;66;9

Table 1. Restriction polymorphisms of the RNAtGlu/cytochrome b fragment in sturgeons (fragments length is represented in base pairs).

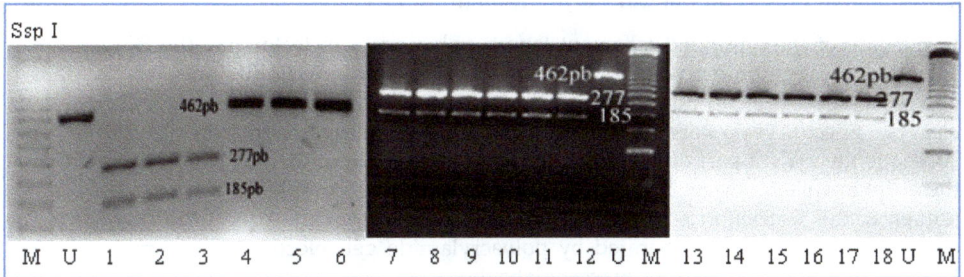

Fig. 2. The electrophoretic profile obtained after the digestion of the RNA$_t$Glu/cytochrome b fragment with *SspI*. M – molecular size ladder 50bp; U–uncut amplicon; 1,2,3–*H. huso*; 4, 5,6–*A. stellatus*; 7,8,9–*A. ruthenus*; 10,11,12–*A. gueldenstaedtii*; 13,14–*A. baeriiXA. gueldenstaedtii* (aquaculture); 15,16–*A. gueldenstaedtii* (aquaculture); 17,18-*A. baerii* (aquaculture).

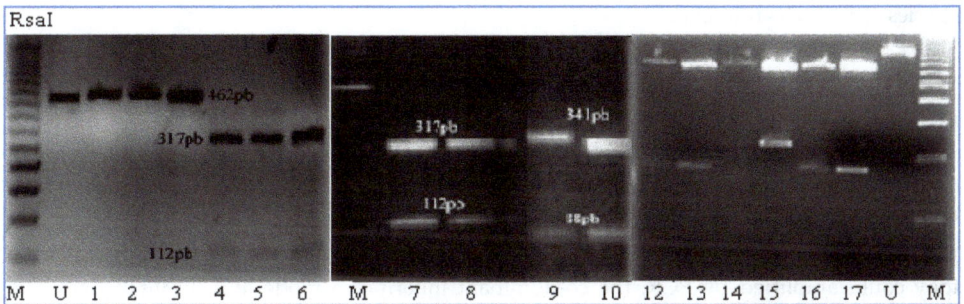

Fig. 3. The electrophoretic profile obtained after the digestion of the RNA$_t$Glu/cytochrome b fragment with *RsaI*. M - molecular size ladder 50bp; U - uncut amplicon; 1,2, 3–*A. stellatus*; 4,5,6 –*H. huso*; 7, 8 –*A. gueldenstaedtii*; 9, 10 –*A. ruthenus*; 11, 12 –*A. baerii* X *A. gueldenstaedtii* (aquaculture); 13, 14 –*A. gueldenstaedtii* (aquaculture); 15, 16 - *A. baerii* (aquaculture).

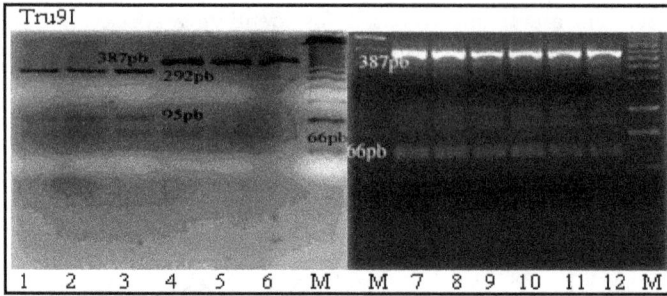

Fig. 4. The electrophoretic profile obtained after the digestion of the RNA$_t$Glu/cytochrome b fragment with *RsaI*. M - molecular size ladder 50bp; 1,2, 3–*A. stellatus*; 4,5,6–*H. huso*; 7, 8 –*A. gueldenstaedtii*; 9,10–*A. baerii*X*A. gueldenstaedtii* (aquaculture);11,12–*A. baerii* (aquaculture).

4.2 Amplified Fragment Length Polymorphism (AFLP)

In the case of the amplified fragment length polymorphism technique, the DNA chain is subjected to digestion by two restriction enzymes, for instance, *EcoRI* and *MseI*. *EcoRI* has a 6bp recognition site and rarely cleaves, and *MseI* has a 4bp recognition site and frequently cleaves. Specific adaptors are bound to the resulting restriction fragments and, subsequently, only a part of the restriction fragments are amplified by PCR. The selection of these fragments is done by using primers which are complementary to the adaptor sequence, the restriction site and several nucleotides from the restriction fragment. The amplified fragments are separated by polyacrylamide gels electrophoresis and visualized through autoradiography or fluorescent labeling.

The markers detected through the AFLP technique present a multiple band profile profile for each individual and in some cases researchers have observed that this profile is specific to one of the genders or a certain species. In the case of sturgeons some studies have been undertaken by Congiu *et al.* (2001, 2002) in order to identify the hybrids that appear in the natural environment or as a result of selective crossbreeding in aquaculture. Apparently, the profile of the bands allows the identification of hybrids when the profiles of each genitor species are used as reference.

4.3 Microsatellites analysis

Microsatellites (also called VNTR - Variable Number of Tandem Repeats) are short repetitive sequences, of 2-9bp, dispersed in the entire genome and they have a high degree of polymorphism. These markers are numerous in all vertebrate species, and in fish they appear at approximately every 10 kB (Wright, 1993).

Due to some of their characteristics, such us the relatively reduced size, easiness with which they could amplified by PCR, the co-dominant inheritance mode and the high degree of polymorphism, microsatellites are very useful markers. They may be used in numerous studies from varied fields of molecular biology, including molecular epidemiology, population genetics and genetic mapping. In the aquaculture and fish genetics studies, microsatellites are used for the genetic characterization of stocks, specimen selection for breeding, linkage map building, mapping and gene identification for QTL and various applications in the assisted breeding programs.

Microsatellite analysis results in individual genetic profiling (genetic fingerprinting). It may be used to establish the interrelations between various individuals and assess their allele frequency within the populations. Taking into account the molecular methods currently available, it is possible to assess length polymorphisms in a great number of individuals for intra- and inter-population genetic analyses. Certain microsatellites possess a high number of allele per locus and they are, therefore, appropriate for the identification of genitors, and their descendents, respectively, in mixed populations. Other microsatellites possess a lower number of alleles and are recommended in the genetic and phylogeny studies of populations (O'Connell & Wright, 1997; Estoup & Angers, 1998).

Currently, most laboratories analyzing microsatellites use the genotyping by PCR amplification with primers that bind to the sequences flanking the repetitive motifs. The resulting amplification products are separated depending on size by polyacrylamide gel electrophoresis or capillary electrophoresis in automated genetic analyzers.

In the case of sturgeons, the studies based on microsatellite markers analysis were initiated for the North American species. Currently, each of these species was studied from the point of view of genetic diversity, despite the difficulties in collecting the biological samples. These difficulties are reflected in the reduced number of analyzed specimens. Taking into account the fact that microsatellites contain important genetic information for the comparison of sturgeon populations, the research directions were especially aimed at the development of disomic microsatellites in the North American sturgeon species and paddlefish (May et al., 1997; McQuown et al., 2000; King et al., 2001; Pyatskowit et al., 2001; Heist et al., 2002; Henderson- Arzapalo & King, 2002; Welsh et al., 2003; Welsh & May, 2006). The identification and characterization of new microsatellite loci in sturgeons is difficult due to the polyploidy of these fish. Many potential microsatellites were eliminated from research studies since they were polysomic and they tended to complicate the interpretation of the inheritance mode and the intra- and inter- population genetic variation. Nevertheless, for the North American species, researchers have developed a set of disomic loci and microsatellites have begun to be used increasingly in population genetics studies. Researchers have identified various levels of genetic diversity, population structures and sturgeon stocks using different molecular markers in different species.

Another application of microsatellites in sturgeons was focusing on the use of these markers for species identification. Thus, these molecular markers with Mendelian inheritance represent a viable solution for the detection of hybridization. They also allow the characterization of individuals as pertaining to pure species or hybrids. Microsatellites can be very useful in the identification of sturgeon hybrids, by taking into consideration the fact that the alleles of an individual represent a combination of the parental alleles.

In the case of the species that spawn in the Danube, we tested various microsatellite loci. Finally, we selected eight of the microsatellite loci tested (LS19, LS34, LS39, LS54 - May et al., 1997; Aox27 – King et al., 2001; AoxD234 – Henderson – Arzapalo & King, 2002; AnacE4, AnacC11 – Forlani et al., 2008). These eight microsatellite loci have demonstrated a very good amplification, result repeatability and interspecific polymorphism in all of the four investigated species.

Allele frequencies were calculated and private alleles were identified (alleles that appear only in one of the analyzed species) given that they are considered diagnostic alleles. These

Locus	Alleles size (bp)	A. stellatus	H. huso	A. gueldenstaedtii	A. ruthenus
LS19	133	0.5238	0.0238	0.1538	0
	136	0.0952	0.0714	0.359	0.5
	139	0.1905	0.0952	0.0641	0.4375
	140	0	0	**0.0128**	0
	142	0.1667	0	0.1282	0.0625
	145	**0.0238**	0	0	0
	148	0	0.0714	0.1923	0
	151	0	**0.0476**	0	0
	154	0	**0.0952**	0	0
	157	0	0.119	0.0897	0
	160	0	**0.4286**	0	0
	163	0	**0.0476**	0	0
LS34	139	0	0	0	0.0313
	142	0	0.9762	1	0
	145	0	0	0	**0.9688**
	148	1	0.0238	0	0
Aox27	**114**	0	0	**0.4605**	0
	118	0	0	0	**1**
	122	0	**0.1905**	0	0
	126	0.0476	0.1429	0.0132	0
	130	0.0714	0.4048	0	0
	134	0.1905	0.2619	0.2895	0
	138	0.3333	0	0.2105	0
	142	**0.1429**	0	0	0
	146	**0.1905**	0	0	0
	150	0	0	**0.0263**	0
LS54	**140**	**0.0476**	0	0	0
	152	0	0	0	**0.9375**
	160	0.1905	0	0	0.0625
	164	**0.0952**	0	0	0
	172	**0.0476**	0	0	0
	176	**0.0952**	0	0	0
	184	**0.2143**	0	0	0
	188	**0.2619**	0	0	0
	192	**0.0476**	0	0	0
	196	0	0	**0.2308**	0
	204	0	0	**0.2051**	0
	208	0	0	**0.2308**	0
	212	0	0	**0.3077**	0
	215	0	0	**0.0128**	0
	220	0	**0.0952**	0	0
	232	0	**0.0714**	0	0
	236	0	**0.2619**	0	0
	240	0	**0.0714**	0	0
	244	0	0.1429	0.0128	0
	248	0	**0.0952**	0	0
	252	0	**0.2619**	0	0

LS39	**96**	1	0	0	0
	102	0	0	0	**0.0625**
	105	0	0.2143	0.7368	0.625
	108	0	0.6667	0.2632	0.3125
	111	0	**0.0714**	0	0
	114	0	**0.0476**	0	0
AnacE4	326	0	0	0.1795	0.4688
	328	0	0	0.4744	0.4375
	332	0	0.119	0	0.0938
	334	0	0.5952	0	0
	336	0	0.0238	0.0385	0
	338	0	0.0476	0.0769	0
	346	**0.1905**	0	0	0
	348	**0.2143**	0	0	0
	350	0.1905	0	0.0128	0
	352	0.1667	0	0.0128	0
	354	**0.1905**	0	0	0
	358	0.0476	0.0714	0	0
	360	0	0.1429	0.2051	0
AoxD234	**195**	0	**0.0238**	0	0
	199	0	**0.0238**	0	0
	203	0.1429	0	0	0
	207	0.0952	0.0476	0.2361	0
	211	0	**0.0238**	0	0
	215	0	0.2381	0.3056	0
	219	0	**0.0238**	0	0.1563
	223	0.1905	0	0	0
	227	0.1429	0.2381	0.1667	0
	231	0	0	**0.0278**	0
	235	0.2857	0	0	0
	239	0	**0.0476**	0	0
	243	0	**0.0714**	0	0
	247	0.1429	0.0952	0.0833	0
	251	0	0.1667	0.0833	0.0938
	255	0	0	0.0417	0.25
	259	0	0	0	0.2813
	271	0	0	0.0556	0.2188
AnacC11	**145**	0	**0.4048**	0	0
	153	0.125	0	0	0
	161	0	0	0	**0.25**
	165	0.1	0.1667	0.027	0
	169	0	0	0.1892	0.2188
	173	0.225	0.4286	0	0
	177	0.425	0	0.5676	0.2188
	185	0.1	0	0	0.25
	189	0	0	**0.2162**	0
	193	0.025	0	0	0
	201	0	0	0	**0.0625**

Table 2. Allele frequencies calculated with the Genetix software (Belkhir *et al.*, 2002) for the Danube sturgeon populations. The allele frequencies in bold indicate the potential diagnostic alleles.

diagnostic alleles are distributed in a wide frequency range, from 0.01 to 1 (fixed alleles for that particular population) (Table 2). Taking into account the co-dominant inheritance mode, these private alleles may be used as markers for the identification of hybrids. This method assumes the analysis of a significant number of specimens from the parental species in order to obtain the confidence of an accurate diagnostic.

4.4 Sequencing

The analysis of nucleotide sequences and of the existing degree of variation at their level allows the identification of sturgeon species, subspecies or populations. The sequences recorded in the various data bases such as GenBank or EMBL, for a significant number of markers from various species, are available for researchers for comparison and analysis. As mentioned earlier, sequencing also represents a validation method of the results obtained through various methods (for instance, RFLP or PCR amplification with species-specific primers). Sequencing constitutes the most precise diagnostic method but it also has an important disadvantage by being, for the moment, an expensive technique, which takes a lot of time and, as a consequence, is not suitable for analyzing numerous samples. Currently, due to the development of the Next Generation Sequencing Systems, this technique has become more accessible and it allows the analysis of an extended set of markers for a significant number of species and/or specimens.

It is preferable that, before the development of mitochondrial or nuclear markers for the identification of sturgeon species, to take into consideration the phylogenetic relations established at the level of this group. The establishment of phylogenetic relations in sturgeons is important in order to understand the genetics of these species with an ancient origin, since there are considered genuine "living fossils". Furthermore, taking into account the fact that many of the sturgeon species are on the brink of extinction, it is important to establish the phylogenetic relations in this group of fish because there is a risk that in the future such studies would be focused only on a reduced number of species / specimens from museum collections.

One interesting aspect to consider is the fact that the phylogenetic relations within the *Acipenseriformes* order, as established based on molecular data, present some differences comparing to those established as a result of morphological analyses.

For the phylogenetic classification of the sturgeon species belonging to the *Acipenseriformes* order, which live in the Black Sea and spawn in the Lower Danube, we have sequenced the cytochrome b (1141bp), cytochrome oxidase I (678bp) and cytochrome oxidase II (667bp) mitochondrial markers. In this phylogenetic analysis we have included, apart from our own sequences, 26 sequences already recorded in GenBank (25 from sturgeon species, and as outgroup we have selected the *Polypterus ornatipinnis* species, as a representative of the closest taxon to sturgeons – the *Polypteriformes* order). The phylogenetic tree was built by using the Neighbor-Joining (NJ) method implemented in the MEGA4 software (Tamura *et al.*, 2007), based on the Tamura-Nei plus gamma evolutionary model established with ModelTest (Posada & Crandall, 1998) (Figure 5).

The tree obtained shows a separation in the *Acipenseriformes* order into two monophyletic clades: *Polyodontidae* (represented by *Polyodon spathula* and *Psephurus gladius*) and *Acipenseridae*. The American Atlantic sturgeon (*A. oxyrhinchus*) and the common European

Atlantic sturgeon (*A. sturio*) form a monophyletic group and occupy a basal position in the *Acipenseridae* family. This conclusion confirms other molecular phylogeny studies (Krieger *et al.*, 2008), but it is in contradiction with morphometric data, which places in a basal position the group formed by the species of the *Scaphirhynchus* genus (Mayden & Kuhajda, 1996; Birstein & DeSalle, 1998).

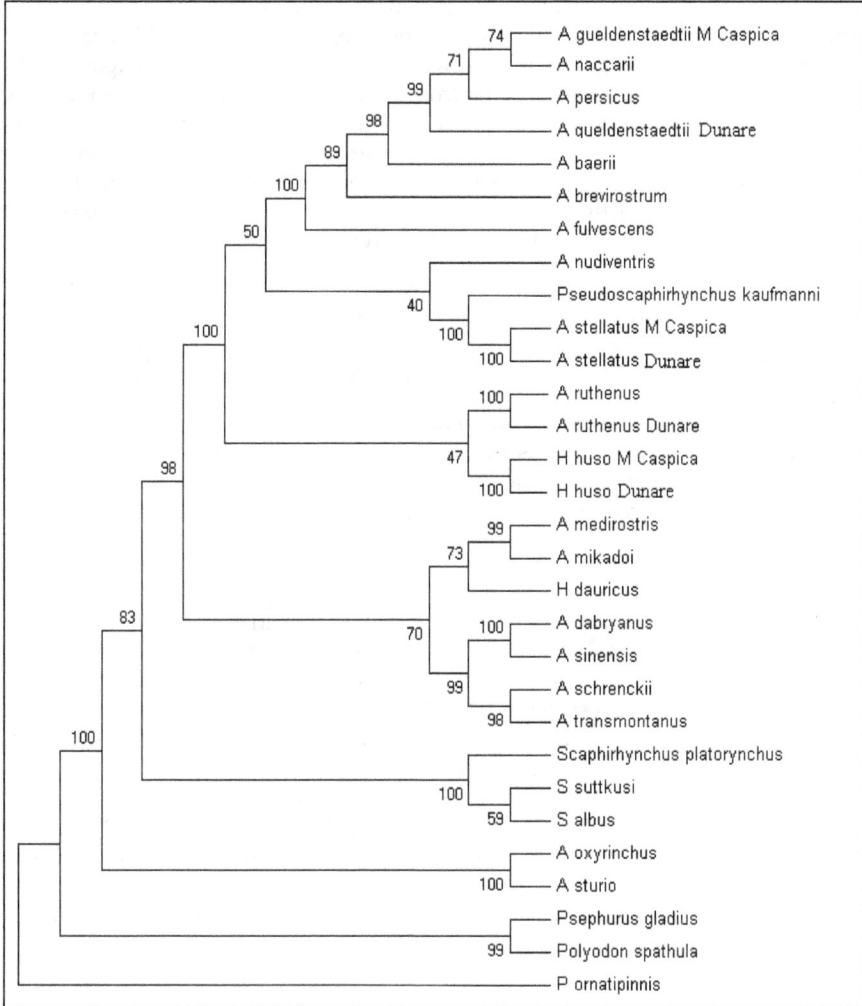

Fig. 5. Phylogenetic trees built through the NJ method based on the cytochrome b + cox I + cox II (2486bp) sequences.

Within the trees obtained, in the Acipenseridae family, two major clades were observed: the Atlantic clade (*A. ruthenus, A. nudiventris, A. stellatus, A. fulvescens, A. brevirostrum, A. baerii, A. persicus, A. gueldenstaedtii, A. naccarii* and *H. huso*) and the Pacific clade (*A. sinensis, A. dabryanus, A. medirostris, A. mikadoi, A. transmontanus, A. schrenckii* and *H. dauricus*). The

Pontic-Caspian species analyzed, among which we included the sturgeon species spawning in the Lower Danube, were distributed in the Atlantic clade.

The NJ analyses based on concatened mitochondrial sequences have shown that *H. huso* and *A. ruthenus* form a cluster, as directly related species. This classification is surprising, given the fact that *H. huso* and *A. ruthenus* present very different traits from a morphological and ecological point of view. The hypothesis that *H. huso* and *A. ruthenus* are closely related is supported by the easiness with which they hybridize, since their hybrids are viable and fertile (Birstein et al., 1997). The results we have obtained may be explained by an introgression of the mitochondrial genome from *H. huso* to *A. ruthenus* which took place in the remote past. It is possible that the analyzed specimens, although not hybrids, could have derived from a mitochondrial line of hybrid origins. Recently, the "bottleneck" phenomenon has been observed within the populations. This phenomenon encourages the fixation of a certain mitochondrial line or the introgression phenomenon (for instance a haplotype from *A. ruthenus* may become dominant in *H. huso* or the reverse may be true).

Under these circumstances, the interrelation between *H. huso* and *A. ruthenus* is artificial and it results from the history of the genes rather than from the history of the species. Similar results which place *H. huso* and *A. ruthenus* in a monophyletic group have also been obtained by Birstein & DeSalle (1998); Birstein et al., (2000) and Krieger et al., (2008).

The two species of the *Huso* genus are distributed among the species of the *Acipenser* genus, which means that the *Huso* genus is not monophyletic and it does not form a separate taxonomic unit. *A. stellatus* forms a monophyletic group with *P. kaufmanni*, and *A. gueldenstaedtii* with the Euroasian species *A. baerii*, *A. persicus* and *A. naccarii*.

4.5 DNA barcoding

"DNA barcoding" is a technique used in order to characterize species using short and standardized DNA sequences. The process involve the organization of a reference collection, obtaining DNA sequences and introducing them into a data base and including in a standardized analysis. The best candidates for the identification of animal species are mitochondrial genes which encode for proteins, these being preferred to those encoding for rRNA16S and rRNA12S. These genes present more interspecific variations and as a result they are able to differentiate even between closely related species. Moreover, the analysis of genes encoding for proteins is facilitated by the fact that, generally, they do not present mutations of the insertion and/or deletion type which occur frequently in the mitochondrial genes for rRNA. One region of 650bp from the 5' end of the gene for CO I was proposed as a potential "barcode" for species identification, taking into consideration the following advantages:

a. The easiness of amplification, using a reduced number of primers for different taxa;
b. The efficiency in discriminating among closely related species in a wide variety of invertebrates and vertebrates.

Once a data base containing the CO I sequences from various species was created and the nucleotides with a key role have been identified, the short species-specific oligonucleotides may be synthesized and placed on a DNA chip (similarly to a "microarray" analysis). In the case of sturgeons the implementation of such a technique may be extremely useful for the identification of species, especially in correlation with the accurate labeling of caviar origin.

5. Conclusions

The accurate identification of sturgeon species and hybrids is very important both from a commercial point of view as well as from the perspective of protecting these extremely valuable species. The decision to initiate repopulation programs in order to recover the natural sturgeon populations must take into account not only the social and economic aspects, but also the assessment of genetic diversity and the correct diagnostic of the species involved.

The diagnostic of species based on mitochondrial markers through PCR-RFLP or sequencing may be applied only in the case of pure species. Hybrids cannot be identified through the analyses of these markers, since they display restriction profiles and mitochondrial haplotypes similar to those of the maternal genitor species. This fact is not surprising due to the exclusively maternal inheritance trait of mtDNA. The analysis of mitochondrial markers is useful in determining the origin of the maternal species involved in the hybridization. For the identification of hybrids and of the genitor species involved in hybridization nuclear markers as microsatellites seems more appropriate. An accurate diagnostic of the sturgeon species and hybrids is very important if we take into consideration their fragile status and such a diagnostic can result only from both morphological and molecular data.

6. Aknowledgements

This work was supported by the National Authority for Scientific Research, Bilateral project Romania-France 484/2011 "Diversité génétique et hybridation des esturgeons de Roumanie et de France" and CNMP, grant PNII 52-167 "Evaluation of distribution and characterization of genetic diversity in salmonids from Romania.

7. References

Antipa, G. (1909). *Fauna ichtiologica a Romaniei*. Acad. Rom. Publ Fond. Adamachi (In Romanian), Bucharest.

Arefyev, V.A. (1997). Sturgeon hybrids: Natural reality and practical prospects. *Aquaculture Magazine*, 23, pp.53–58.

Arefyev, V.A. (1998). Sturgeon hybrids: Natural reality and practical prospects. *Aquaculture Magazine*, 24, pp. 44–50.

Arnold, M. L. (1997). *Natural Hybridization and Evolution*, Oxford University Press, ISBN: 0195099745, New York.

Artyukhin, E.N. (1995). On biogeography and relationships within the genus Acipenser. *The Sturgeon Quarterly*, 3(2), pp. 6-8.

Bacalbasa-Dobrovici, N. (1999). Endangered migratory sturgeons of the lower Danube River and its Delta. *Environmental Biology of Fishes*, 48 (1 – 4), pp. 201- 207, ISSN 0378-1909.

Belkhir, K.; Borsa, P.; Chikhi, L.; Raufaste, N. & Bonhomme, F. (2002). GENETIX 4.05, logiciel sous Windows TM pour la génétique des populations. Laboratoire Génome, Populations, Interactions, CNRS UMR 5171, Université de Montpellier II, Montpellier (France).

Bemis, W.E. & Kynard, B. (1997). Sturgeon rivers: An introduction to Acipenseriform biogeography and life history. *Environmental Biology of Fishes*, 48, pp.167-184, ISSN 0378-1909.

Bemis, W.E.; Finedis, E.K. & Grande, L. (1997). An overview of Acipenseriformes. *Environmental Biology of Fishes*, 48, pp. 25 -71, ISSN 0378-1909.

Billard, R. & Lecointre, G. (2001). Biology and conservation of sturgeon and paddlefish. *Reviews in Fish Biology and Fisheries*, 10, pp. 355-392, ISSN: 0960-3166.

Birstein, V. J. & DeSalle, R. (1998). Molecular phylogeny of Acipenserinae, *Molecular Phylogenetics and Evolution*, 9, pp. 141-155.

Birstein, V. J.; Doukakis, P. & DeSalle, R. (2000). Polyphyly of mtDNA lineages in the Russian sturgeon, Acipenser gueldenstaedtii: forensic and evolutionary implications. *Conservation Genetics*, 1, pp. 81–88, ISSN: 1566-0621.

Birstein, V.J. & Bemis, W.E. (1997). How many species are there within the genus Acipenser? *Environmental Biology of Fishes*, 48, pp. 157-163, ISSN 0378-1909.

Brown, J. R.; Beckenbach, K.; Beckenbach, A.T. & Smith, M. J. (1996). Length variation, heteroplasmy and sequence divergence in the mitochondrial DNA of four species of sturgeon (Acipenser), *Genetics*, 142, pp. 525–535, ISSN: 0016-6731.

Burgener, M & Hubner, P. (1998). Mitochondrial DNA enrichment for species identification and evolutionary analysis. *Zeitschrift fur Lebensmittelunntsersuchung und-Forschung*, 207, pp. 261-263.

Congiu, L.; Dupanloup, I. & Patarnello, T. (2001). Identification of interspecific hybrids by amplified fragment length polymorphism: the case of sturgeon. *Molecular Ecology*, 10, pp. 2355-2359, ISSN: 0962-1083.

Congiu, L.; Fontana, F.; Patarnello, T.; Rossi, R. & Zane. L. (2002). The use of AFLP in sturgeon identification, *Journal of Applied Ichthyology*, 18, pp. 286–289, ISSN: 0175-8659.

Ene, F.; Ene, C. & Suciu, R. (1996) Researches on the marine migratory sturgeons of the Danube River (II) Fin abnormalities of individuals in the region of the Danube River mouths. *Scientific Annals of the Danube Delta Institute*, IV, pp. 95 – 101, ISSN 1583-6932.

Estoup, A. & Angers, B. (1998). Microsatellites and minisatellites for molecular ecology: theoretical and empirical considerations. In: *Advances in molecular ecology*, Ed. Carvalho, G., pp. 55-86, IOS Press. (NATO ASI series), Amsterdam.

Estoup, A.; Largiader, C.R.; Perrot, E.&Chourrout, D. (1996) Rapid one tube DNA extraction for reliable PCR detection of fish polymorphic markers and transgenes, *Molecular Marine Biology and Biotechnology*, 5, pp. 295-298.

Forlani, A; Fontana, F. & Congiu, L. (2008). Isolation of microsatellite loci from the endemic and endangered Adriatic sturgeon (Acipenser naccarii), *Conservation Genetics*, 9, pp. 461–463, ISSN: 1566-0621.

Giurescu, C. C. (1964). *Istoria pescuitului si a pisciculturii*, Editura Academiei R.S.R (In Romanian), Bucharest.

Heist, E.J.; Nicholson, J.T.; Sipiorski & Keeney, D.B. (2002). Microsatellite markers for the paddlefish (Polyodon spathula), *Conservation Genetics*, 3, pp. 205-207, ISSN: 1566-0621.

Henderson-Arzapalo, A. & King, T.L. (2002). Novel microsatellite markers for Atlantic sturgeon (Acipenser oxyrinchus) population delineation and broodstock management. *Molecular Ecology Notes*, 2, pp. 437–439.

Hensel, K. & Holcick, J. (1997). Past and current status of sturgeons in the upper and middle Danube. *Environmental Biology of Fishes*, 48, pp.185–200.

King, T.L.; Lubinski, B.A & Spidle, A.P. (2001). Microsatellite DNA variation in Atlantic sturgeon (Acipenser oxyrinchus oxyrinchus) and cross-species amplification in the Acipenseridae, *Conservation Genetics*, 2, pp. 103-119, ISSN: 1566-0621.

Kiss, J.B. (1997). *Cartea Deltei*. Editura Fundatiei Aves (In Romanian), Odorheiul Secuiesc.

Kocher, T.D.; Thomas, W.K.; Meyer, A.; Edwards, S.V.; Paabo, S.; Villablanca, F.X. & Wilson, A.C. (1989). Dynamics of mitochondrial DNA evolution in animals: amplification and sequencing with conserved primers, *Proceedings of the National Academy of Sciences USA*, 86, pp. 6196-6200.

Krieger, J.; Hett , A. K.; Fuerst, P. A.; Artyukhin, E. & Ludwig, A. (2008). The molecular phylogeny of the order Acipenseriformes revisited, *Journal of Applied Ichthyology*, 24 (1), pp. 36–45.

Kynard, B.; Zhuang P.; Zhang T. & Zhang L. (2002): Ontogenetic behavior and migration of Volga River Russian sturgeon, Acipenser gueldenstaedtii, with a note on adaptive significance of body color. *Environmental Biology of Fishes*, 65, pp. 411–421, ISSN 0378-1909.

Larkin, M.A., Blackshields, G. & Brown, N.P. (2007) ClustalW and ClustalX version 2.0. Bioinformatics, 23, 2947–2948.

Ludwig, A.; Congiu, L.; Pitra, C. & Fickel, J. (2003). Nonconcordant evolutionary history of maternal and paternal lineages in Adriatic sturgeon, *Molecular Ecology*, 12, pp. 3253- 3264, ISSN: 0962-1083.

Ludwig, A.; Debus, L. & Jenneckens, I. (2002). A molecular approach for trading control of black caviar, *International Review of Hydrobiology*, 87, pp. 661-674, ISSN: 1434-2944.

Ludwig, A.; Lippold, S.; Debus, L. & Reinartz, R. (2009). First evidence of hybridization between endangered sterlets (Acipnser ruthenus) and exotic Siberian sturgeons (Acipenser baerii) in the Danube River, *Biological Invasions*, 11, pp. 753-760, ISSN: 1387-3547.

Manea, G.I. (1980). *Sturionii (Acipenseridae). Taxonomie, biologie, sturionicultura si amenajari sturionicole.* Ed. Ceres (in Romanian), Bucharest.

May, B.; Krueger, C. C. & Kincaid, H. L. (1997). Genetic variation at microsatellite loci in sturgeon: primer sequence homology in Acipenser and Scaphirhynchus, *Canadian Journal of Fisheries and Aquatic Sciences*, 54, pp. 1542- 1547, ISSN: 1205-7533.

Mayden, R.L. & Kuhajda, B.R. (1996). Systematics, taxonomy and conservation status of the endangered Alabama sturgeon, Scaphyrhinchus suttkusi, *Copeia*, 2, pp. 241-273.

McDowall, R. M. (1988). *Diadromy in fishes: migrations between freshwater and marine environments.* Croom Helm, ISBN-10: 0881921149, London.

McDowall, R.M. (1992). Diadromy: origins and definitions of terminology. *Copeia* , pp. 248–251, ISSN 0045-8511.

McQuown, E. C.; Sloss, B. L.; Sheehan, R. J.; Rodzen, J.; Tranah, G. & May, B. (2000): Microsatellite analysis of genetic variation in sturgeon: new primers sequences for Scaphirhynchus and Acipenser, *Transactions of the American Fisheries Society*, 129, pp. 1380–1388, ISSN: 0002-8487.

Miracle, A. & Campton, D.E. (1995). Tandem repeat sequence variation and length heteroplasmy in the mitochondrial DNA D-loop of the threatened Gulf of Mexico sturgeon, Acipenser oxyrhynchus detsoi, *Journal of Heredity*, 86, pp. 22-27, ISSN: 0022-1503.

Navodaru I., M. Staras & R. Banks (1999)[b]. Management of sturgeon stocks of the lower Danube River system. *Proceedings of* "The Delta`s: State-of art protection and management", Tulcea, Romania, 26-31 July 1999.

Năvodaru, I.; Constantinescu, A.& Munteanu, I. (1999)[a]. Reproducerea speciilor comerciale de pești de apă dulce în zona Deltei Dunării. *Scientific Annals of the Danube Delta Institute*, VII, pp. 159-164.

O'Connell, M. & Wright, J.M. (1997). Microsatellite DNA in fishes, *Reviews In Fish Biology And Fisheries*, 7, pp. 331–363, ISSN: 0960-3166.

Oksiyuk, O. P.; Zhuravlev, L. A.; Lyashenko, A. V.; Bashmakova, I. Kh.; Karpezo, Yu. I. & Ivanov, A. I. (1992). Water pollution of the Danube River in the Ukraine: general indices. *Gidrobiologicheskiy Zhurnal*, 28, pp. 3–11. (in Russian; English translation:

Otel, V. (2007). *Atlasul Peștilor Din Rezervația Biosferei Delta Dunării.* Editura de Informare Tehnologică Delta Dunării (In Romanian). ISBN978-973-881117-0-6, Tulcea.

Posada, D. & Crandall, K.A. (1998). Modeltest: testing the model of DNA substitution, *Bioinformatics* , 14 (9), pp. 817-818, ISSN: 1367-4803.

Pringle, C.; Vellidis, G.; Heliotis, F.; Bandacu, D. & Cristofor, S. (1993). Environmental problems of the Danube Delta. *Amer. Sci.*, 81, pp. 350–360.

Pyatskowit, J. D.; Krueger, C. C.; Kincaid, H. L. & May, B. (2001). Inheritance of microsatellite loci in the polyploid lake sturgeon (Acipenser fulvescens), *Genome*, 44, pp. 185–191, ISSN: 0831-2796.

Rochard, E.; Castelnaud, G. & Lepage M. (1991). Sturgeons (Pisces Acipenseridae); Threats And Prspects. *Journal of Fish Biology*, 37, pp. 123-132, ISSN 0022-1112.

Shubina, T.N; Popova, A.A.& Vasil'ev, V.V. (1989). Acipenser stellatus Pallas, 1771. In: *The Freshwater Fishes of Europe*, Holcík J. (ed)., pp. 395-443, Aula-Verlag, ISBN: 3891040407, Wiesbaden.

Sokolov, L.I. &Vasil'ev, V.P. (1989). Acipenser ruthenus Linnaeus, 1758. . In: *The Freshwater Fishes of Europe*, Holcík J. (ed)., pp 227 – 262, Aula-Verlag, ISBN: 3891040407, Wiesbaden.

Suciu, R. (2008). Sturgeons of the NW Black Sea and Lower Danube River countries. *Proceedings of* : „International Expert Workshop on CITES Non-Detriment Findings ", Cancun, Mexico, 17-22 November 2008.

Szalanski, A.L.; Bischof, R. & Holland, R. (2001). Mitochondrial DNA variation in pallid and shovelnose sturgeon. *Transactions Nebraskan Academy of Sciences* (Affil. Soc.), 26, pp. 19-21.

Taggart, J.B.; Hynes, R. A.; Prodöuhl, P. A. & Ferguson, A. (1992). A simplified protocol for routine total DNA isolation from salmonid fishes. Journal of Fish Biology, 40(6), pp. 963-965, ISSN: 0022-1112.

Tamura, K.; Dudley, J.; Nei, M. & Kumar, S. (2007). MEGA4: Molecular Evolutionary Genetics Analysis (MEGA) software version 4.0. *Molecular Biology and Evolution*, 24, pp.1596-1599, ISSN: 0737-4038.

Tranah, G.; Campton, D.E. & May, B. (2004). Genetic evidence for hybridization of pallid and shovelnose sturgeon, *Journal of Heredity*, 95, pp. 474-480, ISSN: 0022-1503.

Waldman, J.R.; Grunwald, C. & Stabile, J. (2002) Impacts of life history and biogeography on the genetic stock structure of Atlantic sturgeon Acipenser oxyrinchus oxyrinchus, Gulf sturgeon A-oxyrinchus desotoi, and shortnose sturgeon A-brevirostrum, *Journal of Applied Ichthyology* , 18, pp. 509-518, ISSN: 0175-8659.

Welsh, A. & May, B. (2006). Development and standardization of disomic microsatellite loci for lake sturgeon genetic studies. Journal of Applied Ichthyology, 22, pp. 337–344.

Welsh, A.B.; Blumberg, M. & May, B. (2003). Identification of microsatellite loci in lake sturgeon, Acipenser fulvescens, and their variability in green sturgeon, A. medirostris. *Molecular Ecological Notes*, 3, pp. 47–55.

Wolf, C.; Hübner, P. & Lüthy, J. (1999). Differentiation of sturgeon species by PCR-RFLP, *Food Research International*, 32, pp. 699-705, ISSN: 0963-9969.

Wright, J.M. (1993). DNA fingerprinting in fishes. In: *Biochemistry and Biology of Fishes, Vol. 2*, ed. Hochachka, P.W.& Mommsen, T., pp. 57-91, Elsevier, Amsterdam.

Zaitsev, Yu.P. (1993). Impact of eutrophication on the Black Sea fauna. In: *Studies and Reviews*, General Fisheries Council for the Mediterranean, 64, part 2, pp. 63–86.

14

Aquatic Introductions and Genetic Founder Effects: How do Parasites Compare to Hosts?

April M. H. Blakeslee and Amy E. Fowler
Long Island University, CW Post Campus
Smithsonian Environmental Research Center
USA

1. Introduction

Aquatic parasites have intrigued researchers over the past several decades due to their often unique and complex life cycles, which can require multiple hosts to progress from larval to adult reproductive stages (Shoop, 1998). Parasites are also integral in community and ecosystem functioning and have the potential to impact community structure through direct (e.g., affecting host growth, reproduction, and survivorship) or indirect (e.g., influencing host predation and/or competition) means (Lafferty & Morris, 1996; Torchin et al., 2002; Blakeslee et al., 2009). Recently, parasites have become recognized not only as interesting biological/model species, but also as useful indicator species and biological tools for resolving ecological questions. For example, parasites can be indicators of ecosystem health (Huspeni & Lafferty, 2004) or even utilized to more accurately resolve questions surrounding cryptic species invasions (Blakeslee et al., 2008) or biogeographic movements of hosts (Criscione et al., 2006). Even with these recent developments in aquatic parasite research, and although parasites are known to represent a fundamental component of aquatic systems worldwide (Kuris et al., 2008), genetic diversity patterns of aquatic parasites are much less understood than they are for free-living species. This is especially true for hosts/parasites with broad habitat ranges across bioregions and those introduced to new locations through anthropogenic transport. We believe these knowledge gaps exist for two major reasons: 1) parasites are less visible than free-living species and 2) parasites are logistically more challenging to study (i.e., often requiring destructive sampling, knowledge of parasite taxonomy, and parasite specific genetic markers). Even still, parasites have numerous interesting and important ecological, evolutionary, and conservation implications, including those related to their population genetics in introduced versus native regions. Aquatic parasites thereby represent an important, but overlooked, ecological group. In addition, aquatic invasions are on the rise in recent years (Carlton & Geller, 1993; Ruiz et al., 2000); yet the importance of parasites in those invasions (which have increased both in frequency and in distribution) is often less understood and/or tracked. Therefore, for this chapter, we focus on aquatic parasites, closely exploring how species introductions may affect genetic diversity patterns differently in parasites versus their free-living hosts.

Theoretically, anthropogenic introductions are believed to result in apparent founder effect signatures, whereby the introduced population(s) are subjected to an extreme genetic bottleneck, resulting in significantly lower genetic diversity in the introduced population(s) compared to native populations (Grosberg & Cunningham 2000). Moreover, in some systems, genetic bottlenecks have been correlated with detrimental fitness effects in individuals in the non-native population (Reed & Frankham 2003). However, a recent review by Roman & Darling (2007) found that many introduced populations retained high levels of genetic diversity (i.e., no genetic bottleneck or apparent founder effect signatures), which was counter to theoretical expectations for known introduced populations. The authors suggested that this 'genetic paradox' was likely due to the inherent complexity and individuality of each invasion pathway, which is strongly affected by the number of introduced individuals and the frequency of introduction events (i.e., propagule pressure), as well as invasion timing, effective population size, and the type of introduction vector (Voisin et al., 2005, Roman & Darling, 2007, Darling et al., 2008). Thus, population genetics signatures will be strongly influenced by each species' particular invasion pathway (Roman, 2006; Roman & Darling 2007; Geller et al., 2010).

Not included in the review by Roman & Darling (2007), however, were parasites which, as discussed above, are integral members of aquatic communities. Thus, it remains unclear how genetic diversity patterns across parasite groups could be affected by their hosts' individual invasion pathways. In particular, one important factor to consider is that only a subset of all hosts introduced to a new location will harbor parasites. In addition, parasite life cycles can be highly complex, often requiring multiple hosts (Shoop, 1988). As such, parasite founding populations will be small and subject to evolutionary forces that tend to reduce genetic variability in small populations (i.e., genetic drift). In addition, Allee effects may serve to further reduce variability in newly formed parasite populations if individuals cannot successfully mate and pass on their genes (Chang et al., 2011). In fact, a parasite's complex life cycle could serve to heighten Allee effects since successful reproduction in many aquatic parasite species requires transmission through multiple hosts, and at some stages in the life cycle, host species can be highly specific, if not obligate. This, in turn, may result in inherent challenges for a parasite's successful establishment in a new region (especially if gene flow is limited), which could potentially further reduce genetic diversity and influence its ability to find and infect appropriate hosts vital to its life cycle. In fact, introduced parasite species with complex life stages are expected to show divergent evolutionary patterns at local versus geographic scales (for example, native versus introduced ranges) due their distinct life histories as well as due to the mobility of their hosts (Jarne & Theron, 2003; Prugnolle et al., 2005). For example, highly mobile definitive hosts can play an important role in reducing the genetic variation within populations of parasites by disseminating eggs over long distances (Blouin et al., 1995; Kennedy, 1998; McCoy et al., 2003; Gittenberger et al., 2006; Louhi et al., 2010), potentially outside the area where the parasite's other hosts are present.

As a result of the important evolutionary and ecological distinctions of parasite life cycles compared to their free-living hosts (briefly summarized above), we were curious how genetic diversity patterns may differ between introduced parasites and hosts, and whether parasites would be more likely to exhibit theoretical genetic founder effect signatures than their hosts. Therefore, we ask the following questions: will aquatic parasites exhibit different genetic diversity signatures than their hosts in introduced versus native ranges? And more specifically, will they be more likely to conform to genetic bottlenecks and apparent founder effect signatures than their hosts? Here, we present both a review of the literature and a specific case study to explore these questions, and determine whether general patterns may be observed across systems.

2. Literature review

Below, we present a review of freshwater and marine studies that explores genetic diversity patterns in native and introduced populations in parasite-host systems. We searched for studies using multiple databases and the following keywords: "parasite" "genetic diversity" "introduction/invasion" "host" and "marine" OR "freshwater". Our goal was to compile data and information that could provide insight into our questions above, as well as a general understanding of what is currently known about source and founding population genetics of marine and freshwater hosts and parasites.

2.1 Freshwater systems

While numerous freshwater parasites have rapidly expanded their ranges due to introduced hosts (Taraschewski 2006), studies explicitly focused on the patterns of freshwater parasite and host genetic diversity in native and introduced populations are rare. Those studies that do exist appear to support reductions in genetic diversity in introduced parasite populations, possibly due to single introduction events of host species and subsequent genetic drift in isolated and small populations (e.g., as observed in the Japanese eel swim bladder nematode, *Anguillicola crassus*; Wielgloss et al., 2007, 2008). In some cases, phylogeographic patterns may be affected by mobile definitive hosts which prevent genetic isolation of freshwater parasite populations on a local scale, but at the global scale, parasite populations are genetically isolated and strong genetic differentiation exists among populations (as observed in a tapeworm (*Ligula intestinalis*) introduced to Australia, New Zealand, and North Africa (Bouzid et al., 2008)). However, such a scenario for isolation and genetic differentiation is not always apparent; for example, populations of the introduced eel parasite, *Gyrodactylus anguillae*, in three separate continents exhibit similar genetic structures and identities, hinting at the existence of multiple independently introduced populations from one source population (Hayward et al., 2001). This particular species is unusual, however, in that it has a direct life cycle, increasing their potential to establish new populations with few propagules (Hayward et al., 2001).

The introduction of exotic parasite species in freshwater systems are mainly known from fish species that have been traditionally used for human consumption (e.g., infections in commercial fish by the cestode, *Bothriocephalus acheilognathi* (Font, 1998); monogeneans of the fish genera, *Pseudodactylogyrus* and *Cichlidogyrus* and *Gyrodactylus*; the nematode *Anguillicola crassus* (Ashworth & Blanc, 1997); the tapeworm *Ligula intestinalis* (Bouzid et al., 2008); the leech *Myzobdella lugubris* (Font, 1998), and the heterophyid trematode *Centrocestus formosanus* (Martin, 1958)). However, relatively little is understood about parasite genetic population structure in these parasites compared to that of their hosts (but see Dybdahl & Lively 1996; Criscione & Blouin 2004; Stohler et al., 2004; Rauch et al., 2005; Keeney et al., 2007) and even less is known about how genetic diversity patterns are modified when introductions occur. Several studies have suggested the existence of strong genetic bottlenecks in introduced freshwater parasite populations (e.g. Weekes & Penlington, 1986; Dove, 2000; Tompkins & Poulin, 2006), but these remain to be empirically tested. However, a recent study has found no evidence for genetic bottlenecks during the Ponto-Caspian invasion of the amphipod crustacean *Dikerogammarus villosus* or its associated microparasites (Wattier et al., 2007). An even more complicated parasite system is that of the amphipod, *Crangonyx pseudogracilis*, which exhibits a reduction in post-invasion genetic diversity while its associated microparasites do not (Slothouber-Galbreath et al. 2010). Our review therefore suggests that much remains to be understood regarding post-invasion freshwater parasite systems, and there is great potential for comparable global studies in native and introduced freshwater populations that would help resolve questions regarding host versus parasite genetic diversity patterns. In particular, further understanding of how parasite "spillover" (i.e., when an introduced host transmits its parasites to susceptible native hosts) and parasite "spillback" (i.e., when introduced species are susceptible to endemic parasites in the invaded range) affect host versus parasite genetic diversity patterns would be highly informative, given the effects it would have on parasite reproduction and life cycle transmission, especially in small, isolated populations (Dieguez-Uribeonodo & Soderhall, 1993; Barton, 1997; Dunn & Dick, 1998; Rauque et al. 2003; Torchin et al., 2003; Prenter et al., 2004; Münderle et al., 2006; Kelly et al., 2009).

2.2 Marine systems

Like freshwater systems, studies empirically focused on the differences in genetic diversity of both parasites and their hosts in native versus introduced regions in marine environments remain scarce. Recent studies have focused on resolving the cryptic status of many introduced parasites. For example, Kruse & Hare (2007) utilized molecular techniques to test for genetic homogeneity between native (Gulf of Mexico) and introduced (western Atlantic coast) populations of *Loxothylacus panopaei*, a rhizocephalan parasite of mud crabs. Their results demonstrated a high rate of southward expansion of this introduced parasite species on a scale of tens of kilometers per generation, and they were able to more accurately pinpoint the location of the source population to the western Gulf of Mexico. Parasites have also been used to explain unresolved or cryptic statuses of their hosts in introduced populations to unravel invasion histories. For example, Burreson et al. (2000) resolved the identity of a parasite (*Haplosporidium nelsoni*) in introduced Californian populations of the oyster *Crassostrea gigas*, and were able to show that this parasite species was introduced into Californian and Atlantic waters with native *C. gigas* populations from Japan. Burreson et al. (2000) also demonstrated an example of host-switching in that *H. nelsonii* imported with *C. gigas* to the mid-Atlantic have successfully parasitized previously uninfected populations of native *Crassostrea virginica*. Finally, Blakeslee et al. (2008) explored host and parasite genetics to resolve the cryptogenic (=origin uncertain) status of a highly abundant marine intertidal snail, determining that host and parasite invaded the east coast of North America together.

Studies empirically comparing the genetic structure of native and introduced populations of parasite-host associations in marine systems are also rare, but those few studies that have studied this question provide some support for reductions in genetic diversity in introduced habitats, for both parasites and hosts. For example, Muira et al. (2006) compared the genetic structure of native and introduced populations of the Asian horn snail *Batillaria attramentaria* and its associated parasites and observed reductions in genetic diversity in introduced populations for the snail and one lineage of a cryptic trematode parasite (*C. batillariae*) with the other lineage showing no reductions. Blakeslee et al. (2008) also observed reductions in genetic diversity in the introduced range for a snail host (*Littorina littorea*) and its most common trematode parasite (*Cryptocotyle lingua*); however, the magnitude of these reductions was not explored. Another ecologically important invader, the European green crab (*Carcinus maenas*) has shown reductions in some of its globally invasive populations, but in others, diversity is not significantly reduced (Darling et al., 2008). At the present time, nothing is known about the genetic structure of its most common trematode parasite, *Microphallus similis*, which infects *C. maenas* in both its native and introduced ranges (Torchin et al., 2001; Blakeslee et al., 2009). It would be interesting to see how genetic diversity in this trematode (which uses the crab as its second-intermediate host) compares to its crab host, considering its moderately high prevalence across native and introduced populations. *Microphallus similis* also uses native snails, *Littorina obtusata* and *L. saxatilis*, as first-intermediate hosts; thus in eastern North America, it is completing its life cycle through native (*L. obtusata*, *L. saxatilis*) and non-native (*C. maenas*) hosts, which may potentially affect its genetic diversity patterns in the introduced region.

2.3 Literature review discussion

Our review of the freshwater and marine literature suggests extensive knowledge gaps in studies of host and parasite genetic diversity in introduced populations. In many cases, hosts and parasites have been explored independently and comparisons between them are not always made, thus it is difficult to determine general patterns from the literature review with the present paucity of appropriate studies. However, those studies that do compare host and parasite genetic diversity in introduced and native ranges tend to suggest reductions in both hosts and their parasites in introduced populations; however, the extent of these reductions is not always clear. In some cases, hosts showed little to no reductions in genetic diversity (as was observed in numerous free-living examples in Roman & Darling (2007)), while in other cases the reductions were more extensive. Many of the parasite examples we found suggested extensive reductions in diversity; however, in a few cases they did not. Our review also found studies from numerous parasite groups, but no trends according to group were apparent, except perhaps for the note regarding direct life cycles versus complex life cycles – the former transmitting through one host versus the latter through multiple hosts. It might be expected that hosts with more complex life cycles could show greater reductions in genetic diversity in introduced populations due to inherent difficulties in completing life cycles; e.g., appropriate hosts may be in low abundance or lacking. Further insight into patterns across parasite groups will require more extensive research of introduced populations (the literature for which is presently sparse), focusing on introduced versus native populations of hosts and parasites, in order to understand whether parasite groups in general are more likely to exhibit theoretical signatures of genetic founder effects than their hosts, or whether certain types of parasites (e.g., multi-host) could be more likely to exhibit significant reductions.

3. Case study

Here, we present a case study to provide further insight into understanding whether founder effect signatures would be more apparent in introduced aquatic parasites than their hosts. Our case study focuses extensively on a prominent North American intertidal species, *Ilyanassa obsoleta*, and its trematode parasites (see Figure 1 for a typical trematode life cycle) in native and introduced regions, as well as comparisons with two *Littorina* sp. snails and parasites in native and introduced populations.

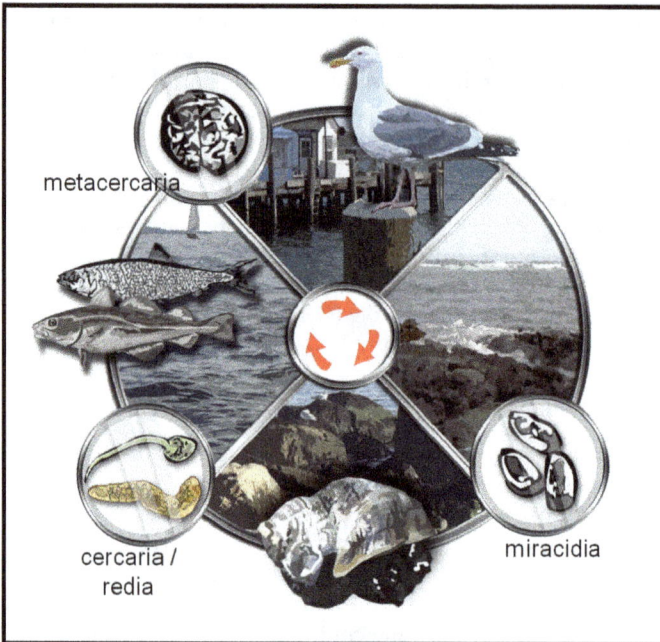

Fig. 1. A typical three-host infection cycle for a trematode parasite. Trematodes asexually reproduce in their first-intermediate snail hosts; then seek out and encyst within a second-intermediate host (e.g., fish); and sexually reproduce within their definitive host (e.g., seabird). This figure represents a typical life cycle of a prevalent trematode species, *Cryptocotyle lingua*, which infects *Littorina littorea* (common periwinkle snail) as its first-intermediate snail host.

3.1 Background information

Ilyanassa obsoleta is native to the east coast of North America, and its range extends from the Gulf of St Lawrence, Canada to northern Florida, USA (Bousfield, 1960; Abbott, 1974). The snail inhabits soft sediment, estuarine, and marine environments and often reaches extremely high abundances in both native and introduced populations (Scheltema, 1961; Brown, 1969; Curtis & Hurd, 1981; Blakeslee et al., 2011; A.M.H.B., pers. obs.). *Ilyanassa obsoleta* was accidentally introduced to the North American west coast in the early 1900's from the mid-Atlantic region of the east coast with commercial shipments of the eastern oyster, *Crassostrea virginica*, for aquacultural purposes (Carlton, 1992). While the intentional introduction of the eastern oyster failed, numerous organisms associated with the oyster successfully established in San Francisco Bay and other areas along the west coast (Carlton, 1979; Carlton, 1992; Miller, 2000), including *I. obsoleta* and several of its parasites (Blakeslee et al., 2011). Presently, *I. obsoleta* is found in three major populations on the west coast: San Francisco Bay (SFB), California, USA (first noted in 1907); Willapa Bay (WB), Washington, USA (first noted in 1945); and Boundary Bay (BB), Washington, USA and British Columbia, Canada (first noted in 1952) (Demond, 1952; Carlton, 1992). Nine trematode parasites infect *I. obsoleta* in its native east coast range (Curtis, 1997), and recent work (Blakeslee et al., 2011) has discovered a total of five trematodes infecting the snail in its introduced populations on the west coast (all five in San Francisco Bay; three in Willapa Bay; and two in Boundary Bay). This recent work also found a significant reduction in trematode parasite diversity and abundance in numerous populations on the west coast compared to native east coast populations. This observation of *parasite escape* is a signature that has been noted in numerous introductions worldwide (Torchin et al., 2003), and is another consequence of species invasions that could potentially abet introduced hosts in becoming highly successful in their new habitats (Keane & Crawley, 2002; Torchin & Mitchell, 2004; Liu & Stiling, 2006). The resulting reduction of parasite species richness and abundance as a result of parasite escape may affect both host and parasite population genetics, but as of yet, such evolutionary effects of the host-parasite invasion have not been investigated. Here, we present new, unpublished genetic data for the snail and four of its trematode parasites in native and introduced regions to understand the effects of the invasion on genetic diversity in the snail and its trematode parasites.

Another prominent marine intertidal snail on the North American east coast is *Littorina littorea* (common periwinkle), which is found from Labrador to Delaware Bay (Steneck & Carlton, 2001), its introduced region. The snail's native range is in Europe, where it inhabits coastlines from the White Sea to Portugal (Reid, 1996). Recent empirical work has suggested the snail was accidentally introduced to the North American east coast in the 1800s when vessel traffic between the British Isles (its purported source area) and North America was high. In fact, a study by Brawley et al. (2010) found congruence between *shipping records* from the British Isles to the Canadian harbor, Pictou, where the snail was first noted in the mid-1800s (Steneck & Carlton, 2001), and *genetic data* that also pinpointed the snail's likely origin of introduction to the British Isles. Rock ballast, prominently used at the time, was suggested as the likely introduction vector for the snail's North American invasion (Brawley et al., 2011). Furthermore, *Littorina littorea* also demonstrated characteristic signatures of parasite escape, in that it showed a significant reduction in parasite diversity in North America versus Europe (Blakeslee & Byers, 2008) when compared to two of its congeners, *Littorina saxatilis* (rough periwinkle) and *Littorina obtusata* (smooth periwinkle), both of which are native throughout the North Atlantic. Moreover, Blakeslee et al. (2008) discovered congruent patterns of genetic diversity reductions in introduced versus native regions for *L. littorea*, as well as its most common trematode parasite, *Cryptocotyle lingua* (Figure 1), finding both the snail and its associated parasite to show a genetic bottleneck in eastern North America and also suggesting a joint introduction of the snail and its parasite to North America. However, the study did not explore the comparative magnitude of these reductions and whether the reduction was more profound for the parasite than the snail. Therefore, we use the data from this study to compare genetic diversity in *L. littorea* and *Cr. lingua*, and additionally include (previously unpublished) diversity data for another common trematode species that infects the snail in its native and non-native ranges, *Cercaria parvicaudata*.

The third snail species we include in our case study is the rough periwinkle, *Littorina saxatilis*, which has a cosmopolitan native range across the North Atlantic (including populations throughout Europe and eastern North America) and an introduced population on the US West Coast in San Francisco Bay. The snail was first noted in SFB in 1993, and it is believed to have been introduced through the live trade vector, specifically baitworms (blood and sand worms) and live lobsters (*Homarus americanus*) from Maine and other areas of New England, which are packed in intertidal seaweed and shipped to locations around the globe (Carlton & Cohen, 1998). Since 1993, the snail has spread rapidly throughout the Bay and is highly abundant in numerous populations (Blakeslee et al., 2011). The snail is infected by fourteen trematode parasites in its native East Coast range (Blakeslee & Byers, 2008) but only three (from a single population) are found in SFB (Blakeslee et al., 2011), again representing a significant reduction in parasite diversity and abundance in the introduced range. While genetic data exists for the snail's native and introduced populations (Brown, 2007; Brown, Geller, Blakeslee, unpublished), there is no available genetic data for its introduced parasites because of the extremely low abundance and richness of trematodes infecting *L. saxatilis* in SFB (<0.5% throughout the Bay; Blakeslee et al., 2011). However, because the snail is a prominent host to trematode parasites in its native range, and because extensive genetic data exists for the snail in both its native and introduced regions, we include *L. saxatilis* here to complement the data we have for the other two snails in order to provide a better understanding of genetic diversity patterns in general for native versus introduced first-intermediate gastropod hosts. Continued sampling of the snail in its introduced range may reveal sufficient parasite data in the future (especially because the snail's introduction vector remains active), which could further support the results we present here based on three snail hosts and six trematode parasites.

3.2 Methodology

Included in our comparative study are three first-intermediate snail hosts, *Ilyanassa obsoleta*, *Littorina littorea*, and *Littorina saxatilis*, and six trematode parasites from native and introduced regions. The genetic data used in this case study are from previously published and unpublished data, and all are from mitochondrial markers (cytochrome b and cytochrome oxidase I genes). The information for genes, primers, sample sizes, and published studies can be found in Table 1.

We used the genetic data to obtain haplotype (=genetic) diversity values in both native and introduced regions for each snail and parasite individually, and then collectively. We also focused on *I. obsoleta* and its four parasites in a more extensive exploration of haplotype identities, frequencies, and connections across populations and within subregions and regions. For the latter, we obtained frequency data for each individual population and then also combined populations into larger subregions, which included: "North" – those populations found in Maine, New Hampshire, and Massachusetts; "Long Island Sound (LIS)" – those populations located along Long Island Sound; "DELMARVA" – those populations located along the Delaware, Maryland, Virginia (DELMARVA) peninsula; "South" – those populations from North Carolina, South Carolina, and Georgia; "BB" – those populations found in Boundary Bay, British Columbia; "WB" – those populations found in Willapa Bay, Washington; and "SFB" – those populations found in San Francisco Bay, California. The LIS and DELMARVA subregions are both known to be areas where oysters were collected for shipments west (Miller, 2000) and thus are likely source areas for *I. obsoleta's* and its parasites' introductions to the North American west coast (Blakeslee et al., 2011).

Because sampling effort was not equal across populations and regions, we also employed rarefaction techniques to find expected total haplotype richness for each species. This was especially important for the trematode parasites, where sampling was impacted by locating

Species	Mitochondrial Gene(s)	Total base-pairs	Primers	# Samples Native region	# Samples Introduced region	# Sites Native region	# Sites Introduced region	Published study for data
Snails								
Ilyanassa obsoleta	COI	546	1) TCGTGCTGAACTTGGACAAC 2) CCCCAGCTAATACAGGCAAA	250	173	15	11	previously unpublished
Littorina littorea	Cytb & COI	1197	1) CCTTCCCGCACCTTCAAATC 2) ATGAGAAATTTTCAGGGTC 3) CTCTCCTGGGAGATGACCAG 4) TTCTGGGTGACCGAAGAATC	187	183	22	29	Blakeslee et al. (2008)
Littorina saxatilis	COI	757	1) GGGGAGGAGACCCTATTCT 2) GCTCCTGTTTCAGGTGCATT	322	326	23	12	Brown (2007); Geller et al. (in prep)
ALL SNAILS -- TOTAL				759	682			
Trematode Parasites								
Austrobilharzia variglandis (IO)	COI	571	1) CGCCTCTGTCGTTGTTGAA 2) AAACCCAACACTACCACAA	11	19	3	4	previously unpublished
Himasthla quissetensis (IO)	COI	522	1) CTGCGTCGGTTTGTTTAGGT 2) TCCCAAACACACAATAGCC	46	30	7	3	previously unpublished
Lepocreadium setiferoides (IO)	COI	514	1) CCCCCTTGTCGAGTGGGGAT 2) TGCAGTATGCACATCCAAACCCACC	62	8	11	4	previously unpublished
Zoogonus rubellus (IO)	COI	535	1) CCGCCTTTATCTTCTGTGGA 2) TATGCACATCCAAACCAACC	109	9	12	4	previously unpublished
Cryptocotyle lingua (LL)	COI	450	1) TTTTTGGGCATCCTGAGGTTTAT 2) TAAAGAAAGAACATAATGAAAATG	98	98	16	20	Blakeslee et al. (2008)
Cercaria parvicaudata (LL)	COI	450	1) TTTTTGGGCATCCTGAGGTTTAT 2) TAAAGAAAGAACATAATGAAAATG	36	29	9	9	previously unpublished
ALL TREMATODES -- TOTAL				362	193			

Table 1. Genetic data used in the case study of three snail hosts (*Ilyanassa obsoleta, Littorina littorea,* and *Littorina saxatilis*) and six trematode parasites (*Austrobilharzia variglandis, Himasthla quissetensis, Lepocreadium setiferoides, Zoogonus rubellus, Cryptocotyle lingua,* and *Cercaria parvicaudata*). First-intermediate hosts of the six trematode parasites are identified by 'IO' for those infecting *Ilyanassa obsoleta* and 'LL' for those infecting *Littorina littorea*. The second column describes the mitochondrial gene used for each species (COI = cytochrome oxidase I; cytb = cytochrome b). The third column lists the total number of base-pairs for each mitochondrial gene fragment. The fourth column lists the forward and reverse primers used in PCR reactions. Columns five through eight describe the number of sequences sampled in native and introduced regions, as well as the number of sites included in each region. The last column lists the published study for the genetic data or whether the data are 'previously unpublished'.

infected snails among all snails (which can only be determined through destructive sampling). Thus, species-specific prevalence can be very heterogeneous (see Blakeslee et al., 2011). As a result, particular trematode species may have been sufficiently prevalent, in low abundance, or completely absent from a site, making it challenging to control for sampling abundance at each site, especially in introduced populations where prevalence was already lower than in native populations (Blakeslee et al., 2011). However, the rarefaction techniques allowed us to obtain *estimates* of expected haplotype richness at each site for native and introduced regions. We could then analyze data using both observed and expected values. We performed these rarefaction analyses using ESTIMATES 8.2 (Colwell, 2009) and selected the non-parametric estimator, Jack2, for calculating expected richness values because this estimator has been shown to perform well in numerous richness studies in terms of bias, precision, and accuracy (e.g., Canning-Clode et al., 2008).

We analyzed snails and trematode parasites in terms of proportional genetic diversity for native and introduced regions. We also explored haplotype richness for each individual species, as well as collectively ('snails' versus 'trematodes'), to determine whether general differences existed among hosts and their parasites. Finally, we used a single-factor ANOVA to look for significant differences between regions for hosts and parasites, using both observed and expected values (based on rarefaction analysis described above).

3.3 Results and discussion

Overall, we believe our study is suggestive that founder effect signatures are more apparent in the introduced parasites we featured here than their hosts. This is because we found trends for greater reductions in genetic diversity among introduced parasites at both the individual level and collectively than their snail hosts, which we describe in detail below.

Focusing first on our exploration of *Ilyanassa obsoleta* and its four parasites, we found genetic diversity to be reduced in the introduced region for both the snail and its parasites, but this reduction was much greater among the parasites than for the snail. Figure 2 explores haplotype frequencies as a series of pie charts at a biogeographic scale: within populations (smaller pies) and within subregions (larger pies). *Ilyanassa obsoleta* displayed much greater levels of diversity in terms of both shared (colored pie slices) and unshared (gray pie slices) haplotypes in its introduced region than did its trematode parasites. On average ($\pm$SE), native east coast *I. obsoleta* populations had about 17 ($\pm$0.6) haplotypes per site compared to 16 ($\pm$0.2) haplotypes per site in the introduced west coast. In contrast, the snail's trematodes had on average about 17 ($\pm$4.4) haplotypes per site from east coast populations versus only 8 ($\pm$3.5) haplotypes per site from introduced populations. At the subregional level, east coast *I. obsoleta* had on average about 63 ($\pm$6.1) haplotypes per subregion compared to 58 ($\pm$25) haplotypes per subregion on the west coast. In contrast, east coast trematodes had on average 57 ($\pm$12) haplotypes per subregion compared to 22 ($\pm$12) haplotypes per subregion on the west coast. Altogether, these analyses suggest that introduced populations of *I. obsoleta* trematodes tended to have about one-half to one-third the haplotype diversity of native populations/subregions, while the differences for the snail in native versus introduced populations and subregions were nearly nonexistent.

This trend for substantial differences in comparative genetic diversity between *I. obsoleta* and its parasites was also found in our exploration of diversity at the regional level. While *I.*

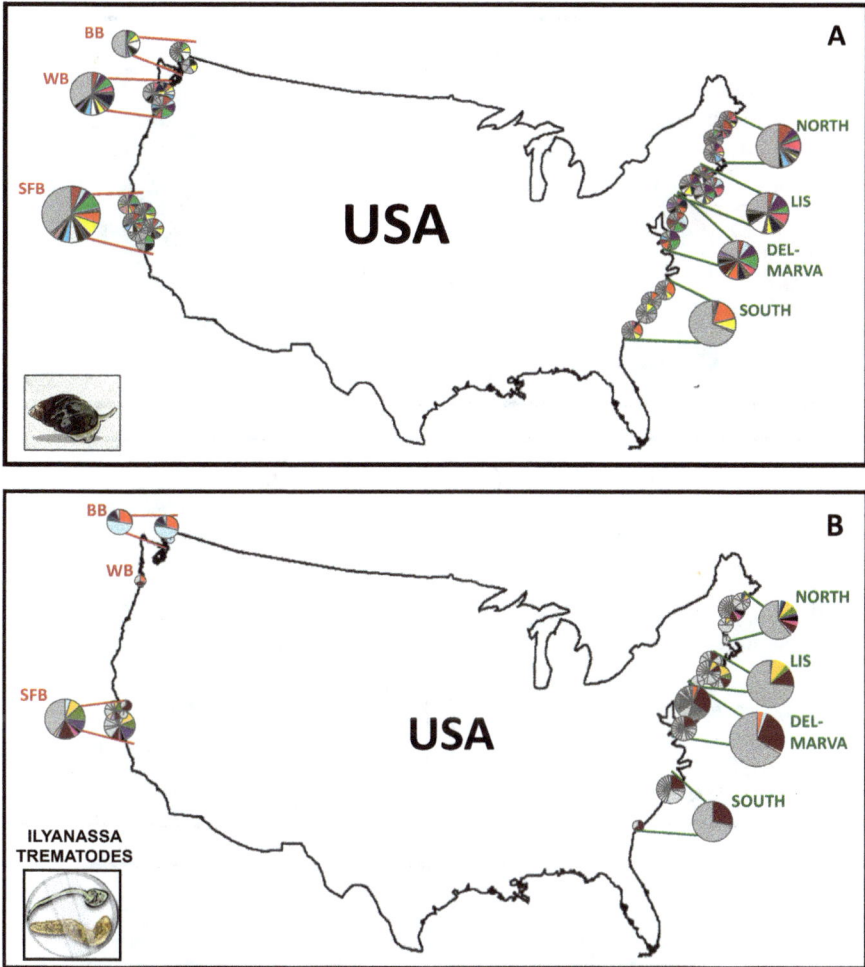

Fig. 2. Haplotype frequencies in native and introduced populations (small pies) and subregions (large pies) on the east and west coasts of North America for *Ilyanassa obsoleta* (A) and its trematode parasites (B). Larger subregions are as follows: "North" – those populations found in Maine, New Hampshire, and Massachusetts; "Long Island Sound (LIS)" – those populations located along Long Island Sound; "DELMARVA" – those populations located along the Delaware, Maryland, Virginia (DELMARVA) peninsula; "South" – those populations from North Carolina, South Carolina, and Georgia; "BB" – those populations found in Boundary Bay, British Columbia; "WB" – those populations found in Willapa Bay, Washington; and "SFB" – those populations found in San Francisco Bay, California. Colored pie pieces represent shared haplotypes between the native east coast and introduced west coast. Gray pie pieces represent unshared haplotypes. Pie charts are relatively sized based on sample size at a site or subregion. *Ilyanassa obsoleta* demonstrates a substantial amount of genetic diversity in its introduced west coast populations, while its trematodes show declines in diversity in introduced west coast populations compared to native east coast populations.

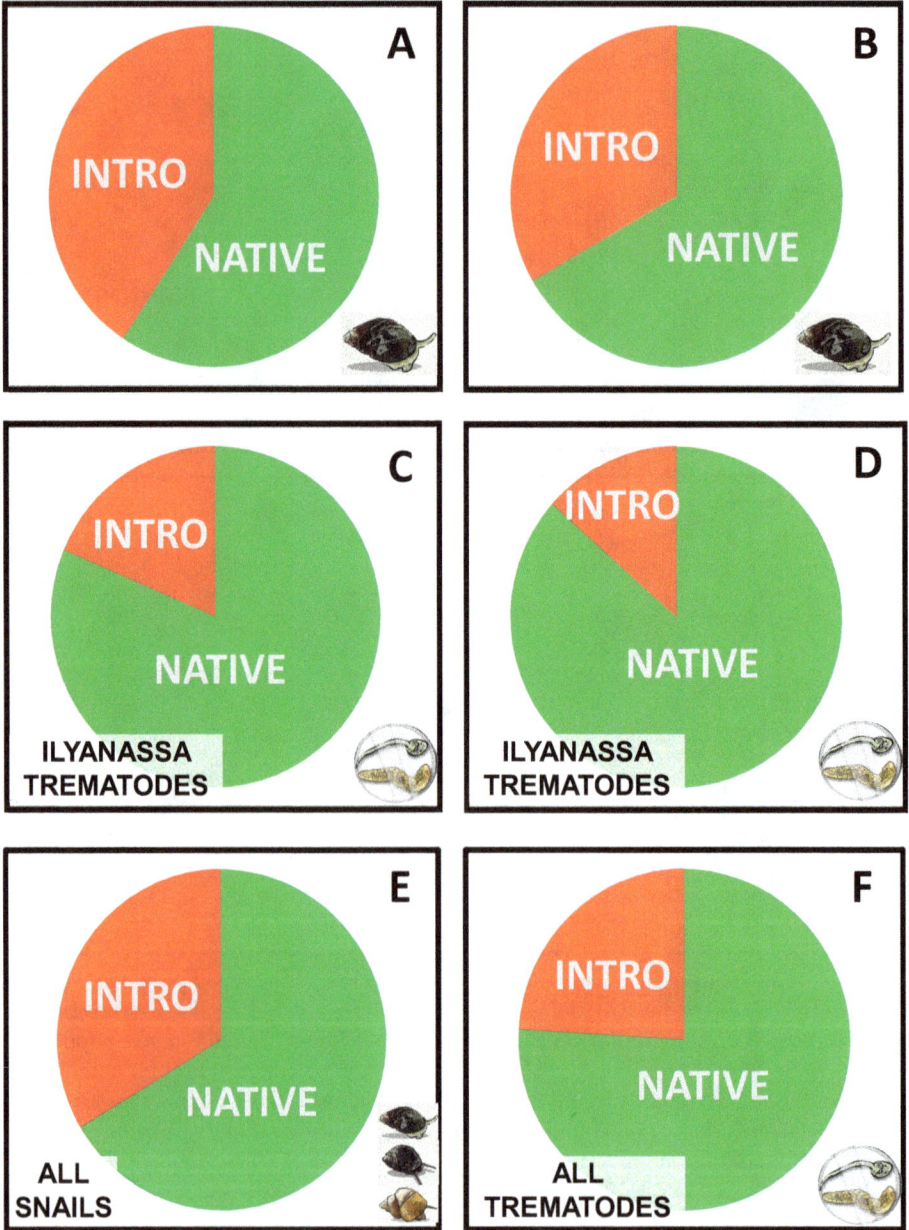

Fig. 3. Measures of regional genetic diversity in *Ilyanassa obsoleta* (A-B), its trematode parasites (C-D), all snail hosts (E), and all trematode parasites (F) for observed (A,C,E,F) and expected (B&D) total diversity in native and introduced (INTRO) regions. In general, snails tended to show a greater proportion of their overall genetic diversity to be 'Introduced' than did parasites.

obsoleta showed less diversity in its introduced (38-40%) compared to its native (60-62%) range (for both the observed and expected analyses; Figure 3A-B), its trematodes showed substantially less diversity in the introduced (13-19%) compared to native (81-87%) ranges (for both the observed and expected analyses; Figure 3C-D). This suggests that the differential between native and introduced genetic diversity is much lower for *I. obsoleta* than for its trematode parasites.

In our analyses where we included all three snails (*I. obsoleta*, *Littorina littorea*, and *L. saxatilis*) and all six trematode parasites, we found similar patterns as those described above for just *I. obsoleta*. In particular, the snails showed less of a reduction in genetic diversity in their introduced (33-34%) versus native (66-67%) regions [for both observed and expected analyses; Figure 3E (because observed and expected results are so similar, only observed values are shown in the figure)], compared to their trematode parasites, where the decline in genetic diversity was more substantial in the introduced (21-24%) versus native (76-79%) region (for both the observed and expected analyses; Figure 3F).

These differences were also observed at the individual species level. Snail hosts tended to show less reduction in diversity in their introduced versus native regions than their trematode parasites (Figure 4A-B), though variability between and among species was very apparent. This variability is not unexpected, however, given the different life histories (reproduction, life span, dispersal ability, etc.) of each species, as well as their different invasion histories, which would affect the number, identity, and frequency of haplotypes being carried over, as well as the likelihood of becoming established and maintained in the population. Even still, it is interesting that when we explored snail hosts and trematodes collectively, we found some general patterns, especially in regards to diversity reductions in introduced versus native regions in snails versus parasites. For example, when grouped, we found that the introduced snails had on average about half the genetic diversity of their native region (this difference was non-significant; p=0.30); whereas for trematodes, introduced diversity was about one-third that of native diversity (Figure 4C), and the reduction in introduced versus native diversity was marginally significant (p=0.07) [note: an analysis using expected haplotype diversity revealed similar p-values: 0.06 for trematodes and 0.27 for snails].

Altogether, these results continue to support a more profound reduction in genetic diversity for the introduced trematode parasites than their hosts and also are suggestive of general patterns (at least for the three snails and six trematodes from which we collected genetic data) – in that the snails (especially *I. obsoleta*) appear to conform to the "genetic paradox" observed in numerous species in the Roman & Darling (2007) study (where a significant reduction in genetic diversity was not apparent in the introduced region). In contrast, the parasites do show significant reductions in haplotype diversity in the introduced versus native regions, and thus appear to demonstrate founder effect signatures. However, sampling for trematodes was challenging because they are more difficult to locate than their snail hosts (given the logistic nature of having to destructively sample enough snails to obtain an adequate amount of parasite DNA), especially in introduced populations where parasite abundance is already lower (Blakeslee et al., 2011). As a result, we attempted to control for some of the sampling variation in our study (through rarefaction analyses); in addition, we explored collective parasite data in many cases to enhance sample and effect sizes. Therefore, we believe the patterns we observed have merit and suggest that inherent

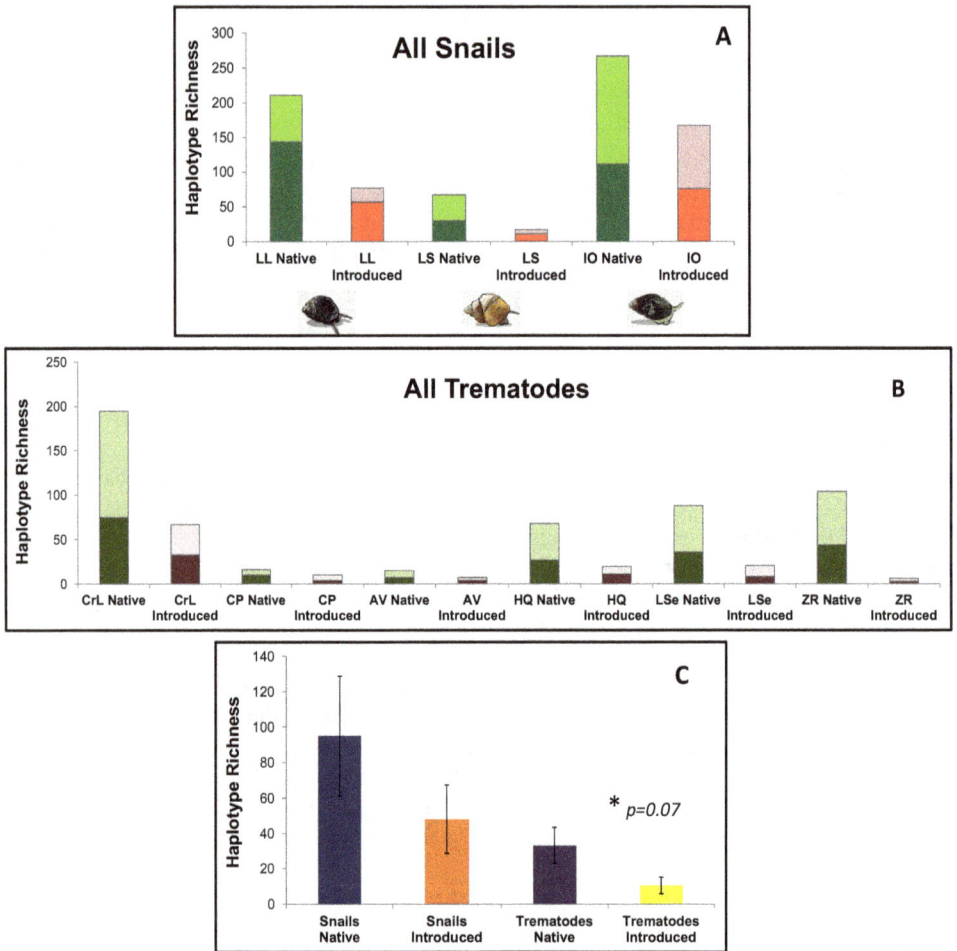

Fig. 4. Native and introduced haplotype richness in the: (A) three snail hosts, *Littorina littorea* (LL), *Littorina saxatilis* (LS), and *Ilyanassa obsoleta* (IO); (B) trematode parasites infecting *Littorina littorea* and *Ilyanassa obsoleta*; (C) snails combined and trematodes combined. While all three snails (individually and collectively) showed some level of genetic diversity reductions in introduced versus native regions, the magnitude of the reduction was less than that of their parasites both individually and combined. In the combined treatment (C), the reduction for the parasites was marginally significant (p=0.07) using a single-factor ANOVA compared to a non-significant reduction (p=0.30) in the snails. Lighter shading in bar graphs of (A) and (B) represent the differential between expected haplotype richness and observed haplotype richness (i.e., observed richness is represented by darker shading and the light shading represents haplotypes that were estimated to have been missed in the sampling). CrL=*Cryptocotyle lingua*; CP=*Cercaria parvicaudata*; AV=*Austrobilharzia variglandis*; HQ=*Himasthla quissetensis*; LSe=*Lepocreadium setiferoides*; ZR=*Zoogonus rubellus*.

differences in invasion pathways and life histories of snail hosts and their parasites affects the likelihood for observing founder effect signatures.

Interestingly, we found consistent patterns across parasites even with differences in host invasion histories – which (depending on the particular introduction vector) may result in a greater or lesser likelihood for parasite and genetic transfer to introduced populations. For example, *I. obsoleta's* particular introduction vector would have likely provided numerous opportunities for gene flow and parasite transfer between native east coast populations and those areas on the west coast (in particular San Francisco Bay) receiving shipments of eastern oysters and associated biota (including *I. obsoleta*) (Blakeslee et al., 2011). This is likely for several reasons. First, shipments of oysters to the west coast occurred on a massive scale sustained over numerous years and resulting in billions of oysters being transported across the country (Miller, 2000), increasing the likelihood for associated organisms like *I. obsoleta* to be transferred with the oysters. This would have also resulted in multiple introductions of the snail, enhancing genetic diversity in the introduced region. In addition, the main harvesting method for oyster extraction – dredging – was very unselective (Ingersoll, 1881; Carlton, 1992), thus numerous individuals from numerous source populations could have been captured in the process, resulting in a greater diversity of alleles being transferred. Finally, commercial oysters were packaged for transcontinental shipping in a manner that ensured their survival, also benefiting the survival of hitchhiking organisms (including parasites) (Carlton, 1979); as such, the loss of alleles and parasites due to mortality during the transfer process would have been reduced. Altogether, this information suggests that *I. obsoleta* could have transferred a substantial amount of the genetic diversity of its source populations (which span a large area, from Virginia to Connecticut; Blakeslee et al., 2011) to its introduced populations. Additionally, this introduction vector would have allowed for some level of parasite gene flow between native and introduced populations as well. However, due to the inherently lower number of parasites being transferred (since only a subset of invading snails would be infected) and the trematode's complex life history requiring multiple hosts (e.g., Figure 1), the movement of trematode genes to the introduced populations would have automatically been lower and more subject to forces reducing diversity, like genetic drift (given the smaller introduced population) and Allee effects. Altogether, this would have resulted in much greater reductions in genetic diversity in the introduced populations of the parasites than their snail host.

Our genetic data appear to support these theories. In particular, genetic diversity in introduced *I. obsoleta* populations was not significantly depressed, and in some west coast populations (especially in San Francisco Bay), genetic diversity was similar if not larger than native east coast populations. We also observed that a substantial amount of the snail's introduced diversity originated from putative source populations along Long Island Sound and the Delmarva peninsula (Figure 2A). For parasites, on the other hand, all of *I. obsoleta's* trematodes showed signatures of genetic founder effects. This was also the case for *Littorina littorea* and its trematode parasites, though the latter snail showed a greater reduction in diversity in its introduced population than *I. obsoleta*. This difference between the snail hosts could be because of major distinctions between their invasion pathways. In particular, *L. littorea's* introduction vector (purportedly rock ballast in ships; Brawley et al., 2009) probably did not allow for as much gene flow between source and founding populations as did *I. obsoleta's* introduction vector due to longer transit times and harsher transfer conditions for *L. littorea* versus *I. obsoleta*. Even still, we found parasites of both snails to exhibit more apparent founder effect signatures than their snail hosts.

Altogether, these snails may benefit from the lack of a considerable genetic bottleneck in their introduced populations, especially for *I. obsoleta*, which exhibited substantial amounts of genetic diversity in many of its introduced populations. This is because small, genetically depauperate populations can be subject to detrimental fitness effects often associated with extreme bottlenecks, including inbreeding depression and loss of diversity through genetic drift (Roman & Darling, 2007). Avoiding such impacts may have assisted *I. obsoleta* in its successful establishment and spread throughout San Francisco Bay and other west coast bays, where it is presently highly abundant (A.M.H.B., pers. obs.).

4. Conclusion

As we have seen throughout the chapter, characteristics associated with a specific introduction vector strongly affect the likelihood for a species' successful introduction to a new location, and its likelihood for transferring a substantial subset of the genetic diversity of its native range. In particular, the number of host individuals transported, frequency of transport events, and transport conditions are important characteristics that will affect the entrainment, transfer, and ultimate success of a species and its ability to establish and maintain a significant level of genetic diversity in its new environment (Miller & Ruiz, 2009; Roman & Darling, 2007). In particular, for successful establishment and spread, invasive species must be able to survive and reproduce in their novel environment, and for a multi-host parasite, this depends on the presence and abundance of suitable hosts, which serve as their biological habitats (Blakeslee et al., 2011). Thus, hosts and parasites are likely to be impacted differently by the invasion process, ultimately affecting their genetic diversity patterns. This is supported by our case study results, which suggest that, for the most part, invasive parasites will be more likely to exhibit genetic bottleneck and founder effect signatures (in terms of genetic diversity loss) in introduced ranges than their hosts, and this is likely a result of the inherent differences in their propagule pressures, life histories, and invasion pathways.

Furthermore, it is important to note that because parasites and hosts are fundamentally intertwined, they will also greatly impact one another's evolutionary processes. For example, parasites may affect host genetic structure if there is differential reproductive success of infected versus uninfected host genotypes. Thus, if parasites reduce host reproduction or survival and if host genotypes differ in their susceptibility to parasitic infection, then parasite-mediated changes in gene frequencies can occur. As a result, co-evolutionary processes will be highly important to the genetic structure of both groups; i.e. the Red Queen hypothesis (Haldane 1949; Jaenike 1978; Hamilton 1980; Bell 1982), which states that parasites and hosts are both under strong selection pressures as a result of one another – the parasite to adapt to infect locally common host genotypes, and the host to adapt to be genetically unique (e.g., possessing rare genotypes) to avoid fitness-reducing infections. As hosts and parasites are constantly co-evolving, this can generate a time-lag in selection of both parasite and host genotypes and produce oscillations in gene frequencies (Clarke 1976; Hutson and Law 1981; Nee 1989, Dybdahl and Lively 1998; Jokela et al. 2009; Wolinska and Spaak 2009) all over a relatively short evolutionary time period (Koskella and Lively 2009; Morran et al. 2009; Paterson et al. 2010, Schulte et al. 2010). Parasites may also differ from their hosts in being genetically structured over relatively small spatial scales due to their complex life histories that depend on hosts for habitat and successful reproduction (Prugnolle et al., 2005). Thus, there are many ecological and evolutionary bases for strong differences in genetic diversity and genetic structure between parasites and hosts in natural situations; anthropogenic movements of species and genes complicates understanding further and has resulted in many unanswered questions regarding the effects of invasion on host and parasite population genetics. Our study has explored some of these questions, though much still needs to be investigated and resolved.

Our chapter has therefore provided important preliminary clues regarding the differences between host and parasite genetic structure and diversity patterns in aquatic systems as a result of anthropogenic introductions. Continued research in these areas will further our understanding of the roles of each and how they may impact one another and affect evolutionary change, especially in recently introduced founding populations. Furthermore, detailed invasion history information is required to understand the effects of the invasion process on hosts and parasites, especially considering the impact of the invasion pathway on the transfer of propagules and alleles across geographic barriers. Genetic diversity data is also important in understanding parasite contributions to aquatic communities, especially for those parasites with human health effects (such as schistosome trematodes). Moreover, at the conservation level, detailed genetic information can help track the movements of genotypes locally and also at the global scale, providing managers with important information on source, timing, and introduction vector which can aid in prevention or mitigation measures. Finally, we have identified significant knowledge gaps in parasite research for aquatic introductions in this chapter through our literature review, where very few studies existed comparing host and parasite genetic diversities in founding populations; therefore, our chapter also provides impetus for continued research on comparative host-parasite studies to determine whether patterns could be found across systems and across parasite groups.

5. Acknowledgments

We thank Irit Altman, Jeb Byers, Linsey Haram, and Kristin Maglies for field and/or lab help. We thank the Smithsonian Institution's Marine Science Network for funding and support. We thank Greg Ruiz and Whitman Miller for advice and mentoring. We thank Chris Brown for use of genetic data.

6. References

Abbott, R.T. (1974) *American seashells: the marine Mollusca of the Atlantic and Pacific coasts of North America.* Van Nostrand Reinhold, New York.

Ashworth, S. T., and Blanc, G., (1997) *Anguillicola crassus*, a recently introduced aggressive colonizer of European eel stocks. *Bulletin Francais de la Peche et de la Pisciculture*, 344/ 345, 335-342.

Barton, D.P. (1997) Introduced animals and their parasites: the cane toad, *Bufo marinus*, in Australia. *Australian Journal of Ecology*, 22: 316–24.

Bell, G. (1982) The masterpiece of nature: the evolution and genetics of sexuality. Univ. of California Press, Berkeley, CA.

Blakeslee, A.M.H., Altman, I., Miller, A.W., Byers, J.E., Hamer, C.E. & Ruiz, G.M. (2011) Parasites and invasions: a biogeographic examination of parasites and hosts in native and introduced ranges. *Journal of Biogeography*. DOI: 10.1111/j.1365-2699.2011.02631.x

Blakeslee, A.M.H., Keogh, C.L., Byers, J.E., Kuris, A.M., Lafferty, K.D. & Torchin, M.E. (2009) Differential escape from parasites by two competing introduced crabs. *Marine Ecology Progress Series*, 393, 83–96.

Blakeslee, A.M.H., Byers, J.E. & Lesser, M.P. (2008) Resolving cryptogenic histories using host and parasite genetics: the resolution of *Littorina littorea's* North American origin. *Molecular Ecology*, 17, 3684-3696.

Blakeslee, A.M.H. & Byers, J.E. (2008) Using parasites to inform ecological history: comparisons among three congeneric marine snails. *Ecology*, 89, 1068-1078.

Blouin, M.S., Yowell, C.A., Courtney, C.H., Dame, J.B. (1995) Host movement and the genetic structure of populations of parasitic nematodes. *Genetics*, 141, 1007-1014.

Bouzid, W., Stefka, J., Hypsa, V., Lek, S., Scholz, T., Legal, L., Hassine, O.K.B., Loot, G. (2008) Geography and host specificity: Two forces behind the genetic structure of the freshwater fish parasites *Lingua intestinalis* (Cestoda: Diphyllobothriidae). *International Journal for Parasitology*, 38, 1465-1479.

Brawley, S.H., Coyer, J.A., Blakeslee, A.M.H., Olsen, J.L., Hoarau, G., Johnson, L.E., Byers, J.E. & Stam, W.T. (2009) Historical invasions of the intertidal zone of Atlantic North America associated with distinctive patterns of trade and emigration. *Proceedings of the National Academy of Sciences USA*, 106, 8239-8244.

Brown, C.W. (2007) Genetic variation of the invasive gastropod *Littorina saxatilis* (Olivi) in San Francisco Bay. M.S. Thesis, San Francisco State University, San Francisco, CA.

Brown, S.C. (1969) The structure and function of the digestive system of the mud snail *Nassarius obsoletus* (Say). *Malacologia*, 9, 447-500.

Bousfield, E.L. (1960) *Canadian Atlantic Sea Shells*. Department of Northern Affairs and National Resources, National Museum of Canada, Ottawa, 72 pp.

Burreson, E.M., Stokes, N.A., Friedman, C.S. (2000) Increased Virulence in an Introduced Pathogen: *Haplosporidium nelsoni* (MSX) in the Eastern Oyster *Crassostrea virginica*. *Journal of Aquatic Animal Health* 12, 1–8.

Canning-Clode, J., Valdivia, N., Molis, M., Thomason, J.C. & Wahl, M. (2008) Estimation of regional richness in marine benthic communities: quantifying the error. *Journal of Limnology and Oceanography Methods*, 6, 580– 590.

Carlton, J.T. (1979) *History, biogeography, and ecology of the introduced marine and estuarine invertebrates of the Pacific coast of North America*. PhD Thesis, University of California, Davis, CA.

Carlton, J.T. (1992) Introduced marine and estuarine mollusks of North America: an end-of-the-century perspective. *Journal of Shellfish Research*, 11, 489-505.

Carlton, J.T. & Cohen, A.N. (1998) Periwinkle's progress: the Atlantic snail *Littorina saxatilis* (Mollusca: Gastropoda) establishes a colony on a Pacific shore. *Veliger*, 41, 333-338.

Carlton, J.T. & Geller, J.B. (1993) Ecological roulette: the global transport of nonindigenous marine organisms. *Science*, 261, 78-82.

Chang, A.L., A.M.H. Blakeslee, A.W. Miller & G.M. Ruiz (2011) Why do invasions fail? A case history of *Littorina littorea* in California, USA. *PLOS One*, 6, e16035.

Curtis, L.A. & Hurd, L.E. (1981) Nutrient procurement strategy of a deposit-feeding estuarine neogastropod, *Ilyanassa obsoleta*. *Estuarine Coastal and Shelf Science*, 13, 277-285.

Clarke, B. (1976) The ecological relationships of host-parasite relationships. In A. E. R. Taylor, & R. Muller, eds. *Genetic aspects of host-parasite relationships*. Blackwell, Oxford. Pp. 87–103.

Criscione, C.D. & Blouin, M.S. (2004) Life cycles shape parasite evolution: comparative population genetics of salmon trematodes. *Evolution*, 58, 198–202.

Criscione, C.D., Cooper, B., Blouin, M. (2006) Parasite genotypes identify source populations of migratory fish more accurately than fish genotypes. *Ecology*, 87, 823–826.

Darling, J.A., Bagley, M.J., Roman, J., Tepolt, C.K., Geller, J.B. (2008) Genetic patterns across multiple introductions of the globally invasive crab genus *Carcinus*. *Molecular Ecology*, 17, 4992–5007.

Dieguez-Uribeondo J. & Soderhall, K. (1993) *Procambarus clarkia* (Girard) as a vector for the crayfish plague fungus, *Aphanomyces ascaci* (Schikora). *Aquatic Fisheries Management*, 2,:761–765.

Dove, A.D.M. (2000) Richness patterns in the parasite communities of exotic poeciliid fishes. *Parasitology*, 120, 609–23.

Dunn, A.M. & Dick, J.T.A. (1998) Parasitism and epibiosis in native and non-native gammarids in freshwater in Ireland. *Ecography*, 21, 593–598.

Dybdahl, M.F. & Lively, C.M. (1996) The geography of coevolution:comparative population structures for a snail and   its trematode parasite. *Evolution*, 50, 2264–2275.

Dybdahl, M.F. & Lively, C.M. (1998) Host-parasite coevolution: evidence for rare advantage and time-lagged selection in a natural population. *Evolution*, 52, 1057–1066.

Font, W.F. (1998), Parasites in paradise: patterns of helminth distribution among streams and fish hosts in Hawai'i. *Journal of Helminthology*, 72, 307-311.

Geller, J.B., Darling, J.A. & Carlton, J.T. (2010) Genetic Perspectives on marine biological invasions. *Annual Review of Marine Science*, 2, 367–393.

Gittenberger, E., Groenenberg, D.S.J., Kokshoorn, B., Preece, R.C. (2006) Molecular trails from hitch-hiking snails. *Nature*, 439, 409.

Grosberg R, Cunningham CW (2000) Genetic structure in the sea from populations to communities. In: Marine Community Ecology (eds Bertness M, Gaines S, Hay M), pp. 61–84. Sinauer Associates, Sunderland, Massachusetts.

Haldane, J. B. S. (1949) Disease and evolution. *Ricerca Scientifica*. 19, 68–76.

Hamilton, W. D. (1980) Sex versus non-sex versus parasite. *Oikos*, 35, 282-290.

Hayward, C.J., Iwashita, M., Ogawa, K. Ernst I. (2001) Global spread of the eel parasite *Gyrodactylus anguillae* (Monogenea). *Biological Invasions*, 3, 417-424.

Huspeni, T.C. & Lafferty, K.D. (2004) Using larval trematodes that parasitize snails to evaluate a saltmarsh restoration project. *Ecological Applications*, 14, 795-804.

Hutson, V., & Law, R. (1981) Evolution of recombination in populations experiencing frequency-dependent selection with time delay. *Proceedings of the Royal Society of London* B 213, 345–359.

Jaenike, J. (1978) An hypothesis to account for the maintenance of sex within populations. *Evolutionary Theory*, 3, 191–194.

Jarne, P. & Theron, A. (2003) Genetic structure in natural populations of flukes and snails: a practical approach and   review. *Parasitology*, 123, 27–40.

Jokela, J., Dybdahl, M. F., Lively, C. M. (2009) The maintenance of sex, clonal dynamics, and host-parasite coevolution in a mixed population of sexual and asexual snails. *American Naturalist*, 174, S43-S53.

Keane, R.M. & Crawley, M.J. (2002) Exotic plant invasions and the enemy release hypothesis. *Trends in Ecology and Evolution*, 17, 164-170.

Keeney, D.B., Waters, J.M., Poulin, R. (2007) Diversity of trematode genetic clones within amphipods and the timing of same-clone infections. *International Journal for Parasitology*, 37, 351-357

Kelly, D.W., Paterson, R.A., Townsend, C.R., Poulin, R., Tompkins, D.M. (2009) Parasite spillback: a neglected concept in invasion ecology? *Ecology* 90, 2047-2056.

Kennedy, C.R. (1998) Aquatic birds as agents of parasite dispersal: a field test of the effectiveness of helminth colonisation strategies. *Bulletin of the Scandinavian Society for Parasitology*, 8, 23-28.

Koskella, B. & Lively, C.M. (2009) Evidence for negative frequency-dependent selection during experimental coevolution of a freshwater snail and a sterilizing trematode. *Evolution*, 63, 2213-2221.

Kruse, I. & Hare, M.P. (2007) Genetic diversity and expanding nonindigenous range of the rhizocephalan *Loxothylacus panopaei* parasitizing mud crabs in the western North Atlantic. *Journal of Parasitology*, 93, 575-582.

Kuris, A.M., Hechinger, R.F., Shaw, J.C., Whitney, K.L., Aguirre-Macedo, L., Boch, C.A., Dobson, A.P.,

Dunham, E.J., Fredensborg, B.L., Huspeni, T.C., Lorda, J., Mababa, L., Mancini, F.T., Mora, A.B.,

Pickering, M., Talhouk, N.L., Torchin, M.E. & Lafferty, K.D. (2008) Ecosystem energetic implications of parasite and free-living biomass in three estuaries. *Nature*, 454, 515-518.

Lafferty, K.D. & Morris, A.K. (1996) Altered behavior of parasitized killifish increases susceptibility to predation by bird final hosts. *Ecology*, 77, 1390-1397.

Louhi, K.-R., Karvonen, A., Rellstab, C., Jokela, J. 2010. Is the population genetic structure of complex life cycle parasites determined by the geographic range of the most motile host? *Infection, Genetics and Evolution,* 10, 1271-1277.

Liu, H. & Stiling, P. (2006) Testing the enemy release hypothesis: a review and meta-analysis. *Biological Invasions*, 8, 1535-1545

Martin, W. E. (1958) The life histories of some Hawaiian heterophyid trematodes. *Journal of Parasitology* 44, 305-323.

McCoy, K.D., Boulinier, T., Tirard, C., Michalakis, Y. (2003) Hostdependent genetic structure of parasite populations: differential dispersal of seabird tick host races. *Evolution*, 57, 288-296.

Miller, A.W. (2000) *Assessing the importance of biological attributes for invasion success: Eastern oyster* (Crossostrea virginica) *introductions and associated molluscan invasions of Pacific and Atlantic coastal systems.* PhD Thesis, University of California Los Angeles, Los Angeles, CA.

Miller, A.W. & Ruiz, G.M. (2009) Differentiating successful and failed invaders: species pools and the importance of defining vector, source and recipient regions. *Biological invasions in marine ecosystems: ecological, management, and geographic perspectives* (ed. by G. Rilov and J. Crooks), pp. 153-170, Springer Verlag, Berlin.

Miura, O., Torchin, M.E., Kuris, A.M., Hechinger, R.F., Chiba, S. (2006) Introduced cryptic species of parasites exhibit different invasion pathways. *Proc Natl Acad Sci USA*, 103, 19818-19823.

Morran, L. T., Parmenter, M. D., Phillips, P. C. (2009) Mutational load and rapid adaptation favour outcrossing over self-fertilization. *Nature*, 462, 350-352.

Münderle, M., Taraschewski, H., Klar, B. Chang, C. W. , Shiao, J. C. , Shen, K. N. , He, J. T. , Lin, S. H. , Tzeng, W. N.   (2006) Occurrence of *Anguillicola crassus* (Nematoda: Dracunculoidea) in Japanese eels *Anguilla japonica* from a river and an aquaculture unit in southwest Taiwan. *Diseases of Aquatic Organisms* 71, 101-108.

Nee, S. (1989) Antagonistic coevolution and the evolution of genotypic randomization. *Journal of Theoretical Biology*, 140, 499-518.

Paterson, S., Vogwill, T., Buckling, A. D., Benmayor, R., Spiers, A. J., Thomson, N. R., Quail, M., Smith, F., Walker, D., Libberton, B., Fenton, A., Hall, N., Brockhurst, M. A. (2010) Antagonistic coevolution accelerates molecular evolution. *Nature*, 464, 275-278.

Prenter, J., MacNeil, C., Dick, J.T.A., Dunn, A.M. (2004) Roles of parasites in animal invasions. *Trends in Ecology and   Evolution*, 19, 385-390.

Prugnolle, F., Liu, H., de Meeus, T., Balloux, F. (2005) Population genetics of complex life-cycle parasites: an illustration with trematodes. *International Journal of Parasitology*, 35, 255-263.

Rauch G, Kalbe M, Reusch TBH (2005) How a complex life cycle can improve a parasite's sex life. *Journal of Evolutionary Biology*, 18, 1069-1075.Rauque, C.A., Viozzi, G.P., Semenas, L.G. (2003) Component population study of *Acanthocephalus tumescens* (Acanthocephala) in fishes from Lake Moreno, Argentina. *Folia Parasitology* 50, 72-78.

Reed, D.H. & Frankham, R. (2003) Correlation between Fitness and Genetic Diversity. *Conservation Biology*, 17, 230-237.

Reid, D.G. (1996) *Systematics and evolution of Littorina* (ed. by D.G. Reid), pp. 278-340. The Ray Society, Andover, UK.

Roman, J. (2006) Diluting the founder effect: cryptic invasions expand a marine invader's range. *Proceedings of the Royal Society B: Biological Sciences*, 273, 2453-2459.

Roman, J. & Darling, J.A. (2007) Paradox lost: genetic diversity and the success of aquatic invasions. *Trends in Ecology and Evolution*, 22, 454-464.

Ruiz, G.M., Fofonoff, P., Carlton, J.T., Wonham, M. & Hines, A.H. (2000) Invasion of coastal marine communities in North America: apparent patterns, process, and biases. *Annual Review of Ecology and Systematics*, 31, 481-531.

Scheltema, R.S. (1961) Metamorphosis of the veliger larvae of *Nassarius obsoletus* (Gastropoda) in response to bottom sediment. *Biological Bulletin*, 120, 92-109.

Schulte, R. D., Makus, C., Hasert, B., Michiels, N. K., Schulenburg, H. (2010) Multiple reciprocal adaptations and rapid genetic change upon experimental coevolution of an animal host and its microbial parasite. *Proceedings of the National Academy of Sciences*, 107, 7359-7364.

Shoop, W.L. 1988. Trematode transmission patterns. *Journal of Parasitology*, 74, 46-59.

Slothouber-Galbreath, J. G. M., Smith, J. E., Becnel, J. J., Butlin, R. K., and Dun, A. M. (2010) Reduction in post-invasion genetic diversity in Crangonyx pseudogracilis; a genetic bottleneck or the work of hitchhiking, vertically transmitted microparasites? *Biological Invasions*.12 191-209.

Steneck, R. & Carlton J.T. (2001) Human alterations of marine communities: students beware!. In: Marine Community Ecology (eds Bertness M, Gaines S, Hay M), pp. 445–468. Sinauer Associates, Sunderland, Massachusetts.

Stohler, R. A., Curtis, J., Minchella, D. J. (2004) A comparison of microsatellite polymorphism and heterozygosity among field and laboratory populations of *Schistosoma mansoni. International Journal of Parasitology*, 34, 595–601.

Taraschewski, H. (2006) Hosts and parasites as aliens. *Journal of Helminthology*, 80, 99-128.

Tompkins, D.M. & Poulin, R. (2006). Parasites and biological invasions. In: *Biological Invasions in New Zealand* (eds Aleen, R.B. & Lee, W.G.). Springer-Verlag, Berlin, pp. 67–82.

Torchin, M.E., Lafferty, K.D., Kuris, A.M. (2001) Release from parasites as natural enemies: increased performance of a globally introduced marine crab. *Biological Invasions* 3, 333–45.

Torchin, M.E., Lafferty, K.D., Dobson, A., McKenzie, V. & Kuris, A.M. (2003) Introduced species and their missing parasites. *Nature*, 421, 628–630.

Torchin, M.E. & Mitchell, C.E. (2004) Parasites, pathogens, and invasions by plants and animals. *Frontiers in Ecology and the Environment*, 2, 183–190.

Voisin, M., Engel, C.R. & Viard, F. (2005) Differential shuffling of native genetic diversity across introduced regions in a brown alga: aquaculture vs. maritime traffic effects. *Proceedings of the National Academy of Sciences USA*, 102, 5432-5437.

Wattier, R. A., Haine, E. R., Beguet, J., Martin, G., Bollache, L., Musko, I. B., Platvoet, D., and Rigaud, T. (2007). No genetic bottleneck or associated microparasite loss in invasive populations of a freshwater amphipod, *Oikos* 116, 1941–1953.

Wielgoss, S., Sanetra, M., Meyer, A., Wirth, T. (2007) Isolation and characterization of short tandem repeats in an invasive swimbladder nematode, parasitic in Atlantic freshwater eels, *Anguillicola crassus. Molecular Ecology Notes*, 7, 1051–1053.

Wielgloss, S., Taraschewski H., Meyer, A., Wirth, T. (2008) Population structure of the parasitic nematode *Anguillicola crassus*, an invader of declining North Atlantic eel socks. *Molecular Ecology*, 17, 3478-3495.

Wolinska, J., & Spaak, P. (2009) The cost of being common: evidence from natural *Daphnia* populations. *Evolution*,63, 1893-1901.

15

Molecular Biodiversity Inventory of the Ichthyofauna of the Czech Republic

Jan Mendel et al.*
Institute of Vertebrate Biology
Academy of Sciences of the Czech Republic, Brno
Czech Republic

1. Introduction

The term biological diversity (biodiversity) covers the variability of life on Earth. In 1989, the World Wildlife Fund defined biodiversity as "the richness of life on Earth – millions of plants, animals and microorganisms, including the genes which they carry, and complex ecosystems that create the environment" (Primack et al., 2001).

The degree of biodiversity was dynamically changing during the centuries and millennia, according to available resources it was gradually growing. In the course of evolution the periods of intensive speciation and relative "speciation rest" were alternating, with five major episodes of mass extinction (Wilson, 1989; Raup, 1992). The worst extinction event took place at the end of the Permian period, 250 million years ago (Primack et al., 2001), the most recent and also the best known mass extinction event occurred in the late Cretaceous period, i. e. 65 million years ago (Freeland, 2005). From then on, the rate of speciation was in equilibrium with the rate of extinction, or was even higher. Currently, the rate of extinction is 100-1000 times higher, namely almost exclusively in consequence of human activity. Therefore a lot of experts call the present situation the sixth mass extinction (Primack et al., 2001). At present, only approximately 1.5 million species are described, of which the majority is represented by insects and plants (Wilson, 1992). According to various estimates the number of non-described species amounts to 10 million or even 30-150 million (Hammond, 1992).

In recent years, when the effort to map and at the same time to protect global biodiversity has been intesifying, it is becoming clear that the existing morphological approach to species classification is not sufficient. Therefore, interest has focused on the methods of molecular genetics.

In 2003, a global project aimed at the mapping and protection of global biodiversity using a new taxonomic method called "DNA barcoding" was started. The key personality in this

* Eva Marešová[1], Ivo Papoušek[1], Karel Halačka[1], Lukáš Vetešník[1], Radek Šanda[2],
Milena Koníčková[1] and Soňa Urbánková[1]
[1]*Institute of Vertebrate Biology, Academy of Sciences of the Czech Republic, v.v.i. Brno*
[2]*Department of Zoology, National Museum, Prague*
Czech Republic

initiative was Canadian professor Paul D. N. Hebert from the Biodiversity Institute of Ontario (BIO), who proposed the creation of a global public library of DNA barcodes for all living organisms, also known as BoLD (Barcode of Life Database). The Barcode of Life Data Systems is an informatics workbench for the acquisition, storage, intercontinental comparison and publication of DNA barcode records (Fig. 1).

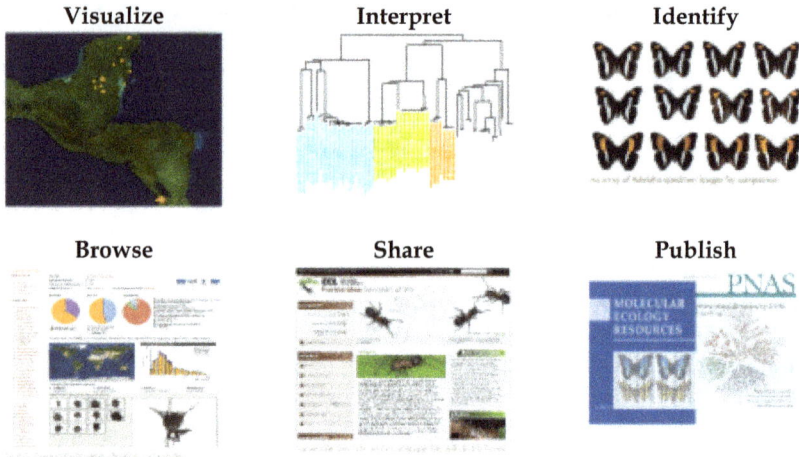

Fig. 1. Multifunctionality of the BoLD database (adopted from http://ibol.org/resources/barcode-library/)

At the same time it bridges a long-standing bioinformatics chasm between molecular, morphological and distributional data. BoLD is freely available to anyone interested in these problems, the identification and preservation of all life on Earth.

2. DNA barcoding

The basic assumption behind this method is that every biological species has a short sequence in its genome which is unique to that species, like a fingerprint to every human being. Its mutation rate should be sufficiently fast to enable the divergence of the sequences of closely related species and at the same time slow enough to enable the minimization of the differences between members of the same species. The implementation of a comparative sequence analysis of this carefully selected sequence should enable mutual identification of the individual species.

The term DNA barcoding is based on the analogy with EAN barcodes on goods which enable safe identification of the individual products (Hebert et al., 2003a). EAN barcode (Fig. 2) is composed of 10 digits at 11 positions, which gives 10^{11} possible combinations. The genetic code has only 4 letters (DNA bases) for one position, but the length of that code is incomparably longer. Only with the sequence length of 15 nucleotides, theoretically 4^{15} various combinations are available, which is a number of unique combinations far exceeding even the boldest estimates of the number of biological species living on Earth (Hebert et al., 2003a).

The technique of DNA barcoding provides a standardized method based on the mapping of a single gene in all the species on Earth (Hebert et al., 2003a). At present, the fragment of the gene for the first cytochrome c oxidase (COI) subunit appears to be the major marker for species of the animal kingdom. This segment is approximately 650 bp long and the protein it codes is part of the respiration chain. The results suggest that this global standard can also be easily obtained from phylogenetically very distant taxa by using a relatively small set of primers and that it is effective in the differentiation of closely related animal species belonging to a great number of groups of invertebrates and vertebrates (Hajibabaei et al., 2007; Hebert et al., 2003a, 2003 ; Hebert & Barett, 2005; Ivanova et al., 2007).

Fig. 2. EAN barcode (adopted from Brian Krueger).

Hebert et al. (2003b) proposed a limit of a 2% deviation in the barcode sequence for the identification of intraspecific variability. The intraspecific deviation only rarely exceeds 2% and in most cases does not reach 1% (Avise, 2000). They also proposed an experiential rule that the divergence between the species sequences should be 10 times higher than that within one species (Hebert et al., 2004).

For the identification of samples with degraded DNA, Hajibabaei et al. (2006) and Meusnier et al. (2008) proposed using shorter COI segments, the so-called "universal mini-barcodes" of a length smaller than 150 bp. Thus they provided a reliable solution for the identification of older tissues, badly preserved museum specimens, partly digested pieces of food in animal stomachs, etc. The results indicate that although the selection of a concrete position of the mini-barcode plays an important role concerning the success of identification, generally it can be stated that the mini-barcodes analysis provides virtually the same results as the specific full-barcode analysis.

2.1 Global progression of DNA barcoding

In April 2004, the Consortium for Barcode of Life (CBOL) was established (Schindel & Miller, 2005). It is an international collaborative effort of more than 160 member organizations from more than 50 countries on six continents (http://barcoding.si.edu, Hebert et al., 2004). CBOL´s mission is to develop the potential of DNA barcoding as a practical and financially affordable tool for taxonomic research, the study and preservation of biodiversity and the development of applications which will utilize the taxonomic information for the benefit of science and society.

Since 2004, a number of campaigns and projects striving to map global biodiversity of all life on Earth have been started (http://ibol.org/about-us/campaigns/):

* **Formicidae Barcode of Life** – a campaign aimed at barcoding all of the world's more than 12,000 ant species.
* **Bee Barcode of Life Initiative (Bee-BOL)** – a global effort to assembly the barcodes of all 20,000 bee species.

- **All Birds Barcoding Initiative (ABBI)** – the aim is to collect DNA barcodes of all of the approximately 10,000 known bird species. The genetic analyses made within the project show that there are hundreds of bird species which have not been described yet.
- **Trichoptera Barcode of Life** – the aim of the project is to barcode the world's approximately 13,000 species of caddisflies.
- **Coral Reef Barcode of Life** – a detailed study of fishes living at one site of the Great Barrier Reef.
- **Fish Barcode of Life Initiative (FISH-BOL)** – a project aimed at the mapping of global ichthyofauna involving 31,200 known fish species (Ward et al., 2009). This initiative also includes the Czech project „Molecular biodiversity inventory of the ichthyofauna of the Czech Republic" which has been introduced here.
- **All Fungi Barcoding** – a project associating initiatives mapping the global diversity of fungi.
- **HealthBOL** – an initiative coordinating the barcoding of vectors, pathogens, and parasites for the improvement of human health around the world.
- **Lepidoptera Barcode of Life** – a campaign mapping all butterflies, with an associated sub-campaign for selected families of Australia and North America.
- **Mammal Barcode of Life** – a project studying global mammal fauna is part of the initiative encompassing all vertebrates.
- **Mosquito Barcoding Initiative** – an international effort aimed at the identification of approximately 3,200 known mosquito species, the disease-bearing species being the priority.
- **Marine Barcode of Life campaign (MarBOL)** – a joint project of the Consortium for the Barcode of Life (CBOL) and the Census of Marine Life (CoML) is aimed at the comprehensive mapping of the diversity of the world's oceans.
- **Polar Barcode of Life campaign** – a campaign studying the biodiversity of the Arctic and Antarctic. It includes marine, freshwater and terrestrial ecosystems.
- **Shark Barcode of Life** – a project aimed at the barcoding of sharks. It plans for 1,000 marine and 100 freshwater shark species.
- **Sponge Barcoding Project** – it is the first global project studying diploblastic species using DNA barcoding. It covers the complete taxonomic range of Porifera.

The existing projects proceed successfully and have already brought the first results regarding various groups of plants and animals, e. g., birds (Hebert et al., 2004), gorillas (Thalmann et al., 2004), tropical beetles (Monaghan et al., 2005), spiders (Barrett & Hebert., 2005), ants (Fisher, 2006), flowering plants (Kress et al., 2005), amphibians (Vences et al., 2005), fishes (Ward et al., 2005), etc.

Currently, as many as 1,371,809 DNA barcodes are identified and stored, which corresponds to approximately 113,435 denominated species (as of September 30, 2011). The collaborating scientific teams have set a goal by 2015 to collect 5 million barcodes/specimens for 500,000 species.

2.1.1 Fish DNA barcoding

In 2005, the FISH-BOL initiative was established. It is an expression of the global effort to develop and coordinate a standardised library of sequences for all fish species. For this purpose a guideline containing DNA barcodes, photographs of the studied specimens, geospatial coordinates of the locations of the finds, etc. was issued. It also contains references to type specimens, information on species distribution, nomenclature, author taxonomic information, literature citations, etc. Thus, it complements and enhances the information resources available so far – the genome databases (GenBank, EMBL, DDBJ), including the internet encyclopaedia FishBase.

Barcoding in fish has multiple usages, of which we select only some:

- Identifying endangered and protected species (Ward et al., 2008)
- Identifying historical and museum material (Meusnier et al., 2008)
- Identifying new and cryptic species and possible fusions of existing taxa, and insight into phylogenetic relationships (Pyle et al., 2008)
- The development of a reference library for known species (Ratnasingham & Hebert, 2007)

Fishes are the most diversified group of vertebrates; at present, there are about 30,000 known species, including 15,758 (53.3%) marine species, 13,779 (46.4%) freshwater species and 86 (0.3%) brackish species (Ward et al., 2005). By September 2011, DNA barcodes were obtained from 8,293 species (27%) in the world, and from 503 species (25%) in Europe (http://www.fishbol.org/progress.php).

Some of the best-known projects utilizing the DNA barcoding in fishes are:

1. "Barcoding Marine Fishes: A Three-Ocean Perspective Project". The goal of this project is to obtain DNA barcodes of all marine species in the Pacific, Indian and Atlantic oceans. Participating countries include Canada, Australia, South Africa and Portugal.
2. "Planetary Biodiversity Inventories – Catfishes". It incorporates 300 participants from 40 countries whose work is taking an inventory of catfishes (Siluriformes).
3. "The Cypriniformes Tree of Life Initiative" (CToL), an American initiative aimed at the collection and analysis of DNA samples and reference specimens of nearly all North American freshwater species (~ 1100 species).

2.1.2 The state of the molecular biodiversity inventory of the ichthyofauna of the Czech Republic and introduction of the project

The study and recognition of genetic diversity in the Czech Republic has an approximately twenty-year history. Although the methods of biochemical genetics which were used in the past brought significant knowledge, only the application of molecular-genetic methods allowed significant detailed recognition and disclosure of the genetic diversity of species at the level of populations (Lusk & Hanel, 2008). Existing information about the genetic diversity of the individual species of the native ichthyofauna of the Czech Republic was and still is insufficient. The result of the research project VaV – SM/6/3/05 "Genetic diversity of endangered fish species - the essential basis for effective biodiversity protection" brought a breakthrough in the recognition of genetic diversity, especially of the protected and endangered species. The project was conducted in 2005-2007 in cooperation between our department and the research workers from the Institute of Animal Physiology and Genetics, v.v.i., Liběchov. This allowed the presentation of the first taxonomic approaches to some species of the Czech ichthyofauna (Bartoňová et al., 2008; Mendel et al., 2005; Mendel et al.,

2008a; Papoušek et al., 2008a, 2008b; Vetešník et al., 2007). Other taxonomic studies of our team are included in the proceedings "Biodiversity of fishes in the Czech Republic VII", which paid attention to the problems of the genetic diversity of some protected and rare fish species of the Czech Republic. Apart from the study of actual intra-species diversity, attention in this book was also paid to the distribution or specification of the occurrence of the studied species (Bartoňová et al., 2008; Lusk et al., 2008a, 2008b; Lusková et al., 2008a, 2008b; Mendel et al., 2008b, 2008c; Papoušek et al., 2008b).

The main goal of the presented project is the mapping and inventory making of all Czech ichthyofauna using the DNA barcoding technique. In addition to that, in the case of archived specimens forming new museum type series a comprehensive approach was used (morphology and nuclear genome analysis).

A study of this type has been lacking in the Czech Republic. It is unique due to its comprehensiveness using both the classical and the newest taxonomic tools and due to its future potential – the intercontinental mapping of the diversity of life on Earth. Thanks to our cooperation with the Biodiversity Institute of Ontario the Czech Republic became actively involved in the international iBOL project (International Barcode of Life Project), namely in its part which was already mentioned above – FISH-BOL (Fish Barcode of Life Initiative). Thus, the Czech study took the side of already finished or currently running studies which take place within whole countries and continents, e. g.:

- Fishes of Australia (Ward et al., 2005)
- Barcoding of Canadian freshwater fishes (Hubert et. al., 2008)
- DNA barcoding of fish of the Antarctic Scotia Sea (Rock et al., 2008)
- Freshwater fishes from Mexico and Guatemala (Valdez-Moreno et al., 2009)
- Aquarium Imports (Steinke et al., 2009)
- Fishes of Alaska and the Pacific Arctic (Mecklenburg et al., 2011)
- Amazon fishes (Ardura et al., 2010)
- DNA Barcoding of Indian Marine Fish (Lakra et al., 2011)
- Fishes of Japan (Zhang & Hanner 2011)
- Freshwater Fishes of North America (April et al., 2011)
- Barcoding Fishes of Eastern Nigeria (Nwani et al., 2011)
- Fishes of Argentina (running project)
- Fishes from South China Sea (running project)
- Etc.

More information can be found at: http://www.boldsystems.org/views/projectlist.php

3. Material and methods

The inventory and subsequent cataloguing were focused on recent indigenous and non-indigenous fishes and lampreys living in the natural waters of the Czech Republic. It concerned 11 orders, 17 families and 72 species. The selection of collection localities covered the main distribution areas of the studied taxa according to comprehensive surveys and the most recent taxonomic and phylogenetic studies (Baruš & Oliva, 1995; Hanel & Lusk, 2005; Kottelat & Freyhof, 2007; Janko et al., 2007; Mendel et al., 2008d, etc). The collection of samples was divided into three levels:

- classification of the first level is based on the hydrologic position of the Czech Republic (Fig. 3)

Concerning hydrologic division, the Czech Republic belongs to three sea drainage areas (the North Sea, Baltic Sea and Black Sea drainage areas).

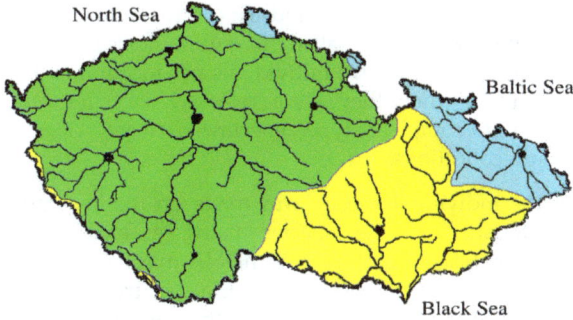

Fig. 3. Collection localities – 1st level of classification.

- classification of the second level (Fig. 4)

The hydrologic network of the Czech Republic was further divided into six areas, which for the most part correspond with the division of state administration: I. the Ohře River basin (the North Sea drainage area, the actual Ohře River basin and parts of other tributaries of the Elbe), II. the Vltava River basin (the North Sea drainage area, without the Ohře and the Elbe River basins), III. the Elbe River basin (the North Sea drainage area, without the Ohře and the Vltava River basins), IV. the Dyje River basin and the lower course of the Morava River, V. the Morava River basin (the Black Sea drainage area, the upper and middle courses of the Morava River), VI. the Odra River basin (the Baltic Sea drainage area).

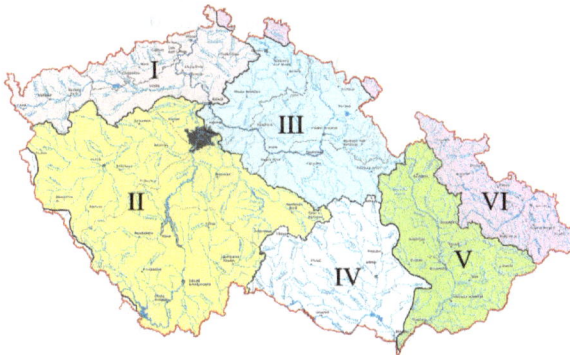

Fig. 4. Collection localities – 2nd level of classification.

- classification of the third level

The selection of the sampling localities for the given species was further specified according to up-to-date information concerning its distribution, according to the specific features of the given species (for example, the Natura 2000 localities, etc.) and with regard to the elimination of "family sampling". In the case of threatened and critically endangered fish species (*Pelecus cultratus, Zingel zingel, Zingel streber*, etc.) the database of the tissue samples collected by our workplace in the past was used.

The capturing of adult specimens was done using electro-fishing gear (direct pulsed current) in collaboration with the Czech and Moravian Fishing Unions and the representatives of the Agency for Nature Conservation and Landscape Protection of the Czech Republic. Approximately 1,400 specimens were taken for subsequent molecular-genetic analyses. From the captured specimens (about five individuals per location) a part of the pectoral fin (fin border) was taken and then they were released back into the stream. Only an indispensable number of complete type specimens were taken for morphological description and museum collections.

The identification of the individual species was made on the background of the study of morphological characteristics of the species using a multilocus and two-genome comparative approach (analysis of both mt and nDNA markers). Thus some limitations of the DNA barcoding method were eliminated, e. g. a) occurrence of hybrids; b) maternal contribution to COI marker as only one part of the information necessary for a valid description of a species. For reasons of saving space we do not present here the morphological characteristics of the individual species and the detailed results of the analyses of the nuclear marker – 1st intron of the S7 r-protein (RP1).

The preparation of the DNA barcodes conformed to the standardized protocols developed by the CCDB Centre, the FISH-BOL initiative and the BoLD database (http://www.ccdb.ca/pa/ge/research/protocols; Ratnasingham & Hebert, 2007; Hajibabaei et al., 2005; deWaard et al., 2008).

DNA isolation was performed by using a commercial kit - the Genomic DNA Mini kit (KRD) – according to the instructions for use. Cocktails of primers from the genetic study (Ivanova et al., 2007) were used for PCR amplification of the COI gene fragment (Fig. 5). For sequencing PCR, mainly M13F and M13R primers (Messing, 1983) were used, in a small number of cases the primers described by Ivanova et al. (2007) were used.

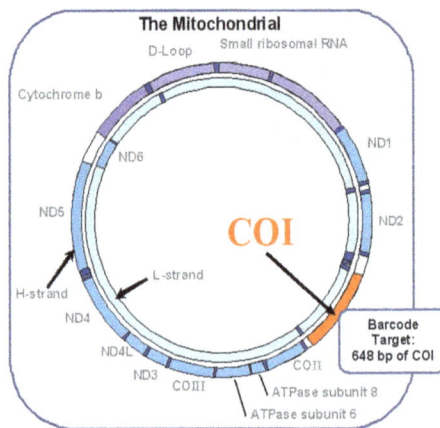

Fig. 5. Mitochondrial DNA – the area of the COI marker – adopted from http://www.barcodeoflife.org.

At the Canadian workplace, the subsequent steps were performed according to the above mentioned protocol. At the Czech workplace, the PCR products were re-purified by precipitation with PEG/Mg/NaAc solution (26% Polyethylene glycol, 6.5 mM $MgCl_2$. $6H_2O$, 0.6 M $NaAc.3H_2O$) and subsequently used for the sequencing reaction using BigDye™ Terminator Cycle Sequencing kit v 1.1. (Applied Biosystems), performed according to the manufacturer's instructions. The products of the sequencing reaction were re-purified with ethanol precipitation. For the COI sequencing, the instrumentation of the Canadian BIO workplace (ABI 3730XL; Applied Biosystems) was used and for test and complementary COI sequencing and nuclear marker analysis also the instrument ABI PRISM 310 (Applied Biosystems) at the Czech workplace was used. Also, commercial sequencing (Macrogen, Korea) was used in optimized samples. The analyzed fragment of mitochondrial DNA from each sample was sequenced from both directions. Chromatograms were adjusted and developed using the computer module SeqManII (part of the program package Lasergene v. 6.0; DNASTAR Inc.) and MEGA software.

The developed sequences were visually checked for errors and subsequently compared using the ClustalW algorithm. The correct taxonomic classification of the sequences was confirmed using comparison with the GenBank and BoLD databases. An electropherogram depicting the order of nucleotides in the sequence was obtained through sequencing. With its subsequent processing the DNA barcode was produced (Fig. 6).

Fig. 6. DNA barcode.

The individual steps of the barcoding process are shown in Figure 7.

The morphologic characteristics were identified in selected adult specimens to enable the establishment of a reference type series. A holotype was selected as a nomenclatoric type and other specimens from the described group were assigned to it as the documentary material (paratypes).

For the purpose of the cataloguing of the taxa of the ichthyofauna of the Czech Republic using the DNA barcoding method, the basic data concerning each specimen were collected and completed, including the body length (SL), weight and sex, photo documentation, locality description (including the GPS coordinates), date of collection and the person who caught or identified the specimen, the catalogue number and place of storage. The COI sequences of a length of at least 500 bp and the sequences of used PCR primers were recorded as well. For further details see the instructions of the National Museum and the standardized protocol of the BoLD database (http://www.boldsystems.org/docs/boldmas.html; Ratnasingham & Hebert, 2007).

In all archived specimens of each species a sequencing of RP1 marker from nuclear genome was also performed in order to exclude the presence of hybrid specimens. The correct nuclear identification is ensured by sequence comparison with the results of various studies from both domestic and foreign workplaces (e. g., He et al., 2008; Mayden et al., 2008; Mendel et al., 2008d; etc.) and international initiatives (Cypriniformes Tree of Life; Assembling the Tree of Life; etc.)

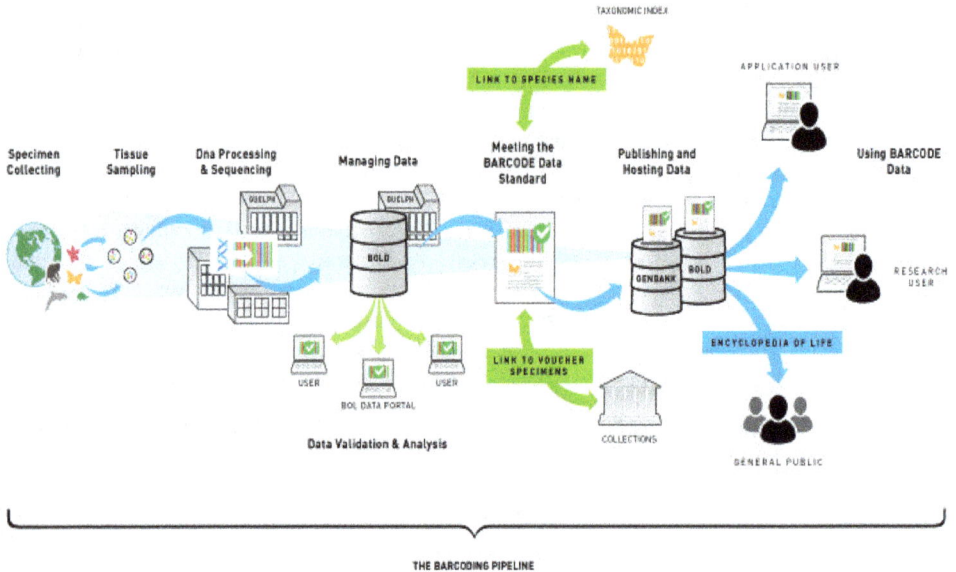

THE BARCODING PIPELINE

Fig. 7. The barcoding pipeline (http://www.barcodeoflife.org/content/about/what-dna barcoding).

The platform of the BoLD system (http://www. barcodinglife.org) with three identification-statistical modules - MAS, IDS and ECS - was used for the organization of the results. The obtained sequences are saved by the CBOL consortium primarily in the BoLD datasystem and then automatically sent, within INSDC (International Nucleotide Sequence Database Collaboration), to the interconnected databases: GenBank, EMBL and DDBJ. For more information see the standardized protocol of the CBOL consortium: http://www.dnabarcoding.ca/pa/ge/research/protocols/.

4. Results

An analysis of the COI gene sequences of 500-652 bp in length was performed on 820 individuals from 67 species, which represent 93.1% of the Czech ichthyofauna (Table 1). The map of sampling distribution (Fig. 8) clearly shows that we have evenly sampled the whole territory of the Czech Republic. Altogether, 109 localities were sampled. The number of individuals per species ranged from 1 to 31, with an average number of 12 individuals per species. Out of a total number of 72 species, the following species, which occur very rarely, were not analysed: *Micropterus salmoides*, *Romanogobio banaticus*, *Romanogobio belingi*, *Ameirus melas* and *Acipenser baeri*. Only one or two individuals were analysed in some critically endangered (*Zingel zingel*, *Zingel streber*, *Sabanejewia balcanica*, *Eudontomyzon mariae*, *Pelecus cultratus*, *Ballerus sapa*, *Gymnocephalus schraetser*), and endangered (*Misgurnus fossilis*) species. Neither insertions/deletions nor stop codons were found, thus encouraging the view that all amplified sequences represent functional COI gene sequences.

Family	Species	N	No of haplotypes	Intraspecies difference range (%)
	Acipenser gueldenstaedtii	3 (+1)	1 (+1)*	0 (4.484)
Acipenseridae	*Acipenser ruthenus*	4	1	0
	Acipenser stellatus	5	1	0
Anguillidae	*Anguilla anguilla*	5	5	0.154 – 3.494
Balitoridae	*Barbatula barbatula*	27	13	0 – 5.967
Centrarchidae	*Lepomis gibbosus*	10	1	0
	Cobitis sp.	3	2	0 – 0.677
Cobitidae	*Misgurnus fossilis*	1	1	-
	Sabanejewia balcanica	1	1	-
Cottidae	*Cottus gobio*	21	4	0 – 0.773
	Cottus poecilopus	15	6	0 – 2.681
	Abramis brama	29	4	0 – 0.617
	Alburnoides bipunctatus	14	5	0 – 0.617
	Alburnus alburnus	26	4	0 – 0.308
	Aspius aspius	14	2	0 – 0.308
	Ballerus ballerus	3	1	0
	Ballerus sapa	2	1	0
	Barbus barbus	25	2	0 – 0.163
	Blicca bjoerkna	14 (+1)	4 (+1)**	0 – 0.308 (4.316)
	Carassius carassius	16	5	0 – 1.266
	Carassius gibelio	27	6	0 – 0.621
	Carassius langsdorfii	3	1	0
	Ctenopharyngodon idella	3	1	0
	Cyprinus carpio	10	2	0 – 0.31
	Gobio gobio	27 (+1)	3 (+1)***	0 – 0.308 (11.420)
Cyprinidae	*Gobio obtusirostris*	5	3	0 – 0.308
	Hypophthalmichthys molitrix	3	1	0
	Chondrostoma nasus	25	5	0 – 0.308
	Leucaspius delineatus	5	2	0 – 3.808
	Leuciscus idus	13	4	0 – 0.621
	Leuciscus leuciscus	21 (+5)	2 (+1)†	0 – 0.161 (0.933)
	Pelecus cultratus	2	1	0
	Phoxinus phoxinus	25	10	0 – 3.973
	Pseudorasbora parva	22	7	0 – 1.084
	Rhodeus amarus	18	7	0 – 1.244
	Romanogobio vladykovi	7	3	0 – 0.618
	Rutilus rutilus	31	7	0 – 0.795
	Scardinius erythrophthalmus	16	1	0
	Squalius cephalus	28	4	0 – 5.252
	Tinca tinca	24	3	0 – 1.521
	Vimba vimba	4	1	0
Esocidae	*Esox lucius*	22	4	0 – 0.772
Gasterosteidae	*Gasterosteus aculeatus*	2	1	0
Gobiidae	*Neogobius melanostomus*	5	1	0
	Proterorhinus semilunaris	6	2	0 – 1.087
Ictaluridae	*Ameiurus nebulosus*	2	1	0
Lotidae	*Lota lota*	16	3	0 – 0.308

Percidae	Gymnocephalus baloni	1	1	-
	Gymnocephalus cernua	16	5	0 – 0.798
	Gymnocephalus schraetser	2	1	0
	Perca fluviatilis	29	5	0 – 0.772
	Sander lucioperca	16	1	0
	Sander volgensis	2	1	0
	Zingel streber	1	1	-
	Zingel zingel	1	1	-
Petromyzontidae	Eudontomyzon mariae	2	2	0.197
	Lampetra planeri	14	3	0 – 0.308
Salmonidae	Coregonus maraena	4 (+1)	1 (+1)††	0 (1.558)
	Coregonus peled	5	4	0 – 0.618
	Hucho hucho	2	1	0
	Oncorhynchus mykiss	24	5	0 – 0.308
	Salmo salar	3	2	0 – 0.154
	Salmo trutta	31	4	0 – 1.087
	Salvelinus fontinalis	16	4	0 – 0.617
	Thymallus thymallus	16	2	0 – 0.155
Siluridae	Silurus glanis	11	1	0
Umbridae	Umbra krameri	5	1	0

Table 1. List of species analysed in the current study. N - number of studied individuals (further individuals with haplotypes shared with another species in parentheses). No of haplotypes - number of COI haplotypes found (further haplotypes shared with another species in parentheses). * - shared haplotype with *Acipenser ruthenus*. ** - shared haplotype with *Abramis brama*. *** - shared haplotype with *Romanogobio vladykovi*. † - shared haplotype with *Leuciscus idus*. †† - shared haplotype with *Coregonus peled*. Intraspecies difference range - range of intraspecific genetic distances excluding haplotypes shared with another species (maximum genetic distance including shared haplotypes in parentheses).

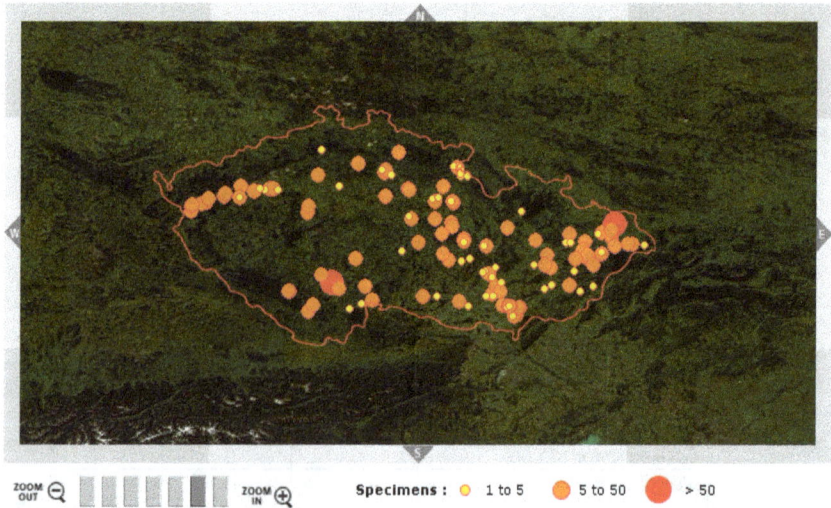

Specimens :   ○ 1 to 5   ● 5 to 50   ● > 50

Fig. 8. Overview of sampling localities in the Czech Republic. The borders of the Czech Republic are schematically drawn in red.

As we expected, we found a growing genetic divergence with raising taxonomical level. On the intra-species level, the distances ranged 0-5.97% (Fig. 9), not including inter-specific shared haplotypes (hybrids, misidentifications etc.). Intra-specific diversities exceeding 5% were detected in *Barbatula barbatula* (the highest value of 5.97% was found here), and in *Squalius cephalus* (0-5.25%), values exceeding 2% were found in *Phoxinus phoxinus* (0-3.97%), *Anguilla anguilla* (0.15-3.49%), *Leucaspius delineatus* (0-3.81%) and *Cottus poecilopus* (0-2.68%). In these six mentioned species, there was an apparent deep divergence of the lineages that had been assigned to a single species. Two or three (*B. barbatula, S. cephalus*) lineages were distinguished in each species, both from the phylogenetic tree (Fig. 10) and the list of variable sites (Fig. 11). Two lineages can also be distinguished in the following species, although they possess lower genetic distances: *Tinca tinca* (0-1.52%) and *Carassius carassius* (0-1.27%). In *Rhodeus amarus,* a single individual diverges from all other, which raises the mean intra-species diversity from 0.46 to 1.09%.

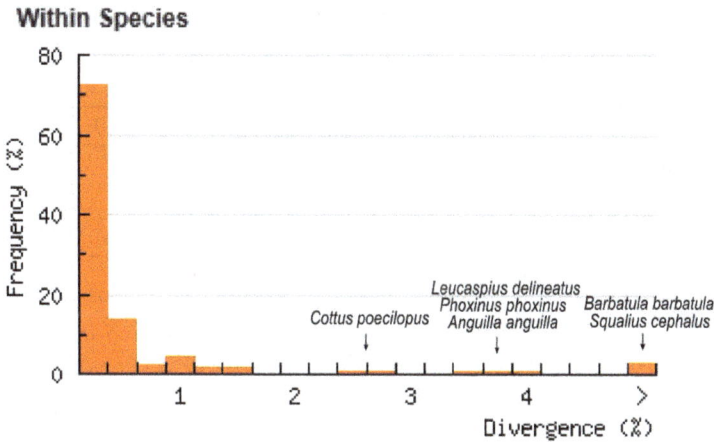

Fig. 9. Plot of an average intraspecific divergence. Species that exceeded 2% are listed.

An analysis of distribution of the nearest-neighbour distance (NND, the minimum distance between a species and it closest – usually congeneric – relative, Table 2) has shown that only a single species pair exhibited NND lower than 1%, that being namely: *Leuciscus leuciscus* and *L. idus*. The NND averaged 9.67%, which was 40 times higher than the mean within species distance (ca. 0.25%) and 14 times higher than the mean maximum intraspecific distance (0.70%).

We have detected altogether nine cases of shared barcode haplotypes: one *Acipenser gueldenstaedtii* individual possessed an *A. ruthenus* haplotype, one *Blicca bjoerkna* individual possessed an *Abramis brama* haplotype, one *Gobio gobio* individual possessed a *R. vladykovi* haplotype, one *Coregonus maraena* possessed a *C. peled* haplotype, and five *Leuciscus leuciscus* carried a *L. idus* haplotype.

A phylogenetic tree based on all sequences longer than 500 bp, employing the Kimura-2-parametr (K2P) model is available on http://www.ivb.cz/projects-molecular-biodiversity-inventory-of-the-ichthyofauna-of-the-czech-republic.html. Subsequently, we selected the representatives of all species or intra-species lineages with intra-species distance at least three times higher (1.176%) than the mean distance within the species (0.392%; Hubert et al., 2008). Based on these sequences, a reduced phylogenetic tree was constructed and put through a bootstrap analysis of 1000 replications. The reduced phylogenetic tree (Fig. 10) documents strong bootstrap support both on the level of genera and on family level. Only the most numerous family Cyprinidae has exhibited moderate support. The *Neogobius* and *Proterorhinus* genera were not supported in forming a single clade of the Gobiidae family.

The phylogenetic tree supports the most recently proposed changes in the scientific taxonomical nomenclature. According to recent taxonomical opinions and studies (Froese & Pauly, 2010; Kottelat & Freyhof, 2007; Perea et al., 2010), *A. sapa* and *A. ballerus* have been currently sorted into the genus *Ballerus*; *Abramis bjoerkna* now belongs again to the *Blicca* genus, *L. cephalus* into *Squalius* and it is suggested that *Aspius aspius* be included into *Leuciscus*. All these proposales are reflected in the tree, as the above mentioned species always cluster together.

Order	Family	N	< 0.1	0.1 - 1.0	1.0 - 2.7	> 2.7
Acipenseriformes	Acipenseridae	3				3
Cypriniformes	Cobitidae	3				3
	Cyprinidae	30		2*	2**	26
	Balitoridae	1				1
Scorpaeniformes	Cottidae	2				2
Salmoniformes	Salmonidae	8			2***	6
Petromyzontiformes	Petromyzontidae	2				2
Esociformes	Umbridae	1				1
	Esocidae	1				1
Perciformes	Gobiidae	2				2
	Percidae	8				8
	Centrarchidae	1				1
Gadiformes	Lotidae	1				1
Anguilliformes	Anguillidae	1				1
Siluriformes	Siluridae	1				1
	Ictaluridae	1				1
Gasterosteiformes	Gasterosteidae	1				1
Totals:		67	0	2	4	61

Table 2. Summary of Czech ichthyofauna diversity and distribution of genetic distance of each species analysed to nearest neighbour (in per cent). N - number of species in family.
* - *Leuciscus idus* and *Leuciscus leuciscus* (0.62%). ** - *Blicca bjoerkna* and *Vimba vimba* (2.51%).
*** - *Coregonus peled* and *Coregonus maraena* (1.4%).

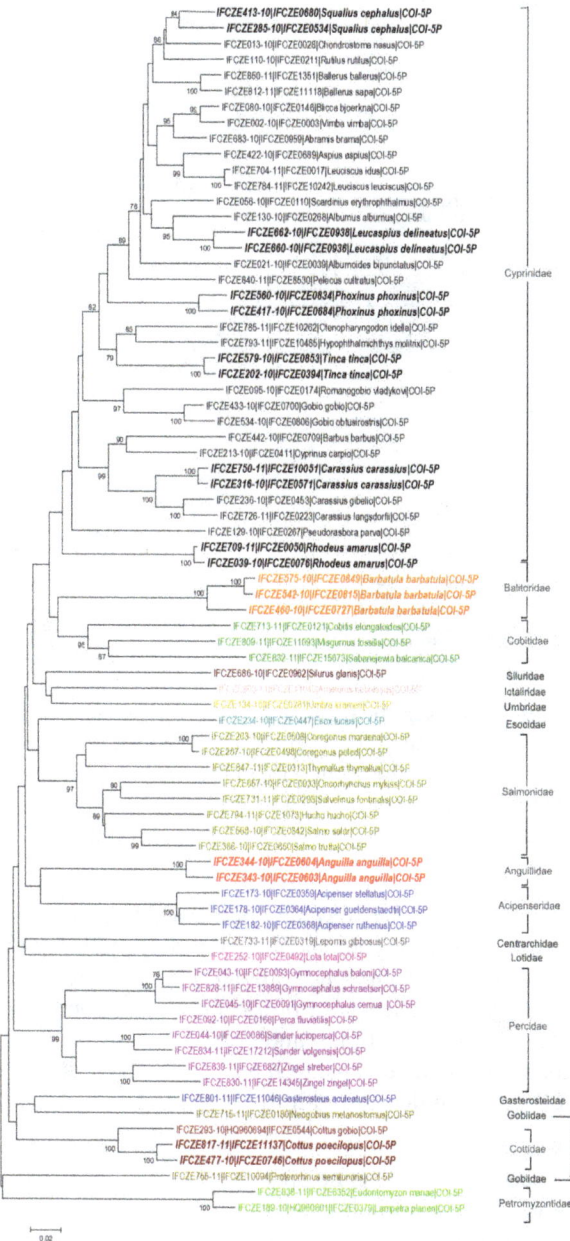

Fig. 10. K2P/NJ tree of representatives of each species, or lineage within species. Bootstrap values are listed near the nodes, only values ≥75% are shown. Families are listed in colour. Species showing an internal divergence more than three times higher than the mean distance within the species are listed in bold italics.

```
[                    1222222 2223333333 4444445555 5555555666 66]        [                          1222333 3444445556 666]
[                    2389223345 6890112379 0113680015 6667899011 22]      [                          2247128114 7013802380 012]
[                    5150035870 5081362408 9250645810 2581038469 25]      [                          5605796030 0059426001 438]
#IFCZE0815|Barbatula_barbatula  ACCCACCGCG GCCTAGGAAC CGTAAGGATC TGCACCCGCG GA   #IFCZE0248|Anguilla_anguilla AGCTCAACAA GCTCGTCTGG CTA
#IFCZE0556|Barbatula_barbatula  .......... ........G. .......... ...... ..       #IFCZE0603|Anguilla_anguilla GA........ .......... A..
#IFCZE0802|Barbatula_barbatula  .......... .......... ...G...... .T.... ..       #IFCZE0771|Anguilla_anguilla .A........ .......... A..
#IFCZE0809|Barbatula_barbatula  .......... .......... ...G...... ...... ..       #IFCZE0604|Anguilla_anguilla ..TCT..TGG ATCTACTCAA .CG
#IFCZE11111|Barbatula_barbatula .......... ...C...... .......... ...... ..       #IFCZE0772|Anguilla_anguilla ..TCTGGTG. ATCTACTCAA .CG
#IFCZE0555|Barbatula_barbatula  G........T ....GA..GT ...G..A... .......... A.
#IFCZE0835|Barbatula_barbatula  ........T ....GA..GT ...G..A... .......... A.    [                          2223333 344445556]
#IFCZE0849|Barbatula_barbatula  ........T ....GA..GT ...G..A... .......... A.    [                          1560490147 702394561]
#IFCZE11031|Barbatula_barbatula ........T ....GA..GT ...G..A... .......G.... A.   [                          0548181300 397067329]
#IFCZE11071|Barbatula_barbatula ........T A..GA..GT ...G..A... .......... A.      #IFCZE0187|Cottus_poecilopus AAGAAAAATA CATATTGCC
#IFCZE0726|Barbatula_barbatula  .TGTCTAT.A ATA..AAG.. TAC..A.GCT CTTCT.TAGA A.   #IFCZE1132|Cottus_poecilopus G......... .........
#IFCZE0727|Barbatula_barbatula  .TGTCTAT.A ATA..AAG.. TAC..A.GCT CTTCT.TAGA AG   #IFCZE11133|Cottus_poecilopus .......... T........
#IFCZE0730|Barbatula_barbatula  .TGTCTAT.A ATA..AAG.. T.C..A.GCT CTTCT.TAGA A.   #IFCZE11137|Cottus_poecilopus .......... .........
                                                                                 #IFCZE0746|Cottus_poecilopus .CAG.GGGAG .GG.CCAAT
[                    122222233 3333333444 4444445555 56666]                       #IFCZE0750|Cottus_poecilopus .CAG.GGGAG .GGGCCAAT
[                    4600144700 2223457000 0368991235 71235]
[                    7028747447 2581020036 9334797023 10242]              [                          111223345]
IFCZE0830|Squalius_cephalus CAGAATTATG GGAGACACTA GGAACCAATT ACGAA              [                          089041210]
IFCZE0680|Squalius_cephalus .......... ...A..G... A.....G... G....              [                          076249259]
IFCZE0785|Squalius_cephalus .......... ...C...... .......... .....             #IFCZE0264|Tinca_tinca TACGTAGCG
IFCZE0534|Squalius_cephalus TCAGGCCGAA AA.AGT.TCG .AGGTAGGCC .TAGG              #IFCZE0853|Tinca_tinca ........T
                                                                                 #IFCZE0394|Tinca_tinca CGTACGAT.
[                    111 2222333333 4444455556 6]
[                    3448899099 3679144456 0389903561 4]                  [                          2222334 56]
[                    7032512836 2515903627 3643655953 3]                  [                          2550277256 54]
#IFCZE0682|Phoxinus_phoxinus GGCGGCCTGT CAATATCCCA CCGATCTTTT C           [                          0563917550 09]
#IFCZE0684|Phoxinus_phoxinus A...A..... T....CA... ..A....... .          #IFCZE0048|Rhodeus_amarus CTCCCACCCC GC
#IFCZE0686|Phoxinus_phoxinus A...A..... T.G..CA... ..A....... .          #IFCZE0050|Rhodeus_amarus G......... ..
#IFCZE0767|Phoxinus_phoxinus ....A..... T...GCA..G ..A....... .          #IFCZE0494|Rhodeus_amarus GCT....... ..
#IFCZE0805|Phoxinus_phoxinus A...AT.... T....CA... ..A....... .          #IFCZE0633|Rhodeus_amarus G......... A.
#IFCZE0816|Phoxinus_phoxinus AA..A..... T....CA... ..A....... .          #IFCZE0822|Rhodeus_amarus G...T..... ..
#IFCZE0127|Phoxinus_phoxinus ..TTATT.A. TG.CGCGTT. TT.CCGCCA. T          #IFCZE10047|Rhodeus_amarus G.......G. ..
#IFCZE0255|Phoxinus_phoxinus ..TCATTCA. TG.CGCGTT. TT.CCGCCAC T          #IFCZE0076|Rhodeus_amarus G..T.GTT.T .T
#IFCZE0551|Phoxinus_phoxinus ..TT.TT.A. TG.CGCGTT. TT.CCGCCAC T
#IFCZE0834|Phoxinus_phoxinus ..TTATT.A. TG.CGCGTT. TT.CCGCCAC T          [                          1223344]
#IFCZE0914|Phoxinus_phoxinus ..TTATT.AG TG.CGCGTT. TT.CCGCCAC T          [                          66272428]
                                                                         [                          20012077]
[                    1111222 2333334455 5556]                             #IFCZE0571|Carassius_carassius CATATTGC
[                    4780037013 7122895612 4572]                          #IFCZE10150|Carassius_carassius G.......
[                    9926938378 7958877640 1015]                          #IFCZE0392|Carassius_carassius ..CGCCAT
#IFCZE0936|Leucaspius_delineatus GGTGTCCCGT CGGGGATAAA CTAA              #IFCZE0821|Carassius_carassius ..C.CCAT
#IFCZE0938|Leucaspius_delineatus AACACTATAG TAAAAGCGTG TAGG              #IFCZE10051|Carassius_carassius .GCGCCAT
```

Fig. 11. Summary of variable sites in the COI gene in haplotypes of nine species showing an internal divergence more than three times higher than the mean distance within the species. Numbers show variable position of nucleotide in whole 652 bp sequence. Dots represent nucleotide identical with first sequence.

5. Discussion

This study presents the first molecular screening of the whole Czech ichthyofauna. Indeed, the position of our project, „The Ichthyofauna of the Czech Republic" (abbreviated IFCZE in the BoLD database), is in a certain sense unique. Table 3, which compares genetic divergences on various taxonomic levels in projects spanning countries across continents from North America to Asia, Australia and Africa, contains only a single project representing the European ichthyofauna – Czech project. It is the first and currently the only public project in BoLD database dealing with the barcoding of European ichthyofauna on this scale.

Among the analyzed species, there was no species endemic to the Czech Republic. Aside from indigenous species, further 18 non-indigenous species introduced in the past to the Czech ichthyofauna were also tested. One species is considered to be regionally extinct in the wild, 12 species are considered to be critically endangered, five to be endangered and six to be vulnerable according to the IUCN classification (Lusk et al., 2011). Also included in the project was the *Umbra krameri*, while its indigenity to the Czech Republic is currently disputed.

The amplification of the 5′ region of the COI gene (up to 652 bp) was successful for all 820 tested individuals, while utilizing a single primer set for amplification of all tested species. No nuclear copies of the COI gene were detected, neither frame shifts nor mutations causing the emergence of stop codons. By means of DNA barcoding, all tested species could be successfully distinguished. The DNA barcodes were acquired for 67 species. An example of such a barcode record can be seen in Figure 12.

	CAN	MEX	NorthAm.	IND	PHL	NGA	AUS	CZE	
min%									
within species	0	0	0	0	0	0	0	**0**	
within genus	0	0	0	0.10	10.0	1.80	0	**0**	
within family	2.67	1.804	0.185	0.20	5.50	0.46	1.39	**0**	
within order	14.25	17.24	14.53	8.00	16.20	15.52	9.55	**18.07**	
within class	17.49	17.88	16.42	-	-	18.46	14.33	**18.36**	
mean dist%									
within species	0.27	0.45	2.52	0.30	0.60	0.30	0.39	**0.39**	
within genus	8.37	5.10	14.10	6.60	11.70	1.90	9.93	**5.00**	
within family	15.38	13.57	16.00	9.91	17.67	17.19	15.46	**14.85**	
within order	20.60	23.38	21.50	16.00	24.80	21.62	22.18	**24.62**	
within class	24.57	25.34	24.80	-	-	25.40	23.27	**24.87**	
max dist%									S.E.
within species	7.42	2.11	15.45	0.30	9.30	22.38	14.80	**11.61**	0.013
within genus	19.33	16.67	23.68	11.90	15.30	36.61	20.63	**11.20**	0.067
within family	23.22	28.48	29.92	23.10	22.60	24.82	35.7	**24.46**	0.011
within order	29.44	28.77	30.95	23.40	31.50	28.12	37.52	**28.84**	0.013
within class	31.20	31.16	37.50	-	-	31.90	37.39	**30.77**	0.004
species	190	76	605	98	23	59	207	**61**	
genus	85	56	134	78	21	33	122	**51**	
family	28	32	36	53	17	19	?	**17**	
order	20	11	18	24	9	8	?	**11**	
class	2	2	2	4	-	2	?	**2**	

Table 3. Comparison of genetic divergences (K2P model used) within various taxonomic levels among selected freshwater and marine fishes BoLD projects. (CAN-Barcoding of Canadian freshwater fishes, MEX-Freshwater Fishes of Mexico, NorthAm-Freshwater Fishes of North America, IND-Marine Fishes of India, PHL-Freshwater Fish of Taal Lake, Philippines, NGA- Barcoding Freshwater Fishes of Eastern Nigeria, AUS-Marine Fishes of Australia, CZE-Ichthyofauna of the Czech Republic).

The barcoding of the Czech ichthyofauna has not ended yet: several further species (*M. salmoides*, *R. banaticus*, *R. belingi*, *A. baerii*, and *A. melas*) are currently being tested as well. Therefore, the analyzed sample set currently covers 67 of 72 species, or 93.1% of the Czech ichthyofauna. Certain species are present in all six investigated regions and were therefore analysed in greater numbers and in some cases exhibited greater haplotype variability (e.g. *B. barbatula*, 27 individuals/13 haplotypes or *P. phoxinus*, 25 individuals/10 haplotypes. For more details see Table 1.

Overall, the mean congeneric divergence was ca. 13 times higher than the mean conspecific divergence (5.00 vs 0.39%, Table 3). These values are in good concordance with the results of Valdez-Moreno et al. (2009), which studied the diversity of ichthyofauna in Mexico and Guatemala (5.10 vs 0.45%, Table 3). The divergence ratio was higher in studies dealing with the ichthyofauna of Australia (25x, Ward et al., 2005), Canada (31x, Hubert et al., 2008) and the Philippines (19x, Aquilino et al., 2011) (Table 3). Such a low divergence in our project is probably caused by overall more recent speciation of the freshwater species in comparison to their marine counterparts (which are included in the mentioned projects). This is also supported by the results of the marine project from India (27x, Lakra et al., 2011), China (50x, a running project) and Japan (59x, Zhang & Hanner, 2011).

The overlapping of conspecific and congeneric levels of divergence was minimal: only two genera have shown inter-specific distance between the nearest neighbours lower than 2.7%: *Leuciscus* – *L. leuciscus* and *L. idus* (0.62%) and *Coregonus* – *C. peled* and *C. maraena* (1.4%). Interestingly, in one case an inter-specific distance lower than 2.7% occurred between two non-congeneric species: *B. bjoerkna* and *Vimba vimba* (2.51%, Table 2). Zardoya & Doadrio (1999) and Cunha et al. (2002) observed the same effect, based on the sequencing of a different marker. All congeneric species always clustered together in the phylogenetic analysis. Furthermore, several species pairs (not always congeneric) have been shown to possess shared haplotypes. In most cases, a single individual of one species bore a haplotype typical of a close (sister) species. Such was the case of a single *B. bjoerkna* among *A. brama*, *G. gobio* among *R. vladykovi*, *C. maraena* among *C. peled*, and *A. gueldenstaedtii* among *A. ruthenus*. All these cases can be easily explained as simple hybridization, in the case of *Coregonus* even human-driven intentional hybridization, (with the second species being the maternal one due to the maternal inheritance of mtDNA), which was confirmed by the sequencing of the nuclear marker (first intron of the S7 ribosomal protein gene).

The last species pair, *L. leuciscus* and *L. idus*, represented an exception. We found five individuals classified as *L. leuciscus* bearing the haplotype of *L. idus*. These two species also presented the lowest mutual genetic distance from all analyzed species pairs (0.62%). We therefore conclude that in these two species, not only hybridization (as confirmed by the sequencing of S7) but also probably recent speciation might have taken place in the genetic similarity of both species.

Analyses have revealed a significant deep intraspecies divergence in several taxa. In most cases, those are species with large areas (Kottelat & Freyhof, 2007), such as *B. barbatula*, *L. delineatus*, *P. phoxinus* or *S. cephalus*.

The most pronounced example is the stone loach (*B. barbatula*). The DNA barcoding technique has revealed high intra-specific divergence (up to 5.97%, Table 1). We have tested 27 individuals from six regions and we found 13 haplotypes diverging into three lineages (Table 1, Fig. 10, Fig. 11). Šedivá et al. (2008) has described the presence of the lineage V in the Labe River basin and the sublineage V_a in the Morava River basin, both of the lineages belonging to the Danubian clade. Our results are in good concordance with these findings. Furthermore, we have distinguished a third, the most divergent lineage, occurring in the Odra River basin. This lineage possibly belongs to the Eastern clade similarly to previously detected individuals from the same basin in Poland (Šedivá et al., 2008).

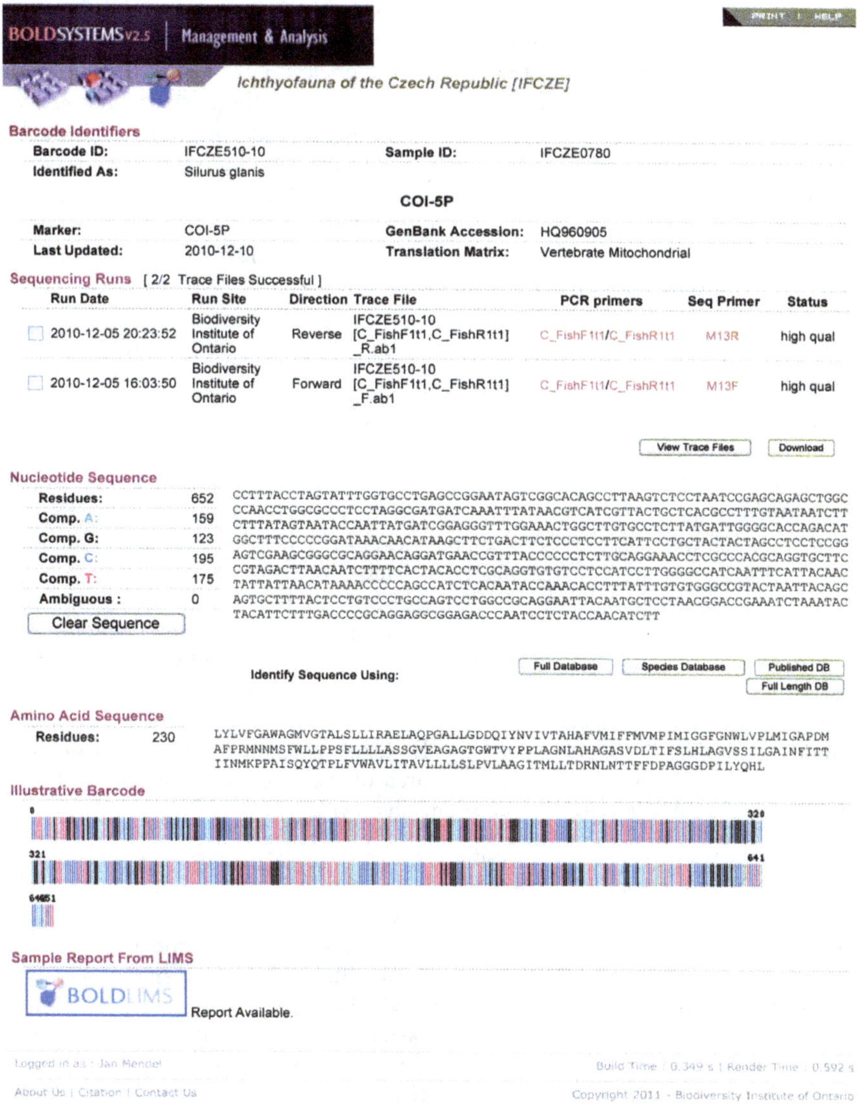
BOLDSYSTEMS v2.5 | Management & Analysis
PRINT | HELP
Ichthyofauna of the Czech Republic [IFCZE]
Barcode Identifiers
Barcode ID: IFCZE510-10
Sample ID: IFCZE0780
Identified As: Silurus glanis
COI-5P
Marker: COI-5P
GenBank Accession: HQ960905
Last Updated: 2010-12-10
Translation Matrix: Vertebrate Mitochondrial
Sequencing Runs [2/2 Trace Files Successful]
Run Date | Run Site | Direction | Trace File | PCR primers | Seq Primer | Status
2010-12-05 20:23:52 | Biodiversity Institute of Ontario | Reverse | IFCZE510-10 [C_FishF1t1,C_FishR1t1] _R.ab1 | C_FishF1t1/C_FishR1t1 | M13R | high qual
2010-12-05 16:03:50 | Biodiversity Institute of Ontario | Forward | IFCZE510-10 [C_FishF1t1,C_FishR1t1] _F.ab1 | C_FishF1t1/C_FishR1t1 | M13F | high qual
View Trace Files
Download
Nucleotide Sequence
Residues: 652
Comp. A: 159
Comp. G: 123
Comp. C: 195
Comp. T: 175
Ambiguous : 0
Clear Sequence
CCTTTACCTAGTATTTGGTGCCTGAGCCGGAATAGTCGGCACAGCCTTAAGTCTCCTAATCCGAGCAGAGCTGGC
CCAACCTGGCGCCCTCCTAGGCGATGATCAAATTTATAACGTCATCGTTACTGCTCACGCCTTTGTAATAATCTT
CTTTATAGTAATACCAATTATGATCGGAGGGTTTGGAAACTGGCTTGTGCCTCTTATGATTGGGGCACCAGACAT
GGCTTTCCCCCGGATAAACAACATAAGCTTCTGACTTCTCCCTCCTTCATTCCTGCTACTACTAGCCTCCTCCGG
AGTCGAAGCGGGCGCAGGAACAGGATGAACCGTTTACCCCCCTCTTGCAGGAAACCTCGCCCACGCAGGTGCTTC
CGTAGACTTAACAATCTTTTCACTACACCTCGCAGGTGTGTCCTCCATCCTTGGGGCCATCAATTTCATTACAAC
TATTATTAACATAAAACCCCCAGCCATCTCACAATACCAAACACCTTTATTTGTGTGGGCCGTACTAATTACAGC
AGTGCTTTTACTCCTGTCCCTGCCAGTCCTGGCCGCAGGAATTACAATGCTCCTAACGGACCGAAATCTAAATAC
TACATTCTTTGACCCCGCAGGAGGCGGAGACCCAATCCTCTACCAACATCTT
Identify Sequence Using:
Full Database
Species Database
Published DB
Full Length DB
Amino Acid Sequence
Residues: 230
LYLVFGAWAGMVGTALSLLIRAELAQPGALLGDDQIYNVIVTAHAFVMIFFMVMPIMIGGFGNWLVPLMIGAPDM
AFPRMNNMSFWLLPPSFLLLLASSGVEAGAGTGWTVYPPLAGNLAHAGASVDLTIFSLHLAGVSSILGAINFITT
IINMKPPAISQYQTPLFVWAVLITAVLLLLSLPVLAAGITMLLTDRNLNTTFFDPAGGGDPILYQHL
Illustrative Barcode
0
320
321
641
Sample Report From LIMS
BOLDLIMS
Report Available.
Logged in as : Jan Mendel
Build Time : 0.349 s | Render Time : 0.592 s
About Us | Citation | Contact Us
Copyright 2011 - Biodiversity Institute of Ontario

Fig. 12. An example of barcode record of the species.

The second deepest divergence has been detected in chub (*S. cephalus*). We have tested 28 individuals and found four haplotypes, diverging into two lineages (Table 1, Fig. 10, Fig. 11), separated by 5.25%. The less abundant lineage occurred only in the North Sea drainage area (Labe and Vltava River basins), the second, more numerous lineage was found across all river basins. Seifertová et al. (2008) in their large study of phylogenetic diversity of chub across its European range distinguished four different haplogroups, corresponding to four lineages, i.e. the Western, Eastern, Aegean and Adriatic (*sensu*

Durand et al., 1999) ones, with the populations collected in the Czech Republic belonging to the Western Lineage. It is probable that our more prevalent lineage corresponds to the Western lineage, sharing its distribution pattern. It is unclear, but possible that the second of our lineages corresponds to the Eastern lineage. However, the Eastern lineage was not found by Seifertova et al. (2008) in the North Sea basin, but in populations from the Vistula River basin in Poland and in Finland. Therefore we cannot exclude the possibility that we have found a new, fifth lineage. A definite conclusion about the identity of lineages can not be drawn due to different markers utilized in both studies. The existence of other deeply divergent lineages of chub is also evident e.g. in Perea et al. (2010), who found two lineages of chub in their study of phylogenetic relationships in the Circum-Mediterranean subfamily Leuciscinae. It is clear that the taxonomy of *S. cephalus* needs to be further addressed; both on the level of additional samples, which we have already collected, and by the unification of utilized markers.

The DNA barcoding further discriminated two deep lineages in two other cyprinid species, *P. phoxinus* (lineages separated by 3.97%) and *L. delineatus* (lineages separated by 3.81%). In the most recent Red List of the Czech Republic, the minnow (*P. phoxinus*) is classified as vulnerable (Lusk et al., 2011). The priority task in the protection of minnow lies in the long-term sustenance of water quality, the segmentation of river bed and banks together with thoughtful fishing and angling management (Hanel & Lusk 2005). We have tested 25 individuals from six regions and we distinguished 10 haplotypes diverging into two lineages (Table 1, Fig. 10, Fig. 11). Lineage I is located in water courses flowing into the North and Black Sea drainage areas and the lineage II is located in the Baltic Sea drainage area (including the Bečva River). While the Bečva River currently belongs to the Black Sea drainage area, here we can consider an ancient proximity or identical origin of the two populations, as during the Pleistocene (e.g. the Elster glaciation) the waters of the Odra flowed towards the south into the drainage area of the river Bečva (Czudek, 1997). The systematics of the genus *Phoxinus* has not been solved yet.

The belica (*L. delineatus*) is currently classified among the critically endangered species in the Red List of the Czech Republic (Lusk et al., 2011). Apart from food competition and the continuous forcing out of original localities by invasive species, its vanishing is also probably influenced by the spread of a disease caused by the pathogen *Sphaerothecum destruens*, which came to Europe through an invasion the invasive fish species *Pseudorasbora parva* with the import of herbivorous fish fry (Gozlan et al., 2005; Hanel & Lusk 2005). Interestingly, both our lineages of the belica coexist in the same locality. Those five tested individuals came from one fish hatchery in the Ohře River basin. Currently, there is no detailed phylogenetic study of this species. Only Perea et al. (2010) dealt with the taxonomy and systematics of this genus, but based on two individuals only and in the context of general phylogenetic relationships and biogeographical patterns in the Circum-Mediterranean subfamily Leuciscinae. Our studies of this species continue in order to collect individuals from further localities. We also aim to evaluate hybrid events in the scope of uncovered lineages based on a nuclear marker.

Two separate lineages were also detected in eel (*A. anguilla*) (3.49%) and Siberian bullhead (*C. poecilopus*) (2.68%). The intra-specific diversity of eel on a molecular level has not been studied yet. Both of our two lineages were found in North Sea and Baltic Sea drainage areas. However, when performing a comparison of our sequences with other sequences stored in

BoLD, one of these lineages is identical to samples of the American eel (*A. rostrata*). Kottelat & Freyhof (2007) mention that the American eel has been stocked in Eastern Europe; our findings confirm their suspicion that the presence of isolated individuals of American eel cannot be ruled out in our free waters. In order to confirm these findings, further samples will be evaluated.

In the Siberian bullhead, Lusk et al. (2008a) distinguished two complexes based on the sequencing of the cyt *b* gene and mitochondrial control region: the "baltica" complex, which consists of the populations from the Baltic Sea drainage area, and the "danubialis" complex, which consists of the populations from the Black Sea drainage area. An existence of the lineages uncovered in this study further supports these findings: our tested populations from the Odra River basin fall into the "baltica" complex, the individuals from the Morava and Dyje River basins fall into the "danubialis" complex. Lusk et al. (2008a) also further discuss the necessity of a more complex genetic analysis (utilizing a nuclear marker in combination with morphological characteristics) in order to properly evaluate the possible taxonomic status of these complexes.

All these findings (visualized in a bootstrapped NJ tree, Fig. 10) are based on the sequencing of a single marker. Thus they are not *per se* a sufficient tool and criterion for a definitive taxonomical and phylogenetic consequences. For this reason, our project has employed a more complex approach: based on these findings, a sequencing analysis of a nuclear marker must be performed, and comparisons of the morphological characteristics of the lineages must be evaluated. Such complex evaluations will subsequently serve for the definitive confirmation of taxonomic conclusions, and also for the reference collection of the fish which will be stored in the National Museum of Natural History, Prague.

6. Conclusion

The presented study provides a clear example of the usefulness and suitability of DNA barcoding for the cataloguing of the biodiversity of the Czech freshwater fishes. The lineages newly uncovered in our study are being subjected to further detailed research in order to evaluate their correct taxonomical status and systematic position. Definitive results of the project, once obtained, can probably serve as a good starting point for subsequent comparatory phylogenetic studies in an Euroasian context. The presented project also enabled us to focus our attention also on taxa which have never been in the scope of molecular-genetic investigation (e.g. *L. delineatus*, *P. phoxinus*, *Gasterosteus aculeatus*, *A. anguilla*, etc.).

New taxonomically and systematically important knowledge may (and hopefully will) contribute to the updating of the background information for the monitoring of the NATURA 2000 system. It also brings crucial information and recommendations to the Agency for Nature Conservation and Landscape Protection of the Czech Republic. A new design of the species collection of type individuals, constructed on the basis of most recent molecular methods and including morphological data, as well as organizing a unified voucher, is also a great benefit for the National Museum of Natural Science in Prague.

By employing a complex approach (analysis of both the nuclear and mitochondrial genomes plus morphological analyses) we were able to eliminate the danger of including a hybrid individual as a representative of a particular species into the type series. Combined preservation approaches (whole individuals stored in formaldehyde, fin clippings in ethanol, DNA isolates) targeted at individuals from the new type collections will also enable possible comparative DNA analyses in future, whenever needed. All the results are clearly and concisely assembled in the BoLD database, which allows very simple and quick on-line access to informations for the broadest scientific and for the general public.

The UN has declared the year, 2010, to be the International Year of Biodiversity. Let us all wish much success to all future projects and initiatives dealing with biodiversity; for every successful project will bring us closer to the ultimate goal of the knowledge and understanding of the biodiversity.

7. Acknowledgement

This study was carried out within the framework of research projects no. M200930901 supported by the Program of Internal Support for International Collaborative Projects of the Academy of Sciences of the Czech Republic and no. 206/09/P608 supported by the Grant Agency of the Czech Republic. We would like to thank our colleagues from the Biodiversity Institute of Ontario, University of Guelph for their financial, laboratory and informatics support, the Czech and Moravian Fishing Unions, the representatives of the water-resources authorities (pond management companies, river basin management agencies), the management of the Protected Landscape Areas, our colleagues from universities and other people for their help with sample collection.

Data accessibility

Cytochrome c oxidase subunit I gene sequences, trace files, digital images of fish and other metadata were submitted to Barcoding of Life Data systems (BoLD – http://www.barcodinglife.com) and were given BoLD Process ID's IFCZE001-10 to IFCZE701-10 and IFCZE702-11 to IFCZE852-11, all being part of the project „The ichthyofauna of the Czech Republic (IFCZE). The complete NJ/K2P tree can be found on the website of IVB on http://www.ivb.cz/projects-molecular-biodiversity-inventory-of-the-ichthyofauna-of-the-czech-republic.html. Acquired DNA sequences are also being stored in GenBank with accession numbers HQ960417-HQ961093.

8. References

April, J., Mayden, R. L., Hanner, R. H. & Bernatchez, L. (2011). Genetic calibration of species diversity among North America`s freshwater fishes. Proceedings of the National Academy of Sciences of the United States of America, *In Press*

Aquilino, S. L., Tango, J. M., Fontanilla, I. C., Pagulayan, R. C., Basiao, Z. U., Ong, P. S. & Quilang, J. P. (2011). DNAbarcoding of the ichthyofauna of Taal Lake, Philippines. *Molecular Ecology Resources*, Vol. 11, No. 4, pp. 612–619, ISSN 1755-0998

Ardura, A., Linde, A. R., Moreira, J. C. & Garcia-Vazquez, E. (2010). DNA barcoding for conservation and management of Amazonian commercial fish. *Biological Conservation*, Vol. 143, No. 6, pp. 1438-1443, ISSN 0006-3207

Avise, J. C. (2000). *Phylogeography. The history and formation of species.* Harvard University Press, ISBN 978-0674666382, Cambridge

Barrett, R. D. H. & Hebert, P. D. N. (2005). Identifying spiders through DNA barcodes. *Canadian Journal of Zoology*, Vol. 83, No. 3, pp. 481-491, ISSN 1480-3283

Bartoňová, E., Papoušek, I., Lusková, V., Koščo, J., Lusk, S., Halačka, K., Švátora, M. & Vetešník, L. (2008). Genetic diversity and taxonomy of *Sabanejewia balcanica* (Osteichthyes: Cobitidae) in the waters of the Czech Republic and Slovakia. *Folia Zoologica*, Vol. 57, No. 1-2, pp. 60-70, ISSN 0139-7893

Baruš, V. & Oliva, O. (1995). *Mihulovci a Ryby*. Academia, ISBN 80-200-0501-3, Praha

Cunha, C., Mesquita, N., Dowling, T. E., Gilles, A. & Coelho, M. M. (2002). Phylogenetic relationships of Eurasian and American cyprinnids using cytochrome *b* sequences. *Journal of Fish Biology* Vol. 61, No. 4, pp. 929-944. ISSN 1095-8649

Czudek, T. (1997). *Reliéf Moravy a Slezska v kvartéru*. Sursum, ISBN 80-85799-27-8, Tišnov, (in Czech)

deWaard, J. R., Ivanova, N. V., Hajibabaei, M. & Hebert, P. D. N. (2008). Assembling DNA barcodes. Analytical protocols. *Methods in Molecular Biology*, Vol. 410, pp. 275-293, ISSN 1064-3745

Durand, J. D., Persat, H. & Bouvet, Y. (1999). Phylogeography and postglacial dispersion of the chub (*Leuciscus cephalus*) in Europe. *Molecular Ecology*, Vol. 8, No. 6, pp. 989-997, ISSN 0962-1083

Fisher, B. L. (2006). DNA Barcoding: Tomorrow Is Too Late. *California Wild*, Vol. 59, No. 1, pp. 1-3

Freeland, J. R. (2005). *Molecular ecology*. John Wiley & Sons Ltd, ISBN 978-0-470-09062-6, Chichester

Froese, R. & Pauly, D. (2010). FishBase. World Wide Web electronic publication. Available from: www.fishbase.org

Gozlan, R. E., St-Hilaire, S., Feist, S. W., Martin, P. & Kent, M. L. (2005). Biodiversity: Disease threat to European fish. *Nature*, Vol. 435, No. 7045, p. 1046, ISSN 0028-0836

Hajibabaei, M., deWaard, J. R., Ivanova, N. V., Ratnasingham, S., Dooh, R. T., Kirk, S. L., Mackie, P. M. & Hebert, P. D. N. (2005). Critical factors for assembling a high volume of DNA barcodes. *Philosophical Transactions of the* Royal *Society B: Biological Science*, Vol. 360, No. 1462, pp. 1959-1967, ISSN 1471-2970

Hajibabaei, M., Smith, M. A., Janzen, D. H, Rodriguez, J. J., Whitfield, J. B. & Hebert, P. D. N. (2006). A minimalist barcode can identify a specimen whose DNA is degraded. *Molecular Ecology Notes*, Vol. 6, No. 4, pp. 959–964, ISSN 1471-8286

Hajibabaei, M., Singer, G. A. C., Hebert, P. D. N. & Hickey, D. A. (2007). DNA barcoding: how it complements taxonomy, molecular phylogenetics and population genetics. *Trends in Genetics*, Vol. 23, No. 4, pp. 167-172, ISSN 0168-9525

Hammond, P. M. (1992). Species inventory, In: *Global Biodiversity: status of the Earth's living resources*, B. Groombridge (ed.), pp. 17–39, Chapman and Hall, ISBN 0-41247240-6, London

Hanel, L. & Lusk, S. (2005). *Ryby a mihule České republiky, rozšíření a ochrana*. ČSOP Vlašim, ISBN 80-86327-49-3,Vlašim

He, S., Mayden, R. L., Wang, X., Wang, W., Tang, K. L., Chen, W. J. & Chen, Y. (2008). Molecular phylogenetics of the family Cyprinidae (Actinopterygii: Cypriniformes) as evidenced by sequence variation in the first intron of S7 ribosomal protein-coding gene: Further evidence from a nuclear gene of the systematic chaos in the family. *Molecular Phylogenetics and Evolution,* Vol. 46, No. 3, pp. 818-829, ISSN 1055-7903

Hebert, P. D. N., Cywinska, A., Ball, S. L & deWaard, J. R. (2003a). Biological identifications through DNA barcodes. *Proceedings of the Royal Society B: Biological Science,* Vol. 270, pp. 313-321

Hebert, P. D. N., Ratsingham, S. & Dewaard, J. R. (2003b). Barcoding animal life: cytochrome c oxidase subunit 1 divergences among closely related species. *Proceedings of the Royal Society B: Biological Science,* Vol. 270 (Suppl 1), pp. 96-99, ISSN 0962-8452

Hebert, P. D. N., Stoeckle, M. Y., Zemlak, T. S. & Francis, Ch. M. (2004). Identification of Birds through DNA Barcodes. *PLoS Biology,* Vol. 2, No. 10, pp. e312, ISSN 1545-7885

Hebert, P. D. N. & Barrett, R. D. H. (2005). Reply to the comment by L. Prendini on "Identifying spiders through DNA barcodes". *Canadian Journal of Zoology,* Vol. 83, No.3, pp. 505–506, ISSN 1480-3283

Hubert, N., Hanner, R., Holm, E., Mandrak, N. E., Taylor, E., Burridge, M., Watkinson, D., Dumont, P., Curry, A., Bentzen, P., Zhang, J., April, J. & Bernatchez, L. (2008). Identifying Canadian Freshwater Fishes through DNA Barcodes. *PloS One,* Vol. 3, No. 6, pp. e2490, ISSN 1932-6203

Ivanova, N. V., Zemlak, T. S., Hanner, R. H. & Hebert, P. D. N. (2007). Universal primer cocktails for fish DNA barcoding. *Molecular Ecology Notes,* Vol. 7, No. 4, pp. 544-548, ISSN 1471-8286

Janko, K., Flajšhans, M.., Choleva, L., Bohlen, J., Šlechtová, V., Rábová, M.., Lajbner, Z., Šlechta, V., Ivanova, P., Dobrovolov, I., Culling, M., Persat, H. Kotusz, J. & Ráb, P. (2007). Diversity of European spined loaches (genus Cobitis L.): an update of the geographic distribution of the Cobitis taenia hybrid complex with a description of new molecular tools for species and hybrid determination. *Journal of Fish Biology,* Vol. 71, No. 2, pp. 387- 408, ISSN 1095-8649

Kottelat, M. & Freyhof, J. (2007). *Handbook of European freshwater fishes.* Kottelat, Cornol, Switzerland and Freyhof, ISBN 978-2-8399-0298-4, Berlin

Kress, W. J., Wurdack, K. J., Zimmer, E. A., Weigt, L. A. & Janzen, D. H. (2005). Use of DNA barcodes to identify flowering plants. *PNAS,* Vol. 102, No. 23, pp. 8369-8374, ISSN 0027-8424

Lakra, W., Verma, M. & Goswami, M. (2011). DNA barcoding Indian marine fishes. *Molecular Ecology Resources,* Vol. 11, No. 1, pp. 60–71, ISSN 1755-0998

Lusk, S., Bartoňová E., Lusková, V., Lojkásek, B. & Koščo, J. (2008a). Vranka pruhoploutvá *Cottus poecilopus* – rozšíření a genetická diverzita v povodí řek Morava, Odra (Česká republika) a Hornád (Slovensko) [Siberian sculpin *Cottus poecilopus*: distribution and genetic diversity in the Morava, Odra (Czech Republic), and Hornád (Slovakia) drainage areas], In: *Biodiverzita ichtyofauny ČR (VII),* S. Lusk, V. Lusková (Ed.), pp. 67-80, Ústav biologie obratlovců AV ČR, v.v.i, ISBN 978-80-87189-01-6, Brno, (in Czech with English summary)

Lusk, S. & Hanel, L. (2008). Foreword. In: *Biodiverzita ichtyofauny ČR (VII),* S. Lusk, V. Lusková (Ed.), pp. 3-5, Ústav biologie obratlovců AV ČR, v.v.i, ISBN 978-80-87189-01-6, Brno

Lusk, S., Papousek, I., Lusková, V., Halačka,K. & Koščo, J. (2008b). Výskyt a genetická diverzita drska menšího a drska většího v povodí Moravy (Česká republika)

[Occurrence and genetic diversity of Streber *Zingel streber* and Zingel *Zingel zingel* in the Morava river drainage area (Czech Republic)], In: *Biodiverzita ichtyofauny ČR (VII)*, S. Lusk, V. Lusková (Ed.), pp. 81-87, Ústav biologie obratlovců AV ČR, v.v.i, ISBN 978-80-87189-01-6, Brno. (in Czech with English summary)

Lusk, S., Lusková, V., Hanel, L., Lojkásek, B. & Hartvich, P. (2011). Červený seznam mihulí a ryb České republiky – verze 2010, [Red List of lampreys and fishes of the Czech Republic – Version 2010], In: *Biodiverzita ichtyofauny ČR (VII)*, S. Lusk, V. Lusková (Ed.), pp. 68-78, Ústav biologie obratlovců AV ČR, v.v.i, ISBN 978-80-87189-08-5, Brno, (in Czech with English summary)

Lusková, V., Bartoňová, E. & Lusk, S. (2008a). Ostrucha křivočará *Pelecus cultratus* – vzácný objekt ochrany v České republice [The Ziege *Pelecus cultratus*, a rare object of protection in the Czech Republic], In: *Biodiverzita ichtyofauny ČR (VII)*, S. Lusk, V. Lusková (Ed.), pp. 38-45, Ústav biologie obratlovců AV ČR, v.v.i, ISBN 978-80-87189-01-6, Brno. (in Czech with English summary)

Lusková, V., Bartoňová, E. & Lusk, S. (2008b). Karas obecný *Carassius carassius* Linnaeus, 1758 v minulosti obecně rozšířený a v současnosti ohrožený druh v České republice [Crucian carp, *Carassius carassius* Linnaeus, 1758, a common species in the Czech Republic in the past yet endangered at present], In: *Biodiverzita ichtyofauny ČR (VII)*, S. Lusk, V. Lusková (Ed.), pp. 46-53, Ústav biologie obratlovců AV ČR, v.v.i, ISBN 978-80-87189-01-6, Brno. (in Czech with English summary)

Mayden, R. L., Tang, K. L., Wood, R. M., Chen, W.-J., Agnew, M. K., Conway, K. W., Yang, L., Simons, A. M., Bart, H. L., Harris, P. M., Li, J., Wang, X., Saitoh, K., He, S., Liu, H., Chen, Y., Nishida, M. & Miya, M. (2008). Inferring the Tree of Life of the order Cypriniformes, the earth's most diverse clade of freshwater fishes: Implications of varied taxon and character sampling. *Journal of Systematics and Evolution*, Vol. 46, No. 3, pp. 424-438, ISSN 1759-6831

Mecklenburg, C. W., Moller, P. R. & Steinke, D. (2011). Biodiversity of arctic marine fishes: taxonomy and zoogeography. *Marine Biodiversity*, Vol. 41, No. 1, pp. 109–140, ISSN 1867-1624

Mendel, J., Lusková, V., Halačka, K, Lusk, S. & Vetešník, L. (2005). Genetic diversity of *Gobio gobio* populations in the Czech Republic and Slovakia, based on RAPD markers. *Folia Zoologica*, Vol. 54, No. 1, pp. 13-24, ISSN 0139-7893

Mendel, J., Lusk, S., Halačka, K., Lusková, V., Vetešník, L., Ćaleta M. & Ruchin, A. (2008b). Genetická diverzita a poznámky k výskytu ouklejky pruhované *Alburnoides bipunctatus* [Genetic diversity of Spirlin, *Alburnoides bipunctatus*, with notes on its occurrence]. In: *Biodiverzita ichtyofauny ČR (VII)*, S. Lusk, V. Lusková (Ed.), pp. 25-37, Ústav biologie obratlovců AV ČR, v.v.i, ISBN 978-80-87189-01-6, Brno. (in Czech with English summary)

Mendel, J., Lusk, S., Koščo, J., Vetešník, L., Halačka, K. & Papoušek, I. (2008a). Genetic diversity of *Misgurnus fossilis* populations from the Czech Republic and Slovakia. *Folia zoologica*, Vol. 57, No. 1-2, pp. 90-99, ISSN 0139-7893

Mendel, J., Lusk, S., Lusková, V., Koščo, J., Vetešník, L., Halačka, K. (2008c). Druhová pestrost hrouzků rodů *Gobio* a *Romanogobio* na území České republiky a Slovenska [The species diversity of gudgeon of the genera *Gobio* and *Romanogobio* in the territory of the Czech Republic and Slovakia], In: *Biodiverzita ichtyofauny ČR (VII)*, S.

Lusk, V. Lusková (Ed.), pp. 17-24, Ústav biologie obratlovců AV ČR, v.v.i, ISBN 978-80-87189-01-6, Brno, (in Czech with English summary)

Mendel, J., Lusk, S., Vasil'eva, E. D., Vasil'ev, V. P., Lusková, V., Ekmekci, F. G., Erk'akan, F., Ruchin, A., Koščo, J., Vetešník, L., Halačka, K., Šanda, R., Pashkov, A. N. & Reshetnikov, S. I. (2008d). Molecular phylogeny of the genus Gobio Cuvier, 1816 (Teleostei: Cyprinidae) and its contribution to taxonomy. *Molecular Phylogenetics and Evolution*, Vol. 47, No. 3, pp. 1061-1075, ISSN 1055-7903

Messing, J. (1983). New M13 vectors for cloning. *Methods in Enzymology*, Vol. 101, pp. 20-78

Meusnier, I., Singer, G. A. C., Landry, J. F., Hickey, D. A., Hebert, P. D. N. & Hajibabaei, M. (2008). A universal DNA mini-barcode for biodiversity analysis. *BMC Genomics*, Vol. 9, pp. 214, ISSN 1471-2164

Monaghan, M. T., Balke, M., Gregory, T. R. & Vogler, A. P. (2005). DNA-based species delineation in tropical beetles using mitochondrial and nuclear markers In: *Philosophical Transactions of the* Royal *Society* B: *Biological Science*, Avalible from: doi:10.1098/rstb.2005.172.4

Nwani, D. W., Becker, S., Braid, H. E., Ude, E. F., Okogwu, O. I. & Hanner R. (2011). DNA barcoding discriminates freshwater fishes from southeastern Nigeria and provides river system-level phylogeographic resolution within some species. *Mitochondrial DNA, In Press*

Papoušek, I., Lusková, V., Koščo, J., Lusk, S., Halačka, K., Povž, M. & Šumer, Z. (2008a). Genetic diversity of *Cobitis* spp. (Cypriniformes: Cobitidae) from different drainage areas. *Folia Zoologica* , Vol. 57, No. 1-2, pp. 83–89, ISSN 0139-7893

Papoušek, I., Lusková, V., Lusk, S. & Koščo, J. (2008b). Výskyt a genetická diverzita ježdíka žlutého a ježdíka dunajského v České republice. [Occurrence and genetic diversity of Stripped ruffe *Gymnocephalus schraetser* and Danube ruffe *Gymnocephalus baloni* in the Czech Republic], In: *Biodiverzita ichtyofauny ČR (VII)*, S. Lusk, V. Lusková (Ed.), pp. 88-95, Ústav biologie obratlovců AV ČR, v.v.i, ISBN 978-80-87189-01-6, Brno, (in Czech with English summary)

Perea, S., Böhme, M., Zupančič, P., Freyhof, J., Šanda, R., Özuluğ, M., Abdoli, A. & Doadrio, I. (2010). Phylogenetic relationships and biogeographical patterns in Circum-Mediterranean subfamily Leuciscinae (Teleostei, Cyprinidae) inferred from both mitochondrial and nuclear data. *BMC Evolutionary Biology*. Vol. 10, pp. 265, ISSN 1471-2148

Primack, R. B., Kindlmann, P. & Jersáková, J. (2001). Biological principles of nature conservation, Portál, ISBN 80-7178-552-0, Praha

Pyle, R. L., Earle, J. L. & Greene, B. D. (2008). Five new species of the damselfish genus Chromis (Perciformes: Labroidei: Pomacentridae) from deep coral reefs in the tropical western Pacific. *Zootaxa*, Vol. 1671, pp. 3–31, ISSN 1175-5334

Ratnasingham, S. & Hebert, P. D. N. (2007). BOLD: the Barcode of Life Data System (www.barcodinglife.org). *Molecular Ecology Notes*, Vol. 7, No. 3, pp. 355-364, ISSN 1471-8286

Raup, D. M. (1992). *Extinction: bad genes or bad luck?* W. W. Norton & Company, ISBN: 978-0393309270, New York

Rock, J., Costa, F. O., Walker, D. I., North, A. W., Hutchinson, W. F. & Carvalho, G. R. (2008). DNA barcodes of fish of the Scotia Sea, Antarctica indicate priority groups for taxonomic and systematice focus. *Antarctic Science*, Vol. 20, No. 3, pp. 253–262, ISSN 1365-2079

Schindel, D. E. & Miller, S. E. (2005). DNA barcoding a useful tool for taxonomists. *Nature*, Vol. 435, pp. 17, ISSN 1476-4687

Seifertová, M., Vyskočilová, M., Morand, S. & Šimková, A. (2008). Metazoan parasites of freshwater cyprinid fish (Leuciscus cephalus): testing biogeographical hypotheses of species diversity. *Parasitology* Vol. 135, No. 12, pp. 1417-35, ISSN 0031-1820.

Steinke, D., Zemlak, T. S. & Hebert, P. D. N. (2009). Barcoding Nemo: DNA-Based identifications for the ornamental fish Trade. *PLos One*, Vol. 4, No. 7, pp. e6300, ISSN 1932-6203

Šedivá, A., Janko, K., Šlechtová, V., Kotlík, P., Simonović, P. Delic. A. & Vassilev, M. (2008). Around or across the Carpathians: colonization model of the Danube basin inferred from genetic diversification of stone loach (*Barbatula barbatula*) populations. *Molecular Ecology*, Vol. 17, No. 5, pp. 1277-1292, ISSN 0962-1083

Thalmann, O., Hebler, J., Poinar, H. N., Pääbo, S. & Vigilant, L. (2004). Unreliable mtDNA data due to nuclear insertions: a cautionary tale from analysis of humans and other great apes. *Molecular Ecology*, Vol. 13, No. 2, pp. 321–335, ISSN 0962-1083

Valdez-Moreno, M., Ivanova, N. V., Elías-Gutiérrez, M., Contreras-Balderas, S. & Hebert, P. D. N. (2009). Probing diversity in freshwater fishes from Mexico and Guatemala with DNA barcodes. *Journal of Fish Biology*, Vol. 74, No. 2, pp. 377-402, ISSN 1095-8649

Vences, M., Thomas, M., van der Meijden, A., Chiari, Y. & Vieites, D. R. (2005). Comparative performance of the 16S rRNA gene in DNA barcoding of amphibians. *Frontiers in Zoology*, Vol. 2, pp. 5, ISSN 1742-9994

Vetešník, L., Papoušek, I., Halačka, K., Lusková, V. & Mendel, J. (2007). Morphometric and genetic analysis of the *Carassius auratus* complex from an artificial wetland in floodplain of the Morava River (Czech Republic). *Fisheries Science*, Vol. 73, No. 4, pp. 817-822, ISSN 1444-2906

Ward, R. D., Zemlak, T. S., Innes, B. H., Last, P. R. & Hebert, P. D. N. (2005). DNA barcoding Australia's fish species. *Philosophical Transactions of the Royal Society B: Biological Science*, Vol. 360, pp. 1847-1857, ISSN 1471-2970

Ward, R. D., Holmes, B. H. & Yearsley, G. K. (2008). DNA barcoding reveals a likely second species of Asian seabass (barramundi) (Lates calcarifer). *Journal of Fish Biology*, Vol. 72, No. 2, pp. 458–463, ISSN 1095-8649

Ward, R. D., Hanner, R. & Hebert, P. D. N. (2009). The campaign to DNA barcode all fishes, FISH-BOL. *Journal of Fish Biology*, Vol. 74, No. 2, pp. 329–356, ISSN 1095-8649

Wilson, E. O. (1989). Threats to biodiversity. *Scientific American*, Vol. 261, pp. 108-116, ISSN 0036-8733

Wilson, E. O. (1992). *The Diversity of Life*. The Belknap Press of Harvard University Press, ISBN 978-0393319408, Cambridge

Zardoya, R. & Doadrio, I. (1999). Molecular Evidence on the Evolutionary and Biogeographical Patterns of European Cyprinids. *Journal of Molecular Biology*, Vol. 49, pp. 227-237, ISSN 0022-2836

Zhang, J. & Hanner, R. (2011). DNA barcoding is a useful tool for the identification of marine fishes from Japan. *Biochemical Systematics and Ecology*, Vol. 39, No. 1, pp. 31–42, ISSN 0305-1978

16

Shark DNA Forensics: Applications and Impacts on Genetic Diversity

Luis Fernando Rodrigues-Filho[1], Danillo Pinhal[2],
Davidson Sodré[1] and Marcelo Vallinoto[1,3]
[1]Federal University of Pará, Campus of Bragança,
Institute of Coastal Studies (IECOS), Bragança, Pará
[2]Department of Genetics, Biosciences Institute
Sao Paulo State University (UNESP), Botucatu, SP
[3]CIBIO/UP, Research Center in Biodiversity and Genetic Resources
Campus Agrário de Vairão, University of Porto
[1,2]Brazil
[3]Portugal

1. Introduction

Despite their worldwide distribution, an increasing conservation appeal and humans' great fascination with them, many aspects of the biology and life history of sharks remain enigmatic. Shark-focused investigations are important because these animals are prominent elements in the marine ecosystem, and the majority of species play pivotal roles as apex predators in the food web.

Over the past decades, several studies have shown the remarkable and accelerated depletion of the natural stocks of many sharks worldwide; population reductions that range from 50% to 89% have been documented (Kotas et al., 1995, Vooren, 1997; Musick et al., 2000; Baum et al., 2003; Baum & Myers, 2007; Dulvy et al., 2008). This decline has been explained to be a consequence of the extensive, unregulated exploitation of wild stocks by fisheries coupled with the restrictive biological characteristics that are intrinsic to the majority of shark species, such as slow growth, high longevity, late sexual maturity and relatively low fecundity (Branstetter, 1990). Moreover, the pressure that is experienced by shark populations is augmented by accidental capture (by-catch) during fisheries that traditionally exploit bony fish species (e.g., tuna and billfishes).

According to Dulvy et al. (2003), the major forces that drive the extinction of shark species are habitat destruction and exploitation by fisheries. Regarding the fisheries, it is well known that the majority of shark populations may not be able to recover after periods of high fishing pressure, which can lead to a permanent loss of genetic diversity in their wild populations (Stevens et al., 2000).

Currently, several coastal and oceanic sharks have reached the status of threatened species (i.e., vulnerable, endangered or critically endangered) according to data that were released by the Red List of threatened fauna of the International Union for Conservation of Nature

(IUCN, 2011). Assorted events are responsible for the current adverse condition that has been reported for diverse shark species worldwide. These events include the lack of governmental policies that regulate and control fishing activity in several countries; the inefficiency in creating priority areas for conservation; the high commercial value of shark fins, which are utilised as food and for pharmaceutical and medical purposes; and the extremely negative stereotypes, which have been acquired over the years due to relatively rare instances of attacks on swimmers, divers and surfers.

A remarkable example of overexploitation involves the daggernose shark *Isogomphodon oxyrhynchus* (Muller & Henle, 1839) (Figure 1), which is a coastal species with a low reproductive capacity (Bigelow & Schroeder, 1948) and a fragmented distribution between Venezuela and Northern Brazil (Lessa, 1999). Although it is protected by law in Brazil (MMA-IBAMA, 2004), *I. oxyrynchus* continues to be illegally caught by fisheries and is currently listed as critically endangered in the Red List of IUCN.

Fig. 1. A daggernose shark *Isogomphodon oxyrhynchus* specimen from the Guiana coast, Atlantic. This photo is courtesy of Dr. Keiichi Matsuura.

In response to the recognised low sustainability of shark fishing on a global scale, the Food and Agriculture Organization (FAO) developed an international plan of action, and a few countries have implemented equivalent national plans for the management and conservation of shark stocks. The primary recommendation of these plans is the collection of shark-catch and trade data on a species-specific basis because different shark species display a broad array of unique biological characteristics, which ultimately reflects their variable susceptibility to exploitation by fisheries (Heist & Gould, 1999). For this reason, precise species identification is required to determine genetic stock structure and population subdivision (Morgan et al., 2011).

Overall, the biodiversity of sharks is high, although species-specific information that can be used to assist in sustainable resource exploitation is scarce (Ovenden et al., 2011). A major recurrent impediment that is encountered in species monitoring is the historical lack of long-term data on shark landings and commerce on a species-by-species basis; this absence virtually prevents proper inferences regarding population status and future trends (Bonfil, 1994; Castro et al., 1999). In addition, identification is prejudiced due to the illegal finning practice, in which the head and fins, which are key morphological characters for the identification of sharks, are removed at sea. Therefore, unidentifiable landed carcasses are generically recorded as "sharks", which prevents the tracking of the composition of species and detection of species that are protected by law in fisheries (Figure 2). Moreover, the difficulty of accurate species identification is exacerbated by the fact that many exploited sharks, such as those in the orders Carcharhiniformes and Lamniformes (Last & Stevens, 1994; Naylor & Marcus, 1994), are morphologically very similar. This makes the desirable taxonomic identification to the species level a challenge even for experienced taxonomists (Castro, 1993; Stevens & Wayt, 1998; Heist, 2004).

Based on this scenario, research projects that are devoted to the conservation of genetic diversity have sought simple, rapid and low-cost tools that can be immediately implemented. Diverse DNA-based techniques have been developed primarily to assist in the identification of species and the monitoring of shark fishing and trade (Lavery, 1992; Heist & Gold, 1999). These molecular approaches are a resourceful way to identify shark body parts, such as fins, carcass, meat, and processed shark subproducts, which are sold in forms such as extracts and pills (Sebastian et al., 2008; Wong et al., 2009).

Due to the emergence of forensic genetics, individuals from numerous species that were previously not easily identifiable by traditional taxonomy can now be accurately traced back to their geographical region and population of origin using diminutive amounts of DNA. Accordingly, the identification of shark species by molecular methods has contributed to control programs that are directed to fisheries management, trade monitoring and market surveillance, and it has especially facilitated the tracking of endangered and/or heavily exploited shark species (Pank et al., 2001; Clarke et al., 2006; Rodrigues-Filho et al., 2009).

In the present chapter, we review the most important molecular markers and the associated technologies that are employed to resolve species identities and their impacts on the genetic diversity of natural shark stocks.

2. Molecular markers and approaches for shark species identification

Current DNA technologies that have been developed for shark species identification are based on one- or two-step approaches that are applied to polymorphic regions of the nuclear genome, mitochondrial genome or both genomes combined. Among the most popular and well-established methodologies for the discovery and validation of informative molecular markers are polymerase chain reaction (PCR) and nucleotide sequencing.

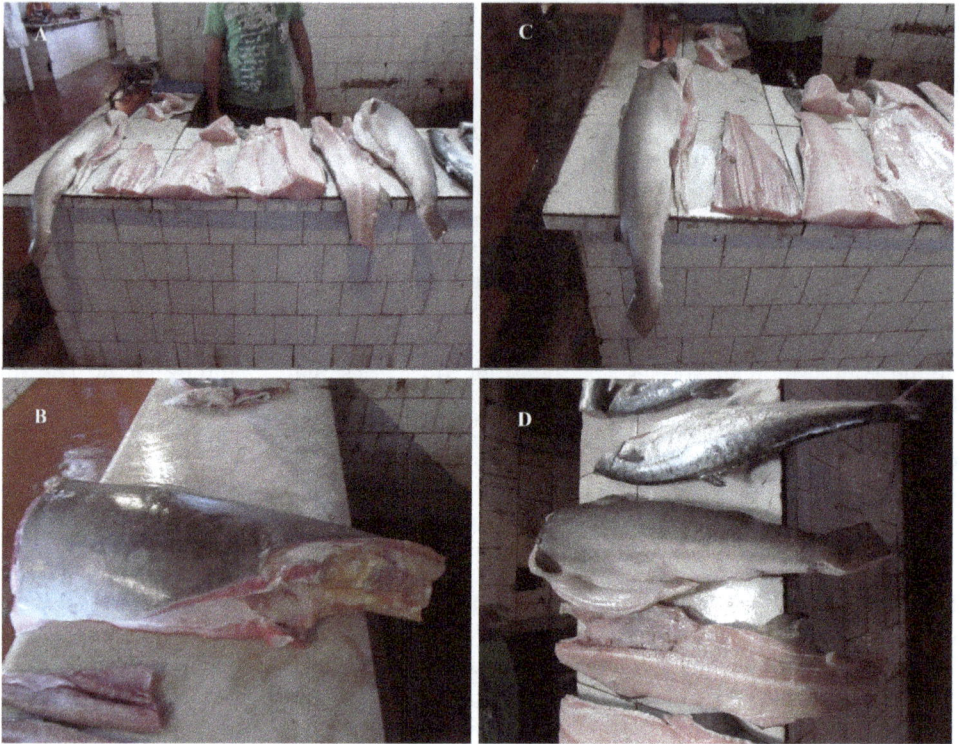

Fig. 2. Headless and finless shark carcasses that were traded in the fish market at Bragança, Pará, Brazil.

2.1 PCR and associated methodologies

PCR (Mullis & Faloona, 1987) has been extensively used for species identification due to its simplicity, specificity and sensitivity. Among molecular markers that have been studied via PCR, the 5S rDNA has gained prominence for species identification by providing consistent results and utilising fast, low-cost strategies (Asensio et al., 2001; Imsiridou et al., 2007). The 5S rDNA is characteristically organised into multiple tandemly arrayed repeats that are distributed on one or several chromosomes in the genome. Structurally, this nuclear ribosomal region consists of highly conserved 120-base pair (bp) coding sequences that are isolated from each other by a variable nontranscribed spacer (NTS) (Long & David, 1980). Specifically, 5S rDNA is a highly informative region for studies that use the PCR technique due to the following features: (a) the 5S rRNA gene is widely conserved even in distantly related species, which makes it less laborious to isolate the whole repeat from different species by PCR; (b) the repeats do not exceed the range of PCR amplification; and (c) the sequences can be isolated from low-quality DNA because they are small in size and are tandemly organised in a large number of copies (Martins & Wasko, 2004; Pinhal et al., 2009, 2011). Although the 5S rRNA genes are conserved even among unrelated taxa, their NTSs show extensive variations in length and nucleotide composition among species; these properties make these sequences highly useful as species-specific genetic markers (Wasko et al., 2001; Martins & Wasko, 2004; Pinhal et al., 2009). Such differences in the NTS architecture are useful for the PCR amplification of species-specific 5S rDNA profiles that can be directly discriminated with agarose gels.

Diverse studies have demonstrated the potential of 5S rDNA as a genetic marker for identifying shark species. For example, Pinhal et al. (2008) utilised 5S rDNA to discriminate eight shark species (*Carcharhinus leucas, C. acronotus, C. limbatus, C. obscurus, Galeocerdo cuvier, Sphyrna lewini, Isurus oxyrinchus* and *Alopias superciliosus*) by producing differently sized amplicons by simple PCR and analysing them in agarose gels. Later, the same genetic marker was employed to discriminate two closely related species of the genus *Rhizoprionodon* (*Rhizoprionodon porosus* and *R. lalandii*); however, in this case, the diagnostic required an extra step of enzymatic digestion in addition to PCR (Pinhal et al., 2009). A representative diagnostic 5S rDNA profile of three distinct shark species is shown below (Figure 3).

The multiplex PCR approach, which is an extension of traditional PCR, utilises a reaction in which multiple primers are simultaneously used for PCR; this permits the simultaneous amplification of various DNA fragments. This approach is used to generate amplicons that are of variable lengths between different species (i.e., unique DNA profiles), which will ultimately permit the identification of one species from another (Rasmussen & Morrissey, 2008; Carrier et al., 2010). Additional advantages of the multiplex approach include the ease of data interpretation, the feasibility of application for high-throughput screening and the overall low cost per analysed sample. Moreover, this approach does not require the sequencing of all of the samples or an extra step of enzymatic digestion to discriminate species (Magnussen et al., 2007).

Studies that have been performed on sharks have employed the multiplex PCR approach in several distinct genomic regions, such as the nuclear internal transcribed spacer 2 locus (ITS 2) and the mitochondrial genes cytochrome oxidase I (COI) and NADH dehydrogenase subunit 2 (ND2) (Pank et al., 2001; Shivji et al., 2002 ; Chapman et al., 2003; Abercrombie et al., 2005; Shivji et al., 2005; Clarke et al., 2006 ; Magnussen et al., 2007 ; Mendonça et al., 2009; Farrell et al., 2009).

The multiplex PCR technique that uses ITS 2 is the most well-established approach for the recovery of a shark's identity. For sharks, sets of species-specific primers (SSP) are used in conjunction with universal primers (positive controls) in a single reaction. Pank et al. (2001) developed SSPs for multiplex PCR using the ITS 2 region to discriminate two shark species, *Carcharhinus obscurus* and *C. plumbeus*. This work was shortly expanded by Shivji et al. (2002), who incorporated five additional species. The same approach was used by Chapman et al. (2003) and subsequently by Shivji et al. (2005) in samples with the generic name of "shark" that were obtained in Asian markets. Both of the studies proved that the meat and fins present among the samples traded were of the great white shark (*Carcharodon carcharias*), a species whose capture and trade is prohibited in many countries. In another study by the same group, Abercrombie et al. (2005) validated new SSPs to identify three large-body-sized species of hammerhead sharks (*Sphyrna lewini, S. zygaena* and *S. mokarran*), and their results confirmed the worldwide commerce of these species. Subsequently, Clarke et al. (2006) estimated the proportion of each taxon in the total shark fins and meat that was sold in Asian markets by using statistical approaches coupled with molecular diagnosis by multiplex PCR.

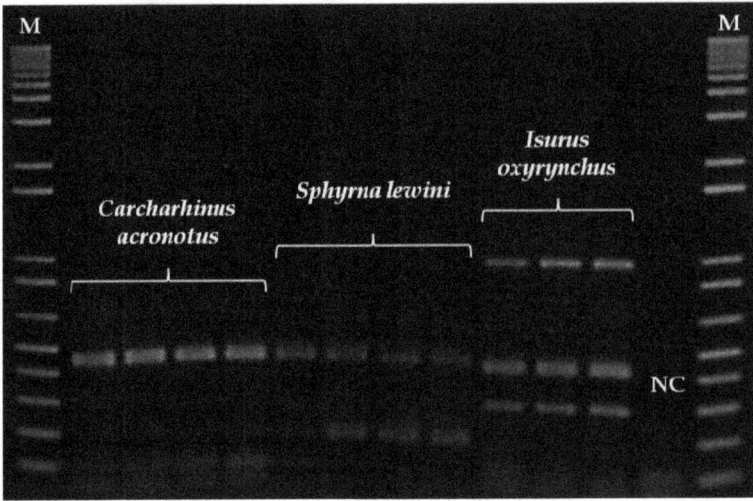

Fig. 3. The 5S rDNA profile for Carcharhiniform (*C. acronotus* and *S. lewini*) and Lamniform (*I. oxyrynchus*) species.

Although the multiplex PCR approach was originally designed to identify large pelagic and oceanic sharks, it has also has been applied to genetically profile globally distributed coastal shark species that are intensively exploited by fisheries (Pinhal et al. in preparation). In this work, the authors designed and validated seven species-specific PCR primers in a multiplex format that could simultaneously discriminate body parts from the seven known sharpnose shark species (genus *Rhizoprionodon*). Unlike the majority of shark species, *Rhizoprionodon* species exhibit life-history characteristics (rapid growth, early maturity and annual reproduction) that suggest that they could be fished in a sustainable manner if an investment in monitoring, assessment and careful management was made. Therefore, the usefulness of the developed assay is noteworthy because the acquisition of catch and trade data for species with a large quantity of landed individuals, which is the case for sharpnose sharks, will require high-throughput profiling. Currently, approximately 35 ITS2-based SSPs are available for shark species identification, and SSPs for an additional two species are currently being validated (Nachtigall et al. in preparation).

Recently a new multiplex PCR-based tool that utilises real-time PCR (RT-qPCR) and high-resolution melting (HRM) technologies to identify shark species was developed. Using a multiplex RT-HRM PCR assay, Moore et al. (2011) distinguished among three closely-related sharks (*Carcharhinus tilstoni, C. limbatus* and *C. amblyrhynchoides*) based on mutations in the mtDNA ND4 gene. Although promising, this technology still requires more sophisticated equipment, reagents and trained personnel before it can be implemented in routine species identification.

2.2 Nucleotide sequencing

DNA sequencing is a powerful and sensitive method for species identification, and it has gained prominence as a rapid, accessible methodology owing to technological advances and the decrease in the cost per sample (Wong et al., 2009).

When sequencing samples for species identification, it is important to correctly choose the genomic portion to be analysed. Distinct regions of the genome evolve at variable rates; therefore, to avoid errors in species diagnostics, the chosen portion should display high interspecific and low intraspecific variability (Bossier, 1999).

For example, Heist & Gold (1999) sequenced mitochondrial DNA (mtDNA) to discriminate among 11 species of Carcharhiniformes. Douady et al. (2003) conducted a study with a 2.4-kb fragment of the mitochondrial genome, which extended from the 12S rRNA gene to the 16S rRNA gene. They examined the phylogenetic relationships among the orders of sharks, which demonstrated that this region may be useful for resolving relationships at lower taxonomic levels. Subsequently, Greig et al. (2005) extended on this analysis by discriminating 35 shark species in the North Atlantic based on the high intraspecific and low interspecific divergence of these mitochondrial gene sequences.

Recently, Rodrigues-Filho et al. (2009) utilised the same marker to discriminate shark species that are exploited by fisheries in Northern Brazil. They revealed the occurrence of 11 species among morphologically indistinguishable landed specimens (Figure 4). Among the species that were identified, the species *Carcharhinus porosus* and sharks of the genus *Rhizoprionodon* were the most frequently caught. The former species is not listed by IUCN; however, local *C. porosus* stocks have been depleted, which has caused this species to be included in the Brazilian Red List of endangered fauna (MMA, 2004). In addition, Rodrigues-Filho et al. (2009) used genetic screening to show that the previously abundant *Isogomphodon oxyrhynchus* was not present among species that were caught by fisheries.

A different strategy that is widely used to identify species is the sequencing of the DNA barcode region (Hebert et al., 2003). In this approach, the identification of a given species is based on a profile that is generated with an approximately 650-base pair fragment of the mitochondrial COI gene. Studies have shown that this portion of COI displays low intraspecific and high interspecific variability, which is important for the discrimination of species on the basis of permanent nucleotide differences that exist among them. An advantage of this approach is the availability of a global reference databank, which is the result of international efforts that were mediated by the Consortium for the Barcode of Life (CBOL; http://www.barcoding.si.edu). The primary databank is called BOLD (Barcode of Life Data System, http://www.barcodinglife.org/views/login.php) (Ratnasingham & Hebert, 2007), which has currently assembled approximately 100,000 barcoded species. BOLD, which is the host of the FISH-BOL databank (Fish Barcode of Life campaign, http://www.fishbol.org) is dedicated to the assembly of barcode sequences from marine and freshwater species of fishes. Currently, 556 species of sharks and rays have been barcoded; this represents approximately 50% of the total number of elasmobranches that have been described (FISH-BOL, accessed in July/2011).

Some of the primordial studies that used DNA barcodes were performed by Ward et al. (2005), who sequenced the COI gene of 61 species and standardised this region for the identification of elasmobranches. Other studies that have used the DNA barcode approach have efficiently identified shark species (Ward et al., 2007; Moura et al., 2009; Ward et al., 2008; Holmes et al., 2009; Wong et al., 2009; Barbuto et al., 2010). In a wide screen, Ward et al. (2008) evaluated the variability of COI sequences from 210 species of Chondrichthyes (123 species of sharks) and discriminated 99% of the species that were screened. In addition, Holmes *et al.* (2009) identified and quantified the relative abundance of 20 shark species by analysing dried fins that were confiscated from illegal fisheries. Similarly, the DNA barcoding approach was also useful for the detection of fraudulent mislabelling in Italian fish markets, in which 80% of the commercialised sharks did not corresponded to the referred species that were sold (Barbuto et al., 2010).

Despite of the large number of studies that have demonstrated the potential of DNA barcode techniques for species identification, its use is controversial. Many authors argue that a single, short marker is not reliable as a taxonomic tool, and traditional systematics utilises a large number of characteristics to delimit species (Dasmahapatra & Mallet, 2006). In addition, the limitations of the DNA barcode approach include its inability to detect hybridisation (which originated from gene flow between distinct species) due to the maternal inheritance of mitochondrial DNA (Moritz & Cicero, 2004). Therefore, to gauge the confidence of the DNA barcode technique for species identification, scientists have recommended the examination of a large number of individuals, especially of globally distributed species, to confirm that the genetic divergence of COI is higher between species than within species (see Shivji, 2010).

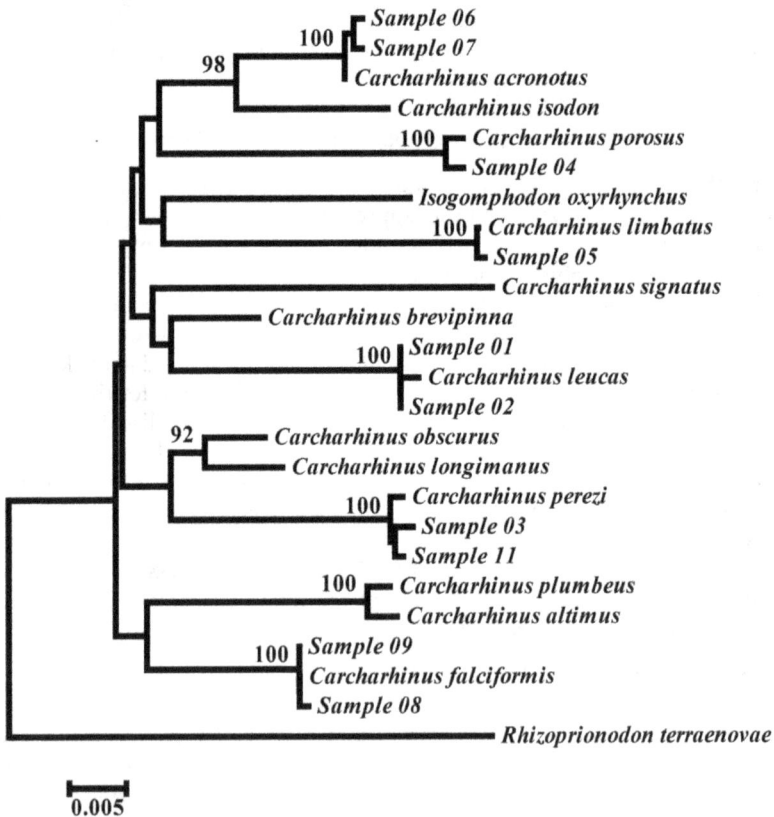

Fig. 4. Maximum likelihood tree of the mitochondrial genes 12S-16S that was used to discriminate shark samples (genus *Carcharhinus*).

2.3 Genetic diversity measured by molecular markers and implications for shark management

Among the large migratory fish species, sharks perhaps represent some of the more challenging taxa to trace demographically because they are rarely observed in the open ocean where they feed, breed and otherwise interact with conspecifics (Bradshaw et al., 2007). Despite the advances in mark–recapture and other tagging strategies that are used to estimate survival rates and long-range movements, these methods still have restrictions for deriving population estimates and exchange (Kohler & Turner, 2001), and they cannot provide a quantitative measure of the genetic diversity of the stocks.

Genetic techniques have been used with increasing frequency to inform conservation and management (Crandall et al., 2000; Allendorf et al., 2004), and they are especially useful for studying species for which ecology, demographics or behaviour limit the traditional field methods that are used to estimate the population trends.

The decreasing fish stocks and near extinction of several species will impact the economy and the whole environment. By measuring the variability of genetic stocks, it is possible to determine the real impact of fisheries on shark populations.

Because genetic diversity is correlated with population fitness (Reed & Frankham, 2003), understanding patterns of gene flow and dispersal is crucial for deriving how shark populations are distributed and if they constitute distinct genetic stocks. When this occurs populations are alleged to be structed. Therefore, the presence of genetic structure is indicative of low levels of migration among distinct stocks and indicates the need for the adoption of independent management measures for each stock (Schrey & Heist, 2003). This information is particularly useful for large free-living marine animals as sharks and can greatly assist for the conservation of wildlife in oceanic landscapes.

Sharks and other large marine fishes with a broad distribution frequently exhibit little population structure (Graves, 1998; Heist, 2004; Curley & Gillings, 2009). Exceptions to this pattern have been attributed to viviparous reproduction, sedentary behaviour and disjunct distributions, but they have been principally attributed to reproductive philopatry (Palumbi 1994; Pratt & Carrier, 2001; Waples et al., 2001; Keeney et al., 2003; Hueter et al., 2004; Keeney et al., 2005; Duncan et al., 2006; Karl et al., 2010). Philopatry is defined as the tendency of an individual to return to or stay in its home area, natal site, or another adopted locality; this is in opposition to nonreturning roaming behaviour or simple dispersal away from home areas (Hueter et al., 2004). Philopatric tendencies can be strong or weak for a given species, and special patterns of this behaviour can include natal philopatry (returning to the natal nursery area) and sex-specific philopatry (where one sex is more philopatric than the other, as in many male birds and female mammals) (Hueter et al., 2004). Females that exhibit philopatric behaviour return to coastal embayments that are called "nursery areas" to give birth (Pardini et al., 2001; Feldheim et al., 2002; Schrey & Heist, 2003). Nursery areas are usually located in shallow coastal waters and offer the advantages of low predation rates, prey abundance and spatial segregation from adults, which use geographically distant feeding areas, to neonates and juveniles (Morrissey & Gruber, 1993; Merson & Pratt, 2001). As a result, philopatric behaviour may clearly directly influence the levels of population structure and genetic diversity of a species, and molecular markers can be employed to detect such pattern.

Currently, a wide set of genetic markers has been utilised to infer population genetic structure in marine organisms (Baker et al., 1990; Baker et al., 1996; Amos et al., 1993; Innes et al., 1998). Due to high polymorphism and mutation rates (among other features), the molecular markers that are commonly used are those that involve the analysis of mitochondrial DNA and nuclear microsatellite loci.

In the mitochondrial genome, the non-coding control region (CR) is the molecular marker that is frequently utilised (Brown et al., 1985; Warn et al., 1986; Avise, 2000; Warn, 2000) due to its high evolutionary rates (Li, 1997; Warn, 2000). Specifically, the control region contains portions of its sequence in which the variability is even greater, which makes the region a highly informative target for population studies.

Studies that have used the CR as molecular marker for sharks are somewhat scarce in comparison to other marine species (for review, see Shivji et al., 2010). This implies incipient information regarding migration rates, dispersion movements, gene flow and the composition of genetic stocks for the majority of species. Indeed, knowledge of the genetic structure and diversity of shark populations is critical for maintaining species reproductive strength, resistance to diseases and the ability to adapt to environmental changes (Frankham et al., 2002).

Overall, most of the studies that have been performed on sharks have shown that distinct species possess lower genetic diversity when compared to other taxonomic groups (FAO). Nevertheless, despite the extensive exploitation by fisheries, some studies have reported that species still carrying high levels of genetic diversity (Pardini et al., 2001; Bowen et al., 2004; Keeney et al., 2005; Duncan et al., 2006; Martinez et al., 2006; Natoli et al., 2006) (Table 1).

Species	Pi	H	N	HAP	References
Carcharhinus limbatus (Terra Ceia Bay)	0.0106	0.720	45	8	Keeney *et al.* (2003)
Carcharhinus limbatus (Yankeetown)	0.0134	0.796	45	9	Keeney *et al.* (2003)
Carcharhinus limbatus (Belize City)	0.00077	0.680	13	4	Keeney *et al.* (2005)
Carcharhinus limbatus (Dangriga)	0.00049	0.526	19	2	Keeney *et al.* (2005)
Cetorhinus maximus	0.0013	0.720	62	6	Hoelzel *et al.* (2006)
Carcharias taurus	0.003	0.717	26	4	Stow *et al.* (2006)
Sphyrna lewini	0.013	0.8	271	24	Duncan *et al.* (2006)
Rhincodon typus	0.011	0.974	69	44	Castro *et al.* (2007)
Triakis semifasciata	0.0067	-	169	5	Lewallen *et al.* (2007)
Galeorhinus galeus	0.0071	0.92	116	38	Chabot *et al.* (2009)

Table 1. Genetic diversity of the mitochondrial control region from several shark species. **Pi** – Nucleotide diversity. **H** – Haplotype diversity. **N** – sampling. **HAP** – Number of observed haplotypes.

However, a few studies, such as the study that was published by Keeney *et al.* (2003), reported a low genetic diversity for the CR sequences from some populations of *C. limbatus* in the Atlantic coast. Other species that displayed low levels of diversity were *Cetorhinus maximus*, which is a worldwide distributed species that is highly susceptible exploitation by fisheries, and *Sphyrna tiburo*, which is heavily exploited in the Gulf of Mexico (Chapman et al., 2004; Hoelzel et al., 2006).

Within the nuclear genome, microsatellites are among the most utilised markers. These sequences can correspond to more than 20 alleles and carry heterozygosity above 95% (Heist, 2004). Other advantages of microsatellites are that they are diploid, codominant and analysable from small amounts of DNA; they also display high and constant mutational rates per generation (Jarne & Lagoda, 1996). The PCR allows access to different alleles at microsatellite loci and requires small amounts of DNA.

Recently, significant advances have been made in developing new statistical tools to analyse specific microsatellite loci (Pritchard et al., 2000). Such specificity in the analysis makes microsatellites powerful markers, and their applicability has been proven by studies in fish, such as in genetic mapping and paternity tests (Kocher et al., 1998; Keeney & Heist, 2003, Feldheim et al., 2004). Additionally, microsatellites are extensively used in studies that evaluate genetic variability among species and populations, evolutionary history, population structure and conservation and natural resource management (Bruford & Wayne, 1993; Warn, 2010). The variation in the number of repeated units in each microsatellite locus is the basis for studies that use these markers.

Currently, microsatellite markers are employed to study several shark species, and these markers can be useful for estimating genetic structure and gene flow in wild populations. In fact, DNA markers with different modes of inheritance, such as the maternally inherited mitochondrial DNA (mtDNA) and bi-paternally inherited nuclear microsatellites, are the markers that are most commonly used to infer the genetic diversity and structure of shark populations (e.g., gummy shark *Mustelus antarcticus*, Gardner & Ward 1998; great white shark *Carcharodon carcharias*, Pardini et al., 2001; blacktip sharks *Carcharhinus limbatus*, Keeney *et al.*, 2003, 2005; shortfin mako *Isurus oxyrinchus*, Schrey & Heist 2003; whale shark, *Rhincodon* typus, Castro et al., 2007; bull sharks *Carcharhinus leucas*, Karl et al., 2010). Several shark studies have used a combination of both markers, which has facilitated the obtainment of improved data regarding historical and contemporary events that drive population dynamics.

3. Conclusion

Oceans are a bountiful but not unlimited source of fishes that have been extensively harvested for decades. As a consequence, several top predators, such as the majority of shark species, are now endangered or live on the edge of extinction. Given the worldwide depletion of shark populations, which has been largely documented globally, there is an urgent need to increase the knowledge of the life history traits and other fundamental aspects of shark biology to promote conservation.

Therefore, several molecular technologies and methodologies that are summarised and discussed in the present chapter are extremely relevant to monitor the exploitation and prevent the loss of genetic diversity of sharks. It is important to note that molecular identification using distinct markers can be applied to virtually all living shark species to prevent extinction and to assist in their sustainable use as a natural resource. Furthermore, the correct monitoring of genetic stocks is extremely important to ensure the possibility of the full recovery of depleted stocks and the maintenance of genetic diversity.

4. References

Abercrombie, D.L.; Clarke, S.C. & Shivji, MS (2005) Global-scale genetic identification of hammerhead sharks: Application to assessment of the international fin trade and law enforcement. *Conservation Genetics, 6*, 775-788

Allendorf, F.W.; Leary, R.F.; Hitt, N.P.; Knudsen, K.L.; Lundquist, L.L. & Spruell, P. (2004) Intercrosses and the U.S. Endangered Species Act: should hybridized populations be included as Westslope Cutthroat Trout? *Conservation Biology, 18*, 1203–1213

Amos, B.; Schlotterer, C. & Tautz, D. (1993) Social structure of pilot whales revealed by analytical DNA profiling. *Science, 260*, 670–672

Asensio, L.; González, I.; Fernández, A.; Céspedes, A.; Rodríguez, M.; Hernández, P. & García, T. (2001) Identification of Nile perch (*Lates niloticus*), grouper (*Epinephelus guaza*), and wreck fish (*Polyprion americanus*) fillets by PCR amplification of the 5S rDNA gene. *Journal of AOAC International, 84*, 777-781

Avise, J.C. (2010) Conservation genetics enters the genomics era. *Conservation Genetetics, 11*, 665-669

Avise, J.C. (2000) Phylogeography: the history and formation of species. Cambridge, MA: Harvard University Press

Avise, J.C.; Helfman, G.S.; Saunders, N.C. & Stanton, L.H. (1986) Mitochondrial DNA differentiation in north atlantic eels: Population genetic consequences of an unusual life history pattern. *Proceedings of the National Academy of Science, 83*, 4350-4354

Baker, C.S.; Palumbi, S.R. & Lambertsen, R.H. (1990) Influence of seasonal migration on geographic distribution of mitochondrial DNA haplotypes in humpback whales. *Nature, 344*, 238–240

Baker, C.S.; Cipriano, F. & Palumbi, S.R. (1996) Molecular genetic identification of whale and dolphin products from commercial markets in Korea and Japan. *Molecular Ecology, 5*, 671–685

Barbuto, M.; Galimberti, A.; Ferri, E.; Labra, M.; Malandra, R.; Galli, P. & Casiraghi, M. (2010) DNA barcoding reveals fraudulent substitutions in shark seafood products: The Italian case of. *Food research international, 43*, 376-381

Baum, J.; Myers, R.; Kehler, D.; Worm, B.; Harley, S. & Doherty, P. (2003) Collapse and conservation of shark populations in the Northwest Atlantic. 299, 389-392

Baum, J.K. & Myers, R.A. (2004) Shifting baselines and the decline of pelagic sharks in the Gulf of Mexico. *Ecol. Lett., 7*, 135–145

Bigelow, H.B. & Schroeder, W.C. (1948) Fishes of the western north Atlantic. *Memoirs of the Sears Foundation for Marine Research* 1, 282-292

Bonfil, R. 1994. *Overview of world elasmobranch fisheries.* FAO Fisheries Technical Paper no. 341, Rome

Bossier, P. (1999) Authentication of seafood products by DNA patterns. *Journal of food science, 64*, 189-193

Bowen, B.W.; Bass, A.L.; Chow, S.M.; Bostrom, M.; Bjorndal, K.A.; Bolten, A.B.; Okuyama, T.; Bolker, B.M.; Epperly, S.; Lacasella, E.; Shaver, D.; Dodd, M.; Hopkinsmurphy, S.R.; Musick, J.A.; Swingle, M.; Rankin-Baransky, K.; Teas, W.; Witzell, W.N. & Dutton, P.H. (2004) Natal homing in juvenile loggerhead turtles (*Caretta caretta*). *Molecular Ecology*, 13, 3797-3808

Bradshaw, C.J.A. (2007) Swimming in the deep end of the gene pool: global population structure of an oceanic giant. *Molecular Ecology*, 16, 5111-5113

Branstetter, S. (1990) Early life-history implications of selected carcharhinoid and lamnoid sharks of the northwest Atlantic. *NOAA Technical Reports NMFS*, 90, 17-28Brown WM (1985) The mitochondrial genome of animals. In: Maclntvre, R.J. (ed). Molecular Evolutionary genetics. Plenum Press, NY, p 95-130

Bruford MW, Wayne RK (1993) Microsatellites and their application to population genetic studies, *Current Opinion in Genetics and Development*, 3, 939–943

Castro, A.L.; Stewart, S.; Wilson, G.; Hueter, R.E.; Meekan, M.G.; Motta, P.J.; Bowen, B.W. & Karl, S. (2007) Population genetic structure of Earth's largest fish, the whale shark (*Rhincodon typus*). *Molecular Ecology*, 16, 5183-5192

Castro, J.I. (1993) *A field guide to the sharks commonly caught in commercial fisheries of the southeastern United States*, US Dept. of Commerce.; National Oceanic and Atmospheric Administration.; National Marine Fisheries Service.; Southeast Fisheries Science Center, Miami

Castro, J.I.; Woodley, C.M. & Brudek, R.L. (1999) *A preliminary evaluation of the status of shark species*,Fisheries technical paper 380. FAO, Rome.

Chapman, D.D.; Abercrombie, DL.; Douady, C.J.; Pikitch, E.K.; Stanhopen, M.J. & Shivji, M.S. (2003) A streamlined, bi-organelle, multiplex PCR approach to species identification: Application to global conservation and trade monitoring of the great white shark, *Carcharodon carcharias*. *Conservation Genetics*, 4, 415-425

Chapman, D.D.; Prodöhl, P.A.; Gelsleichter, J.; Manire, C.A. & Shivji, M.S. (2004) Predominance of genetic monogamy by females in a hammerhead shark, *Sphyrna tiburo*: implications for shark conservation. *Molecular Ecology*, 13, 1965-1974

Clarke, S.C.; Magnussen, J.E.; Abercrombie, D.L.; McAllister, M.K. & Shivji, M.S. (2006) Identification of shark species composition and proportion in the Hong Kong shark fin market based on molecular genetics and trade records. *Conservation Biology*, 20, 201-211

Crandall, K.A.; Bininda-Emonds, O.R.P.; Mace, G.M. & Wayne, R.K. (2000) Considering evolutionary processes in conservation biology. *Trends in Ecology and Evolution*, 15, 290-295.

Curley, B.G. & Gillings, M.R. (2009) Population connectivity in the temperate damselfish *Parma microlepis*: analyses of genetic structure across multiple scales. *Marine Biology*, 156, 381-393

Dasmahapatra, K. & Mallet, J. (2006) Taxonomy: DNA barcodes: recent successes and future prospects. *Heredity*, 97, 254-255

Douady, C.J.; Dosay, M.; Shivji, M.S. & Stanhope, M.J. (2003) Molecular phylogenetic evidence refuting the hypothesis of Batoidea (rays and skates) as derived sharks. *Molecular Phylogenetics and Evolution*, 26, 215-221

Dulvy, N.K; Sadovy, Y. & Reynolds, J.D. (2003) Extinction vulnerability in marine populations. *Fish and Fisheries*, 4, 25–64

Dulvy, N.K.; Baum, J.K.; Clarke, S.; Compagno, L.J.V.; Cortes, E.; Domingo, A.; Fordham, S.; Fowler, S.; Francis, M.P. & Gibson, C. (2008) You can swim but you can't hide: the global status and conservation of oceanic pelagic sharks and rays. *Aquatic Conservation: Marine and Freshwater Ecosystems*, 18, 459-482

Duncan, K.; Martin, A.; Bowen, B. & De Couet, H. (2006) Global phylogeography of the scalloped hammerhead shark (*Sphyrna lewini*). *Molecular Ecology*, 15, 2239- 2251

FAO (2000) Fisheries Management: 1. Conservation and Management of Sharks. FAO Technical Guidelines for Responsible Fisheries 4 (Supplement 1), pp 37

Farrell, E.D.; Clarke, M.W. & Mariani, S. (2009) A simple genetic identification method for Northeast Atlantic smoothhound sharks (*Mustelus* spp.). *ICES Journal of Marine Science: Journal du Conseil*, 66, 561

Feldheim, K.A.; Gruber, S.H. & Ashley, M.V. (2002). Breeding biology of lemon sharks at a tropical nursery lagoon. *Proceedings of the Royal Society of London B*, 269, 1655-1662

Feldheim, K.A.; Gruber, S.H. & Ashley, M.V. (2004) Reconstruction of parental microsatellite genotypes reveals female polyandry and philopatry in the lemon shark, *Negaprion brevirostris*. *International Journal of Organic Evolution*, 58(10), 2332-2342

Frankham, R.; Ballou, J.D. & Briscoe, D.A. (2002) *Introdução para genética da conservação*, Cambridge University Press, ISBN Cambridge

Gardner, M.G. & Ward, R.D. (1998) Population structure of the Australian gummy shark (*Mustelus antarcticus* Gunther) inferred from allozymes, mitochondrial DNA and vertebrae counts. *Marine Freshwater Research*, 49, 733-745

Graves, J.E. (1998) Molecular insights into the population structure of cosmopolitan marine fishes. *Journal of Heredity*, 89, 427-437.

Greig, T.; Moore, M.; Woodley, C. & Quattro, J. (2005) Mitochondrial gene sequences useful for species identification of western North Atlantic Ocean sharks. *Fishery bulletin-national oceanic and atmospheric administration*, 103, 516

Hebert, P.D.N.; Cywinska, A.; Ball, S.L. & DeWaard, J.R. (2003) Biological identifications through DNA barcodes. *Proceedings of the Royal Society of London. Series B: Biological Sciences*, 270, 313

Heist, E. (2004) Genetics of sharks, skates, and rays. In: Biology of Sharks and Their Relatives, Carrier, J.C.; Musick, J.A. & Heithaus, M.R., pp. 485. CRC Press, New York.

Heist, E. & Gold, J. (1999) Genetic identification of sharks in the US Atlantic large coastal shark fishery. *Fishery bulletin-national oceanic and atmospheric administration*, 97, 53-61

Hoelzel, A.; Shivji, M.; Magnussen, J. & Francis, M. (2006) Low worldwide genetic diversity in the basking shark (*Cetorhinus maximus*). *Biology Letters*, 2, 639-642

Holmes, B.H.; Steinke, D. & Ward, R.D. (2009) Identification of shark and ray fins using DNA barcoding. *Fisheries Research*, 95, 280-288

Hueter, R.E.; Heupel, M.R.; Heist, E.J. & Keeney, D.B. (2004) Evidence of philopatry in sharks and implications for the management of shark fisheries. *Journal of Northwest Atlantic Fishery Science*, 35, 7-17

Imsiridou, A.; Minos, G.; Katsares, V.; Karaiskou, N. & Tsiora, A. (2007) Genetic identification and phylogenetic inferences in different Mugilidae species using 5S rDNA markers. *Aquaculture Research*, 38, 1370-1379

Innes, B.H.; Grewe, P.M. & Ward, R.D. (1998) PCR-based genetic identification of marlin and other billfish. *Marine Freshwater Research*, 49, 383–388

IUCN (2011) 2011 IUCN Red List of Threatened Species. <www.iucnredlist.org>. Downloaded on 07 May 2011

Jarne, P. & Lagoda P.J.L. (1996) Microsatellites, from molecules to populations and back. *Trends in Ecology and Evolution*, 11, 424-429

Karl, S.A.; Castro, A.L.F.; Lopez, J.A.; Charvet, P. & Burgess, G.H. (2010) Phylogeography and conservation of the bull shark (*Carcharhinus leucas*) inferred from mitochondrial and microsatellite DNA. *Conservation Genetics*, DOI 10.1007/s10592-010-0145-1

Keeney, D.; Heupel, M.; Hueter, R. & Heist, E. (2005) Microsatellite and mitochondrial DNA analyses of the genetic structure of blacktip shark (*Carcharhinus limbatus*) nurseries in the northwestern Atlantic.; Gulf of Mexico.; and Caribbean Sea. *Molecular Ecology*, 14, 1911-1923

Keeney, D.B. & Heist, E.J. (2003) Characterization of microsatellite loci isolated from the blacktip shark and their utility in requiem and hammerhead sharks. *Molecular Ecology Notes*, 3, 501-504

Kocher, T.D., Lee, W.J., Sobolewska, H., Penman, D. & McAndrew, B. (1998) A genetic linkage map of a cichlid fish, the tilapia (*Oreochromis niloticus*). *Genetics*, 148, 1225-1232

Kohler, N.E. & Turner, P.A. (2001) Shark tagging: a review of conventional methods and studies. *Environmental Biology of Fishes*, 60, 191-223

Kotas, J.E.; Gamba, M.R.; Conoly, P.C.; Hostim-Silva, M.; Mazzoleni, R.C. & Pereira, J.P. (1995) A Pesca de emalhe direcionada aos Elasmobrânquios com desembarques em Itajaí e Navegantes/SC. Reunião Do Grupo De Trabalho E Pesquisa De Tubarões E Raias Do Brasil, VII, Rio Grande, novembro, 1995

Last, P. & Stevens, J. (1994) Sharks and Rays of Australia.; 513 pp. Australia Commonwealth Scientific and Industrial Research Organization. Melbourne

Lavery, S.; Moritz, C. & Fielder, D. (1996) Genetic patterns suggest exponential population growth in a declining species. *Molecular Biology and Evolution*, 13, 1106-1113

Lessa, R.; Santana, F.M.; Rincón, G.; Gadig, O.B. & El-Deir, A.C.A. (1999) *Biodiversidade de Elasmobrânquios no Brasil. Programa Nacional da Diversidade Biológica* (PRONABIO). 125p. (http://www.bdt.org.br/workshop/costa/elasmo/)

Li, W.H. (1997) Molecular Evolution. Sinauer Associates, Inc., Publishers, Sunderland. pp 284

Long, E.O. & David, I.B. (1980) Repeated genes in Eukaryotes. *Annual Review of Biochemestry*, 49, 727-764

Magnussen, J.; Pikitch, E.; Clarke, S.; Nicholson, C.; Hoelzel, A. & Shivji, M. (2007) Genetic tracking of basking shark products in international trade. *Animal Conservation*, 10, 199-207

Martins, C. & Wasko, A. (2004) Organization and evolution of 5S ribosomal DNA in the fish genome. *Focus on Genome Research. Nova Science Publishers, Hauppauge*, 289-319

Martinez, P.; Gonzalez, E.G.; Castilho, R. & Zardoya, R. (2006) Genetic diversity and historical demography of Atlantic bigeye tuna (*Thunnus obesus*). *Molecular Phylogenetics and Evolution*, 39, 404-416

Mendonça, F.; Hashimoto, D.; Porto Foresti, F.; Oliveira, C.; Gadig, O. & Foresti, F. (2009) Identification of the shark species *Rhizoprionodon lalandii* and *R. porosus* (Elasmobranchii, Carcharhinidae) by multiplex PCR and PCR RFLP techniques. *Molecular Ecology Resources*, 9, 771-773

Merson, R.R. & Pratt, Jr. H.L. (2001) Distribution, movements and growth of young sandbar sharks, *Carcharhinus plumbeus*, in the nursery grounds of Delaware Bay. *Environmental Biology of Fishes*, 38, 167-174

MMA – IBAMA. Estatística da Pesca 2002 Brasil: Grandes Regiões e Unidades da Federação. *IBAMA-CEPENE*, 15 agosto 2004, www.ibama.gov.br

MMA (Ministério do Meio Ambiente) (2004) Instrução Normativa n° 5, de 21 de Maio de 2004: Anexo I – LISTA NACIONAL DE ESPÉCIES DE INVERTEBRADOS AQUÁTICOS E PEIXES AMEAÇADAS DE EXTINÇÃO. Diário Oficial da União – seção 1 n°102, sexta-feira, 28 de Maio de 2004

Morgan, J.A.T.; Welch, D.J.; Harry, A.V.; Street, R.; Broderick, D. & Ovender, J.R. (2011) A mitochondrial species identification assay for Australian blacktip sharks (*Carcharhinus tilstoni, C. limbatus* and *C. amblyrhynchoides*) using real-time PCR and high-resolution melt analysis. *Molecular Ecology Resources*, doi: 10.1111/j.1755-0998.2011.03023.x

Moritz, C. & Cicero, C. (2004) DNA barcoding: promise and pitfalls. *PLoS Biology*, 2, e354

Morrissey, J.F. & Gruber, S.H. (1993) Home range of juvenile lemon sharks *Negaprion brevirostris*. *Copeia*, 2, 425-434

Moura, T.; Silva, M.C.; Figueiredo, I.; Neves, A.; Muñoz, P.D.; Coelho, M.M. & Gordo, L.S. (2008) Molecular barcoding of north-east Atlantic deep-water sharks: species identification and application to fisheries management and conservation. *Marine and Freshwater Research*, 59, 214-223

Musick, J.; Burgess, G.; Cailliet, G.; Camhi, M. & Fordham, S. (2000) Management of sharks and their relatives (Elasmobranchii). *Fisheries*, 25, 9-13.

Natoli, A.; Cañadas, A.; Peddemors, V.; Aguilar, A.; Vaquero, C.; Fernández-Piqueras, P. & Hoelzel, A. (2006) Phylogeography and alpha taxonomy of the common dolphin (*Delphinus* sp.). *Journal of Evolutionary Biology*, 19, 943-954

Naylor, G.J.P. & Marcus, L.F. (1994) *Identifying isolated shark teeth of the genus Carcharhinus to species: Relevance for tracking phyletic change through the fossil record*, American Museum of Natural History, New York

Ovenden, J.R.; Morgan, J.A.T.; Street, R.; Tobin, A.; Simpfendorfer, C.; Macbeth, W. & Welch, D. (2011) Negligible evidence for regional genetic population structure for two shark species *Rhizoprionodon acutus* (Ruppell, 1837) and *Sphyrna lewini* (Griffith & Smith, 1834) with contrasting biology. *Marine Biology*, 158, 1497–1509

Palumbi, S.R. (1994) Genetic divergence, reproductive isolation, and marine speciation. *Annual Review of Ecology and Systematics*, 25, 547–572

Pank, M.; Stanhope, M.; Natanson, L.; Kohler, N. & Shivji, M. (2001) Rapid and simultaneous identification of body parts from the morphologically similar sharks *Carcharhinus obscurus* and *Carcharhinus plumbeus* (Carcharhinidae) using multiplex PCR. *Marine Biotechnology*, 3, 231-240

Pardini, A.T.; Jones, C.S.; Noble, L.R.; Kreiser, B.; Malcolm, H.; Bruce, B.D.; Stevens, J.D.; Cli, G.; Scholl, M.C.; Francis, M.; Duvy, C.A.J. & Martin, A.P. (2001) Sex-biased dispersal of great white sharks. *Nature*, 412, 139–140

Pinhal, D.; Yoshimura, T.S.; Araki, C.S. & Martins, C. (2011) The 5S rDNA family evolves through concerted and birth-and-death evolution in fish genomes: an example from freshwater stingrays. *BMC Evolutionary Biology*, 11:151 doi:10.1186/1471-2148-11-151

Pinhal, D.; Araki, C.; Gadig, O. & Martins, C. (2009a) Molecular organization of 5S rDNA in sharks of the genus Rhizoprionodon: insights into the evolutionary dynamics of 5S rDNA in vertebrate genomes. *Genetics Research*, 91, 61-72

Pinhal, D.; Gadig, O.B.F. & Martins, C. (2009b) Genetic identification of the sharks *Rhizoprionodon porosus* and *R. lalandii* by PCR-RFLP and nucleotide sequence analyses of 5S rDNA. *Conservation Genetetics Research*, 1, 35-38.

Pinhal, D.; Gadig, O.; Wasko, A.; Oliveira, C.; Ron, E.; Foresti, F. & Martins, C. (2009b) Discrimination of Shark species by simple PCR of 5S rDNA repeats. *Genetics and Molecular Biology*, 31, 361-365

Pratt, H.L. & Carrier, J.C. (2001) A review of elasmobranch reproductive behavior with a case study on the nurse shark, *Ginglymostoma cirratum*. *Environmental Biology of Fishes*, 60, 157-188

Pritchard, J.K.; Stephens, M. & Donnelly, P. (2000) Inference of population structure using multilocus genotypic data. *Genetics*, 155, 945-959

Rasmussen, R. & Morrissey, M. (2008) DNA-based methods for the identification of commercial fish and seafood species. *Comprehensive Reviews in Food Science and Food Safety*, 7, 280-295

Ratnasingham, S. & Hebert, P.D.N. (2007) BOLD: The Barcode of Life Data System (http://www. barcodinglife. org). *Molecular Ecology Notes*, 7, 355-364

Rodrigues-Filho, L.; Rocha, T.; Rêgo, P.; Schneider, H.; Sampaio, I. & Vallinoto, M. (2009) Identification and phylogenetic inferences on stocks of sharks affected by the fishing industry off the Northern coast of Brazil. *Genetics and Molecular Biology*. 32(2), 405-413

Sebastian, H.; Haye, P.A. & Shivji M.S. (2008) Characterization of the pelagic shark-fin trade in north-central Chile by genetic identification and trader surveys. *Journal of Fish Biology*, 73, 2293-2304

Shivji, M.; Clarke, S.; Pank, M.; Natanson, L.; Kohler, N.; Stanhope, M. (2002) Genetic identification of pelagic shark body parts for conservation and trade monitoring. *Conservation Biology*, 16, 1036-1047

Shivji, M.S. (2010) DNA Forensic Applications in Shark Management and Conservation. Sharks and Their Relatives II: Biodiversity, Adaptive Physiology, and Conservation. CRC Press, Boca Raton, FL.

Shivji, M.S.; Chapman, D.D.; Pikitch, E.K.; Raymond, P.W. (2005) Genetic profiling reveals 33 illegal international trade in fins of the great white shark, *Carcharodon carcharias*. *Conservation Genetics*, 6, 1035-1039

Schrey, A.W. & Heist, E.J. (2003) Microsatellite analysis of population structure in shortfin mako (*Isurus oxyrinchus*). *Canadian Journal of Fisheries and Aquatic Sciences*, 60, 670-675

Stevens, J. & Wayte, S. (1998) A review of Australia's pelagic sharks resources. Fisheries Research and Development Corporation project 98/107, 64 p. *CSIRO Marine Research, GPO Box 1538, Hobart, Tasmania 7001 Australia.*

Stevens, J.D.; Bonfil, R.; Dulvy, N.K. & Walker, P.A. (2000) The effects of fishing on sharks, rays, and chimaeras (Chondrichthyans) and the implications for marine ecosystems. *ICES Journal of Marine Science*, 57, 476-494

Vooren, C.M. (1997) Elasmobrânquios demersais. *In:* Os ecossistemas costeiro e Marinho do extremo sul do Brasil. Seelinger, U., Oderbrecht, C. & Castello, J.P., pp162, Ed. Ecoscientia, Rio Grande

Waples, R.S.; Gustafson, R.G. & Weitkramp, L.A. (2001) Characterizing diversity in salmon from the Pacific Northwest. *Journal of Fish Biology*, 59, 1-41

Ward, R.; Zemlak, T.; Innes, B.; Last, P. & Hebert, P. (2005) DNA barcoding Australia's fish species. *Philosophical Transactions of the Royal Society. B. BiologicalSciences*, 360, 1847-1857

Ward, R.D.; Holmes, B.H.; White, W.T. & Last, P.R. (2008) DNA barcoding Australasian chondrichthyans: results and potential uses in conservation. *Marine and Freshwater Research*, 59, 57-71

Ward, R.D.; Holmes, B.H.; Zemlak, T.S. & Smith, P. (2007) DNA barcoding discriminates spurdogs of the genus Squalus. *Descriptions of new dogfishes of the genus Squalus (Squaloidea: Squalidae). CSIRO Marine and Atmospheric Research Paper*, 14, 117-130

Wasko, A.; Martins, C.; Wright, J. & Galetti, Jr. P. (2001) Molecular organization of 5S rDNA in fishes of the genus Brycon. *Genome*, 44, 893-902

Wong, E.H.K.; Shivji M.S. & Hanner, R.H. (2009) Identifying sharks with DNA barcodes: assessing the utility of a nucleotide diagnostic approach. *Molecular Ecology Resources*, 9, 243-256

Wynen, L.; Larson, H.; Thorburn, D.; Peverell, S.; Morgan, D.; Field, I. & Gibb, K. (2009) Mitochondrial DNA supports the identification of two endangered river sharks (*Glyphis glyphis* and *Glyphis garricki*) across northern Australia. *Marine and Freshwater Research*, 60, 554-562

17

Estimating the Worth of Traits of Indigenous Breeds of Cattle in Ethiopia

Girma T. Kassie[1], Awudu Abdulai[2] and Clemens Wollny[3]
[1]International maize and Wheat Improvement Center (CIMMYT), Harare
[2]University of Kiel
[3]Bingen University of Applied Sciences, Bingen
[1]Zimbabwe
[2,3]Germany

1. Introduction

Livestock in general and cattle in particular are indispensable components of rural livelihoods in Ethiopia. In semi-arid and arid parts of the country, the pastoral communities depend mainly on their livestock for their livelihoods (Little et al. 2001; Barrett et al. 2003; Ouma et al. 2007). In the more dominant mixed crop-livestock livelihood system, cattle serve in providing traction power, in generating cash, in buffering shocks, as sources of consumables, as sources of prestige, and as the main indicator of wealth. The vital importance of cattle in supporting rural livelihoods against the backdrop of negative effects of climate change and continuing erosion of animal genetic resources (AnGR) justify a thorough analysis of the preferred characteristics of cattle to guide conservation and improvement programs.

Accordingly, proper identification, valuation, and maintenance of different traits of animal genetic resources are necessary to make them available and relevant for future use without compromising their current utilization. The main challenge in this regard is that the economic implications of erosion of genetic diversity and consequently its conservation are not well understood. This is essentially so because the diversity of AnGRs has a quasi-public nature (Scarpa et al. 2003a) and this makes it inadequate to value it in ordinary markets using revealed preference techniques.

Both revealed and stated preference techniques have been employed to analyze the marketing or pricing of livestock in Africa. The revealed preference techniques mainly employ the hedonic pricing method. Previous studies that used this method are Andargachew and Brokken (1993), Fafchamps and Gavian (1997), Jabbar (1998), Barrett et al (2003) and Jabbar and Diedhiou (2003). These studies showed that, in general, weight, age, sex, body condition, body size, coat colour, reason of purchase, season, rainfall pattern, holidays, district location, breed type, market locations, and restrictions such as quarantines determine livestock prices observed in the market.

The stated preference approach has recently become important in analyzing the preferences and economic values of livestock attributes. The significance of this approach in valuing attributes has generated considerable interest and research in the area of AnGR in recent times. After the pioneering work by Sy et al (1997) in Canada, many authors have used stated preference approach to analyze economic values of livestock traits in different parts of the world. Tano et al (2003) analyzed the economic values of traits of indigenous breeds of cattle in West Africa focusing on trypanotolerance by employing conjoint ranking and ordered probit model. Using choice experiments (CE) and mixed logit model, Scarpa et al (2003a) quantified the economic values of different traits of a Creole (local) pig in Yucatan, Mexico. Scarpa et al (2003b) later employed the same method to estimate the values for the traits of indigenous cattle in Northern Kenya. Ouma et al (2007) employed choice experiments to elicit preferences and mixed logit and latent class models to determine the relative values of traits and heterogeneities in trait preferences in the pastoral areas of Northern Kenya and South-Western Ethiopia. Zander (2006) employed conjoint ranking and mixed and multinomial logit models to study the relative values of traits and preference heterogeneities of Borana cattle keeping pastoralists in Northern Kenya and Southern Ethiopia. Roessler et al (2008) employed choice experiments and multinomial logit model to investigate the relative economic weights of pig traits in Vietnam, while Ruto et al (2008) examined the relative values of cattle traits and preference heterogeneities in Northern Kenya using choice experiments and latent class modeling. Recently, Kassie et al (2009) applied the same methodology to estimate the implicit prices of traits of indigenous cows and sources of trait preference heterogeneity.

This study contributes to the scientific literature in two ways. First, it applies both revealed and stated preference methods to analyze the implicit prices of attributes of indigenous cattle. The econometric tools used are advanced and comprehensive in that they reliably represent and predict the reality of the rural markets in study area. Second, unlike most of the past studies whose unit of analysis was the household (see Tano et al. 2003; Zander 2006; Ouma et al. 2007; Ruto 2008), we analyze preferences of cattle buyers at the market level by sampling different representative markets and interviewing market agents who were mainly farmers. This entailed capturing preferences of farmers for different cattle traits during actual market transactions unlike conducting interviews at the farm level when some farmers are not even thinking of selling or buying animals. The remaining part of the paper is organized as follows; next, a description of the study area, and the data generation and management procedures are presented. These are followed by results and discussion. The final section contains conclusions and implications of the results.

2. The approach

2.1 The study area

The study was conducted in the *Danno* district of central Ethiopia, which is located 250 km west of the Ethiopian capital Addis Ababa. The district has an area of about 66,000 hectares, and had a human population of 83,000 in 2005/6. Livestock, particularly cattle, are an important asset of the community. Semi-subsistence crop-livestock mixed farming system is the mainstay of rural livelihoods for the district's human population. The most important livelihood objective of the average household is producing sufficient food for the family each year (Kassie 2007).

The study covered five markets. Four of the markets, namely, *Sayo, Menz, Danno-Roge* and *Awadi-Gulfa*, are situated within the *Danno* district. *Sayo*, the administrative and economic capital of the district, has two different cattle markets that operate on Wednesdays and Saturdays. *Menz* is a small market located at about 12 km north of *Sayo* and is operational on Tuesdays only. *Danno-Roge* is located at the northern tip of the district some 28 km far from *Sayo*. *Danno-Roge market days* are on Thursdays and, unlike in other markets, cows and calves are frequently exchanged. *Awadi-Gulfa* market is located 24 km northeast of *Sayo* and operates on Wednesdays. *Awadi* is mainly a market for male cattle brought from both within and outside the district. The fifth market is *Ijaji* that is located in neighboring *Cheliya* district, very close to the district's border with *Danno*, and it sets on Saturdays. *Ijaji* market was included only because *Danno* farmers mentioned it as a market they visit as frequently as those within the district. All types of cattle are brought to *Ijaji* market and it is the only fenced market of about 30m by 80m area. Comparatively, traders are more frequent in this market than in others. Animals are trekked to and from the markets throughout the year. All cattle markets are managed and run by male buyers and sellers with virtually no women around. All the markets set for half a day mostly in the afternoons.

2.2 Sampling and data generation

2.2.1 Revealed preference analysis

Data were generated through a survey in the five rural markets described above. The survey was conducted in four rounds every three months over a sample of 20 cattle buyers in each season from each of the five rural markets. Given that some of the buyers purchased two animals at a time, the final sample size was 411. The survey focused on the phenotypic traits of the animals, places the animals were brought from, price, and the characteristics of the buyers. The phenotypic characteristics were identified in the initial survey and included color, class, age and body size of the animal bought.

Data collection for each season[1] was carried out over two weeks simultaneously in all of the markets. Season one covers the period from end of February to beginning of March. This is immediately after the major crop harvest where crop prices are normally low and livestock prices are high. Most of the cattle keepers want to sell their animals during this period against the challenge of the imminent feed scarcity in the dry season that follows right away which is evident from Figure 1. Season two covers the period from late May to early June. This is a period when prices of cattle decline, as buyers - predominantly farmers, usually lack money or cash to purchase cattle. Moreover, since it is the beginning of the rainy season, farmers tend to focus on their cropping activities.

The third season spans from late August to early September, a period of serious feed shortage. Prices are normally expected to be low for the animals that are yet to regain weight they lost in the dry season and for those that are subjected to a restricted free grazing during the rainy season. As expected, this is the most favored period by buyers and the least preferred by sellers. The last round of data was collected in late November. This is the beginning of the crop harvest period for early maturing crop varieties and the declining prices of crops. The animals normally recover from the weight losses of the past seasons and farmers can then postpone their selling decisions, if the prices offered are not attractive enough.

---

[1] Season implies the periods in which the market level surveys were conducted.

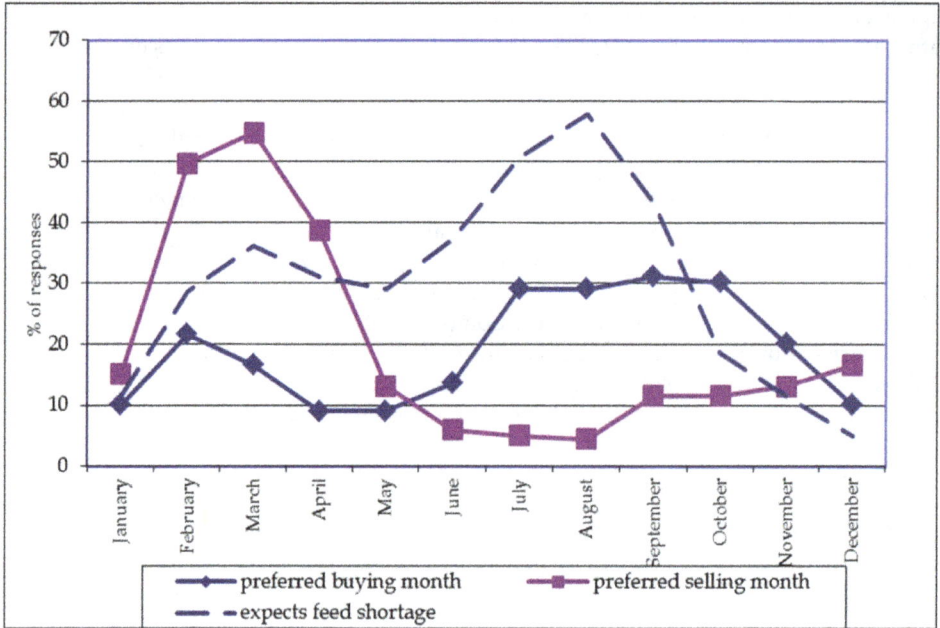

Fig. 1. Annual feed availability calendar and cattle selling and buying season preferences in *Danno*

An important observation in this study is the kind of classification farmers have for cattle and the influence of this classification on cattle prices. Male cattle which have ploughed for more than two seasons and which are sometimes castrated are called *'sangota'*. Non-castrated or intact younger male cattle with plowing experience of less than two seasons are called *'jibbota'*. A cow which has delivered more than once is called *'sa'a'*. Young female cattle that have delivered only a calf or none are called *'gorba'*. Very young female and male cattle with no parturition or plowing experience are called *'jabbota'*. We retain this classification in this paper but use the conventional terms ox for *'sangota'*, bull for *'jibbota'*, cow for *'sa'a'*, heifer for *'gorba'*, and calf for *'jabbota'*. It is worth noting that this classification somehow overlaps and might differ for the buyers and sellers. For instance, a younger cow for a seller might be a heifer for the buyer. The present study adopts the buyers' classification.

2.2.2 Stated preference analysis - Choice experiment

Choice experiment (CE) is a popular stated preference method which is used to elicit preferences for attributes of differentiated goods based on statistically efficient designs of attributes and attribute levels. CE surveys have already become routine in the fields of, *inter alia*, environmental (e.g., Rolfe et al. 2000; Campbell 2007), food and beverage (e.g., Rigby and Burton 2005; Mtimut and Albisu 2006), and plant genetic resources (e.g., Windle and Rolfe 2005; Birol et al. 2006) economics. Application of CE for the valuation of attributes of livestock is very recent and only a few studies (Scarpa et al. 2003a,b; Ouma et al. 2007; Roessler et al. 2008; Ruto et al. 2008, and Kassie et al. 2009) employed it.

The most important issues in designing a CE survey are attribute and attribute level determination, generation of statistically efficient and practically manageable experimental design, and management of the field interview. In this study, trait identification and trait level determination were done after a series of informal and focus group discussions both in the villages and in the markets where people of *Danno* district make a living and undertake cattle transactions. Respondents were asked to mention the attributes they consider when valuing the animals they keep or buy.

Seven traits were identified for cow CE and six traits for bull CE. Age was fixed at 3 years for cows based on the average of the figures indicated by farmers. This is in line with the fact that the average age of a cow at its first parturition is about 3.2 years in this part of the country (Ayalew and Rowlands 2004). For bulls, age was fixed at 4 years, as this is the age at which a bull would have ploughed for a year. The price levels used in the CE are averages of the minimum, average and maximums of the price distributions generated from respondents in the villages and markets for an 'average' cow and 'average' bull – average as perceived by respondents. Table 1 presents the traits and trait levels used in the choice experiments.

Variable	Levels	Reference level
Origin	*Danno*	Danno
	Nearby districts	
	Wellega	
	Keffa	
Body size	Small	Small
	Medium	
	Big	
Fertility	A calf/ 2 years	A calf/2 years
(Cows only)	A calf/ year	
Milk yield (Cows only)	1 liter/day	1 liter/day
	2 liter/day	
	3 liter/day	
Plowing potential (Bulls only)	Poor	Poor
	Good	
Calf vigor	Poor	Poor
	Good	
Disease resistance/ illness frequency	>2 times per year	>2 times per year
	<2 times per year	
Price – cows	Low price = Birr[2] 500.00	Birr 500.00
	Medium price = Birr 700.00	
	High Price= Birr 900.00	
Price - bulls	Low price = Birr 800.00	Birr 800.00
	Medium price = Birr 1000.00	
	High Price= Birr 1200.00	

Table 1. Traits and levels included in the choice experiments

---

[2] Birr is the local currency in Ethiopia. One USD ≈ 8.8 Ethiopian Birr in 2007.

The traits and trait levels were statistically combined in an efficient way to generate the final profiles based on the attributes and attribute levels. The most comprehensive approach to generating statistically efficient design is with SAS algorithm. This was employed in this study as suggested by Kuhfeld (1997, 2005). In addition to orthogonality, statistically efficient designs are characterized with balanced distribution of attribute levels, balanced utility across alternatives, and minimum overlap of levels in a choice set (Huber and Zwerina 1996).

The design for cow traits generated 36 profiles classified into 18 choice sets (two profiles in each set) blocked into 3 so that each respondent could be presented with six choice sets. Similarly, the design for the bull traits CE generated 24 profiles categorized into 12 choice sets blocked into 2 so that each respondent receives six choice sets. In total, each respondent received 12 choice sets. Attributes and attribute levels were described with pictures and sketches which were carefully selected to clearly show the attributes and the differences in the levels of the attributes. The choice experiment was administered by three experienced researchers from the department of livestock improvement at *Bako* Agricultural Research Centre (BARC) and supervised by an agricultural economist.

Completed experiments were 195 for cows and 198 for bulls. Accordingly, the total number of cow choice sets responded to were 1170 and that of bulls were 1188 with three alternatives in each set. The third alternative was an opt-out option included for the purposes of avoiding forced choice and of generating theoretically sound taste parameter estimates. In each market, one or two well-known brokers were identified and briefed about the objectives of the study and the equal opportunity sampling procedure to be employed. Then broker(s) identified respondents from the different spots in the markets. This is a relatively isolated community and the five markets are virtually the only markets where cattle in the district are traded. The sample is therefore believed to be representative of the cattle buyers in *Danno* district.

3. Analytical framework

3.1 Combining revealed and stated preference approaches

Both revealed (RP) and stated (SP) preference analyses have advantages and disadvantages. According to Louviere et al (2003) and Hensher et al (2005) RP data represent the real world scenario, possess inherent relationships between attributes, embody market and personal constraints on the decision maker, have high reliability and face validity, have limitations on alternatives, attributes, and attribute levels, yield one observation per respondent at each observation point, and show attribute level invariance. on the other hand, SP data show virtual decision contexts, allow mapping of utility functions with technologies different from existing ones, can include existing and/or proposed and/or generic choice alternatives, cannot easily represent changes in market and personal constraints, seem to be reliable when respondents understand, are committed to and can respond to tasks, and usually yield multiple observations per respondent at each observation point. The advantages of these approaches can be harnessed and the disadvantages abated if employed systematically combined.

Combining of RP and SP data and analysis is a highly recommended approach (Louviere et al 2003; Bateman et al, 2003). It has also been indicated, however, to be difficult and sometimes impractical (Bateman et al, 2003). The combination can be in terms of merging the RP and SP data sets generated from the same sample or merging the approaches on different sample of the same population. The latter widens the scope of the investigation and generates sets of information that can complement each other. Accordingly, this study employed the two approaches on two different samples of the same population of livestock keepers and consumers in order to explain indigenous cattle trait preferences in the real and hypothetical scenarios.

3.2 Revealed preference analysis

Theoretically, the prices cattle sellers receive reflect the anticipated utility by the buyers and this utility is derived from the attributes of the product as cattle can be considered as quality (attribute) differentiated goods (Lancaster 1966; Rosen 1974). This paper focuses on the main phenotypic attributes that buyers check or verify by inspecting the animals during the bargaining period and purchase. The external features farmers look for and attach value to are age, color, body size, sex, and the place where the animals were brought from.

The different levels of the attributes that differentiate cattle are known to both buyers and sellers, albeit at different levels of detail. The levels considered in this analysis are those perceived by the buyers, despite the possibility of imperfect knowledge and differences in measurement. The buyers and sellers in the markets considered are mainly farmers who raise the cattle. In line with the household modeling literature, where goods are produced, consumed and sold by households, a hedonic model can be employed to value the attributes of the quality differentiated indivisible goods. Therefore, estimation of the relationship between the characteristics of the cattle and their prices can be made through hedonic price analysis.

Under competitive market conditions, implicit prices will normally be related to product attributes alone, without accounting for producer or supplier attributes. However, as widely documented in the literature, rural markets in developing countries, particularly in Sub-Saharan Africa, are rarely competitive (Barrett and Mutabatsere 2007). Several empirical studies have shown that prices are also related to the characteristics of buyers, season and market location (e.g., Oczkowski 1994; Jabbar and Diedhiou 2003). Hence, season, market location and education level of the buyer were included in the models estimated in this research. As mentioned above, cattle price discovery in the rural markets surveyed is done through a one-to-one bargaining with the help of brokers. Brokers are usually invited by buyers - mainly farmers, as they have much less market information about prices and tend to be price takers. Therefore, the bargaining power of the buyer is very important in influencing the price paid. No direct information was gathered on bargaining power, but education level was taken as a proxy to indicate strength in bargaining, under the assumption that higher education increases the bargaining skills of buyers.

Another important issue in estimating hedonic functions is the identification of the appropriate functional form and estimation procedure. The critics of Rosen (1974) model on identification emphasize on Rosen's formulation that attempts to obtain higher order approximations to the utility function by imposing homogeneity across individuals (Brown and Rosen, 1982; Epple, 1987; Bartik, 1987). Ekeland et al (2002) suggest a solution based on transformation models and instrumental variable models, however, their approach only allows for a single dimensional characteristic which must be observed. Ekeland et al (2002) in fact indicated that if the price is constructed to be non-linear (e.g., log transformation), its non-linear variation gives an added piece of information which can help to identify preference parameters. Similarly, Bajari and Benkard (2004) solve the identification problem by allowing each individual to have different utility parameters but relying on parametric restrictions on the utility function. Moreover, they generalize the Rosen's approach by allowing for imperfect competition, a discrete product space with discrete characteristics, and one product characteristic that is not observed by the econometrician.

In general, the functional form of the hedonic price equation is unknown (Haab and McConnel 2002). Parametric, semi-parametric and non-parametric estimation procedures have all been suggested and used in different applications (e.g., Anglin and Gencay, 1996; Parmeter et al. 2007). As this research focuses on the estimation of the relative weights of cattle attributes (first step hedonic analysis), the technical details of these alternative approaches are not of interest.

The estimation strategy adopted in this study is a simple linear model based on the suggestion by Cropper et al (1988) as well as Haab and McConnel (2002). Cropper et al (1988) employed Monte-Carlo simulation analysis to show that the linear and linear-quadratic functions give the smallest mean square error of the true marginal value of attributes. However, when some of the regressors are measured with error or if a proxy variable is used, then the linear function gives the most accurate estimate of the marginal attribute prices. Haab and McConnel (2002) also argue that when choosing a functional form and the set of explanatory variables, the researcher must bear in mind the almost inevitable conflict with collinearity. High collinearity makes the choice of a flexible functional form less attractive, since the interactive terms of a flexible functional form result in greater collinearity. Given these considerations, we begin with a restrictive basic linear model given by

$$\ln(price) = X\beta + \varepsilon \tag{1}$$

where X is the vector of independent variables including the constant term, characteristics of cattle and the socioeconomic variables considered, β is a vector of parameters to be estimated and ε is an independently and identically distributed (iid) error term.

The iid assumption for the error term implies that the conditional distribution of the errors given the matrix of explanatory variables has zero mean [E{ε} = 0], constant variance [V{ε} = σ^2], and zero covariance [V{ε} = σ^2I, where I is the identity matrix]. These assumptions and hence the reliability of the estimates based on such assumptions hardly hold in analyzing survey data. We tested the basic model for specification error and heteroscedasticity. Ramsey's RESET test of the hypothesis of no omitted variables generated F (3, 381) value of 1.54 which is much below the critical value of 2.60 at α = .05 implying non-rejection of the null hypothesis. Both White and Breusch-Pagan tests rejected the hypothesis of homoskedasticity at the one percent level of significance, suggesting the presence of heteroscedastic error terms.

The data analysis employed in this study follows the approach used by Barrett *et al* (2003) in their study of the determinants of price and price variability in Northern Kenya. They applied the well-established concepts of structural heteroscedasticity and GARCH-M models to iteratively estimate price of cattle simultaneously accounting for price variability in the estimation.

Two equations are estimated simultaneously. The first equation regresses the conditional mean of the ln(price) on the independent variables discussed above and the standard deviation of the residual for each observation from the original OLS regression given by:

$$\ln(price) = X\beta + \sigma\gamma + \varepsilon \qquad (2)$$

where σ is the conditional standard deviation of the natural log of price and γ is its coefficient.

The second equation is the regression of σ on selected exogenous variables (Z) in X.

$$\sigma = Z\lambda + \upsilon \qquad (3)$$

where λ is the vector of parameter estimates and υ is an iid error term.

The estimation is conducted such that the predicted values of equation (3) will be substituted into equation (2) in each step until the parameters converge. This simultaneous estimation strategy is suitable for an analysis of price risk and the risk premiums relevant to cattle marketing (Barrett *et al.* 2003).

3.3 Stated choice analysis

Random utility theory (McFadden, 1974) formulates utility (U) as an additive function of deterministic and random components:

$$U_{njt} = X'_{njt}\beta_n + \varepsilon_{njt} \qquad (4)$$

where, X_{njt} is a vector of attributes of alternatives and ε_{njt} is unexplained utility assumed to be independently and identically distributed (iid) across individuals, alternatives and choice sets with extreme value type I distribution. β_n is a conformable vector of the unknown weights the respondent assigns to the explanatory variables.

Given the stochastic component of utility is iid extreme value type I, the probability conditional on β_n (CP$_{njt}$) that the cattle buyer chooses alternative 'j' out of 'm' alternatives in a choice set 't' is a conditional logit (McFadden 1974):

$$CP_{njt}(\beta_n) = \frac{\exp X'_{njt}\beta_n}{\sum_{l=1}^{m} \exp X'_{nlt}\beta_n} \qquad (5)$$

However, this assumes homogeneous preference for traits across all respondents and the taste parameters of each individual (β_n) are known and completely explained by their means only.

Preference heterogeneity is, however, known to be common among cattle producers and consumers (e.g., Ouma et al. 2007, Kassie et al. 2009). A random parameters logit model which accounts for heterogeneity is therefore used here. In random parameters logit (RPL), the β_n's are specified to be random and normally distributed:

$$\beta_n \sim N[\beta, \Sigma_\beta] \tag{6}$$

where β is the mean and Σ_β is the covariance of the distribution of β_n.

The random taste parameters (β_n) are unobserved and so the unconditional probability that a cattle buyer will choose alternative 'j' is estimated by integrating the conditional probabilities over all values of each of the random taste coefficients weighted by its density function. That is

$$P_{njt} = \Pr[y_{nt} = j] = \int \frac{\exp(x'_{njt}\beta_n)}{\sum_{l=1}^{m} \exp(x'_{nlt}\beta_n)} \phi(\beta_n \mid \beta, \Sigma_\beta) d\beta_n \tag{7}$$

where the integral is multidimensional and $\phi(\beta_n \mid \beta, \Sigma_\beta)$ is the multivariate normal density for β_n with mean β and variance Σ_β.

The maximum likelihood estimation then maximizes

$$\ln L_N = \sum_{n=1}^{N} \sum_{j=1}^{m} y_{njt} \ln P_{njt} \tag{8}$$

with respect to β and variance $\Sigma\beta$. This maximization cannot be solved; because, the integral (equation 7) has no closed form solution as its dimension is given by the number of components of β_n that are random, with non-zero variance. Simulated maximum likelihood estimation is, therefore, employed to estimate the unconditional choice probabilities (Train 2003; Cameron and Trivedi 2005). Following Cameron and Trivedi (2005), the integral (equation 7) is replaced by the average of R evaluations of the integrand at random draws of β_n from the $N[\beta, \Sigma_\beta]$ distribution. The maximum simulated likelihood estimator then maximizes

$$\ln \hat{L}_N(\beta, \Sigma_\beta) = \sum_{n=1}^{N} \sum_{j=1}^{m} y_{njt} \ln \left[\frac{1}{R} \sum_{r=1}^{R} \frac{e^{x'_{njt}\beta_n^{(r)}}}{\sum_{l=1}^{m} e^{x'_{njt}\beta_n^{(r)}}} \right] \tag{9}$$

where y_{njt} is 1 if alternative j is chosen and 0 otherwise, and $\beta_n^{(r)}$, $r = 1,2,...,$ R, are random draws from the density $\phi(\beta_n \mid \beta, \Sigma_\beta)$.

4. Results and discussion

4.1 Description of the sample population

Three data sets generated from two samples are used in this study. The revealed preference analysis is based on observations on 400 respondents and their 411 transactions. The sample size for the stated preference analysis is 200 but the valid numbers of observation were 195 for cows and 198 for bulls. The mean age of the respondents is about 36 years for all samples. The average number of male and female family members of households (2.73) is less than that of stated preference sample. Most of the respondents in both samples are either illiterate or have completed elementary school or less. The occupation of the respondents in the SP sample was found to be that nearly 89% of the respondents are either farmers (~47%) or farmer traders (~42%). About 8% of the respondents were full time traders and the remaining 2% were in other occupation such as small restaurant ownership and civil service (Table 2).

Variables	RP sample	SP sample	
		Cows	Bulls
Mean age of respondent	36.19	36.35	36.45
No. of make family size	2.73	3.39	3.40
No. of female family size	2.34	3.34	3.32
Literacy level (%)			
Illiterate	33.40	22.10	23.70
Reading and writing	9.80	9.20	9.10
Religious studies	2.60	2.60	2.50
Elementary school	41.40	51.80	51.00
Secondary school	10.50	13.80	12.60
Above secondary	2.30	0.50	1.00
Occupation (%)			
Farmer		46.70	47.00
Farmer trader		42.60	41.90
Trader		7.70	8.10
Other		3.10	3.00

Table 2. Some characteristics of the sample population

4.2 Revealed preference analysis

Both the mean and standard deviation equations have 411 observations and were found to be statistically significant. The mean equation has 27 parameters and R^2 value of 0.84, whereas, the standard deviation equation has 19 parameters and R^2 value of 0.41.

The estimations show that season, market location, class (age and sex based) of cattle, body size and age are important determinants of cattle prices in the rural markets of central Ethiopia (Table 3). Cattle prices in seasons one and two were found to be similar. However, the prices in season three were significantly lower than those in season one. Season three is the period when farmers would not have harvested their crops and their liquid assets are believed to be low. This implies that they could be forced to sell their cattle to generate cash with the market responding with lower price due to excess supply. Season three is, therefore, the least preferred period to sell cattle by farmers (see Figure 1). Average price in season four was found to be significantly higher than that of season 1. This is expected because during this season (season 4) the farmers can afford to postpone their cattle selling decisions if the prices are not acceptable, since they can easily rely on the recently harvested crop yield to meet subsistence and other needs.

Most of the coefficients of the market dummies were also found to be significantly different from zero, implying price differentials for cattle relative to *Danno*. The frequency of each class of animal is also decisive in this particular estimation. It is only at *Danno-Roge* market that the frequency of the bigger animals – oxen and cows – is less than that of *Sayo*. This clearly undermines the prices in *Danno-Roge* as compared to other markets and hence the negative coefficients. Cattle prices in *Menz* and *Awadi* markets are significantly higher than in *Sayo*. These markets have higher frequency of oxen and cow transactions as compared to others. In addition, *Awadi* is one of the routes out of the district to trek to secondary markets such as *Guder* and Ambo. Traders in *Menz* also trek their cattle to these secondary markets via *Awadi*.

ln(price)	Modified SHM [ln(price)]		Modified SHM [St.dev. ln(price)]	
	Coef.	St. Err.	Coef.	St. Err.
Constant	6.265*	0.079	0.540*	0.038
Season 2	0.017	0.016	0.004	0.018
Season 3	-0.032‡	0.019	0.018	0.019
Season 4	0.101*	0.017	0.033‡	0.018
Menz	0.118*	0.024	0.017	0.023
Awadi	0.088*	0.020	-0.025	0.021
Ijaji	0.008	0.025	-0.015	0.027
Roge	-0.063	0.040	0.095*	0.023
Ox	0.252*	0.053	-0.110*	0.018
Cow	-0.077	0.093	-0.255*	0.023
Heifer	-0.098*	0.037	0.061*	0.026
Bull	0.059†	0.027	-0.103*	0.025
Medium body size	0.028	0.020		
Big body size	0.174*	0.019		
Red coat color	0.036	0.026		
Black coat color	-0.091*	0.029		
White coat coloer	0.021	0.053		
Age	0.181*	0.029		
Age square	-0.011*	0.002		
Neighbor district	-0.036	0.031	-0.075*	0.034
Wellega	0.113‡	0.066	0.110	0.073
Keffa	-0.067	0.056	0.080	0.064
Read and write	-0.037	0.032	0.025	0.035
Elementary	-0.020	0.024	0.002	0.023
Secondary	0.067†	0.028	-0.009	0.031
Above second.	0.074	0.066	-0.054	0.058
Religious study	-0.078	0.050	0.028	0.055
St.dev. ln(price)	-0.155	0.267		

*, †, and ‡ significant at α = 0.01, α = 0.05, and α = 0.05, respectively.

Table 3. Modified SHM model parameter estimates

Farmers' classification of cattle into sex and functional categories was found to be important determinant of prices. For example, oxen have a price premium of about 25% over calves. This is the highest premium followed by that of bull. The heifers were found to have lower prices than the calves. Given the frequency of heifer and calf transactions, the fact that the calves include mainly male young cattle might have inflated the prices for calves over heifers. Coefficient for the cow dummy has the unexpected negative sign. Though the cow dummy coefficient is not statistically significant, the result generally shows that the relative value attached to female cattle is lower. This is essentially due to the fact that milk is not tradable in the district and female cattle are kept mainly for herd replacement purposes.

Body size was found to be very important determinant of cattle prices, with big size having a price premium of about 18% over small ones. This is a clear indication of the interests of cattle keepers/buyers of the area on larger body size, an observation also made by previous studies elsewhere on the topic (Jabbar and Diedhiou 2003; Barrett *et al.* 2003; Scarpa *et al.* 2003a). The most consistent variable in determining the price of cattle in these rural markets was age of the animal. The results show a strong quadratic relationship between age and price of cattle that at younger ages an increase in age increases the price of the animal and as the animals get older increase in age decreases price.

The coat color of cattle is also an attribute buyers normally consider in setting an animal's price. The results reveal that red and white colors attract similar prices (i.e. are equally preferred) as compared to mixed color, which is the base level. However, black coat color, relative to mixed color, has a significant price lowering effect on cattle. The coefficient for black coat color is not only statistically significant but also exhibits the highest value. Specifically, black coated cattle will attract a downward premium of about 9% as compared to mixed color coated cattle. This is attributed to the perception in the community that black coated animals are very susceptible to trypanosomosis that is prevalent in the area. It is an established fact that tsetse flies are attracted by a combination of blue and black colors (Barrass 1960; Roth 1967; Steverding and Troscianko 2004) which makes black-colored animals unsuitable for trypanosome prevalent areas.

Among the cattle origins, only *Wellega* appears to be marginally significant. Cattle from *Wellega* had a price premium of up to 11% above those from within *Danno* district. This is expected as the field surveys revealed that cattle from *Wellega* are considered to be bigger in size, disease free and therefore more marketable. Although statistically insignificant, the location *Keffa* has the expected negative sign as cattle from this zone are considered to be susceptible to diseases. Literacy related variables included as proxies for bargaining power did not significantly influence cattle price.

As expected, the coefficient of the conditional standard deviation of the natural log of price in the natural log price equation is negative, but is statistically insignificant. The negative sign implies the commonly observed phenomenon that as market prices grow more volatile, those who, nonetheless, opt to sell their animals in the markets are somewhat more desperate for cash and so are less able to hold out for a good price from traders (Barrett et al. 2003). The variability of the natural log of price is indicated to be influenced mainly by the age and functional classes of cattle as defined by marketers as well as season, market, and origin of the cattle.

4.3 Stated preference analysis

4.3.1 Cow trait preferences

Choosing a profile in the choice sets, as opposed to opting out, was found to be highly preferred by the respondents as indicated by the significant constant term (Table 4). Fertility, disease resistance, calf vigor, and milk yield were found to be highly significant (P<< 0.001) in influencing the choice of a cow. Body size, price and some locations were found to be statistically insignificant. The signs of all the taste parameters are as expected, except that of medium body size. The model in general is highly significant (P<< 0.001) at 29 degrees of freedom (Table 4).

The magnitude of the parameter estimates shows that fertility – or short calving interval – is more important than all the other attributes considered by cattle buyers. Disease resistance was also found to be more important than calf vigor, milk yield and the origin of the cow. Vigor of the calf was also identified to be very important in influencing cow choice. These findings conform to the basic objectives of rural life in this part of Ethiopia in general and with the specific purposes for which animals are kept.

The primary goal of majority of the households in this part of rural Ethiopia is to produce sufficient food for the family. Secondly, households aim at selling part of their produce to generate cash to pay for other costs of life including food, as food shortage is not uncommon. The main contribution of livestock towards achievement of these goals is through traction power generated from bulls and through livestock sales. Shorter calving interval implies more animals to sell over the lifespan of a breeding female, and higher total number of male calves over the same period, to replace the aging bulls as well as to sell. Disease resistance is so important not only because it assures the herd stays productive but also saves the scarce cash resources (treatment costs) of the rural people. A vigorous calf is described in the area as one that is fast growing, healthy and strong. The high value assigned to larger herd and the medication cost implications show the importance of calf vigor. The relative importance of these traits is comparable to the corresponding findings of studies elsewhere which analyzed preferences for cow traits (Tano et al. 2003; Ouma et al. 2007; Zander 2006) with apparent differences in the relative weights of the attributes.

Production of milk is important attribute of cows. However, the relative weight assigned to milk yield potential of cows in the study area is lower than those for other cattle traits. In *Danno* and the neighboring districts, milk is only produced for household consumption and selling milk is a social taboo that people would rather give it free. Some households milk their cows every other day as they do not have storage facilities, or cannot sell it. This is unlike the high importance attached to milk yield by the latent class of crop-livestock farmers in Kenya (Ouma et al. 2007). Given the fact that recent public livestock development efforts in the highlands of Ethiopia in general and in the study area in particular have focused on improving milk production from dairy cows, the relatively weight attached to milk production capacity of cows shows the considerable disparity between the official public livestock development agenda and rural livelihood objectives. Therefore, genetic improvement efforts targeted at rural settings need to consider the breeding goal of the community.

Variable	Structural Parameters		SD of the parameter distributions	
	Coefficient	St. Error	Coefficient	St. Error
Random parameters				
Medium body size	-0.42	0.30	0.21	1.62
Big body size	0.28	0.47	0.07	3.34
Fertility	1.80*	0.61	1.06‡	0.60
Milk yield	1.00*	0.33	0.60	0.37
Calf vigour	1.05*	0.29	0.11	1.88
Disease resistance	1.59*	0.51	1.45*	0.54
Medium price	-0.20	0.29	0.98	0.99
High price	-0.13	0.32	0.79	0.77
Non-random parameters				
Nearby districts	0.55‡	0.30		
Wellega zone	-0.47	0.32		
Keffa Zone	-0.27	0.29		
Constant	-2.98*	0.65		
Heterogeneity in mean parameters				
Big body*education	0.17‡	0.1		
Fertility* farmer trader	-0.29‡	0.16		
Fertility*family size	-0.09†	0.04		
Milk*trader	-0.51†	0.24		
Disease res.*farmer trader	-0.81†	0.35		
Disease res.*other occupant.	1.00‡	0.58		
High price*trader	-1.00‡	0.56		
High pr.*farmer trader	-0.31	0.30		
High Pr.*other occupant.	1.25‡	0.66		
N = 1170	LL = - 630.47		Pseudo R^2 = 0.51	
χ^2 (df=29)= 1309.80	LL$_{base}$ = -1285.4		Adj. R^2 = 0.50	

*, †, and ‡ significant at α = 0.01, α = 0.05, and α = 0.05, respectively. LL is value of log-likelihood function, LL$_{base}$ is value of the restricted (no coefficient) log likelihood function and χ^2 is chi-squared.

Table 4. Random Parameters logit model parameter estimates for cows

The origin cows are brought from is another important attribute cattle buyers consider, but this is not explicitly related to breed identity of the cows, although both livestock keepers and buyers implicitly recognize that cattle of a common origin share certain typical or characteristic features. People ask for the origin of the cow to judge its adaptability, in addition to examining some phenotypic characteristics which show considerable difference across locations. The regression results show that cows from immediate neighboring districts are preferred to those of the same district. Taste coefficients of Wellega and Keffa zones were found to be negative and statistically insignificant. The negative sign implies that cows from these areas, which are very far, are less preferred, *ceteris paribus*.

The three price levels were entered as categorical variables like all other traits with low price (Birr 500.00) fixed as reference level. The coefficients of the two price levels are statistically insignificant showing that the price levels used in the CE did not significantly influence the choices of alternatives. The respondents apparently considered the price levels too small for most of the profiles presented. Identification of traits (including price) and trait levels was completed four months before the CE survey. In the subsequent four months, the inflation that had been rampant in Ethiopia since May 2005 made the prices identified for the CE quite low.

4.3.2 Heterogeneity in cow trait preferences

This paper also examines preference heterogeneity based on the means and standard deviations of the random parameters, as well as the mean coefficients of the interaction terms. In line with Hensher *et al* (2005) and Train (2003), differential distributional assumptions were tried for random parameters. However, all random preference parameters were assumed to be normal based on the likelihood ratio test.

Preference heterogeneity was evident around the means of fertility and disease resistance. This implies that not all cattle buyers attach equal value to these cow attributes. The estimated means and standard deviations of each of the random taste parameters give information about the share of the population that places positive values or negative values on the respective attributes or attribute levels (Train, 2003). Considering attributes with statistically significant standard deviation estimates, 96% of the respondents prefer the fertility to be good (a calf per year), while 4% of the respondents prefer lower fertility (a calf/ 2 years). Similarly, 86% of the respondents indicated preference for higher disease resistance.

The sources of taste heterogeneity were further investigated by introducing interaction of the attributes and socioeconomic characteristics. As education level increases, the sensitivity towards body size increases. The relatively educated group of respondents is composed of non-farmers who intend to consume the animals than keeping them either for production or reproduction. Higher sensitivity for diseases resistance of cows was also observed among small restaurant owners as compared to farmers. This is essentially because these respondents cannot afford to keep sick animals or take them to clinics after purchase as the animals are to be slaughtered for immediate use. For buyers, other than this group, purchasing sick animals might not be that risky as there is always a one month guarantee with which they can return the cows for the seller in case they are seriously ill. These restaurant and inn owners are also quite sensitive to the high prices of cows as compared to farmers. This is clearly the result of the effective demand of these buyers that they have to purchase the animals to run their businesses and postponing their decisions in case of high prices is less likely.

The results also show that farmer traders are less interested in fertility of cows as compared to farmers. This is intuitive and implies that the marginal utility of fertility is lower for farmer-traders as they mainly intend to resell the animals. Similarly, as family size increases, interest in fertility of cows decreases. This shows that bigger family sizes are of well-established households with possibly less interest in increasing their herd size as compared to smaller families of young households that are expected to intend to increase their herd size. Traders, as compared to farmers, were also found to be less interested in milk yield of cows. This is in line with the peculiar culture of the community that discourages milk selling. Farmer-traders are less interested in disease resistance of the cows. This group of people purchases the cows essentially for reselling and hence is not expected to be interested in diseases resistance as much as farmers do. Traders and farmer-traders were uniformly found to be less interested in high price levels of cows as compared to farmers. These respondents are interested in increasing their marketing margins and are supposed to be keener on paying less than more. As farmers are less informed about the prices across markets, the sensitivity of traders and farmer traders is expected.

4.3.3 Bull trait preferences

Body size was found to be relatively less important trait in influencing bull type choices in these rural markets (Table 5). Negative sign of the medium body size level was, however, unexpected and this might possibly be due to the lack of sufficiently distinct level descriptions in the survey or the levels were too close to differentiate from respondents' perspective. The mixed crop-livestock production system depends very much on the traction power of bulls for all the activities from first plowing to threshing. Only bulls are used for plowing in this area, making traction power a crucial characteristic of a bull. That is essentially what the model results reflect (Table 5). Plowing suitability has the largest taste coefficient with the expected positive sign and high statistical significance, indicating that good plowing potential is a trait that respondents consider when purchasing bulls.

The rural community has multiple objectives in buying and keeping cattle in such a production system. The bulls are bought and kept at least for two purposes - traction and reproduction. The reproductive contribution of bulls is very important as there are no communal or village-owned bulls selected for this purpose. In particular, farmers normally do not take within-the-herd mating for granted and focus on traction suitability only. They usually inquire about the reproductive characteristics of the bull, for which is calf strength is taken as a proxy. The attribute's coefficient is highly significant. The more vigorous the offspring of a bull is, the higher the probability that it will be chosen on the premise that a higher utility can be derived. Disease resistance was also found to be positive and statistically significant, indicating preferences for healthy or disease tolerant animals. With limited resources to employ on medication and hygienic costs for their animals, rural livestock keepers are expected to be very interested in healthy animals.

The RPL estimation resulted in negative and statistically significant coefficients for nearby districts and *Keffa* zone. The negative signs of the coefficients indicate that bulls from both origins are less preferred to those from *Danno* and will result in less probability of choice for a bull. The differences in absolute magnitudes of the structural parameters of the location variables show that the probability of not selecting an animal will be higher if the origin is *Keffa* than neighboring districts. This is an exact reflection of the preference of farmers in *Danno*, as cattle from Keffa region are considered susceptible to trypanosomosis and less adapted to the local conditions at *Danno* district. This again implies that buyers generally give high value to the fact that they know the pedigree of the cattle they buy which could only be possible if the animals were raised in their proximity. Given the lack of accurate information and the uncertainties under which farmers make decisions, it is obvious that cattle buyers in this semi-subsistent farming system would prefer bulls from their own districts.

	Structural Parameters		SD of the parameter distributions	
Variables	Coefficient	St. Error	Coefficient	St. Error
Random parameters				
Medium Size	-0.254†	0.108	0.005	0.300
Big Size	0.836*	0.192	0.655	0.480
Ploughing	1.994*	0.218	1.357*	0.255
Calf strength	0.752*	0.084	0.006	0.300
Illness freq.	0.821*	0.124	0.003	0.307
Price 1 (birr 1000.00)	0.237	0.183	0.003	0.245
Price 2 (birr 1200.00)	-0.267‡	0.170	0.014	0.444
Non-random parameters				
Constant	-2.476*	0.226		
Nearby districts	-0.417‡	0.240		
Wellega zone	0.223	0.130		
Keffa zone	-0.634*	0.193		
N= 1188	LL base = -1305.15		χ^2= 1024.4,	
LL = -792.9	Ps. R^2 = 0.392		df=18	

*, †, and ‡ significant at α = 0.01, α = 0.05, and α = 0.05, respectively. LL is value of log-likelihood function, LL$_{base}$ is value of the restricted (no coefficient) log likelihood function and χ^2 is chi-squared.

Table 5. Random Parameters logit model parameter estimates for bulls

The results also show that both medium (Birr 1000.00) and high (Birr 1200.00) levels of price have no significantly different influence on choice as compared to small price level. These results appear realistic, given that the price levels used during the choice experiment were already low as stated above in relation to prevailing inflation and the low and medium levels of prices were nearly indifferent for the respondents. Even the high level of price was considered quite acceptable for almost all the hypothetical profiles presented in the choice sets.

4.3.4 Willingness to Pay (WTP) values for bull attributes

The marginal rate of substitution between the traits and the monetary coefficient provides estimates of the implicit prices for the traits. These implicit prices are also referred to as willingness to pay (WTP) or willingness to accept. The price volatility prevalent in the study area makes the absolute magnitude of the willingness to pay (WTP) values less important. In order to assess prioritization of traits by the buyers, only the relative magnitudes of the WTP weights should be used. The willingness to pay values computed for each attribute (γ) at the highest price (p) level show that changing the traction potential level from poor to good is valued 2.65, 2.42, and 2.39 times more than a comparable change in offspring vigor, illness frequency (implying disease resistance) and big body size, respectively (Table 6).

Trait	WTP$_Y$ =E(-$\beta \sqrt{/\beta_P}$)	SD = E(-$\delta \sqrt{/\beta_P}$)	Min.	Max.
Medium Body	-0.954	0.018	-0.956	-0.951
Big body	3.134	2.382	1.698	4.842
Ploughing potential	7.476	4.550	-1.624	11.286
Calf strength	2.819	0.021	2.811	2.824
Illness freq.	3.078	0.013	3.070	3.084

Table 6. Willingness to pay for bull traits computed at the highest price level

5. Conclusion and recommendation

This study used both revealed and stated preference approaches to determine the values attached to the different features of indigenous cattle in central Ethiopia. A hedonic model was employed to examine the determinants of cattle prices in the primary rural markets. Transaction level data of cattle farmers and farmer-traders were used in the analyses. Data collected in rural markets to identify cattle price determinants result in estimates with standard errors that are mostly heteroscedastic. We employed SHM estimations to account for heteroscedastic errors. Based on Akaike, Bayesian and log-likelihood criteria of model selection, we found that the modified SHM formulation is very appropriate for examining price functions in such rural markets.

The empirical estimation showed that market place, seasonal differences, sex and function-based classification of cattle, body size, and age were very important factors influencing the market prices cattle sellers receive. The significance of the characteristics of animals in influencing prices paid for the animals reveals the importance of the preferences for traits in the decision-making process related to buying and selling of cattle. These preferences at the farmers and farmer-traders levels are the ones that matter most in shaping up the diversity of animals marketed at farm level. Furthermore, depending on the relative contribution of selling and purchasing of breeding cattle in herd dynamics, the same preferences are expected to have lasting and cumulative influence on the genetic diversity of cattle maintained at farm level. Phenotypic and genetic diversity in the existing cattle genetic resources provides the basis for selecting preferred attributes of breeding stock in the context of the livelihood objectives of the target community. Thus, the cattle breeding strategies and activities should duly consider the preferences expressed through the prices paid for animals in such markets, where the cattle keepers are the main sellers and buyers.

For the stated preference analysis the study employed choice experiments and random parameters logit to elicit and analyze cattle trait preferences of buyers in the semi-subsistence livelihood systems of rural central Ethiopia. The results of the cows CE revealed that in areas where livestock serve multitude of purposes and where the production and marketing system is semi-subsistence, cows have other functions more important than milk production. Fertility, disease resistance and strength of the calves they bear are as much or more important than milk. The breed concept which is very much associated in Ethiopia with the area where the animal is brought from (Ayalew and Rowlands 2004), was found to be less important as such and it appears that farmers are interested in obtaining animals from the district or locations close by. This is essentially because cattle buyers, who are mostly farmers, are more concerned about adaptability and therefore give high value to the fact that they know the pedigree of the cattle they buy.

The results of the CE for bulls indicate that cattle buyers assign high values for good traction potential, disease resistance, calf vigor, and for places of origin when choosing bulls in the market. The preferences cattle buyers have for these attributes do vary essentially due to differences in occupation, education and age. The primary objective of the rural community to produce sufficient food for the family for each year was manifested through the value assigned to traction potential which is more than twice that of disease resistance. These results are consistent with the basic reasons why animals are kept in the area, but appear to be incoherent with the government funded interventions of livestock development. An observation which needs to be emphasized is the consistency of the preferences of the cattle buyers in such a system characterized by lack of information in every aspect. Given the importance of livestock, bulls in particular, for the livelihoods of the communities in rural Ethiopia, such consistent valuation of the traits show that the objectives of the agrarian life are quite clear among the community – farmers, farmer traders, traders, and others – that production and marketing decisions are made on broader considerations than just milk and meat production.

On-going and planned cattle improvement programs for Ethiopian central highlands should take note of the significant livestock and socio-economic attributes that influence the production, marketing and utilization of cattle and cattle products. Current public policies on improvement of cattle production in central Ethiopian highlands promote use through crossbreeding of exotic dairy type breeds considered as improver genotypes mainly to increase milk production. The smallholder community in this part of Ethiopia depends on semi-subsistence agriculture and so livestock development interventions should focus on reproductive and adaptive traits that stabilize the herd structure, rather than focusing on traits that are only important for commercial purposes which are accorded low priority by the cattle keeping farmers. It can also be observed that improving these preferred traits of cows owned by smallholder farmers in the area has a better chance of addressing immediate livelihoods needs of the farmers than can introducing and testing new genotypes from outside.

6. References

Andargachew, K, & Brokken, R F. (1993). Intra-Annual Sheep Price Patterns and Factors Underlying Price Variations in the Central Highlands of Ethiopia. *Agricultural Economics*, Vol. 8, No. 2, pp. 125-138.

Anderson, S. (2003). Animal genetic resources and sustainable development, *Ecological Economics* Vol. 45, pp. 331 – 339.

Anglin, P M., & Gencay, R. (1996). Semi-parametric Estimation of a Hedonic Price Function. *Journal of Applied Econometrics* Vol. 11, pp. 633-648.

Ayalew, W.; van Dorland, A. & Rowlands, J. (2004). *Design, Execution and Analysis of the Livestock Breed Survey in Oromiya Regional State, Ethiopia*. Oromia Agricultural Development Bureau (OADB), Addis Ababa Ethiopia and the International Livestock Research Institute (ILRI), Nairobi, Kenya. 260pp.

Bajari, P. & Benkard, C. L. (2004). *Demand Estimation with Heterogeneous Consumers and Unobserved Product Characteristics: A Hedonic Approach*. Research Paper No. 1842, Stanford University.

Barrass, R. (1960). The settling of tsetse flies Glossina morsitans westwood (Diptera, Muscidae) on cloth screens. *Entomologia Experimentalis et Applicata*, Vol. 3, No. 1, pp. 59-67

Barret, C, & Mutambatsere, E. (2007). Agricultural Markets in Developing Countries. Entry in S. N. Durlauf and L. E. Blume, (Eds.), *The New Palgrave Dictionary of Economics*, 2nd Edition, Palgrave McMillan , London.

Barret, C.; Chabari, F.; Bailey, D.; Little, P. & Coppock D. (2003). Livestock Pricing in the Northern Kenyan Rangelands. *Journal of African Economies*, Vol. 12, No. 2, No. 127 – 155.

Bartik, T. J. (1987). The Estimation of Demand Parameters in Hedonic Price Models. *The Journal of Political Economy*, Vol. 95, No. 1, pp. 81-88.

Bateman, I., Carson, R., Day B., Hanemann M. *et al.* (2003). *Economic Valuation with Stated Preference Techniques*. Cheltenham: Edward Elgar.

Birol, E.; Smale, M, & Gyovai, A. (2006). Using Choice Experiment to estimate farmers' valuation of on Hungarian small farms. *Environmental and Resource Economics*, Vol. 34, pp. 439-469.

Brown, J. N., & Rosen, H. S. (1982). On the Estimation of Structural Hedonic Price Models. *Econometrica, Vol.* 50, No. 3 (May), pp. 765-68.

Cameron, A. C. & Trivedi, P. K. (2005). *Microeconometrics: Methods and Applications* Cambridge University Press.

Campbell, D. (2007). Willingness to pay for Rural Landscape Improvements: Combining Mixed Logit and random Effects Model. *Journal of Agricultural Economics, Vol.* 58, pp. 467-483.

Cropper, M. L.; Deck, L. B, & McConnel, K. E. (1988). On the Choice of Functional Form for Hedonic Price Functions. *The Review of Economics and Statistics*, Vol. 70, No. 4, pp. 668 – 675.

Drucker, A. G.; Gomez, V. & Anderson, S. (2001). The Economic Valuation of Farm Animal Genetic Resources: A Survey of Available Methods. *Ecological Economics*:36:1-18.

Ekeland, I.; Heckman, J. & Nesheim, L. (2002). Identification and Estimation of Hedonic Models. IFS Working Paper CWP07/02.

Epple, D. (1987). Hedonic Prices and Implicit Markets: Estimating Demand and Supply Functions for Differentiated Products. *The Journal of Political Economy*, Vol. 95, No. 1, pp. 59-80.

Fafchamps, M. & Gavian, S. (1997). The Determinants of Livestock Prices in Niger. *Journal of African Economies,* Vol. 6, No. 2, pp. 255-295.

Haab, T. C. & McConnel, K. E. (2002). *Valuing Environmental and Natural Resources: The Econometrics of Non-Market Valuation.* Edward Elgar, Cheltenham, UK, Northampton, MA, USA.

Hensher, D.; Rose, J. & Greene, W. (2005). *Applied Choice Analysis: A Primer.* Cambridge University Press.

Huber, J. & Zwerina, K. (1996). The Importance of utility Balance and Efficient Choice Designs. *Journal of Marketing Research,* Vol. 33, No. 307-317.

Jabbar, M. A. (1998). Buyer Preferences for Sheep and Goats in Southern Nigeria: A Hedonic Price Analysis. *Agricultural Economics, Vol.* 18, No. 1, pp. 21-30.

Jabbar, M. A. & Diedhiou, M. L. (2003). Does Breed Matter to Cattle Farmers and Buyers? Evidence from West Africa? *Ecological Economics,* Vol. 45, pp. 461 – 472.

Kassie, G. T. (2007). *Economic Valuation of the Preferred Traits of Indigenous Cattle in Ethiopia.* Department of Food Economics and Consumption Studies, Christian-Albrechts University of Kiel, Germany. PhD Dissertation.

Kassie, G T.; Abdulai, A. & Wollny C. 2009. Valuing Traits of Indigenous Cows in Central Ethiopia. *Journal of Agricultural Economics,* Vol. 60, No. 2, pp. 386-401.

Kuhfeld, W. F. (1997). *Efficient Experimental Designs using Computerized Searches* Sawtooth Software Research Paper Series. SAS Institute, Inc.

Kuhfeld, W. F. (2005). *Marketing Research Methods in SAS: Experimental Design, Choice, Conjoint, and Graphical Techniques* (SAS Institute Inc., Cary, NC, USA, http://support.sas.com/techsup/technote/ts722.pdf)

Lancaster, K. (1966). A New Approach to Consumer Theory. *Journal of Political Economy,* Vol. 74, No. 132-57.

Little, P.; Smith, K.; Cellarius. B.; Coppock, D., & Barrett, C. (2001). Avoiding Disaster: Diversification and Risk Management among East African Herders. *Development and Change, Vol.* 32, No. 3, pp. 401 – 433.

Long, J. S. & Ervin, L. H. (2000). Using Heteroscedasticity Consistent Standard Errors in the Linear Regression Model. *The American Statistician,* Vol. 54, pp. 217 – 224.

Louviere, J.; Hensher, D. & J. Swait. (2000). *Stated Choice Methods and Analysis.* Cambridge: Cambridge University Press.

McFadden, D. (1974). Conditional Logit Analysis of Qualitative Choice Behavior. In Zarembka, P. (ed.) *Frontiers in Econometrics.* Academic Press, New York.

Mtimet, N. & Albisu, L. M. (2006). Spanish Wine Consumer Behavior: A Choice Experiment Approach. *Agribusiness,* Vol. 22, No. 3, pp. 343–362.

Oczkowski, E. (1994). A Hedonic Price Function for Australian Premium Table Wine. *Australian Journal of Agricultural Economics,* Vol. 38, No. 93 – 110.

Ouma, E.; Abdulai, A. & Drucker, A. (2007). Measuring Heterogeneous Preferences for Cattle Traits amongst Cattle Keeping Households in East Africa. *American Journal of Agricultural Economics,* Vol, 89, No. 4, pp. 1005-1019.

Parmeter, C. F.; Henderson, D. J. & Kumbhakar, S. C. (2007). Nonparametric Estimation of a Hedonic Price Function. *Journal of Applied Econometrics*, Vol. 22, pp. 695 – 699.

Rigby, D. & Burton, M. (2005). Preference heterogeneity and GM food in the UK. *European Review of Agricultural Economics*, Vol. 32, pp. 269-288.

Roessler, R.; Drucker, A.; Scarpa, R.; Markemann, A.; *et al.* (2008). Using choice experiments to assess smallholder farmers' preferences for pig breeding traits in different production systems in North-West Vietnam. *Ecological Economics, Vol.* 66, pp. 184-192

Rolfe, J.; Bennet, J. & Louviere, J. (2000). Choice modeling and its potential application to tropical rainforest preservation. *Ecological Economics, Vol.* 35, No. 2, pp. 289-302.

Roosen, J.; Fadlaoui, A. & Bertaglia, M. (2005). Economic Evaluation for Conservation of farm animal genetic resources. *Journal of Animal Breeding and Genetics*, Vol. 122, pp. 217-228.

Rosen, S. (1974). Hedonic Prices and Implicit Markets: Product Differentiation in Pure Competition. *Journal of Political Economy*, Vol. 82, No. 1, pp. 34 -35.

Roth, H.H. (1967). *White and black rhinoceros in Rhodesia.* Oryx 9, 217-231.

Ruto, E.; Garrod, G. & Scarpa, R. (2008). Valuing animal genetic resources: a choice modeling application to indigenous cattle in Kenya *Agricultural Economics*, Vol. 38, No. 1, pp. 89-98.

Scarpa, R.; Drucker, A.; Anderson, S.; Ferraes-Ehuan, N.; Gomez, V.; Risoparion, C. & Rubio-Leonel O. (2003a). Valuing Animal Genetic Resources in Peasant Economies: The Case of the Box Keken Creole Pig in Yacutan. *Ecological Economics*, Vol. 45, pp. 427-443.

Scarpa, R.; Kristjanson, P.; Ruto, E.; Radeny, M.; Drucker, A. & Rege, J. (2003b). Valuing Indigenous Farm Animal Genetic Resources in Kenya: A Comparison of Stated and Revealed Preference Estimates. *Ecological Economics*, Vol. 45, pp. 409-426.

Steverding, D. & Troscianko, T. (2004). On the role of blue shadows in the visual behavior of tsetse flies. *Proc. R. Soc. Lond. B (Suppl.)*, Vol. 271, pp. S16–S17.

Sy, H. A.; Farminow, M. D.; Johnson, G. V. & Crow, G. (1997). Estimating the Values of Cattle Characteristics Using an Ordered *Probit* Model, *American Journal of Agricultural Economics*, Vol. 79, pp. 463-476.

Tano, K.; Faminow, M. D.; Kamuanga, M. & Swallow, B. (2003). Using Conjoint Analysis to Estimate Farmers' Preferences for Cattle Traits in West Africa. *Ecological Economics*, Vol. 45, pp. 393 –408.

Train, K. (2003). *Discrete Choice Methods with Simulation.* Cambridge University Press. Cambridge.

Verbeek, M. (2004). *A Guide to Modern Econometrics.* Second Edition. John Wiley and Sons Ltd, The Atrium, Southern Gate, Chichester, England.

Windle, J. & Rolfe, J. (2005). Diversification choices in agriculture: a Choice Modeling case study of sugarcane growers, *The Australian Journal of Agricultural and Resource Economics*, Vol. 49, No. 1, pp. 63-74.

Zander, K. K. (2006). *Modeling the Value of Farm Animal Genetic Resources – Facilitating Priority Setting for the Conservation of Cattle in East Africa.* PhD Dissertation; University of Bonn.

Permissions

The contributors of this book come from diverse backgrounds, making this book a truly international effort. This book will bring forth new frontiers with its revolutionizing research information and detailed analysis of the nascent developments around the world.

We would like to thank Prof Dr. Mahmut Calıskan, for lending his expertise to make the book truly unique. He has played a crucial role in the development of this book. Without his invaluable contribution this book wouldn't have been possible. He has made vital efforts to compile up to date information on the varied aspects of this subject to make this book a valuable addition to the collection of many professionals and students.

This book was conceptualized with the vision of imparting up-to-date information and advanced data in this field. To ensure the same, a matchless editorial board was set up. Every individual on the board went through rigorous rounds of assessment to prove their worth. After which they invested a large part of their time researching and compiling the most relevant data for our readers. Conferences and sessions were held from time to time between the editorial board and the contributing authors to present the data in the most comprehensible form. The editorial team has worked tirelessly to provide valuable and valid information to help people across the globe.

Every chapter published in this book has been scrutinized by our experts. Their significance has been extensively debated. The topics covered herein carry significant findings which will fuel the growth of the discipline. They may even be implemented as practical applications or may be referred to as a beginning point for another development. Chapters in this book were first published by InTech; hereby published with permission under the Creative Commons Attribution License or equivalent.

The editorial board has been involved in producing this book since its inception. They have spent rigorous hours researching and exploring the diverse topics which have resulted in the successful publishing of this book. They have passed on their knowledge of decades through this book. To expedite this challenging task, the publisher supported the team at every step. A small team of assistant editors was also appointed to further simplify the editing procedure and attain best results for the readers.

Our editorial team has been hand-picked from every corner of the world. Their multi-ethnicity adds dynamic inputs to the discussions which result in innovative outcomes. These outcomes are then further discussed with the researchers and contributors who give their valuable feedback and opinion regarding the same. The feedback is then collaborated with the researches and they are edited in a comprehensive manner to aid the understanding of the subject.

Apart from the editorial board, the designing team has also invested a significant amount of their time in understanding the subject and creating the most relevant covers. They scrutinized every image to scout for the most suitable representation of the subject and create an appropriate cover for the book.

The publishing team has been involved in this book since its early stages. They were actively engaged in every process, be it collecting the data, connecting with the contributors or procuring relevant information. The team has been an ardent support to the editorial, designing and production team. Their endless efforts to recruit the best for this project, has resulted in the accomplishment of this book. They are a veteran in the field of academics and their pool of knowledge is as vast as their experience in printing. Their expertise and guidance has proved useful at every step. Their uncompromising quality standards have made this book an exceptional effort. Their encouragement from time to time has been an inspiration for everyone.

The publisher and the editorial board hope that this book will prove to be a valuable piece of knowledge for researchers, students, practitioners and scholars across the globe.

List of Contributors

Georgescu Sergiu Emil and Costache Marieta
University of Bucharest, Faculty of Biology, Department of Biochemistry and Molecular Biology, Splaiul Independentei, Bucharest, Romania

Anila Hoda
Agriculural University of Tirana, Albania

Paolo Ajmone Marsan
Università Cattolica del S. Cuore, Piacenza, Italy

Luciana P. B. Machado, Daniele C. Silva, Daiane P. Simão and Rogério P. Mateus
Universidade Estadual do Centro-Oeste – UNICENTRO, Brazil

Alireza Seidavi
Department of Animal Science, Rasht Branch, Islamic Azad University, Rasht, Iran

Mei-Chen Tseng and Yin-Huei Hung
Department of Aquaculture, National Pingtung University of Science & Technology, Pingtung, Taiwan, Republic of China

Chuen-Tan Jean
Department of Physical Therapy, Shu Zen College of Medicine and Management, Kaohsiung, Taiwan, Republic of China

Peter J. Smith
Museum Victoria, Melbourne Victoria, Australia

Alexeia Barufatti Grisolia
Universidade Federal da Grande Dourados, UFGD, FCBA, MS

Vanessa Roma Moreno-Cotulio
Universidade Federal de Alfenas, UNIFAL-MG, ICN, MG, Brazil

Naoto Hanzawa, Ryo O. Gotoh, Hidekatsu Sekimoto, Tadasuke V. Goto and Hidetoshi B. Tamate
Department of Biology, Faculty of Science, Yamagata University, Yamagata, Japan

Satoru N. Chiba
Center for Molecular Biodiversity Research, National Museum of Nature and Science, Tokyo, Japan

Kaoru Kuriiwa
Department of Zoology, National Museum of Nature and Science, Tokyo, Japan

Hideyuki Imai and Misuzu Aoki
University of the Ryukyus, Nara Women's University, Japan

Pooja Deshpande
School of Anatomy and Human Biology, University of Western Australia, Western Australia, Australia

Michaela Lucas
Institute of Immunology and Infectious Diseases, Murdoch University, Western Australia, Australia Department of Health, Western Australia, Australia

Silvana Gaudieri
School of Anatomy and Human Biology, University of Western Australia, Western Australia, Australia Institute of Immunology and Infectious Diseases, Murdoch University, Western Australia, Australia

Enrique Blanco Gonzalez and Tetsuya Umino
Graduate School of Biosphere Science, Hiroshima University, Japan

Lorraine Pariset, Maria Gargani and Alessio Valentini
Department for Innovation in Biological, Agro-Food and Forest Systems (DIBAF), University of Tuscia, Viterbo, Italy

Stephane Joost
Laboratory of Geographic Information Systems (LASIG), School of Architecture Civil and Environmental Engineering (ENAC), Ecole Polytechnique Fédérale de Lausanne (EPFL), Lausanne, Switzerland

Shawn Larson
Seattle Aquarium, United States

Marieta Costache, Andreea Dudu and Sergiu Emil Georgescu
University of Bucharest, Faculty of Biology, Department of Biochemistry and Molecular Biology, Romania

April M. H. Blakeslee and Amy E. Fowler
Long Island University, CW Post Campus, Smithsonian Environmental Research Center, USA

Jan Mendel Eva Marešová, Ivo Papoušek, Karel Halačka, Lukáš Vetešník, Milena Koníčková and Soňa Urbánková
Institute of Vertebrate Biology, Academy of Sciences of the Czech Republic, Brno, Czech Republic

Radek Šanda
Department of Zoology, National Museum, Prague, Czech Republic

Luis Fernando Rodrigues-Filho and Davidson Sodré
Federal University of Pará, Campus of Bragança, Institute of Coastal Studies (IECOS), Bragança, Pará, Brazil

Girma T. Kassie
International maize and Wheat Improvement Center (CIMMYT), Harare, Zimbabwe

Awudu Abdulai
University of Kiel, Germany

Clemens Wollny
Bingen University of Applied Sciences, Bingen, Germany

Danillo Pinhal
Department of Genetics, Biosciences Institute, Sao Paulo State University (UNESP), Botucatu, SP, Brazil

Marcelo Vallinoto
Federal University of Pará, Campus of Bragança, Institute of Coastal Studies (IECOS), Bragança, Pará, Brazil CIBIO/UP, Research Center in Biodiversity and Genetic Resources, Campus Agrário de Vairão, University of Porto, Portugal

www.ingramcontent.com/pod-product-compliance
Lightning Source LLC
Chambersburg PA
CBHW070716190326
41458CB00004B/997